Mesoporous Materials

Edited by

Inamuddin[1], Arwa Alrooqi[2], Amal Alshehri[2], Mohammad Abu Jafar Mazumder[3,4]

[1]Department of Applied Chemistry, Zakir Husain College of Engineering and Technology, Faculty of Engineering and Technology, Aligarh Muslim University, Aligarh-202002, India

[2]Department of Chemistry, Faculty of Science, Al Baha University, Al Baha, Saudi Arabia

[3]Chemistry Department, King Fahd University of Petroleum & Minerals, Dhahran 31261, Saudi Arabia

[4]Interdisciplinary Research Center for Advanced Materials, King Fahd University of Petroleum & Minerals, Dhahran 31261, Saudi Arabia

Published by **Materials Research Forum LLC**
Millersville, PA 17551, USA

Published as part of the book series
Materials Research Foundations
Volume 173 (2025)
ISSN 2471-8890 (Print)
ISSN 2471-8904 (Online)

Print ISBN 978-1-64490-344-5
eBook ISBN 978-1-64490-345-2

Distributed worldwide by

Materials Research Forum LLC
105 Springdale Lane
Millersville, PA 17551
USA
https://www.mrforum.com

Manufactured in the United States of America
10 9 8 7 6 5 4 3 2 1

Table of Contents

Preface

In recent years, there has been a significant increase in the development of inorganic materials for use in a wide range of devices, from supercapacitors to fuel cells and from optoelectronics to catalysts. Mesoporous materials are a special class of inorganic nanomaterials characterized by uniform nanopores ranging from 2 to 50 nm in diameter. These materials include nanoporous silica, nanoporous alumina, porous carbon, organosilica compounds, and functional nanoporous silica. They possess either a crystalline or amorphous network with cage-like or cylindrical pores. These porous materials offer numerous advantages, including a high surface area, thermal stability, moderate catalytic activity, biocompatibility, and low toxicity. Furthermore, mesoporous materials can be easily customized to achieve the desired composition, shape, and morphology of pores for specific properties and applications.

The time has come to highlight the latest advancements in the fabrication and design of inorganic materials, along with discussions on their properties and applications. *Mesoporous Materials: Synthesis, Properties, and Allied Applications* aims to fill this gap by providing a concise introduction to the synthesis of mesoporous materials, followed by an exploration of their intriguing properties and diverse applications.

This book offers an in-depth guide to the world of mesoporous materials, with a particular focus on their advanced applications in the sorption, removal, and degradation of dyes, volatile organic compounds, and other pollutants for effective environmental remediation. Additionally, it examines their role in healthcare and medicine, including tissue engineering and diagnostic applications. The use of mesoporous materials as supports in size exclusion chromatography, owing to their large pore volume and high surface area, is also discussed. Furthermore, cutting-edge applications such as separation techniques, high-performance catalysis, and enzyme immobilization are explored.

With its emphasis on the characteristics and applications of these next-generation materials, this book serves as an important contribution to the ever-growing field of inorganic chemistry and its allied disciplines. The book contains 18 chapters, summarized as follows:

Chapter 1 discusses inorganic mesoporous materials for the sorption of heavy metal ions, including hydrophosphates, hydrous oxides, double hydroxides of multivalent metals, zeolites, oxidized graphene, and its composites with these inorganic compounds. These materials are also used for modifying ion-exchange resins, which, in their swollen state, are related to mesoporous materials.

Chapter 2 explores the synthesis, structural properties, and applications of mesoporous materials for dye adsorption in wastewater treatment. It highlights their high surface area, tuneable pore sizes, and functionalization potential, emphasizing adsorption mechanisms, recent advances, and case studies that showcase their efficiency, sustainability, and industrial relevance.

Chapter 3 provides a detailed study of mesoporous materials' ability to remove pollutants. The fundamental properties of these materials are discussed along with functionalization methods. Techniques for pollutant removal in water, air, and soil treatment are examined, along with kinetic aspects and future perspectives on their applications.

Chapter 4 focuses on mesoporous materials in liquid chromatographic separation. It briefly introduces the fundamentals of liquid chromatography and covers methods for synthesizing and characterizing mesoporous materials, including surface modification techniques. Their commercial applications are also discussed, emphasizing advancements in stationary phases, improved selectivity, and separation efficiency.

Chapter 5 provides a comprehensive overview of bentonite's various applications, focusing on its use in pharmaceutical formulations, cosmetic products, controlled drug release, detoxification, pharmacological activity, cholesterol reduction, and anti-inflammatory action, as well as its potential in treating bone tissue disorders and cancer, among other areas. By exploring its unique properties and versatility, the chapter highlights the fundamental role of this clay in sectors such as health, beauty, and materials technology.

Chapter 6 highlights sustainable mesoporous materials, valued for their high surface area, tuneable pores, and chemical stability. These materials facilitate pollutant adsorption, catalytic purification, drug delivery, and targeted therapies. Furthermore, eco-friendly synthesis approaches are presented, emphasizing their potential to address global environmental and healthcare challenges.

Chapter 7 explores mesoporous materials as photocatalysts for contaminant degradation, discussing principles such as TiO_2- and ZnO-based photocatalysis, factors influencing efficiency (light intensity, pH, catalyst dosage), various synthesis methods such as sol-gel, and a patent analysis that highlights their expanding role in advanced water and air purification technologies globally.

Chapter 8 details various methods for the synthesis and functionalization of mesoporous solid-phase extraction sorbents. The chapter also covers the diverse advantages and applications of these materials, along with a comparison to traditional sorbents.

Chapter 9 explores various synthetic routes for mesoporous materials, including sol-gel and template-assisted techniques. Additionally, the chapter highlights surface modifications based on extensive literature reviews. Various characterization techniques are discussed, along with unique applications such as protein separation, enzyme immobilization, drug delivery, and heavy-metal adsorption.

Chapter 10 examines the use of mesoporous material-based nanocomposites in sensing and wastewater treatment, detailing various synthesis methods, advantages, and limitations. These nanocomposites are applied in electrochemical sensing and photocatalytic degradation, offering high sensitivity, rapid response, and efficient degradation of dyes, drugs, food additives, and toxins.

Chapter 11 discusses mesoporous materials, characterized by their high surface area, adjustable pore size, and chemical stability, as effective adsorbents for removing volatile organic compounds (VOCs). Various mesoporous materials such as silica, carbon, and metal-organic frameworks (MOFs) are explored, along with their structural properties, synthesis techniques, and functionalization strategies that enhance VOC adsorption. The chapter emphasizes ongoing research and innovation to optimize these materials for efficient VOC removal, highlighting their potential to improve air quality and environmental sustainability.

Chapter 12 discusses mesoporous materials as stationary phases in size exclusion chromatography (SEC). Traditional stationary phases used in SEC are reviewed and compared with mesoporous materials based on their unique properties. Applications in protein, polymer, and nanoparticle separation are explored, along with advancements in material design to enhance resolution and efficiency in SEC.

Chapter 13 examines the use of mesoporous materials for enzyme immobilization. Various immobilization techniques are discussed, along with the advantages of immobilized enzymes over free enzymes. The chapter also emphasizes enzyme immobilization on ordered mesoporous carbons and their applications.

Chapter 14 explores mesoporous materials and their applications in biomolecule separation. Highlighting their unique properties, it discusses their role in enzyme immobilization, nucleic acid adsorption, biologically active molecule separation, and drug delivery. Strategies to enhance biomolecule affinity and emerging techniques are emphasized, showcasing their potential in biomedical research and industrial applications.

Chapter 15 illustrates recent advances and current trends in the application of magnetic mesoporous materials in food analysis, discussing the benefits of these developments and raising awareness about their potential in the field.

Chapter 16 details the importance of mesoporous scaffolds in tissue engineering, covering their customizable physical and chemical properties, which make them an integral component for structural and biological frameworks that support cell adhesion and proliferation. The integration of these materials holds significant promise for improving tissue regeneration outcomes and advancing regenerative medicine.

Chapter 17 discusses the application of mesoporous materials in cosmetic formulations. It explores their role in controlled drug release, enhancing skin penetration, and improving product stability. The chapter highlights various types of mesoporous materials, their synthesis methods, and their potential use in cosmetics, addressing both benefits and challenges for future applications in the industry.

Chapter 18 presents the properties of mesoporous materials as highly functional systems designed to address medical challenges in health diagnostics. These materials feature tunable structures, broad surface areas, advanced cargo capabilities, low cost, and low toxicity. The chapter focuses on preparation and characterization methods, as well as their applications in diagnostics.

Materials Research Foundations 173 (2025) 1-39 https://doi.org/10.21741/9781644903452-1

Chapter 1

Inorganic Mesoporous Materials for Sorption of Heavy Metal Ions

Yu.S. Dzyazko, M.O. Chaban

[1]V.I. Vernadskii Institute of General and Inorganic Chemistry of the NAS of Ukraine

dzyazko@gmail.com

Abstract

In this chapter, we consider typical inorganic mesoporous materials, which are widely used for sorption of heavy metal ions: hydrophosphates, hydrous oxides and double hydroxides of multivalent metals, and also zeolites. Oxidized graphene and its composites with these inorganic substances are also in a focus of attention. Composites based on ion exchange resins modified with mentioned inorganic compounds are discussed, since this type of polymers is also related to mesoporous materials, when they are in a swollen state. Embedded inorganic particles (non-aggregated nanoparticles, aggregates or agglomerates) affect the porous structure of the polymers changing their sorption properties.

Keywords

Hydrophosphates of Multivalent Metals, Hydrated Oxides, Zeolites, Polymer-Inorganic Composites, Ion Exchange Resins, Oxidized Graphene

Contents

Materials Research Foundations 173 (2025) 1-39 https://doi.org/10.21741/9781644903452-1

1. Introduction

According to the IUPAC classification of porous solids, materials, porous structure of which involves voids with a diameter of 2–50 nm. Among known mesoporous materials, ordered structures with a narrow pore size distribution are the most interesting from a practical point of view. This structural feature is attributed to crystalline substances, whereas pore size of amorphous materials is in a wide diapason. Chemical composition of mesoporous solids can be different: aluminosilicates, materials based on multivalent metals, synthetic ion exchange polymers (resins and membranes), allotropic carbon modifications and so on. Mesoporous substances are utilized as sensors, catalysts, sorbents, materials for electronics, optics, medicine and nanotechnologies. In the last case, they are used as nanoreactors, where nanosized isolated particles are formed.

In this chapter, both natural and synthetic mesoporous materials are considered as sorbents of heavy metal ions. Namely mesoporosity provides their rather fast sorption rate. It should be stressed that amorphous substances are more preferable compared with crystalline materials, since in the last case, the ion uptake is often accompanied by the formation of a new phase. This sufficiently slows down sorption and complicates regeneration but provides selectivity. Alternately, sorption on amorphous materials is faster compared with crystalline materials, they can be easily regenerated. However, amorphous solids are characterized by lower selectivity.

Here we will consider typical representatives of inorganic mesoporous materials, such as hydrophosphates, hydrous oxides and double hydroxides of multivalent metals, and also zeolites. Graphene oxide (GO) and its composites with these inorganic substances are also in a focus of attention. Hydrophosphates and hydrous oxides of multivalent metals are often used as modifiers of ion exchange resins, since these swollen polymers are also related to mesoporous materials. Embedded inorganic particles affect porous structure of the polymer matrices changing their sorption properties. Sorption ability of the polymer-inorganic composites strongly depends on the state of modifier (non-aggregated nanoparticles, aggregates or agglomerates).

2. Inorganic sorbents

Various industries require ion exchangers, which are characterized by mechanical strength and high resistance against aggressive reagents, elevated temperatures and ionizing radiation. Inorganic ion exchange materials are satisfied by these requirements. Their research began in the middle of the last century with the development of nuclear technologies.

2.1 Hydrophosphates of metals of IV group

One of the well-known types of ion exchangers is hydrophosphates of multivalent metals. In 1964, Clearfield and Stynes [1], and later Alberti and Toracca [2] synthesized finely dispersed crystalline zirconium hydrophosphate (ZHP), namely its α-modification (Fig. 1 [3]). The information about this compound is systemized in [4].

The α-ZHP crystals consist of extended $[Zr(HPO_4)_2]_n$ layers, where zirconium atoms are in an octahedral environment of oxygen atoms of phosphate groups. Only 3 oxygen atoms are involved to the formation of P–O–Zr bonds, and the fourth is oriented in the interplanar space. It is valence bound to hydrogen atom. Functional groups coordinate the water molecule. Later, other modifications were synthesized, which differ in the ratio of Zr : P : H_2O, as well as in an interplanar distance [5-7]. Acidic hydrogen phosphate groups and a continuous network of hydrogen bonds

Materials Research Foundations 173 (2025) 1-39 https://doi.org/10.21741/9781644903452-1

provide the proton conductivity of crystalline ZHP and ion exchange ability due to diffusion through the interlayer space.

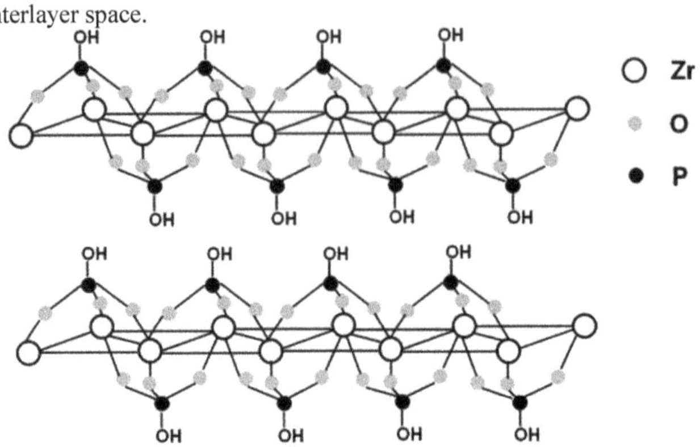

Figure 1. Structure of α-modification of ZHP [3]. Reproduced according to the permission of the Elsevier publisher.

It was established that a significant disadvantage of crystalline HZP is very slow sorption, which is due to the formation of new phases [8, 9]. The incorporation of sorbed ions to the crystal structure complicates the sorbent regeneration. In addition, crystalline modifications of this ion exchanger can be obtained in the form of a finely dispersed powder, which limits their practical application. The size of granules is from 2.5-3.6 μm [10] to several tens microns [11]. As shown in [11], mesopores are absent in hydrogen forms of α-ZHP. Ion exchange leads to an increase of the interlayer distance, as a result, mesoporosity appears. The volume of mesopores depends on the size of ions and increases within the order: Li-form<Na-form<K-form. Mesopores are rather small and regular: the pore size distribution shows a narrow peak at 4 nm.

As opposed to crystalline substances, amorphous materials are characterized by larger and disordered pores. As shown in the example of amorphous titanium hydrophosphate (THP), pores with a size of 25-30 nm dominate (Fig. 2) [12]. Peaks are within the interval of 3-110 nm. Increasing the content of titanium sulfate in the reaction mixture that contains also H_3PO_4 results in an enlargement of the volume of dominating pores. Smaller and more ordered pores (2.5 nm) were obtained for zirconium alkyl phosphates [13]. The size and volume of mesopores of tin hydrophosphate strongly depends on the transformation of as-prepared hydrogel into xerogel [14]. According to the ability to increase the diameter (from 3.5 to 6 nm) and volume of dominating pores, the efficiency of different techniques can be ranged as: mechanochemical treatment < hydrothermal treatment < microwave treatment. However, pores are disordered in this row.

Two types of functional phosphorus-containing groups, namely dihydrophosphate ($-OPO_3H_2$) and hydrophosphate ($(-O)_2PO_2H$) were recognized in amorphous ZHP [15] and THP [16] with ^{31}P NMR spectroscopy. Moreover, potentiometric titration indicated also ≡Me–OH groups (where Me is Zr or Ti) [15, 16]. Due to functional groups, which are able to change H^+ ions, sorption of heavy metal cations is realized according to ion exchange mechanism.

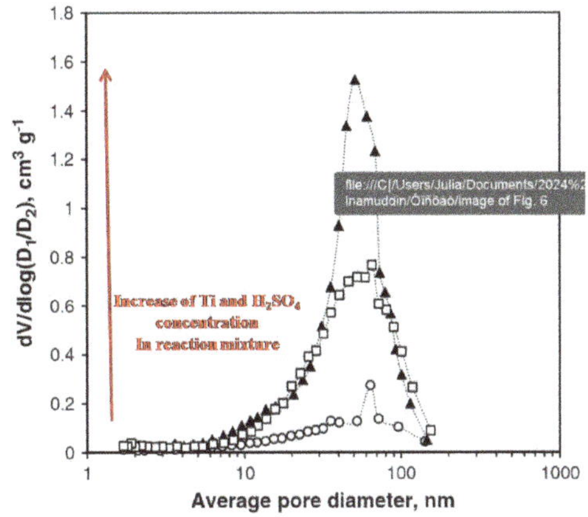

Figure 2. Pore size distributions for amorphous THP samples, which were obtained under different content of components in a reaction mixture [12]. Reproduced according to the permission of Elsevier publisher.

However, ion exchange is accompanied by the complexation of sorbed ions with functional groups as confirmed by the data of UV-VIS [15], ^{31}P NMR [15, 16], FTIR [16, 17] and XPE spectroscopy [17]. Hydrolysis of heavy metal ions directly in the sorbent phase was proved with methods of UV-VIS, ^{31}P NMR and impedance spectroscopy [15]. In this case, the amount of phosphate ligands in the octahedral environment of Ni^{2+} ions was estimated as 2-4. The composition of complexes is shown in Fig. 3. $Ni(OH)^+$ ions are also involved in the complex formation.

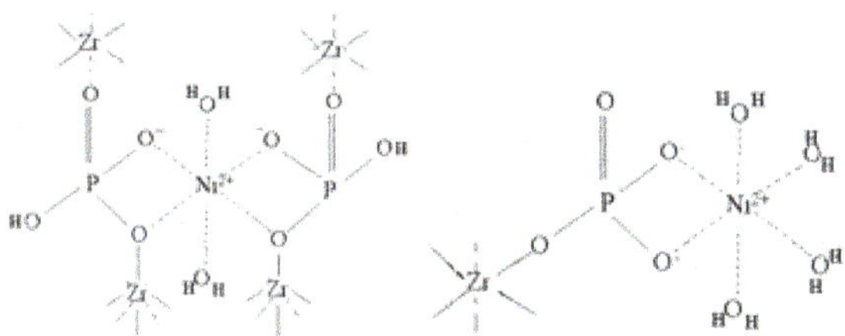

Figure 3. Ni^{2+} complexes with phosphate groups of ZHP.

In some cases, ion exchange can be accompanied by the precipitation of insoluble compounds. More than 50 years ago, the interrelation between the selectivity of THP to one or other cations and the pH of precipitation of their hydroxides was found [18]. Later the formation of insoluble $(UO_2)_3(PO_4)_2{\cdot}4H_2O$ phase as a result of UO_2^{2+} sorption on THP was suggested [19]. The order of sorption capacity was shown to be in agreement with the solubility constants for corresponding phosphates: the smaller this constant, the higher the capacity. Both hydrolysis and precipitation can be a reason for over-equivalent sorption: the content of sorbed ions is higher than the value, which is expected from the number of functional groups.

When colloidal $UO_2(OH)_2$ particles exist in a U(VI)-containing solution under high pH, sorption can be considered as their coagulation on granules. At pH 8, both $UO_2(OH)_2$ nanoparticles and $UO_2(OH)^+$ ions coexist in the solution [20]. During sorption, the pH is shifted to acidic region as a result of $UO_2(OH)^+ \rightarrow H^+$ exchange. It means the transition of $UO_2(OH)_2 \rightarrow UO_2(OH)^+ \rightarrow UO_2^{2+}$, which facilitates sorption. At the same time, the dissociation of functional groups is depressed in acidic media – the pK_a values are 2.2-3.4 and 4.6-5.2 for $-OPO_3H_2$ [15, 16] and $(-O)_2PO_2H$ [16] groups respectively. That is why the dependencies of sorption capacity on pH show extremes (Fig. 4) [20, 21].

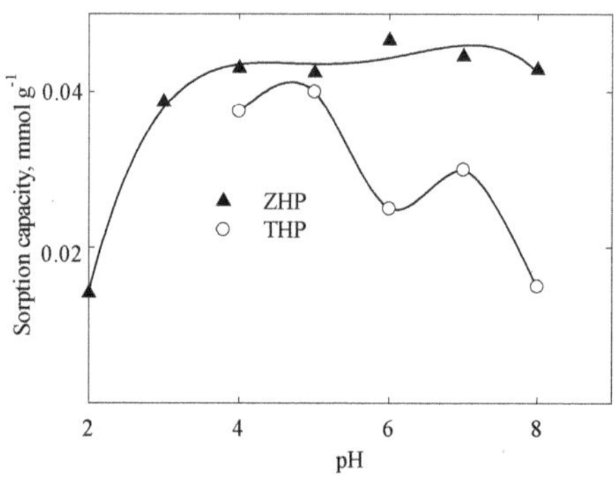

Figure 4. Dependencies of sorption capacity of ZHP and THP towards U(VI)-containing ions on pH. The curves were plotted based on [20, 21].

Some data of sorption of heavy metal ions on phosphate sorbents are given in Table 1. As shown, the equilibrium time of sorption does not exceed 2 h despite rather large granule size. It is interesting that the crystalline and amorphous modifications are characterized by the same sorption rate. A wide interval of sorption capacity towards UO_2^{2+} ions is due to their different initial concentration. High capacity of amorphous [17] and crystalline [22] ZHP towards Pb^{2+} ions should be noted. As suggested in [23], these species are sorbed preferably comparing with Cu^{2+} and Fe^{2+} ions. Preferable sorption of heavy metal ions on amorphous THP compared with Ca^{2+} and Mg^{2+} ions is noted [16]. The order of the value of distribution coefficient is 1000 (Co^{2+}), 100 (Ni^{2+}, Zn^{2+}, Cu^{2+}) and 10 (Mn^{2+}, Ca^{2+}, Mg^{2+}). This difference is explained by the complexation of sorbed ions with functional groups. When ions of different charge are present in a solution simultaneously, the species of higher charge are sorbed preferably. It is also true for solutions to contain ions of only one metal. This regularity was suggested for sorption of ^{234}Th- and ^{238}U-containing ions and $^{134}Cs^+$ [11], Cs^+, Sr^{2+}, UO^{2+} [14], transplutonium ions [24], and a large variety of inorganic cations [25, 26] on amorphous [11, 24] or crystalline ZHP [11, 26], also on amorphous tin [14] and titanium-zirconium hydrophosphates [25].

Table 1. Sorption of heavy metal ions on some hydrophosphates

Compound	Modification	Granule size	Ions	Capacity, mmol g^{-1} (initial pH)	Equilibrium time, min	Reference
Tin hydrophosphate	Amorphous		Cs^+, Sr^{2+}, UO^{2+}	0.25-1.9 (5.2)	–	[14]
ZHP	Amorphous	Several microns - several ten microns	Pb^{2+}, Zn^{2+}, Cd^{2+}	0.8 (5.5)	45-60	[17]
THP	Amorphous	250-500 μm	UO_2^{2+}	4.4 (5.5)	–	[19]
ZHP	Amorphous	800 μm	UO_2^{2+}	0.045 (6)	30	[20]
THP, ZHP	Amorphous	200-2000 μm	UO_2^{2+}	0.3-0.8 (5)	20-120	[21]
ZHP	Crystalline	–	Pb^{2+}	0.9 (6)	25	[22]
	Amorphous	–		0.1 (6)		

Amorphous ZHP prepared with hydrothermal method [27], and THP aerogel, which was obtained with sol-gel technique followed by supercritical drying [28], should be also mentioned. Titanium-zirconium hydrophosphates removes Cd^{2+} ions practically within a few minutes at pH>3, when their initial concentration is lower than 0.1 mmol dm^{-3} [27]. The THP aerogel was applied to Sr^{2+} sorption: the capacity value of 4.3 mmol g^{-3} has been reported [28].

A remarkable property of hydrophosphates of multivalent metals is their chemical stability against mineral acids. This gives a possibility to regenerate them with solutions of mineral [20, 25, 29] or organic acids and even with a NH_4NO_3 solution [25]. It means an opportunity to use these sorbents many times [29].

Despite sufficient advantages, such as high sorption capacity and selectivity towards heavy metal ions, stability against acids and oxidizing reagents, as well as rather high sorption rate, crystalline hydrophosphate materials and their amorphous modifications containing high phosphorus amount are suitable only for the sorption of cations [30]. However, amorphous sorbents comprise also ≡Me–OH groups (where Me is a metal of the IV group) [16], which can provide not only cation exchange, but also anion exchange ability. Unfortunately, literature sources contain few information about the anion sorption on these materials. For instance, sorption of phosphate anions on ZHP was studied, the optimal conditions were found to be realized at pH 2-6. the distribution coefficients are 7-17 depending on coexisting anions (SO_4^{2-}, NO_3^-, Cl^-) [31]. Regarding metal-containing anions, the information is implicit. Uranyl complexes were sorbed on ZHP and THP from acetate media at pH<5 [21]. The acetate complexes are mainly in anionic form ($[UO_2Ac_3]^-$) at pH 3-4 [32].

Inorganic sorbents possessing amphoteric properties are considered further.

2.2 Hydrated simple oxides of multivalent metals

Among known mesoporous adsorbents, oxide materials occupy special position due to their ability to sorb both cations and anions depending on the solution pH [33]. This is due to surface –OH groups, which are dissociated in alkaline media:

$$\equiv Me{-}OH \leftrightarrow \equiv Me{-}O^{-} + H^{+} \quad (1)$$

or protonated in acidic solutions:

$$\equiv Me{-}OH + H^{+} \leftrightarrow \equiv Me{-}OH_2^{+}, \quad (2)$$

Especially hydrated oxides ("oxyhydrates", "globular hydrates") should be distinguished. For these materials, the composition of the cores of primary particles is close to that of anhydrous oxides, considerable amount of functional groups are located on the particle surface (Fig. 5). In fact, hydrated oxides occupy an intermediate position between stoichiometric oxides and hydroxides. A specific surface area plays a key role for uptake, while mesoporous structure provides rather fast sorption.

Hydrated oxides can be divided into amorphous and crystalline materials. For instance, amorphous hydrated oxides of Ti, Zr, Sn, Al and Fe(III) can be synthesized by precipitation from the solutions of their salts [35]. Amorphous manganese dioxide is obtained via $KMnO_4$ reduction [36]. Other approaches to synthesis are sol–gel, solvothermal, hydrothermal, ultrasound or microwave-assisted polyol and microemulsion techniques [35]. Fibrous hydrated titanium oxide has been obtained using a special regime for the gel drying [37]. Alumina electrooxidation was applied to obtain hydrated oxide of this metal in a form of membrane [38-41]. Normally crystalline oxides are obtained by a thermal treatment of hydrated materials [35]. However, crystalline samples of hydrated Fe(III) oxide can be obtained under ambient conditions. For instance, stainless steel rust contains α, β and γ modifications of FeOOH as well as Fe_2O_3 [42].

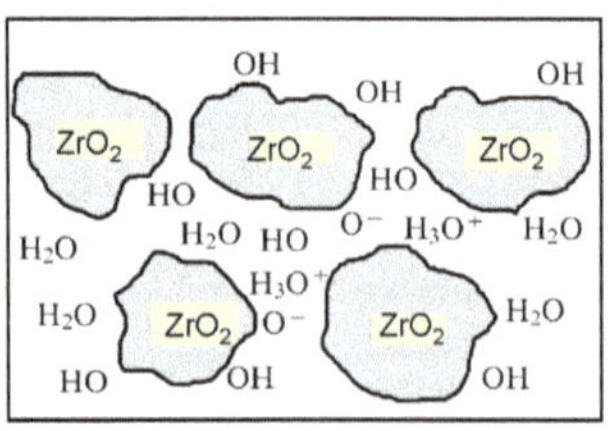

Figure 5. Structure of hydrated zirconium dioxide, adapted from [34].

It should be noted that amorphous hydrated iron (III) oxide shows higher mesoporosity compared with crystalline modifications. A presence of hematite, lepidocrocite and goetite in amorphous phase depresses mesoporosity of oxide sorbents (Fig. 6a) [43].

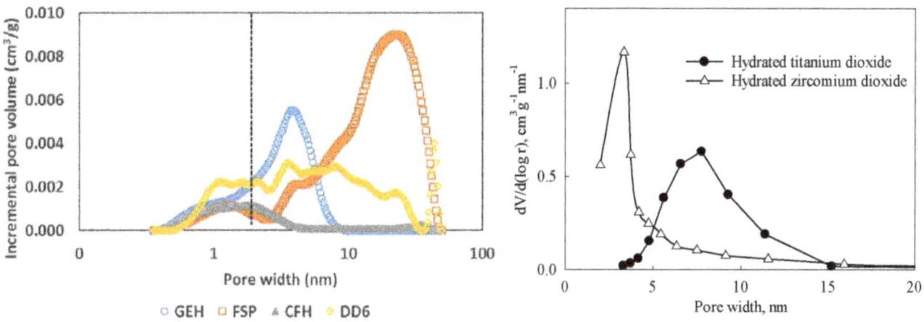

a b

Figure 6. a – pore size distributions for oxide sorbents [43]. Phase composition: GEH – hematite (11%), hydrated iron oxide (89%); FSP – hydrated iron oxide (100 %), CFH goethite/gematite (43%), hydrated iron oxide (57 %); DD6 – hydrated aluminiun oxide (34%), aluminium oxide (66 %). Dashed line divides the regions of micro- and mesopores. b – pore size distributions for hydrated titanium and zirconium dioxides. Reproduced according to the permission of Elsevier publisher (a), plotted according to the data [44] (hydrated titanium dioxide) and [45] (hydrated zirconium dioxide) (b).

As seen from Fig. 6a, sorbents with lower content of amorphous phase show smaller pores. Al_2O_3 containing hydrated aluminium oxide shows a diffuse peak up to 20 nm. It is noted that the adsorption rate is the most independent on the particle size namely for hydrated iron (III) oxide. Expressed mesoporosity is characteristic for hydrated titanium dioxide, where pores of 8 nm dominate (Fig. 6b) [44]. In the case of hydrated zirconium dioxide, smaller pores prevail (3 nm) [45], similar results were also obtained in [46].

An important parameter that affects the sorption properties of oxide sorbents is the solution pH, at which a point of zero charge (PZC) is realized. Under these conditions, the concentration of dissociated and protonated functional –OH groups are equal, the surface charge is 0. For oxides, hydroxides, and hydrated oxides, the Parks review gives a correlation between the pH of PZC and the analogue of the ionic potential (z/R, where z is the metal valency, R is the sum of the radius of the metal ion and the diameter of the oxygen ions, which is 2.8 Å) [47]:

$$pH\ PZC = A - B(z/R) - D\Delta F \qquad (3)$$

Here *A, B* and *D* are the constants, which are similar values for all oxides, hydrated oxides and hydroxides, ΔF is the non-Coulombic contribution of the energy of hydrogen bonds. The *z/R* parameter characterizes charge density of the oxide surface: the larger this value is, the stronger the electrostatic interaction between the sorbed ion and the surface. In other words, sorption is determined by the radius and valence of the metal in oxide material. However, the contribution of the ΔF parameter is considerable. It means, the pH of PZC for different compounds containing the same metal can be sufficiently different. This is shown in Table 2 for various Zr compounds [48]. The PZC strongly depends on the synthesis method (particularly on the solution composition), liquid for washing and aging, since these factors strongly determine the surface composition (ΔF parameter). However, hydrated zirconium dioxide shows close values, which are varied within the interval of 6.0-6.6. Alternately, a wide interval of the pH of PZC has been reported for crystalline sample.

Regarding hydrated oxides, the PZC pH (in brackets) is shifted to alkaline field in the order: Mn(IV) (3.8) < Sn(IV) (4 - 5.2) < Zr (6.0 - 6.6) < Fe(III) (7.9 - 8.3) < Ti (8.1 - 8.5) < Al (8.5-9.0) [48]. Closer values were reported for Sn(IV) (3.9) [49] and hydrated Zr dioxide (≈7) [20]. However, lower value was found for hydrated oxides of Ti (6.2 [49], 6.2 - 6.4 [50] or ≈6 [51]), Al (6.9) or Mn(IV) (2.7) [52]. These disagreements are evidently due to different conditions of the sorbent synthesis and PZC measurements. Hydrated oxides sorb preferably anions from neutral solutions (i.e. close to PZC) [53, 54], however, they show also cation exchange ability under these conditions [52-54]. As shown in the example of UO_2^{2+} species, these materials sorb cations even in acidic media [20, 44, 51].

Table 2. pH of PZC for some Zr-containing compounds (analysis [48])

Compound (structure)	Synthesis method		Electrolyte	pH PZC
Hydrated Zr dioxide (amorphous)	Refluxed at pH 4–7 for 1 day		0.001–0.1 M KNO_3	6.6
	Refluxed at pH 4–7 for 1 days			6.0
	Refluxed at pH 11 for 22 days			6.3
ZrO_2 (crystalline, commercial sample)	Water-washed		0.01 M NaCl	8.0
	No washing			6.1
ZrO_2 (crystalline, prepared under laboratory conditions)	Precipitation from the solution prepared from $ZrOCl_2$, silica template, calcined at 550° C		KOH, HCl	5.0
ZrO_2 (crystalline, commercial sample)	Washed with acid and base		$NaNO_3$, NaBr	7.6
	No washing		0.001 M KCl	6.4
	No washing		0.01-0.1 M $NaNO_3$, NaCl, $NaClO_4$	8.2
	No washing, aged in 0.001 M KNO_3	for 16 h	0.001–0.1 M KNO_3	9.5
		for 50 h		7.0
		For 170 h		6.2

The information about sorption of heavy metal ions on hydrated oxides is given in Table 3, where the data [20, 44, 51, 55-59] are summarized. As seen, rather fast uptake has been reported – in some cases, the equilibrium time does not exceed 1 h despite considerable particle size. This is due to the mesoporosity of sorbents. Very short equilibrium time was found for the hydrogel of hydrated zirconium dioxide (10 min) [56]. Xerogel shows longer equilibrium time (> 1 h).

High capacity (up to several tenths of millimol per gram) can be reached in the PZC region (pH 5 – 7). Moreover, overequivalent uptake was suggested for Cr(VI)-containing ions [55, 56]. This data cannot be explained by the electrostatic model. Considerable cation uptake in acidic media as well as difficult desorption under varying the solution acidity (surface charge) [20, 44, 51] are also outside the framework of the model. Currently, the most widespread models are ion exchange (see equations (1.1-1.2) [60] and the formation of surface complexes [61].

As noted for iron oxide materials, other adsorption mechanisms are possible: (co)precipitation on the surface, incorporation to the structure and mechanical occlusion (Fig. 7) [62]. Surface complexation involves the formation of inner-sphere and outer-sphere complexes. For instance, synergetic effect was found in the case of simultaneous adsorption of Zn^{2+} cations and phosphate anions [63]. The pathway of surface precipitation is as follows: first phosphate anions are bonded with surface hydroxyl groups, then Zn cations are bonded with phosphate ions. In fact, insoluble Zn phosphate is formed on the surface.

Table 3. Adsorption of heavy metal ions on amorphous simple hydrated oxides of multivalent metals

Metals in hydrated oxides	Sorbed ions	Granule size, μm	Sorption capacity, mmol g^{-1} (initial pH)	Equilibrium time	Kinetic model	Reference
Zr	UO_2^{2+}	400	5.1×10^{-2} (pH≥5)	0.5 h	Chemical reaction	[20]
	Pb^{2+}, $HCrO_4^-$	400	0.5 (pH 7)	–	–	[55]
	$HCrO_4^-$	200 – 800	1.8×10^{-1} – $6/5\times10^{-1}$	10 min – > 1 h	Internal diffusion	[56]
Ti	Ni^{2+}	29-36	2.5×10^{-1}– 3.7×10^{-1} (pH 5)	0.5 h	Chemical reaction	[57]
	UO_2^{2+}	900	1.9×10^{-2} – 2.8×10^{-2} (pH 5)	0.5-3 h	Chemical reaction	[44, 51]
Al, Fe(III), Mn(IV)	Zn^{2+}, Cd^{2+}	15-30	2.5×10^{-5}– 2.8×10^{-3} (pH 7)	≈80 days	Internal diffusion	[58]
	Pb^{2+}	15-30	2.0×10^{-2}– 5.5×10^{-2} (pH 7)	30-50 days	Internal diffusion	[59]

The content of sorbed Cd^{2+}, Mn^{2+}, Ni^{2+}, and Pb^{2+} ions in as-precipitated hydrated iron oxide was studied over its aging [64]. The effect of cation concentration on sorption decreased within the row: $Pb^{2+} \gg Ni^{2+} > Mn^{2+} = Cd^{2+}$. During the oxide transformation into the goethite or hematite structure, Cd^{2+} and Ni^{2+} ions were significantly desorbed. At the same time, Pb^{2+} and Mn^{2+} cations were embedded to the structure of crystalline phases. Ions can be incorporated via pore diffusion or lattice diffusion [62]. It was noted that coprecipitation of traces of heavy metal ions simultaneously with hydrated oxides of Fe or Al is much efficient comparing with adsorption [65]. Hydrolysis of cations of alkaline-earth metals on the surface of hydrated zirconium dioxide was suggested [66].

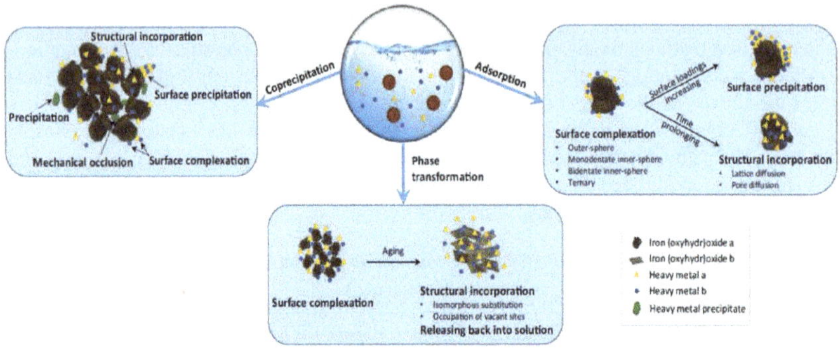

Figure 7. Possible mechanisms of sorption of heavy metal ions on oxide compounds of Fe(III) [62]. Reproduced according to the permission of the Elsevier publisher.

The approach, which allows one to improve functional properties of hydrated oxides, is the development of more complex materials, such as double oxides. They are considered in the next section.

2.3 Double hydroxides of multivalent metals

Crystalline layered double hydroxides are the most well studied, since they are widespread in nature. The first discovered mineral of this class is hydrotalcite, $[Mg_6Al_2(OH)_{16}]CO_3 \cdot 4H_2O$ [67]. The basic structural unit of these materials are octahedra, in the centre of which the metal ion is located, the six vertex angles are occupied by OH^- anions (Fig. 8). Cations in the layer can be replaced by other metal cations, when their radii are close. The interlayer anions can be also substituted by other anions. The flexibility and adjustability of the interlayer space is an unique property of these crystalline materials.

Besides the materials of natural origin, synthetic layered double hydroxides are known. The procedure of the synthesis of mesoporous materials involves co-precipitation of Mg and Al hydroxides followed by the surfactant adsorption [68]. Other approach is hydroxide coprecipitation in a presence of surfactant. Then surfactant is removed via the thermal treatment, then the material is rehydrated. Mesoporosity strongly depends in the surfactant concentration in the solution of metal salts (Fig. 9a).

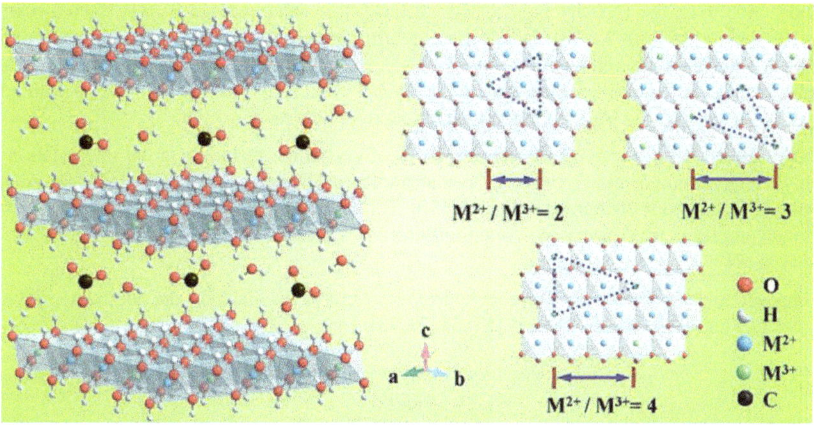

Figure 8. Structure of double layered hydroxides [67]. Reproduced according to the permission of Elsevier publisher.

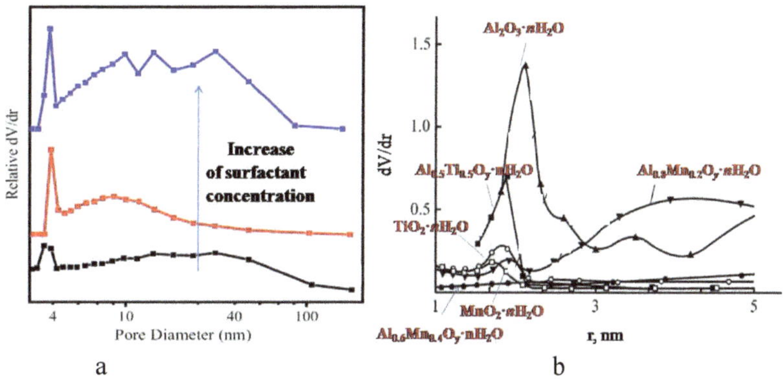

Figure 9. Pore size distributions for crystalline double hydroxides [68] (a) and amorphous [50] simple and double hydrated oxides (b). Here mesoporosity of Mg-Al double hydroxide enhances with increasing of surfactant concentration in the reaction mixture (a). Reproduced according to the permission of Elsevier (a) and Springer (b) publishers.

Among the advantages of crystalline materials, a simple preparation procedure, which involves low-cost reagents, should be mentioned. However, their practical application is limited by a low sorption capacity that is caused by a small amount of functional groups. In order to increase the capacity and adsorption rate, crystalline materials are delaminated [69]. As shown for a number of cations, delaminated samples show very short equilibrium time (in ≈5 min). However, the size of the adsorbent particles is extremely small (≈ 100 nm). Thus, it is difficult to separate solids from liquids. Moreover, the capacity is in a wide interval: this value reaches 0.7 (Fe^{3+}), 0.2 (Cu^{2+}), 1.2×10^{-2} (Zn^{2+}) and even 2.4×10^{-4} (Pb^{2+}) mmol g^{-1}. It is assumed that the sorption behavior of the material is due to a difference in ion radius: these values are 0.70 and 0.73 Å for Fe^{3+} and Cu^{2+} cations respectively. For other ions, they are 0.74 (Zn^{2+}) and 1.19 (Pb^{2+}) Å. Comparing with non-delaminated sorbent, the sorption capacity is higher in 2-3 times.

Another direction to improve sorption properties of crystalline materials is the intercalation of highly selective component. For instance, hexacyanoferrate(II) ions were intercalated to the double layered hydrated oxide containing Zn and Al [70]. The sorbent was applied to the removal of U(VI) hydroxocomplexes from water. As found, adsorption is practically independent on the solution acidity at pH > 3.

As opposed to crystalline layered double hydroxides, amorphous materials are not too sensitive to the size of ions being sorbed (Table 4 [53, 54, 71-73]).

Table 4. Sorption of heavy metal ions on amorphous double hydrated oxides of multivalent metals

Metals in hydrated oxides	Sorbed ions	Granule size, μm	Sorption capacity, mmol g^{-1} (pH),	Equilibrium time	Kinetic model	Reference
Sn-Al	Cd^{2+}, Cu^{2+}, Pb^{2+}	250-500	0.6-1.0	1 h	Internal diffusion	[53]
Zr-Al	$HCrO_4^-$, CrO_4^{2-}	250-500	0.3 (10.5) 0.8 (2.5)	0.7 h	Internal diffusion	[54]
Fe-Mn	Co^{2+}	–	0.4 (6)	2.5–3.5 h	–	[71]
Fe-Mn	Cd^{2+}, Cu^{2+}, Pb^{2+}	–	0.8-1.0 (6)	2-10 h	Chemical reaction	[72]
Fe–Mn	Cd^{2+}	–	0.5 (6)	–	–	[73]

As shown from Table 4, sorption capacity towards cations is within the interval of 0.4-1.0 mmol g^{-1} in neutral media. In general, sorption properties are determined by the solution acidity. For instance, the maximal uptake of $HCrO_4^-$ anions is realized at pH 2.5, the competition of anions of background electrolyte (acid) is expressed under lower pH. The capacity is predictably lower in alkaline media, since oxide materials show mainly cation exchange properties under these conditions.

It should be noted that the porosity of double hydrated oxides occupies intermediate position between these characteristics for simple oxides, as shown in the example of Al-Ti oxide (see Fig. 9b). Conjoining more porous (Al_2O_3•nH_2O) and less porous (MnO_2•nH_2O) materials leads to the development of mesoporosity of resulting double oxide.

It is also possible to regulate the PZC pH by means of control of the sorbent composition. For instance, the addition of MnO_2 to Zr, Sn, Ti hydrated oxides causes a shift of this pH to acidic region [74]. Alternately, the Al_2O_3 addition leads to a shift of the PZC to alkaline region. This means purposeful control of the sorption properties of hydrated oxides.

Materials Research Foundations 173 (2025) 1-39 https://doi.org/10.21741/9781644903452-1

2.4 Zeolites

Among silica-containing sorbents for the removal of heavy metal ions from water, natural and synthetic zeolites are probably the most commonly used materials. The formation of natural zeolites is mainly due to volcanic activity [75]. When an active volcano is located near water bodies, the ejected lava can appear in them. Hot lava interacts with water and salts during thousands of years. As a result, such crystalline materials as zeolites are formed.

All zeolites involve aluminosilicate framework consisting of a tetrahedral arrangement of Si^{4+} cations and Al^{3+} cations, which are surrounded by four O^{2-} anions (Fig. 10a, [76]). Each O^{2-} ion within the Al-O and Si-O bonds connects 2 cations – it is shared between 2 tetrahedrons. Thus, a 3d framework includes AlO_2 and SiO_2 building blocks, a Si:O ratio is 1:2.

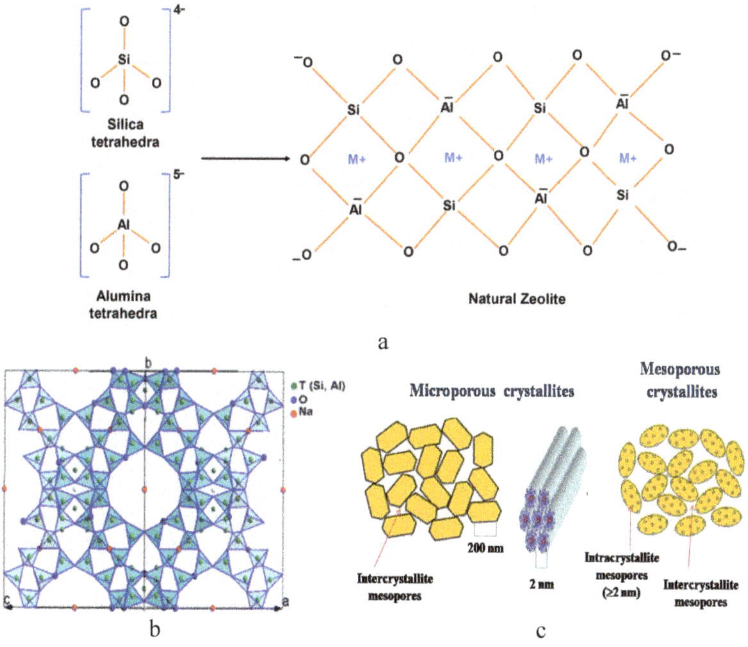

Figure 10. Structural formula of zeolites [76] (a), tetrahedron organization in zeolite microporous monocrystal [77] (b), hierarchical pore system [78, 79] (c). Reproduced according to the permission of Springer (a) and Elsevier (c) publishers or no permission is required (b).

Some Al^{3+} ions substitute Si^{4+} ions. As a result, a net negative charge in the alumosilicate framework is formed. The charge is due to the valency difference between the $(SiO_4)^{4-}$ and $(AlO_4)^{5-}$ tetrahedrons. Normally it is located on one of the O^{2-} anions, which is bonded with Al^{3+} cation. The negative charges are compensated by counterions (alkaline or alkaline earth metals, for instance, Na^+, K^+, Ca^{2+} and Mg^{2+}). They are between tetraedra and bound with them via ionic bonds. The composition of any zeolite can be described by the formula of $Me_2O_n \cdot Al_2O_3 \cdot xSiO_2 \cdot yH_2O$, where n is the valence of the metal cation, x is the mole ratio of SiO_2/Al_2O_3, y is the number of moles of water. The connected tetrahedra form a great variety of structures, which are characterized by the presence of internal crystalline polyhedral voids (Fig. 10b, [77]). The shape and size of the voids in zeolite monocrystal depend on the tetrahedron organization. In general, monocrystals involve a system of pores of angstrom size (micropores), they perform a function of ion/molecular sieves.

From the point of view of porous structure, zeolites can be divided into several types: (i) wide-pore zeolites, which are characterized by wider micropores comparing with regular zeolite structures, (ii) nanosized zeolites possessing only intercrystalline mesopores, (iii) mesoporous zeolite crystals having intracrystalline mesopores (Fig. 10c [78, 79]) [80]. In the last case, intracrystalline micropores are interconnected with mesopores forming hierarchical structure inside crystallites.

In order to obtain materials of these types, they are synthesized in aqueous basic media, which contains Si and Al sources as well as inorganic and/or organic cations, under hydrothermal conditions [81]. For instance, these sources can be a waste residue after the alkaline extraction of lithium from lithium ores [82]. To minimize the diffusion limitation caused by microporous structure, some strategies are applied to the synthesis procedure: decreasing the crystal size, i.e. obtaining nanocrystalline zeolites, and formation of hierarchical porous structure inside crystallites (see Fig. 10c). Some approaches are used for this purpose: dealumination (treatment of zeolite with acids), desilication (treatment with alkali), zeolitization of mesoporous amorphous material, template synthesis.

These techniques give a possibility to obtain materials with high mesoporosity. For instance, the pore size distribution for natural zeolite containing mainly clinoptilolite shows a peak at 2.5 nm that is attributed to the pore diameter of 5 nm (Fig. 11a, [83]). According to [84], the origin of these pores is a cleavage of the primary clinoptilolite particles. They are long flat parallel bars, a length of which is 350–1000 nm, a thickness is 25–100 nm, and a wideness is 200–400 nm. A thickness of the cleavage regions can reach several tens nanometeters, straight ordered voids between these regions (intercrystalline mesopores) probably give a sharp peak. On the other hand, it can be attributed to pores between flat bars. which adjoin each other with their side walls. Ordered pores in a synthetic zeolite are the same size (Fig, 11b, [85]). However, their volume is 2 orders of magnitude higher compared with natural zeolite. As opposed to natural material, the ordered mesopores of synthetic zeolite correspond to intracrystalline voids.

In general, zeolites obtained via template synthesis, are characterized by very high specific surface area (up to 1560 m^2g^{-1} [86]). However, a much lower value (16 m^2g^{-1}) has been reported for the natural clinoptilolite-based material [83]. Nevertheless, the treatment with acid allows one to obtain a sample with more developed surface (114 m^2g^{-1}).

Zeolites are promising sorbents for the removal of heavy metals from water, such as Cd^{2+}, Cr^{3+}, Pb^{2+}, Hg^{2+}, Ni^{2+} [87]. This is possible, since alkaline or alkaline earth metals ions, which occupy the surface of $(SiO_4)^{4-}$ and $(AlO_4)^{5-}$ tetrahedrons, are able to ion exchange. Moreover, slight acidification of equilibrium solution was recorded [88] indicating some amount of H^+ ions in the zeolite matrix. This is possible, when zeolites are preliminarily converted into H-form via the acidic treatment. This approach allows one to improve sorption properties of natural zeolites (other reagents can be alkali or salts).

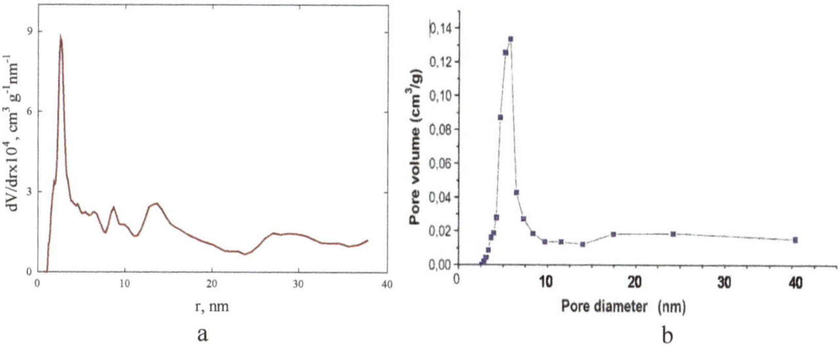

Figure 11. Pore size distributions for natural [84] (a) and synthetic zeolites [85] (b). Reproduced according to the permission of Springer publisher.

Natural zeolite containing mainly clinoptilolite was kept in a 1 M H_3PO_4 solution [84, 89, 90], this procedure results in a leakage of impurities (bernalite) [84]. The treated zeolite shows the improvement of Pb^{2+} and Ca^{2+} sorption under batch conditions evidently due to the removal of admixtures. As mentioned in [91], Pb^{2+} ions are removed practically completely from acidic mine water. Small particles (<1 mm) are more efficient sorbents comparing with large particles (1-2.5 mm). It was found that the capacity is proportional to the initial Pb^{2+} concentration from 50 to 400 mg dm^{-3}. Preferable sorption of Pb^{2+} ions comparing with Ni^{2+} [86, 92, 93], Zn^{2+} [92], Cu^{2+} [86, 93], Cr^{3+} and Cd^{2+} [88] species, was reported for natural zeolite consisting of clinoptilolite/erionite [92] or mordenite [88], as well as for synthetic zeolites [86, 88, 93].

It should be noted that the sorption capacity of zeolites is sufficiently lower compared with hydrophosphates or oxides of multivalent metals. The values of 0.026 - 0.047 mmol g^{-1} (Pb^{2+}) were reported for synthetic zeolites (maximal sorption is achieved at pH>5) [86]. It is higher compared with natural sorbent (0.021 mmol g^{-1}). This disadvantage is compensated by low cost of natural materials or simple preparation from cheap and available regents for synthetic sorbents, Moreover, zeolites can be easily regenerated. For instance, a 0.01 M HCl : HNO_3 solution can be used for this purpose [86]. Sorption is rather fast: for instance, the equilibrium time is 1 h for the sorption of Cd^{2+}, Cu^{2+} and Ni^{2+} ions on natural zeolite (the grain size was 0.2-2 mm) [94]. Shorter time (30 min) was reported for synthetic zeolite (sorption of Zn^{2+} and Cd^{2+} ions) [95], Internal diffusion was found as a rate-determining stage. It should be stressed that the isotherms are fitted by the Dubinin-Radushkevich model. It means that the pores are completely filled with only one ion. As suggested, the pore geometry is especially important for Ni^{2+} sorption [96].

It is noted that Cd^{2+} adsorption from model solution reaches 90-99 % under batch and dynamic conditions [91]. In the case of strong and weak acidic mine water, the removal degree is 86% and 21% respectively. Lower efficiency of zeolites towards Cd^{2+} ions in weakly acidic media is explained by a competition with Ca^{2+}, Mg^{2+} and Mn^{2+} species. The capacity of 0.066-0.089 mmol g^{-1} towards Cd^{2+}ions is reported in [88]. Natural zeolites can be also used for the recovery of Hg^{2+} ions from copper smelting and refining effluents [91]. It is noted that modifying clinoptilolite with complex forming agent gives a possibility to increase the capacity towards these ions up to 0.85 mmol g^{-1}.

Sufficient disadvantage of inorganic sorbents is a strong dependence on the solution pH: in reality, the highest capacity values are achieved in neutral media. Higher uptake can be reached in alkaline solutions, but heavy metal ions form insoluble compounds under these conditions.

In order to solve the problem of the pH effect, a large variety of composites based on inorganic sorbents are developed. Among various additives, such carbon material as GO looks most attractive. GO and its composites are considered further.

3. Graphene oxide and its composites with inorganic sorbents

3.1 Graphene oxide

Graphene oxide (GO) is allotropic carbon modification, which is related to 2d materials: its flakes include nanosheets with a thickness of 0.54 nm, i.e. one atom [97]. GO involves such oxygen-containing groups as carboxyl (–COOH), phenolic (–OH) and epoxy (–C–O–C–) fragments (Fig. 12). Carboxyl groups and partially phenolic fragments are located along the flake perimeter, other part of –OH groups and epoxy groups are placed on planes from 2 sides of flakes.

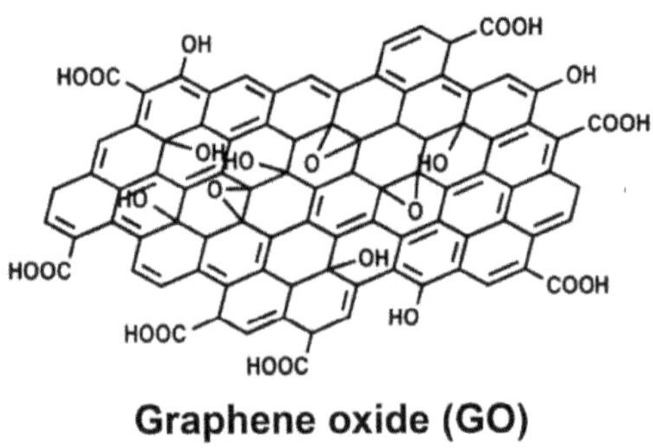

Figure 12. Idealized structure of GO [97]. Reproduced according to the permission of the Elsevier publisher.

The initial substances for the GO synthesis is graphite [98-100] or carbon nanotubes [101-103]. The methods developed by Staudenmaier [98] or Brodie [99] are usually applied to obtaining this material, however, the most well-known approach is the Hummers technique [100]. Graphite or nanotubes are oxidized down to necessary levels by means of a number of aggressive reagents ($KClO_3$, HNO_3. $KMnO_4$, H_2SO_4) [98-101]. Electrochemical method for unzipping carbon nanotubes can be realized under soft conditions [102, 103].

Such feedstock as polysaccharides should be also mentioned [104]. Carbonization in inert atmosphere is necessary to obtain GO with needed functional properties.

GO is characterized by high specific surface area – the values obtained with a BET method (nitrogen adsorption-desorption) are 560 [105] – 900 [106] m^2g^{-1}. Lower magnitude (230 m^2g^{-1}) was reported for partially unzipped carbon nanotubes [45]. This is much less than the theoretical value of specific surface area attributed to completely expended and isolated nanosheets (≈2600 m^2 g^{-1} [107]). This discrepancy is due to curling, agglomeration and overlapping graphene sheets. As a result, N_2 molecules cannot penetrate between them under low temperature. However, a method of standard contact porometry, which provides a usage of octane or water, gives more realistic values (2000-2400 m^2 g^{-1}) due to disjoining pressure of the liquids [108].

Typical integral pore size distributions, which were obtained for GO in aqueous and octane media are given in Fig. 13. When the pore radius is less than 1700 nm, the volume of pores filled with water is larger compared with voids filled with octane [55]. This indicates swelling GO in water – this material behaves similarly to ion exchange polymers, which also show a large volume of hydrophilic pores [109]. Similar results for GO were obtained in [110].

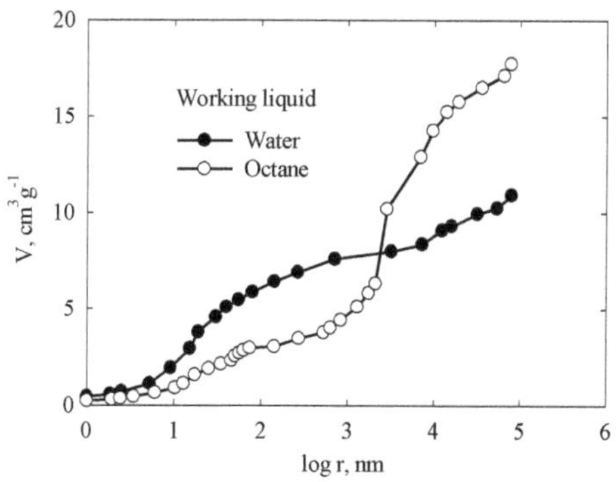

Figure 13. Integral pore size distributions for GO obtained in various media with a method of standard contact porometry. Adapted from [55].

A combination of good dispersibility in water, highly developed surface and large amount of available functional groups allows one to utilize GO as an effective sorbent of heavy metal ions. Mainly –COOH groups are involved in ion-exchange, which is accompanied by surface complexation [111, 112]. Since these groups are weakly acidic (the pH of PZC is 3.9 [112]), the maximal uptake is realized at pH≥4 (Table 5). When the pH exceeds 6, the uptake becomes slightly lower due to the formation of single-charged hydroxocomplexes of ions being sorbed [111]. Sorption was performed even at pH>8–10 [112], however, it is caused mainly by the deposition of insoluble compounds under these conditions. No sufficient effect of ionic strength on the uptake of Cd^{2+} [112, 113, 114], Co^{2+} [112], Cu^{2+}, Ni^{2+} [113] ions was found. Humic acid depresses sorption – the removal degree of Cd^{2+} and Co^{2+} decreases by 10-30 % depending on pH [112]. Aggregation of GO nanoparticles in the presence of Cu^{2+} ions is reported [113].

Since GO is usually in a finely dispersive form, its practical application is limited due to difficult solid-liquid separation. In order to overcome this problem, GO is inserted into a matrix, which also possesses sorption ability towards heavy metal ions. Inorganic materials, such as hydrophosphates and hydrated oxides of multivalent metals as well as zeolites are typical representatives of these matrices.

Table 5. Sorption of heavy metal ions on GO

Sorbed ions	Optimal pH	Sorption capacity, mmol g^{-1}	Isotherms	Kinetic model	Reference
Cu^{2+}, Cd^{2+}, Zn^{2+}, Pb^{2+}	4 – 5	4.6 – 5.4	Langmuir	Pseudo-second order	[111]
Co^{2+}, Cd^{2+}	> 8	1.3 – 2.6	Langmuir, Freundlich	–	[112]
Cd^{2+}	> 6	1.3	Langmuir, Freundlich	–	[113]
Cu^{2+}, Cd^{2+}, Ni^{2+}	7	0.7 – 1.1	Langmuir, Freundlich	Pseudo-second order	[114]
Cu^{2+}	–	0.7	Langmuir	Pseudo-second order	[115]
UO_2^{2+}	4 – 8	1.3 (U(VI))	Langmuir		[116]

3.2 Composites of inorganic sorbents with GO

Synthesis of GO-containing composites, which are based on inorganic sorbents, are mainly performed by framing of previously obtained GO with hydrated oxide [20, 45, 55, 118, 119], hydrophosphate [20, 120] or zeolite [121]. For instance, GO powder was added to the mixed solution of sodium silicate and aluminum sulfate, then the colloidal solution was aged to obtain a gel. This gel was thermally treated in an autoclave at 100° C for several days. A composite consisting of zeolite and GO was obtained by this manner [121]. Zeolite-based composite was synthesized by means of GO adsorption on zeolite powder [122]. A solution prepared from $ZrOCl_2$ [118-120] or sol of Zr hydroxocomplexes [20, 45, 55] were applied to the composite synthesis. The hydrogel granules were deposited with NaOH [20, 45, 55, 119], NaH_2PO_4 [120] or H_3PO_4 [20] solutions.

The composites of GO with magnetite are in a focus of attention, since this inorganic constituent allows one to solve a number of tasks: (i) enlargement of the sorbent particles, (ii) easy separation of solid and liquid, (iii) enhancement of adsorption properties of GO, since magnetite possesses own adsorption ability. A common method for the preparation of the magnetite-GO composites is its precipitation from a suspension containing $FeCl_3$, $FeCl_2$ and GO [123]. Magnetite is formed according to the Elmore reaction [124]:

$$2FeCl_3 + FeCl_2 + 8NH_3\cdot H_2O \rightarrow Fe_3O_4 + 8NH_4Cl + 4H_2O, \quad (4)$$

GO is co-precipitated forming a composite. Another approach has been proposed in [125]. Synthesis was carried out through *in situ* catalytic graphitization (carbothermal reduction) of biochar impregnated with $Fe(NO_3)_3$. Biochar was kept at 900° C in argon atmosphere. Then the

 Materials Research Forum LLC
https://doi.org/10.21741/9781644903452-1

Fe_3O_4 composite was treated at 500–800° C in argon that was preliminarily bubbled through water. In fact, water steam activation was used. As suggested, the composite containing magnetite, graphene and activated carbon is obtained by this manner.

Fig. 14a shows the effect of small GO amount (2 %) on the porous structure of composite based on hydrated zirconium dioxide [117]. No mesopores were found for the inorganic support: only micro- and macroporosity are attributed to it (the integral pore size distribution, which was obtained with a method of standard contact porometry, shows a wide plateau for the pore radius of 1 nm - 1 μm). At the same time, the composite demonstrates mesoporosity – a radius of mesopores exceeds 10 nm. In other words, GO loses the packing particles of hydrated zirconium dioxide, since graphene nanosheets coat them (Fig. 14b). The coating prevents the formation of Zr–O–Zr bonds between the particles protecting them against compaction. Similarly to GO (see Fig. 13), the composite also swells in water – the volume of pores filled with water is larger comparing with pores filled with octane. The specific surface area for the composite was estimated as 360 (octane) and 390 (water) m^2g^{-1} [55]. Increasing the GO content in the composites causes a decrease of their specific surface area that was determined with a method of nitrogen adsorption-desorption. In the case of unzipped carbon nanotubes, this technique gives a value of 230 m^2g^{-1} [20]. The composite based on hydrated zirconium dioxide containing this carbon addition (2 %) shows smaller specific surface area than pure sorbent (280 and 338 m^2g^{-1} respectively). As found for GO-containing composites, this value decreases with increasing its content [45]. The same effect of GO was found for zeolite-based composites [121]. Alternative results were obtained in [119[– GO increases the specific surface area of hydrated zirconium dioxide. However, much lower values were obtained (28 m^2g^{-1} for inorganic constituent and 38 m^2g^{-1} for composite), Unzipped carbon nanotubes increase the specific surface area for ZHP: from 165 (pure ZHP) to 190 m^2g^{-1} (composite) [20]. For magnetic composites, the magnitudes up to 486 m^2g^{-1} were reported [125]. It should be stressed that mentioned values obtained by means of the nitrogen adsorption-desorption technique are understated. Morphology of the composites affected by advanced carbon materials is described in [117, 126].

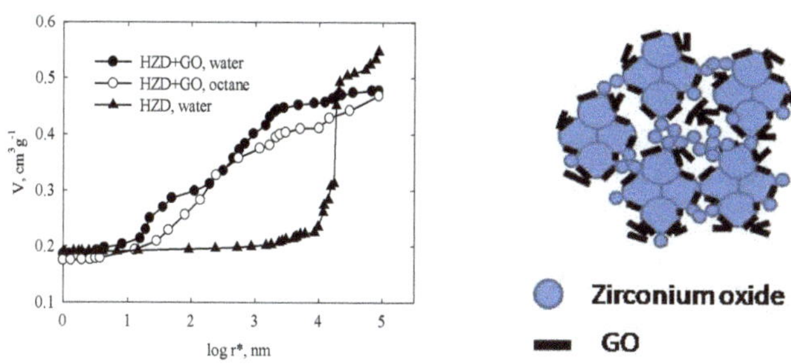

Figure 14. Integral pore size distributions for hydrated zirconium dioxide (HZD) and its composite with GO obtained in water and octane with a method of standard contact porometry (a), loosening particles of inorganic constituent affected by GO (b) [117]. Reproduced according to the permission of the Springer publisher.

GO nanosheets are associated with inorganic sorbents via hydrogen bonds and Lewis acid-base interaction, here Lewis basic sites are due to the π-electron system of graphite planes [127]. The composites include also aggregates and agglomerates of GO [55]. No sufficient change of the pH of PZC due to GO adsorption was found for such crystalline materials as goethite and hematite [128]. At the same time, GO shifts the PZC to acidic region (down to pH 5) [118]. According to the preparation method (precipitation of hydrated zirconium dioxide in GO suspension), the pH of PZC is expected to be ≥6 for pure inorganic sorbent (see Table 2).

GO on the surface of inorganic particles determine sorption properties of composites. For instance, GO enhances cation uptake (Pb^{2+}) and depresses anion sorption ($HCrO_4^-$) on hydrated zirconium dioxide (Fig. 15a) [55]. GO obtained from unzipped carbon nanotubes increases sorption capacity of this inorganic sorbent towards U(VI) [20]. This effect is expressed in the pH interval of 5-8 (Fig. 15b). Alternately, pH 4 is suggested as the most favorable condition [119]. The sorption rate is fitted by the model of chemical reaction of pseudo-second order [20, 119]. The isotherms are fitted with Dubinin-Radushkevich [20] or Langmuir model [119]. In the last case, the maximal sorption capacity of 0.5 mmol g^{-1} is reported.

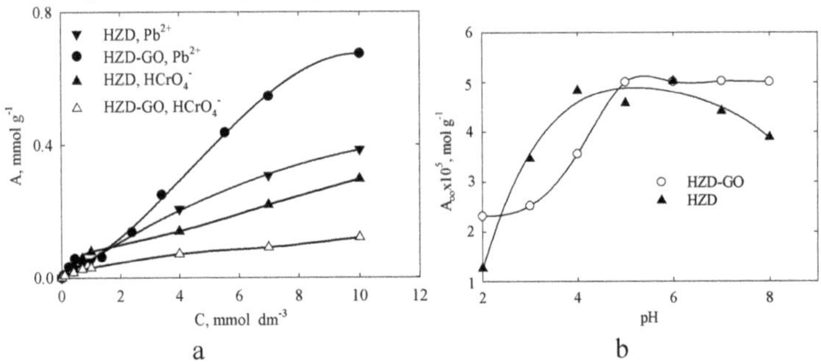

Figure 15. Isotherms of Pb^{2+} and $HCrO_4^-$ sorption on hydrated zirconium dioxide (HZD) and its composite with GO (a), sorption capacity of HZD and its composite towards U(VI) containing ions as a function of pH (b). Adapted from [55] and [20].

Sorption of Pb^{2+}, Cd^{2+} and Hg^{2+} ions on the GO-containing composite based on hydrated zirconium dioxide was investigated in [118]. Langmuir model for sorption isotherms was suggested similarly to [55]. The most favorable conditions for sorption of heavy metal ions are realized at pH 4-6 [118]. The maximal sorption capacity reaches 1.8 (Pb^{2+}), 2.8 (Cd^{2+}) and 1.4 (Hg^{2+}) mmol g^{-1} (lower value of 0.8 mmol g^{-1} is given for Pb^{2+} ions [55]). When the solution contains a mixture of these ions, their removal degree decreases within the order: Hg^{2+}> Cd^{2+}> Pb^{2+}. A rate-determining stage was found to be internal diffusion. The equilibrium is achieved in 1 h.

The values of maximal sorption capacity, which was found for sorption of Pb^{2+} ions on the ZHP-GO composite (1.7 mmol g^{-1} [120]) is very close to that reported in [118] for the material based on hydrated zirconium dioxide. Lower magnitude was found in the case of Cd^{2+} ions (2.1 mmol g^{-1}) [120]. For Cu^{2+} and Zn^{2+} species, the capacity reaches 5.1 and 3.9 mmol g^{-1} respectively. The most favorable conditions are realized at the pH of 3-6, where functional groups of ZHP and GO are dissociated. The complexation of sorbed ions with phosphate groups was proved with a XPS technique. Due to the chemical interaction of sorbed ions with functional groups, the rate of sorption obeys the model of chemical reaction of pseudo-second order. The sorbent regeneration was carried out with a 3 M HCl solution, the desorption degree of 90 % is reported. Such solutions as 0.1 M and 1 M HNO_3 as well as 0.1 M $NaHCO_3$ were used for regeneration [20]. Here the regeneration degree of ZHP-GO composite loaded with U(VI)-containing ions reaches 98, 85 and 69 %. Lower values were obtained for pure ZHP, when it was regenerated with an acid. However, the regeneration degree is 98 % in the case of a $NaHCO_3$ solution. The sorbents are regenerated during 10 min [120] or 2 h [20].

Comparing with composites based on hydrated zirconium dioxide [118] and HZP [120], magnetite containing GO and activated carbon show lower value of the maximal sorption capacity of Pb^{2+} (0.7 mmol g^{-1} at pH 5-7) [125]. However, the advantage of magnetic composite is easy separation from liquid. Magnetite-based composite was also used for the removal of Co^{2+} ions from water [123]. In this case, the maximal capacity of 0.2-0.3 mmol g^{-1} is reported, equilibrium is reached during 5 h. As suggested, the sorption of Co^{2+} ions is independent on ionic strength but is pH-dependent: a rapid increase of sorption capacity was found at pH 6-7.

The composite including zeolites and GO also demonstrates lower capacity towards Pb^{2+} (1.2 mmol g^{-1}) and Cd^{2+} (1.1 mmol g^{-1}) ions [122] than zirconium-containing composites [118, 120]. Sorption of Sr^{2+} and Cs^{+} ions was also investigated, the capacity of 1.3 and 0.8 mmol g^{-1} respectively was reported. Their sorption is rather fast, since the equilibrium time is 30 min. In the case of Pb^{2+} and Cd^{2+} ions, the equilibrium is achieved in 5 h.

The effect of GO additives on the removal of Fe^{3+} ions from water using zeolite-based composite was studied in [121]. As found, the increasing GO amount up to 3 mass % causes a growth of its uptake up to 80 % – the resulting capacity is 1.3 mmol g^{-1} for the sorbent of optimal composition, the equilibrium time is 5-10 min. Alternately, the uptake decreases to 16% with further growth of the GO content.

It should be noted that inorganic sorbents can be used not only as a base for fillers like GO, but also as modifiers of such specific mesoporous materials as ion exchange polymers. The composites of this type are considered further.

4. Composites based on ion exchange polymers containing inorganic sorbents

Inorganic ion-exchangers are used as constituents of the composites based on ion exchange polymers, the most common of which are commercial ion exchange resins. The purpose of modifying polymers with inorganic components is to improve selectivity of strongly acidic or strongly basic resins, which are applied to the removal of heavy metal ions in a wide pH interval.

Though dry ion exchange resins show practically no micro- and mesoporosity, their swollen forms can be related either to mesoporous or to micro-mesoporous materials [109, 129, 130]. This is proved with porometric methods, which are suitable for the investigation of swollen materials (BET water adsorption, standard contact porometry, differential scanning calorimetry, thermoporometry) [130]. The data of porometric measurements for different ion exchange resins and membranes are summarized in [109, 129]. As an example, Fig. 16 illustrates differential pore size distribution for strongly acidic gel-like cation exchange resin, namely Dowex HCR-S (Dow Chemical). The plot shows a considerable fraction of pores within the interval of radius of 1-25 nm, porosity due to micropores and all pores was estimated as 40 and 72 % respectively [131].

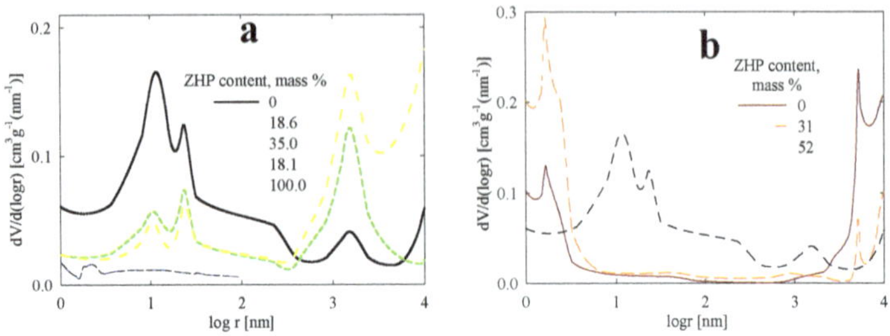

Figure 16. Evolution of porous structure of swollen Dowex HCR-S resin under the influence of ZHP (the modifier content was varied). The composite includes single nanoparticles (a), their aggregates and agglomerates (b) [109]. Reproduced according to the Springer publisher.

As suggested in the middle of last century, any swollen organic ion exchange material is a solution of water in a polymer, i.e. polymer and water phases cannot be divided. Earlier this point of view was common and suggested even in a classical Helfferich monograph [132]. Nevertheless, Hsu and Gierke proved divided phases of polymer and water [133]. They investigated ion exchange Nafion membrane with a small-angle X-ray scattering technique. Based on obtained results, a cluster-channel model of the porous structure of swollen ion exchange polymers has been proposed (Fig. 17 [134]). Despite critique and other points of view [135], the ideas by Hsu and Gierke are currently the most accepted for interpreting the functional properties of ion exchange polymers (both ion exchange resins and membranes).

According to the Hsu-Ghirke model, the elements of porous structure in hydrated ion exchange polymers can be divided into 2 types: gel areas and intergel spaces, where hydrophobic fragments of hydrocarbon chains are located [133]. Gel areas are formed by nanosized clusters, which are interconnected by narrower channels (Fig. 17). Clusters and channels are related to hydrophilic pores, where ion transport is realized. Hydrophobic fragments of polymer chains form pores that belong to the intergel spaces. They are larger compared with clusters. Additionally, the porous structure of ion exchange polymers includes structural defects, a size of which is several microns. A wide range of pores in ion exchange polymers determines a variety of embedded inorganic particles [109, 136, 137]. For instance, pore size distribution for ion exchange resin shows regions corresponding to clusters (pore radius is 10 nm), intergel spaces (23 nm) and structure defects (1.5 μm) (see Fig. 16).

Since the particles of inorganic modifiers contain functional groups, they are additional osmotic centers in ion exchange polymers, and affect their porous structure. For instance, where a polymer contains nanoparticles (in clusters), their aggregates (in intergel spaces) or agglomerates (in structure defects), their influence on porous structure of the polymer matrix is ambiguous (see Fig. 16). Nanoparticles screen clusters, as a result, their volume decreases. At the same time, aggregates and agglomerates stretch intergel voids and structure defects, since they are additional osmotic centres. The size of clusters remains without changes, when the amount of modifier increases, since nanoparticles prevent squeezing from the side of intergel spaces and structure defects. However, when clusters are free from nanoparticles, they become smaller due to squeezing from the side of hydrophobic pores that are filled with aggregates and agglomerates.

Materials Research Forum LLC
https://doi.org/10.21741/9781644903452-1

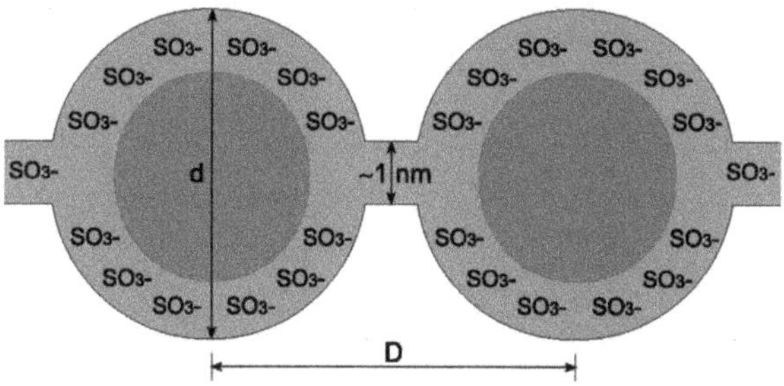

Figure 17. Cluster-channel model of the porous structure of ion exchange polymers [134].

The amount of inorganic modifier and its morphological features determine the functional properties of composites. Table 6 [138-143] illustrates sorption behavior of polymer-inorganic composites in relation to some toxic ions, the extraction of which from water was carried out under dynamic conditions. Besides toxic cations, aqueous solution contained an excess of hardness ions. In the case of $HCrO_4^-$ removal, SO_4^{2-} ions are in a solution, the modifier increases in the breakdown capacity and the degree of extraction of toxic ions. Using the example of arsenate and chromate anions, an increase in the capacity of composites containing particles of hydrated oxides of multivalent metals was recorded in comparison with pristine anion exchange resin [144-148], a synergistic effect was recorded in many cases.

Among non-commercial polymer-inorganic materials, such composite as sulfonic acid – crystalline ZHP [149] should be mentioned. The composite containing ZHP was applied to the removal of radioactive $^{90}Sr^{2+}$ ions from water [150]. The maximal sorption capacity was found to achieve 2.1 mg g^{-1} at pH 4-8. Radionuclides are removed practically completely from the solution containing also Na^+ and K^+ ions under dynamic conditions – their removal degree is up to 10 %. The equilibrium rime was 1.7 h. Increasing temperature from 25° to 45° C causes a slight decrease of capacity (about 5 %). The mechanism of $^{90}Sr^{2+}$ sorption involves not only ion exchange, but also formation of a new phase of the inorganic constituent,

Polymer-inorganic composites, the organic constituent of which possesses no ion exchange ability, were also reported. Such materials as sulfophenol-formaldehyde resin –iron ferrite (magnetic sorbents) [150], polyacrylonitrile – potassium titanate or nickelferrocyanate [151], nylon – ZHP [152], poly(methyl methacrylate) – ZHP [153] (this polymer was also applied to modifying membranes for pressure-driven separation [154]). The composites were applied to sorption of Co^{2+} [150], Ag^+, Sr^{2+} [151], Hg^{2+} [152] Pb^{2+} [153] species. The sorbents reported in [152] and [153] preferably sorb Hg^{2+} [152] and Pb^{2+} [153] ions from binary solutions containing also other heavy metal cations or Mg^{2+} ions.

It should be noted that inorganic sorbents, such as ZHP and hydrated zirconium dioxide, are also applied to modifying membranes for electro- [154] and pressure-driven [155] separation. GO-containing composites based on these inorganic sorbents [156-160] are also used for this purpose. Magnetite is also used for modifying cellulose of plant origin, which is produced from agriculture wastes [161, 162]. These materials are applied not only to the removal of heavy metal ions from water [161], but also to the purification of water surface from oil and oil products [161, 162].

Table 6. Removal of toxic cations from water under dynamic operation

Polymer	Modifier, content, mass. %	Type and size of particles	Ion, initial concentration, mmol dm^{-3}	Break-through capacity, mmol cm^{-3}	Removal degree, %*	Reference
Strongly acidic, gel, 8-10 % of cross-linkage	ZHP, 15	Single, 20-50 nm, aggregates, 70-300 nm	$[UO_2^{2+}OH]^+$, 0.2	3.93×10^{-3} (3.4×10^{-4})	65 (50)	[138]
	ZHP, 50	Dendrite agglomerates, several microns		2.36×10^{-3} (3.4×10^{-4})	87 (50)	
	ZHP, 37	Single, 4-20 nm, aggregates, agglomerates, 50 nm – several microns	Cd^{2+}, 0.035	5,16 $\times10^{-3}$ (1.5×10^{-3})	>99 (19)	[139]
	ZHP, 43	Agglomerates, several microns	Cd^{2+}, 0.1 Ni^{2+}, 0.1	0.24 (0.16) 0.24 (0.16)	99 (90) 99 (95)	[140]
Weakly acidic, macroporous, 8-10 % of cross-linkage	ZHP, 8	Single, 5-15 nm, aggregates, agglomerates, 50 nm – several microns	Ni^{2+}, 0.1	2.1×10^{-2} (7.0×10^{-3})	85 (85)	[141]
Strongly acidic, gel, 2 % of cross-linkage	ZHP, 40	Single, 7-14 nm, aggregates, 30-50 nm, aggregates, agglomerates, up to several microns	Ni^{2+}, 0.3	1.58×10^{-2} (2.7×10^{-3})	>99 (93)	[142]
Strongly basic, 8-10 % of cross-linkage	–	Single, 3-6 nm	$HCrO_4^-$, 0.2	62 (6)	>99 (90)	[143]

* Values for pristine resins are in brackets.

Conclusions

In this chapter, both traditional and unconventional mesoporous materials, which are attractive as sorbents for the removal of heavy metal ions from water, are considered. Synthetic inorganic materials, which show high porosity in a dry state, possess considerable capacity towards these ionic species. It should be noted that their selectivity strongly depends on micropores, while mesopores provide fast sorption. Higher ion uptake is attributed mainly to synthetic sorbents compared with natural materials. However, a sufficient advantage of natural substances is their low cost.

Selectivity to one or other ions is characteristic especially for zeolites, which demonstrate sieving effect – they sorb preferably Pb^{2+}cations. Additionally, synthetic mesoporous substances of this type show rather fast sorption. In general, sorption of ions on crystalline materials is accompanied by the formation of new phases. This slows down sorption and makes regeneration rather difficult. As opposite to crystalline materials, amorphous sorbents demonstrate lower selectivity but rather fast sorption, they can be easy regenerated. The uptake of heavy metal ions is realized according to the mechanism of ion exchange that is accompanied by surface complexation and deposition of insoluble compounds inside granules. Namely the contribution of these mechanisms provides selectivity of amorphous sorbents towards one or other ionic species. All inorganic sorbents show the uptake of cations. The exception is (hydr)oxide materials: amphoteric properties are attributed to them. In order to enhance their ability to sorb cations and depress anion uptake, composites containing GO are prepared. This advanced carbon material possesses its own mesoporosity, which is especially brightly expressed in aqueous media. Moreover, GO-containing composites swell under these conditions as opposed to many inorganic sorbents like hydrophosphates, (hydr)oxides and so on.

Inorganic sorbents are often used for modifying ion exchange polymers, among which ion exchange resins are the most common. The aim of these investigations is to improve their functional properties. As opposite to inorganic sorbents, polymer-based materials are related to unconventional, since they show micro- and mesoporosity in a media of water or non-polar liquids. It should be noted that sorption properties strongly depend on the location of embedded particles, which can be placed both in hydrophilic and hydrophobic pores of polymers. The particle location depends on astate of incorporated particles (non-aggregated nanoparticles, their aggregates and agglomerates) that can be controlled on the synthesis stage.

Surface functionalization of mesoporous materials gives an additional possibility to enhance their sorption properties. Another way is conjoining different mesoporous materials in composites. Systematization of the information in this field can be a subject of further investigations.

References

[1] A. Clearfield, J.A. Stynes, The preparation of crystalline zirconium phosphate and some observation on its ion exchange behaviour, J. Inorg. Nucl. Chem. 26 (1964) 117-129. https://doi.org/10.1016/0022-1902(64)80238-4

[2] G. Alberti, F. Toracca, Crystallyne insoluble salts of polybasic metals. 2. Synthesis of crystalline zirconium and titanium phosphates by direct precipitation, J. Inorg. Nucl. Chem.. 30 (1968) 317-318. https://doi.org/10.1016/0022-1902(68)80096-X

[3] F. Zhang, Y. Xie, W. Lu, X. Wang, S. Xu, X. Lei, Preparation of microspherical a-zirconium phosphate catalysts for conversion of fatty acid methyl esters to monoethanolamides, J. Colloid Interface Sci. 349 (2010) 571-577. https://doi.org/10.1016/j.jcis.2010.05.043

[4] Yu. Cheng, G.K. Chuah, The synthesis and applications of α-zirconium phosphate. Chin. Chem. Let. 31 (2020) 307-310. https://doi.org/10.1016/j.cclet.2019.04.063

[5] S. Yamanaka, M. Tanaka, Formation region and structural model of γ-zirconium phosphate, J. Inorg. Nucl. Chem. 41 (1979) 45-48. https://doi.org/10.1016/0022-1902(79)80391-7

[6] A.M. Krogh, A.P. Norby, J.C. Hanson, J.C. Hanson, T. Vogt, Preparation and characterization of a new 3-dimensional zirconium hydrogen phosphate τ-Zr(HPO4)2. Determination of the complete crystal structure combining synchrotron X-ray single-crystal diffraction and neutron powder diffraction. Inorg. Chem. 37 (1998) 876-881. https://doi.org/10.1021/ic971060h

[7] A. Clearfield, A.I. Bortun, L.N. Bortun, J.R. Garcia, Direct hydrothermal synthesis of zirconium phosphate and zirconium arsenate with a novel basic layered structure in alkaline media. Inorg. Chem. Commun. 1 (1998) 206-208. https://doi.org/10.1016/S1387-7003(98)00055-0

[8] U. Costantino, L. Szirtes, E. Kuzmann, J Megyeri, K Lázár, Exchange of iron ions into layers of α-zirconium phosphate, Solid State Ionics. 141-142 (2001) 359-364. https://doi.org/10.1016/S0167-2738(01)00807-4

[9] Y. Hasegawa, I. Tomita, The ion-exchange behaviour of some cations on crystalline zirconium phosphate, Bull. Chem. Soc. Jap. 4 (1974) 2389-2392. https://doi.org/10.1246/bcsj.47.2389

[10] S.E. Horsley, D.V. Nowell, The particle characteristics of cation exchanged α-zirconium phosphate, J. Colloid Interface Sci. 49(1974) 394-402. https://doi.org/10.1016/0021-9797(74)90384-1

[11] O. Mrad, A. Abdul-Hadi, H. Arsan, Preparation and characterization of three different phases of zirconium phosphate: study of the sorption of 234Th, 238U, and 134Cs, J. Radioanal. Nucl. Chem. 287 (2011) 177-183. https://doi.org/10.1007/s10967-010-0679-1

[12] M.V. Maslova, D. Rusanova, V. Naydenov, O.N. Antzutkin, L.G. Gerasimova, Extended study on the synthesis of amorphous titanium phosphates with tailored sorption properties, J. Non-Crystalline Solids 358 (2012) 2943-2950. https://doi.org/10.1016/j.jnoncrysol.2012.06.033

[13] J.-H. Jung, H.-J. Sohn, Preparation and characterization of mesoporous zirconium phosphates from alkyl phosphates, Micropor. Mesopor. Mater. 106 (2007) 49-55. https://doi.org/10.1016/j.micromeso.2007.02.018

[14] O. Zakutevskyy, S. Khalameida, V. Sydorchuk, M. Kovtun, The effect of hydrothermal, microwave, and mechanochemical treatments of tin phosphate on sorption of some cations, Materials. 16 (2023) 4788, https://doi.org/10.3390/ma16134788. https://doi.org/10.3390/ma16134788

[15] Yu.S. Dzyazko, V.V. Trachevskii, L.M. Rozhdestvenskaya, S.L. Vasilyuk, V.N. Belyakov, Interaction of sorbed Ni(II) ions with amorphous zirconium hydrogen phosphate, Russ. J. Phys. Chem. 87 (2013) 840-845. https://doi.org/10.1134/S0036024413050063

[16] M. V. Maslova, A. S. Chugunov, L. G. Gerasimova, N. V. Konovalova, Acid-base and sorption properties of amorphous titanium phosphate, Radiochem. 55 (2013) 392-398. https://doi.org/10.1134/S1066362213040097

[17] B. Pan, Q. Zhang, W. Du, W. Zhang, B. Pan, Q. Zhang, Z. Xu, Q. Zhang, Selective heavy metals removal from waters by amorphous zirconium phosphate: Behavior and mechanism. Water Res. 41 (2007) 3103-3111. https://doi.org/10.1016/j.watres.2007.03.004

[18] B. N. Laskorin, V.A. Goldobina, V. M. Kopanev, Sorption of ions of non-ferrous metals and iron by inorganic ionites based on zirconium phosphate, Izv. Vyssh. Uchebn. Zaved., Tsvetn. Metallurg. 16 (1973) 22

[19] O. I. Zakutevskyy, T. S. Psareva, V. V. Strelko, Sorption of U(VI) ions on sol-gel-synthesized amorphous spherically granulated titanium phosphates, Russ. J. Appl. Chem. 85 (2012) 1366–1370. https://doi.org/10.1134/S107042721209011X

[20] O.V. Perlova, I.S. Ivanova, Yu.S. Dzyazko, M.O. Danilov, I.A. Rusetskii, G.Ya. Kolbasov, Sorption of U(VI) compounds on inorganic composites containing partially unzipped multiwalled nanotubes, Him. Fiz. Tekhnol. Poverhni. 12 (2021) 18-31. https://doi.org/10.15407/hftp12.01.018

[21] I. Zhuravlev, O. Zakutevsky, T. Psareva, V. Kanibolotsky, V. Strelko, M. Taffet, G. Gallios, Uranium sorption on amorphous titanium and zirconium phosphates modified by Al3+ or Fe3+ ions, J. Radioanal. Nuclear Chem. 254 (2002) 85-89. https://doi.org/10.1023/A:1020893515049

[22] P. Jiang, B. Pan, B. Pan, W. Zhang, Q. Zhang, A comparative study on lead sorption by amorphous and crystalline zirconium phosphates, Colloids Surf. A: Physicochem. Eng. Aspects 322 (2008) 108-112. https://doi.org/10.1016/j.colsurfa.2008.02.035

[23] Y. F. El-Aryan, F.Z. Alqahtany, Removal of some heavy metals from wastewater using hydrothermally synthesized zirconium phosphate as ion exchange material, Russ. J. Phys. Chem. A 95 (2021) S209-S216. https://doi.org/10.1134/S0036024421150085

[24] E.P. Horwltz, The sorption of certain transplutonium ions on amorphous zirconium phosphate, J. Inorg, Nucl. Chem. 28 (1966) 1469-1478. https://doi.org/10.1016/0022-1902(66)80180-X

[25] R. Thakkar, U. Chudasama, Synthesis and characterization of zirconium titanium phosphate and its application in separation of metal ions, J. Hazard. Mater. 172 (2009) 129-137. https://doi.org/10.1016/j.jhazmat.2009.06.154

[26] S. Ahrland, J. Albertsson, Inorganic ion exchangers. III. Equilibrium studies on zirconium phosphate gels, Acta Chem. Scand. 18 (1964) 1861-1878. https://doi.org/10.3891/acta.chem.scand.18-1861

[27] W. Janusz, V. Sydorchuk, E. Skwarek, S. Khalameida, Effect of hydrothermal treatment of zirconium phosphate xerogel on its surface groups properties and affinity adsorption to Cd (II)

ions from acidic solutions, Mater.Res. Bul. 148 (2022) 111674. https://doi.org/10.1016/j.materresbull.2021.111674

[28] Y. Meng, Y. Wang, Z. Ye,, N. Wang,, C. He, Y. Zhu, T. Fujita,, H. Wu, X. Wang, Three-dimension titanium phosphate aerogel for selective removal of radioactive strontium(II) from contaminated waters, J. Environ. Manag. 325 (2023) 116424. https://doi.org/10.1016/j.jenvman.2022.116424

[29] X. Zhang, J. Shen, S. Pan, J. Qian, B. Pan, Metastable zirconium phosphate under nanoconfinement with superior adsorption capability for water treatment, Adv. Funct. Mater. 30 (2020) 1909014. https://doi.org/10.1002/adfm.201909014

[30] M. Pica, Treatment of wastewaters with zirconium phosphate based materials: a review on efficient systems for the removal of heavy metal and dye water pollutants, Molecules 26 (2021) 2392, https://doi.org/10.3390/molecules26082392. https://doi.org/10.3390/molecules26082392

[31] Q. Zhang, Q. Du, T. Jiao, B. Pan, Z. Zhang, Q. Sun, S. Wang, T. Wang, F. Gao, Selective removal of phosphate in waters using a novel of cation adsorbent: Zirconium phosphate (ZrP) behavior and mechanism, Chem. Eng. J. 221 (2013) 315-321. https://doi.org/10.1016/j.cej.2013.02.001

[32] E. H. Bailey, J. F. W. Mosselmans, P. F. Schofield, Uranyl acetate speciation in aqueous solutions - An XAS study between 25°C and 250°C, Geochim. Cosmochim. Acta. 68 (2004) 1711-1722. https://doi.org/10.1016/j.gca.2003.08.024

[33] C. B. Amphlett, Inorganic Ion Exchangers, Elsevier, Amsterdam, 1964.

[34] A.B.Yaroslavtsev, Ion exchange on inorganic sorbents, Russ Chem Rev. 66 (1997) 579-596. https://doi.org/10.1070/RC1997v066n07ABEH000348

[35] L. Wang, C. Shi, L. Wang, L. Pan, X. Zhang, J.-J. Zou, Rational design, synthesis, adsorption principles and applications of metal oxide adsorbents: a review, Nanoscale. 12 (2020) 4790-4815. https://doi.org/10.1039/C9NR09274A

[36] Y. Cai, R. Chua, S. Huang, H. Ren, M. Srinivasan, Amorphous manganese dioxide with the enhanced pseudocapacitive performance for aqueous rechargeable zinc-ion battery, Chem. Eng. J. 396 (2020) 125221. https://doi.org/10.1016/j.cej.2020.125221

[37] L.M. Rozhdestvenska, M.O. Chaban, Yu. S. Dzyazko, O.V. Palchik, O.G. Dzyazko, Formation of lithium-selective sorbent in nanoreactors of the support based on titanium dioxide, Appl. Nanosci. 12 (2022) 1112-1122. https://doi.org/10.1007/s13204-021-01832-5

[38] K.O. Kudelko, L.M. Rozhdestvenska, L.M. Ponomarova, V.M. Ogenko, Anodic aluminum oxide-membrane prepared in electrolyte "oxalic acid - matter with carbon nanodots, Him. Fiz. Tehnol. Poverhni, 14 (2023) 237-248. https://doi.org/10.15407/hftp14.02.237

[39] K. Kudelko, L. Rozhdestvenska, L. Ponomarova, L. Kharkova, O. Dzyazko, Formation of nanopores in anodic oxidized aluminium affected by carbon nanodots, 2023. DOI: 10.1109/NAP59739.2023.10310955, (accessed 14 November 2023). https://doi.org/10.1109/NAP59739.2023.10310955

[40] K.O. Kudelko, L.M. Rozhdestvenska, Y.V. Borysenko, A.M. Mikhnyuk, V.Z. Barsukov, Formation and characterization of porous anodized aluminum oxide synthesized electrochemical with graphene oxide. Technologies and Engineering. 2 (2021) 49-59.

[41] K. Kudelko, L. Rozhdestvenskaya, V. Ogenko, V. Chmilenko, Formation and characterisation of porous anodized aluminum oxide, synthesized electrochemically in the presence of graphene oxide, Appl. Nanosci. 12 (2022) 1967-1976. https://doi.org/10.1007/s13204-022-02457-y

[42] T. Kamimura, S. Hara, H. Miyuki, M. Yamashita, H. Uchida, Composition and protective ability of rust layer formed on weathering steel exposed to various environments, Corrosion Sci. 48 (2006) 2799-2812. https://doi.org/10.1016/j.corsci.2005.10.004

[43] P.S. Kumar, L. Korving, K.J. Keesman, M.C.M. van Loosdrecht, G.-J. Witkamp, Effect of pore size distribution and particle size of porous metal oxides on phosphate adsorption capacity and kinetics, Chem. Eng. J. 358 (2019) 160-169. https://doi.org/10.1016/j.cej.2018.09.202

[44] O.V. Perlova, Yu.S. Dzyazko, O.V. Palchik, I.S. Martovyi, Hydrated titanium dioxide modified with potassium cobalt hexacyanoferrate (II) for sorption of cationic and anionic complexes of uranium(VI), Appl. Nanosci. 12 (2022) 651-663. https://doi.org/10.1007/s13204-021-01721-x

[45] Yu.S. Dzyazko, V.M. Ogenko, L.Ya. Shteinberg, A.V. Bildyukevich, T.V. Yatsenko, Composite adsorbents including oxidized graphene: effect of composition on mechanical durability and adsorption of pesticides, Him. Fiz. Tehnol. Poverhni. 10 (2019) 432-445. https://doi.org/10.15407/hftp10.04.432

[46] T.V. Maltseva, A.V. Pal'chik, E.O. Kudelko, S.L. Vasilyuk, K. A. Kazdobin, Impact of surface properties of hydrated compounds based on ZrO2 on the value of ionic conduction, J. Water Chem. Technol. 37 (2015) 18-24. https://doi.org/10.3103/S1063455X15010051

[47] G.A. Parks, The isoelectric points of solid oxides, solid hydroxides, and aqueous hydroxo complex systems, Chem. Rev. 65 (1965) 177-198. https://doi.org/10.1021/cr60234a002

[48] M. Kosmulski, Isoelectric points and points of zero charge of metal (hydr)oxides:50 years after Parks' review, Adv. Colloid Interface Sci. 238 (2016) 1-61. https://doi.org/10.1016/j.cis.2016.10.005

[49] T.V. Mal'tseva, T.V. Yatsenko, E.O. Kudelko, V.N. Belyakov, The effect of introduction of manganese hydroxide and hydrated aluminum oxide on the pore structure and surface charge of Zr(IV), Ti(IV), and Sn(IV) oxyhydrates, Russ. J. Appl. Chem. 84 (2011) 756-761. https://doi.org/10.1134/S107042721105003X

[50] T.V. Mal'tseva, E.O. Kudelko, T.V. Yatsenko, Specific surface area calculation on the basis of electric double layer capacitance for porous oxyhydrate adsorbents, Rus. J. Appl. Chem. 85 (2012) 359–363. https://doi.org/10.1134/S1070427212030068

[51] O.V. Perlova, Yu.S. Dzyazko, A.A. Malinovska, A.V. Palchik, Peculiarities of U(VI) sorption on composites containing hydrated titanium dioxide and potassium-cobalt hexacyanoferrate(II), Him. Fiz. Tehnol. Poverhni. 12 (2021) 344-357. https://doi.org/10.15407/hftp12.04.344

[52] K. Kudelko, T. Maltseva, V. Belyakov, Adsorption and mobility of Cu(II), Cd(II), Pb(II) ions adsorbed on (hydr)oxide polymer sorbents MxOy·nH2O, M = Zr (IV), Ti(IV), Sn (IV), Mn(IV), Desalination Water Treatment. 35 ((2011)) 295-299.

[53] T.V. Maltseva, E.O. Kudelko, V.N. Belyakov, Adsorption of Cu(II), Cd(II), Pb(II), Cr(VI) by double hydroxides on the basis of Al oxide and Zr, Sn, and Ti oxides, Russ. J. Phys. Chem. A. 83 (2009) 2336-2339. https://doi.org/10.1134/S0036024409130263

[54] E.O. Kudelko, T.V. Mal'tseva, V. N. Belyakov, Sorption of Cr(VI) ions by oxyhydrates of MxAl1- xOy•nH2O composition, where M is Zr(IV), Ti(IV), or Sn(IV), Colloid J. 74 (2012) 313-318. https://doi.org/10.1134/S1061933X12010073

[55] Y.S. Dzyazko, V.M. Ogenko, Y.M. Volfkovich, V.E. Sosenkin, T.V. Maltseva, T.V. Yatsenko, K.O. Kudelko, Composite consisting of hydrated zirconium dioxide and graphene oxide for removal of organic and inorganic components from water. Him. Fiz. Tekhnol. Poverhni. 9 (2018) 417-431. https://doi.org/10.15407/hftp09.04.417

[56] Yu. S. Dzyaz'ko, L. M. Rozhdestvenskaya, S. L. Vasilyuk, Mobility of Cr(VI) anions in the phase of the amphoteric inorganic ion exchanger. J. Water Chem. Technol. 30 (2008) 281-287. https://doi.org/10.3103/S1063455X08050044

[57] S. Debnath, U.C. Ghosh, Nanostructured hydrous titanium(IV) oxide: Synthesis, characterization and Ni(II) adsorption behavior, Chem. Eng. J. 152 (2009) 480-491. https://doi.org/10.1016/j.cej.2009.05.021

[58] P. Trivedi, L. Axe, Modeling Cd and Zn sorption to hydrous metal oxides, Environ. Sci. Technol. 34 (2000) 2215-2223. https://doi.org/10.1021/es991110c

[59] P. Trivedi, L. Axe, A Comparison of strontium sorption to hydrous aluminum, iron, and manganese oxides, J. Colloid Interface Sci. 218 (1999) 554-563. https://doi.org/10.1006/jcis.1999.6465

[60] H. Tamura, Theorization on ion-exchange equilibria: activity of species in 2-D phases, J Colloid Interface Sci. 279 (2004) 1-22. https://doi.org/10.1016/j.jcis.2004.07.010

[61] S. Serrano, P.A. O'Day, D. Vlassopoulos, M.T. García-González, F. Garrido, A surface complexation and ion exchange model of Pb and Cd competitive sorption on natural soils, Geochim. Cosmochim. Acta. 73(2009) 543-558. https://doi.org/10.1016/j.gca.2008.11.018

[62] M. Shi, X. Min, Y. Ke. Z. Lin. Z. Yang, S. Wang, N. Peng, X. Yan, S. Luo, J. Wu, Y. Wei, Recent progress in understanding the mechanism of heavy metals retention by iron (oxyhydr)oxides, Sci. Total. Envir. 752 (2021) 141930. https://doi.org/10.1016/j.scitotenv.2020.141930

[63] J. Liu,, R. Zhu, T. Xu, Y. Xu, F. Ge, Y. Xi, J. Zhu, H. He, 2016. Co-adsorption of phosphate and zinc(II) on the surface of ferrihydrite. Chemosphere 144, 1148-1155. https://doi.org/10.1016/j.chemosphere.2015.09.083

[64] R.G. Ford, P.M. Bertsch, K.J. Farley, Changes in transition and heavy metal partitioning during hydrous iron oxide aging. Environ. Sci. Technol 31 (1997) 2028-2033. https://doi.org/10.1021/es960824+

[65] K.G. Karthikeyan, H.A. Elliott, F.S. Cannon, Adsorption and coprecipitation of copper with the hydrous oxides of iron and aluminum. Environ. Sci. Technol. 31 (1997.) 2721-2725. https://doi.org/10.1021/es9609009

[66] J.B. Stanković, S.K. Milonjić, M.M. Kopečni, T.S. Ćeranić, Sorption of alkaline-earth cations on amorphous zirconium oxide, Colloids Surf. 46 (1990) 283-296. https://doi.org/10.1016/0166-6622(90)80172-Z

[67] X. Feng, R. Long, L. Wang, C. Liu, Z. Bai, X. Liu, A review on heavy metal ions adsorption from water by layered double hydroxide and its composites, Sep. Pur. Technol. 284 (2022) 120099. https://doi.org/10.1016/j.seppur.2021.120099

[68] G. Varga, Z. Somosi, Z. Kónya, Á. Kukovecz, I. Pálinkó, I. Szilagyi, A colloid chemistry route for the preparation of hierarchically ordered mesoporous layered double hydroxides using surfactants as sacrificial templates, J. Colloid Interface Sci. 581 (2021) 928-938. https://doi.org/10.1016/j.jcis.2020.08.118

[69] M.T. Rahman, T. Kameda, S. Kumagai, T. Yoshioka, A novel method to delaminate nitrate-intercalated MgAl layered double hydroxides in water and application in heavy metals removal from waste water, Chemosphere 203 (2018) 281-290. https://doi.org/10.1016/j.chemosphere.2018.03.166

[70] G. N. Pshinko, L.N. Puzyrnaya, B.P. Yatsik, A.A. Kosorukov, Removal of U(VI) from aqueous media with layered double hydroxide of Zn and Al, intercalated with hexacyanoferrate(II) ions, Radiochem. 57 (2015) 616-620. https://doi.org/10.1134/S1066362215060090

[71] L. Jiang, S. Xiao, J. Chen, Removal behavior and mechanism of Co(II) on the surface of Fe-Mn binary oxide adsorbent, Colloids Surf. A: Physicochem. Eng. Aspects 479 (2015) 1-10. https://doi.org/10.1016/j.colsurfa.2015.03.055

[72] W. Xu, H. Lan, H. Wang, H. Liu, J. Qu, Comparing the adsorption behaviors of Cd, Cu and Pb from water onto Fe-Mn binary oxide, MnO2 and FeOOH, Front. Environ. Sci. Eng. 9 (2015) 385-393. https://doi.org/10.1007/s11783-014-0648-y

[73] R. Liu, F. Liu, C. Hu, Z. He, H. Liu, J. Qu, Simultaneous removal of Cd(II) and Sb(V) by Fe-Mn binary oxide: Positive effects of Cd(II) on Sb(V) adsorption, J. Hazard. Mater. 300 (2015) 847-854. https://doi.org/10.1016/j.jhazmat.2015.08.020

[74] T.V. Mal'tseva, T.V. Yatsenko, E.O. Kudelko, V.N. Belyakov, The effect of introduction of manganese hydroxide and hydrated aluminum oxide on the pore structure and surface charge of Zr(IV), Ti(IV), and Sn(IV) oxyhydrates, Russ. J. Appl. Chem. 84 (2011) 756-761. https://doi.org/10.1134/S107042721105003X

[75] M. Moshoeshoe, M.S. Nadiye-Tabbiruka, V. Obuseng, A review of the chemistry, structure, properties and applications of zeolites, Amer. J. Mater. Sci. 7 (2017) 196-221.

[76] V. Yadav, M. Rani, L. Kumar, N. Singh, V. Ezhilselvi, Effect of surface modification of natural zeolite on ammonium ion removal from water using batch study: an overview, Water Air Soil Pollut. 233 (2022) 465. https://doi.org/10.1007/s11270-022-05948-4

[77] E.N. Domoroshchina, V.V. Chernyshev, G.M. Kuz'micheva, A. V. Dorokhov, L. V. Pirutko, G. V. Kravchenko, R. B. Chumakov, Changing the characteristics and properties of zeolite Y

and nano-anatase in the formation of a nano-anatase/Y composite with improved photocatalytic and adsorption properties, Appl. Nanosci. 8 (2018) 19-31. https://doi.org/10.1007/s13204-018-0648-5

[78] Y. Wang, K. Liu, T. He, J. Wu, Y. Fang, Zeolite with trimodal porosity by desilication of zeolite nanocrystals aggregate, J. Solid State Chem. 194 (2012) 416-421. https://doi.org/10.1016/j.jssc.2012.04.028

[79] M. Li, C.K. Ober, Block copolymer patterns and templates, Mater. Today, 9 (2006) 30-39. https://doi.org/10.1016/S1369-7021(06)71620-0

[80] J. Perez-Ramırez, C.H. Christensen, K. Egeblad, C.H. Christensen, J.C. Groen, Hierarchical zeolites: enhanced utilisation of microporous crystals in catalysis by advances in materials design, Chem. Soc. Rev. 37 (2008) 2530-2542. https://doi.org/10.1039/b809030k

[81] R. Srivastava, Synthesis and applications of ordered and disordered mesoporous zeolites: Present and future prospective, Catalysis Today 309 (2018) 172-188. https://doi.org/10.1016/j.cattod.2017.08.017

[82] Y. Lv, B. Ma, Y. Liu, C. Wang, Y. Chen, Adsorption behavior and mechanism of mixed heavy metal ions by zeolite adsorbent prepared from lithium leach residue, Micropor. Mesopor. Mater. 329 (2022) 111553. https://doi.org/10.1016/j.micromeso.2021.111553

[83] Yu.S. Dzyazko, L.M. Rozhdestvenska, K.O. Kudelko, I.V. Fedina, L.M. Ponomaryova, G.M. Nikovska, O.G. Dzyazko, Hydrated iron oxide embedded to natural zeolite: effect of nanoparticles and microparticles on sorption properties of composites, Water, Air Soil Pollut. 233 (2022) 205. https://doi.org/10.1007/s11270-022-05681-y

[84] N. Mansouri, N. Rikhtegar, H.A.Panahi, F. Atabi, B. K. Shahraki, Porosity, characterization and structural properties of natural zeolite - clinoptilolite - as a sorbent, Env. Protect. Eng. 39 (2013) 139-151. https://doi.org/10.37190/epe130111

[85] B. Bensafi, N. Chouat, F. Djafri, High performance of carbon dioxide adsorption of mesoporous mordenite synthesized in the presence of N,N-dimethylaniline, Res. Chem. Intermed. 43 (2017):7443-7456. https://doi.org/10.1007/s11164-017-3085-2

[86] M. Thirumavalavan, Y.-T. Wang, L.-C. Lin, J.-F. Lee, Monitoring of the structure of mesoporous silica materials tailored using different organic templates and their effect on the adsorption of heavy metal ions, J. Phys. Chem. 115 (2011) 8165-8174. https://doi.org/10.1021/jp200029g

[87] L. Velarde, M.S. Nabavi, E. Escalera, M.-L. Antti, F. Akhtar, Adsorption of heavy metals on natural zeolites: A review, Chemosphere 328 (2023) 138508. https://doi.org/10.1016/j.chemosphere.2023.138508

[88] K.M. Wojciechowska, M. Król, T. Bajda, W. Mozgawa, Sorption of heavy metal cations on mesoporous ZSM-5 and mordenite zeolites, Materials 12 (2019) 3271. https://doi.org/10.3390/ma12193271

[89] Y. S. Dzyazko, I.V. Fedina, V.V. Zakharov, Y.O. Kolomiets, K.O. Kudelko, Composite ion-exchangers for the recycling of liquid waste of dairy industry, Ukrainian Chemistry Journal. 86 (2020) 38-52. https://doi.org/10.33609/2708-129X.86.5.2020.38-52

[90] Yu. Dzyazko, L. Ponomarova, L. Rozhdesvenska, K. Kudelko, A. Palchik, T. Yatsenko, Nanocomposite based on natural zeolite containing hydrated iron (iii) oxide for removal of heavy metal ions from water, 2022. DOI: 10.1109/NAP55339.2022.9934432, (accessed 10 November 2022). https://doi.org/10.1109/NAP55339.2022.9934432

[91] L. Velarde, M.S. Nabavi, E. Escalera, M.-L. Antti, F. Akhtar, Adsorption of heavy metals on natural zeolites: A review, Chemosphere 328 (2023) 138508. https://doi.org/10.1016/j.chemosphere.2023.138508

[92] S. Tangkawanit, K. Rangsriwatananon, A. Dyer, Ion exchange of Cu2+, Ni2+, Pb2+ and Zn2+ in analcime (ANA) synthesized from Thai perlite, Micropor. Mesopor. Mater. 79 (2005) 171-175. https://doi.org/10.1016/j.micromeso.2004.10.040

[93] L.-C. Lin, M. Thirumavalavan, Y.-T. Wang, J.-F. Lee, Surface area and pore size tailoring of mesoporous silica materials by different hydrothermal treatments and adsorption of heavy metal ions, Colloids and Surfaces A: Physicochem. Eng. Aspects 369 (2010) 223-231. https://doi.org/10.1016/j.colsurfa.2010.08.032

[94] S. Kocaoba, Y. Orhan, T. Akyüz, Kinetics and equilibrium studies of heavy metal ions removal by use of natural zeolite, Desalination 214 (2007) 1-10. https://doi.org/10.1016/j.desal.2006.09.023

[95] A.A.M. El-Kamash, A.A. Zaki, M. Abed El Geleel, Modeling batch kinetics and thermodynamics of zinc and cadmium ions removal from waste solutions using synthetic zeolite, J. Hazard. Mater. B127 (2005) 211-220. https://doi.org/10.1016/j.jhazmat.2005.07.021

[96] M. Hong, L. Yu, Y. Wang, J. Zhang, Z. Chen, L. Dong, Q. Zan, R. Li, Heavy metal adsorption with zeolites: The role of hierarchical pore architecture, Chem. Eng. J. 359 (2019) 363-372. https://doi.org/10.1016/j.cej.2018.11.087

[97] T.M. McCoy, G. Turpin, B.M. Teo, R.F. Tabor, Graphene oxide: a surfactant or particle? Curr. Opin. Colloid Interface Sci. 39 (2019) 98-109. https://doi.org/10.1016/j.cocis.2019.01.010

[98] L. Staudenmaier, Verfahren zur Darstellung der Graphitsaure. Ber. Deut. Chem. Ges. 31 (1898) 1481-1487. https://doi.org/10.1002/cber.18980310237

[99] B.C. Brodie, Sur le poids atomique du graphite. Ann. Chim. Phys. 59 (1860) 466-472.

[100] W.S. Hummers, R.E. Offeman, Preparation of graphitic oxide. J Am Chem Soc 80 (1958) 1339-1339. https://doi.org/10.1021/ja01539a017

[101] Y. Wang, Z. Wang, Z. Gu, J. Yang, J. Liao, Y. Yang, N. Liu, J. Tang, Uranium(VI) sorption on graphene oxide nanoribbons derived from unzipping of multiwalled carbon nanotubes. J. Radioanalyt. Nucl. Chem. 304 (2015) 1329-1337. https://doi.org/10.1007/s10967-015-3981-0

[102] M.O. Danilov, I.A. Rusetskii, G.I. Dovbeshko, A. Nikolenko, S. Fomanyuk, G.Ya. Kolbasov, Electrochemical synthesis partially unzipped multi-walled carbon nanotubes as electrode materials for fuel cells, ECS Trans. 95 (2019) 273. https://doi.org/10.1149/09501.0273ecst

[103] M.O. Danilov, G.I. Dovbeshko, I.A. Rusetskii, V.I. Pekhnyo, A.S. Nikolenko, G.Ya. Kolbasov, Partially unzipped multi-walled carbon nanotubes-promising material for oxygen electrodes of fuel cells. Appl. Phys. A 126 (2020) 764. https://doi.org/10.1007/s00339-020-03951-2

[104] Y. Dzyazko, V. Ogenko, Polysaccharides: an efficient tool for fabrication of carbon nanomaterials, in: Inamuddin, M.I. Ahamed, R. Boddula, T. Altalhi (Eds.), Polysaccharides: Properties and Applications. Wiley, Hoboken, 2021, pp. 337-366. https://doi.org/10.1002/9781119711414.ch16

[105] R.J. Seresht, M. Jahanshahi, A. Rashidi, A.A. Ghoreyshi, Synthesize and characterization of graphene nanosheets with high surface area and nano-porous structure. Appl. Surf. Sci. 276 (2013) 672-681. https://doi.org/10.1016/j.apsusc.2013.03.152

[106] M.J. McAllister, J.-L. Li, D.H. Adamson, H.C. Schniepp, A.A. Abdala, J. Liu, M. Herrera-Alonso, D.L. Milius, R. Car, R.K. Prud'homme, I.A. Aksay, Single sheet functionalized graphene by oxidation and thermal expansion of graphite. Chem. Mater. 19 (2007) 4396-4404. https://doi.org/10.1021/cm0630800

[107] D.R. Dreyer, S. Park, C.W. Bielawski, R.S. Ruoff, The chemistry of graphene oxide. Chem. Soc. Rev. 39 (2010) 228-240. https://doi.org/10.1039/B917103G

[108] Y.M. Volfkovich, A.Y. Rychagov, V.E. Sosenkin, O.N. Efimov, M.I. Os'makov, Measuring the specific surface area of carbon nanomaterials by different methods, Russ. J. Electrochem. 50 (2014) 1099-1101. https://doi.org/10.1134/S1023193514110111

[109] Yu. Dzyazko, Yu. Volfkovich, O. Perlova, L. Ponomaryova, N. Perlova, E. Kolomiets, Effect of porosity on ion transport through polymers and polymer-based composites containing inorganic nanoparticles (review), Springer Proc. Phys. 222 (2019) 235-253. https://doi.org/10.1007/978-3-030-17755-3_16

[110] Y.M. Shulga, S.A. Baskakov, Y.V. Baskakova, Y.M. Volfkovich, N.Y. Shulga, E.A. Skryleva, Y.N. Parkhomenko, K.G. Belay, G.L. Gutsev, A.Y. Rychagov, V.E. Sosenkin, I.D. Kovalev, Supercapacitors with graphene oxide separators and reduced graphite oxide electrodes, J Power Sources 279 (2015) 722-730. https://doi.org/10.1016/j.jpowsour.2015.01.032

[111] R. Sitko, E. Turek, B. Zawisza, E. Malicka, E. Talik, J. Heimann, A. Gagor, B. Feist, R. Wrzalik, Adsorption of divalent metal ions from aqueous solutions using graphene oxide. Dalton Trans., 42 (2013) 5682-5689. https://doi.org/10.1039/c3dt33097d

[112] Zhao G., Li J., Ren X., Chen Ch., Wang X. Few-layered graphene oxide nanosheets as superior sorbents for heavy metal ion pollution management. Environ. Sci. Technol. 45 (2011) 10454-10462. https://doi.org/10.1021/es203439v

[113] J. Huang, Z. Wu, L. Chen, Y. Sun, Surface complexation modeling of adsorption of Cd(II) on graphene oxides, J. Molec. Liquids 209 (2015) 753-758. https://doi.org/10.1016/j.molliq.2015.06.047

[114] P. Tan, J. Sun, Y. Hu, Z. Fang, Q. Bi, Y. Chen, J. Cheng, Adsorption of Cu2+, Cd2+ and Ni2+ from aqueous single metal solutions on graphene oxide membranes. J. Hazard. Mater. 297 (2015) 251-260. https://doi.org/10.1016/j.jhazmat.2015.04.068

[115] S.-T. Yang, Y. Chang, H. Wang, G. Liu, S. Chen, Y. Wang, Y. Liu, A. Cao, Folding/aggregation of graphene oxide and its application in Cu2+ removal, J. Colloid. Interface Sci. 351 (2010) 122-127. https://doi.org/10.1016/j.jcis.2010.07.042

[116] Z. Li, F. Chen, L. Yuan, Y. Liu, Y. Zhao, Z. Chai, W. Shi, Uranium(VI) adsorption on graphene oxide nanosheets from aqueous solutions, Chem. Eng. J. 210 (2012) 539 -546. https://doi.org/10.1016/j.cej.2012.09.030

[117] Yu.S. Dzyazko, Yu.M. Volfkovich, M.O. Chaban, Composites containing inorganic ion exchangers and graphene oxide: hydrophilic-hydrophobic and sorption properties (review), Springer Proc. Phys. 246 (2020) 93-110. https://doi.org/10.1007/978-3-030-51905-6_8

[118] M. Patel, H. Lanakapati, K. Maheria, Efficient removal of Pb2+, Cd2+, and Hg2+ from wastewater using advanced graphene oxide-zirconium hybrid composite, ACS Appl. Eng. Mater. 2 (2024) 1542−1560. https://doi.org/10.1021/acsaenm.4c00090

[119] S.R. Al-Mhyawi, D.M.D. Bader, M.A. Bajaber, S.M. Abd El Dayem, A.H. Ragab, K. A. Abd El-Rahem, M.A. Gado, B.M. Atia, M.F. Cheira, Zirconium oxide with graphene oxide anchoring for improved heavy metal ions adsorption: Isotherm and kinetic study, J. Mater. Res. Technol. 22 (2023) 3058-3074. https://doi.org/10.1016/j.jmrt.2022.11.121

[120] S. Pourbeyram, Effective removal of heavy metals from aqueous solutions by graphene oxide−zirconium phosphate (GO−Zr-P) nanocomposite, Ind. Eng. Chem. Res. 55 (2016) 5608−5617. https://doi.org/10.1021/acs.iecr.6b00728

[121] I.O. Ali, S.M. El-Sheikh, T.M. Salama, E.K. Abdel-Khalek, M.S. Thabet, M.F. Bakr, M.H. Fodial, Novel composites of multifunctional NaP zeolite/graphene oxide for highly efficient removal of Fe(III) from aqueous solution, J. Inorg. Organomet. Polym. 31 (2021) 577-590. https://doi.org/10.1007/s10904-020-01748-7

[122] D Smržová, L Szatmáry, P Ecorchard, A Machálková, M Maříková, P Salačová, M Straka, Carbon and zeolite-based composites for radionuclide and heavy metal sorption, Heliyon 8 (2022) e12293. https://doi.org/10.1016/j.heliyon.2022.e12293

[123] M. Liu, C. Chen, J. Hu, X. Wu, X. Wang, Synthesis of magnetite/graphene oxide composite and application for cobalt (II) removal, J. Phys. Chem. C 115 (2011) 25234-25240. https://doi.org/10.1021/jp208575m

[124] W.C. Elmore, Ferromagnetic colloids for studying magnetic structures, Phys. Rev. 54 (1938) 309-310. https://doi.org/10.1103/PhysRev.54.309

[125] J. Yan, R. Li, Simple and low-cost production of magnetite/graphene nanocomposites for heavy metal ions adsorption, Sci. Total Envir. 813 (2022) 152604. https://doi.org/10.1016/j.scitotenv.2021.152604

[126] Y. Dzyazko, O. Perlova, I. Martovyi, Advanced carbon nanomaterials and their composites for removal of U(VI) compounds from aqueous solutions, Springer Proc. Phys. 279 (2023) 177-194. https://doi.org/10.1007/978-3-031-18096-5_9

[127] Z. Yang, H. Yan, H. Yang, H. Li, A. Li, R. Cheng, Flocculation performance and mechanism of graphene oxide for removal of various contaminants from water. Water Res. 47 (2013) 3037-3046. https://doi.org/10.1016/j.watres.2013.03.027

[128] G. Sheng, C. Huang, G. Chen, J. Sheng, X. Ren, B. Hu, J. Ma, X. Wang, Y. Huang, A. Alsaedi, T. Hayat, Adsorption and co-adsorption of graphene oxide and Ni(II) on iron oxides: A spectroscopic and microscopic investigation. Environ. Pollut. 233 (2018) 125-131. https://doi.org/10.1016/j.envpol.2017.10.047

[129] Y. Dzyazko, A. Omel'chuk, Porous Ionic Polymers, in: Inamuddin, M.I. Ahamed, R. Boddula (Eds.), Porous Polymer Science and Applications, CRC Press, Boca Raton, London, New York: 2022, pp. 37-59. https://doi.org/10.1201/9781003169604-3

[130] N. Kononenko, V. Nikonenko, D. Grande, C. Larchet, L. Dammak, M. Fomenko, Yu. Volfkovich, Porous structure of ion exchange membranes investigated by various technique. Adv. Colloid Interface Sci. 246 (2017) 196-216. https://doi.org/10.1016/j.cis.2017.05.007

[131] Yu. S. Dzyazko, L. N. Ponomareva, Yu. M. Volfkovich, V.E. Sosenkin, Effect of the porous structure of polymer on the kinetics of Ni2+ exchange on hybrid inorganic-organic ionites, Russ. J. Phys. Chem. 86 (2012) 913-919. https://doi.org/10.1134/S0036024412060088

[132] F. Helfferich, Ion Exchange, second ed., Dover, New Vork, 1995.

[133] W.Y. Hsu, T.D. Gierke, Ion transport and clustering in Nafion perfluorinated membranes. J. Membr. Sci. 13 (1983) 307-326. https://doi.org/10.1016/S0376-7388(00)81563-X

[134] H.E. Andrada, M.B. Franzoni, A.C. Carreras, F.V. Chávez, Dynamics and spatial distribution of water in Nafion 117 membrane investigated by NMR spin-spin relaxation, Int. J. Hydrogen Energy 43 (2018) 8936-8943. https://doi.org/10.1016/j.ijhydene.2018.03.124

[135] K.A. Mauritz, R.B. Moore, State of understanding of Nafion, Chem. Rev. 104 (2004) 4535-4585. https://doi.org/10.1021/cr0207123

[136] Yu.S. Dzyazko, L.N. Ponomareva, Y.M. Volfkovich, V.E. Sosenkin, V.N. Belyakov, Conducting properties of a gel ionite modied with zirconium hydrophosphate nanoparticles, Russ. J. Electrochem. 49 (2013) 209-215. https://doi.org/10.1134/S1023193513030075

[137] Yu.Dzyazko, L. Ponomarova, Yu. Volfkovich, V. Tsirina, V. Sosenkin, N. Nikolska, V. Belyakov, Influence of zirconium hydrophosphate nanoparticles on porous structure and sorption capacity of the composites based on ion exchange resin, Chem. Chem. Technol. 10 (2016) 329-335. https://doi.org/10.23939/chcht10.03.329

[138] Yu.S. Dzyazko, O.V. Perlova, N.A. Perlova, Yu.M. Volfkovich, V.E. Sosenkin, V.V. Trachevskii, V.F. Sazonova, A.V, Palchik, Composite cation-exchange resins containing zirconium hydrophosphate for purification of water from U(VI) cations, Desalination and Water Treatment. 69 (2017) 142-152. https://doi.org/10.5004/dwt.2017.0686

[139] Yu.S. Dzyazko, L.N. Ponomaryova, Yu.M. Volfkovich, V.E. Sosenkin, V.N. Belyakov, Polymer ion-exchangers modified with zirconium hydrophosphate for removal of Cd2+ ions from diluted solutions, Separ. Sci. Technol. 48 (2013) 2140-2149. https://doi.org/10.1080/01496395.2013.794434

[140] Yu.S. Dzyazko, L.N. Ponomaryova, Yu.M. Volfkovich, V.V. Trachevskii, A.V. Palchik, Ion-exchange resin modified with aggregated nanoparticles of zirconium hydrophosphate. Morphology and functional properties, Micropor Mesopor Mater. 198 (2014) 55-62. https://doi.org/10.1016/j.micromeso.2014.07.010

[141] L. Ponomarova, Yu. Dzyazko, Yu. Volfkovich, V.E. Sosenkin, S. Scherbakov, Effect of incorporated inorganic nanoparticles on porous structure and functional properties of strongly and weakly acidic ion exchangers, Springer Proc. Phys. 214 (2018) 63-77. https://doi.org/10.1007/978-3-319-92567-7_4

[142] Y.S. Dzyazko, L.N. Ponomaryova, L.M. Rozhdestvenskaya, S.L. Vasilyuk, V.N. Belyakov. Electrodeionization of low-concentrated multicomponent Ni2+-containing solutions using organic-inorganic ion-exchangers, Desalination. 342 (2014) 52-60. https://doi.org/10.1016/j.desal.2013.11.030

[143] Yu. Dzyazko, E. Kolomyets, Yu. Borysenko, V. Chmilenko, I. Fedina, Organic-inorganic sorbents containing hydrated zirconium dioxide for removal of chromate anions from diluted solutions, Mater. Today: Proc. 6 (2019) 260-269. https://doi.org/10.1016/j.matpr.2018.10.103

[144] Y.O. Kolomiyets, V.N. Belyakov, A.V. Palchik, T.V. Maltseva, Effect of incorporation of hydrated oxides of Sn(IV), Zr(IV) and Fe(III) in a matrix of the anion exchanger Dowex SBR-P on the sorption capacity towards the Arsenic (V) anions, Int. J. Water Wastewater Treat. 2 (2016) http://dx.doi.org/10.16966/2381-5299.120. https://doi.org/10.16966/2381-5299.120

[145] E.O. Kolomiets, T.V. Maltseva, Yu.S. Dzyazko, Organic-inorganic ion-exchangers for selective electrodeionization of diluted aqueous solutions. Voprosy Khimii i Khimicheskoi Tekhnologii. 1 (2019) 18-25. https://doi.org/10.32434/0321-4095-2019-122-1-18-25

[146] T.V. Maltseva, E.O. Kolomiets, Yu.S. Dzyazko, S. Scherbakov, Composite anion-exchangers modified with nanoparticles of hydrated oxides of multivalent metals, Appl. Nanosci. 9 (2019) 997-1004. https://doi.org/10.1007/s13204-018-0689-9

[147] T.V. Mal'tseva, E.A. Kolomiets, S.L.Vasilyuk Hybrid adsorbents based on hydrated oxides of Zr(IV), Ti(IV), Sn(IV), and Fe(III) for arsenic removal. J Water Chem Technol. 39 (2017) 214 - 219. https://doi.org/10.3103/S1063455X17040063

[148] Ye.O. Kolomiyets, E.O. Kudelko, T.V. Maltseva, S.L. Vasilyuk, YuS. Dzyazko, Adsorption of arsenic, chromium, lead, cadmium by adsorbents on the basis of Zr(IV), Ti(IV), Sn(IV), Al(III), Fe(III) oxides. Water and Water Purification Technologies. Scientific and Technical News. 22 (2018) 11-19. https://doi.org/10.20535/2218-93002212018144235

[149] W. Mu, S. Du, Q. Yu, X. Li, H. Wei, Y. Yang, S. Peng, Highly efficient removal of radioactive 90Sr based on sulfonic acid functionalized α-zirconium phosphate nanosheets, Chem. Eng. J. 361 (2019) 538-546. https://doi.org/10.1016/j.cej.2018.12.110

[150] Y.K. Kim, K. J. Lee Synthesis of a magnetic composite resin and its cobalt removal characteristics in aqueous solution, J. Nucl. Sci. Technol. 38 (2001) 785-792. https://doi.org/10.3327/jnst.38.785

[151] J.-K. Moon, K.-W. Kim, C.-H. Jung, Y.-G. Shul, E.-H. Lee, Preparation of organic-inorganic adsorbent beads for removal of radionuclides and heavy metal ions, J. Radioanal. Nucl. Chem. 246 (2000) 299-307. https://doi.org/10.1023/A:1006714322455

[152] Inamuddin, S.A. Khan, W.A. Siddiqu, A.A. Khan, Synthesis, characterization and ion-exchange properties of a new and novel organic-inorganic hybrid cation-exchanger: Nylon-

6,6, Zr(IV) phosphate, Talanta 71 (2007) 841-847. https://doi.org/10.1016/j.talanta.2006.05.042

[153] W.A. Siddiqui, S.A. Khan, Inamuddin, Synthesis, characterization and ion-exchange properties of a new and novel 'organic-inorganic' hybrid cation-exchanger: Poly(methyl methacrylate) Zr(IV) phosphate, Colloids Surf. A: Physicochem. Eng. Aspects 295 (2007) 193-199. https://doi.org/10.1016/j.colsurfa.2006.08.053

[154] L.C.A. Molina, G.A.V. Magalhães-Ghiotto, L. Nichi, Y.S. Dzyazko, R. Bergamasco, Membranes modified with rigid polymer for processing solutions of vegetable proteins, Acta Periodica Technologica 54 (2023) 313-324. https://doi.org/10.2298/APT2354313M

[155] L.M. Rozhdestvenskaya, Y.S. Dzyazko, E.O. Kudelko, S.L. Vasilyuk, V.N. Belyakov, Desalination of glycerol-water solutions by electrodyalysis using the organo-inorganic membranes, J. Water Chem. Technol. 39 (2017) 26-32. https://doi.org/10.3103/S1063455X17010052

[156] Y. Zmievskii, L. Rozhdestvenska, Y. Dzyazko, L. Kornienko, V. Myronchuk, A. Bildukevich, A. Ukrainetz, Organic-inorganic materials for baromembrane separation, Springer Proc. Phys. 195 (2017) 675-685. https://doi.org/10.1007/978-3-319-56422-7_51

[157] Y. Dzyazko, L. Rozhdestvenska, K. Kudelko, V. Ogenko, Y. Kolomiets, Membranes modified with advanced carbon nanomaterials (review), Springer Proc. Phys. 263 (2021) 151-174. https://doi.org/10.1007/978-3-030-74741-1_10

[158] L. Rozhdestvenska, K. Kudelko, Y. Kolomiets, Y. Dzyazko, V. Ogenko, Membranes functionalized with 1d, 2d and 3d carbon materials, Ukrainian Chemistry Journal. 87 (2021) 79-110. https://doi.org/10.33609/2708-129X.87.04.2021.79-110

[159] L. Rozhdestvenska, K. Kudelko, V. Ogenko, O. Palchik, T. Plisko, A.Bildyukevich, V. Zakharov, Yu. Zmievskii, O. Vishnevskii, Filtration membranes containing nanoparticles of hydrated zirconium oxide-graphene oxide, Springer Proc. Phys. 246 (2020) 757-771. https://doi.org/10.1007/978-3-030-51905-6_51

[160] K.O. Kudelko, L.M. Rozhdestvenska, A.A. Verbytska, Y.V. Borysenko, V.Z. Barsukov, Microfiltration membranes modified with hydrated zirconium dioxide containing carbon nanodots. Technologies and Engineering. 5 (2022) 43-55.

[161] Y.O. Kolomiiets, Y.S. Dzyazko, Sorbents based on non-carbonized vegetable raw materials, Ukrainian Chemistry Journal. 88 (2022) 37-68. https://doi.org/10.33609/2708-129X.88.05.2022.37-68

[162] Y.O. Kolomiiets, O.V. Palchik, Yu.S. Dzyazko, T.V. Yatsenko, L.M. Ponomaryova, V.M. Ogenko, Sorbents based on biopolymers of different origin containing magnetite for removal of oil products and toxic ions from water. Him. Fiz. Tehnol. Poverhni. 14 (2023) 121-132. https://doi.org/10.15407/hftp14.01.121

Materials Research Foundations 173 (2025) 40-59
https://doi.org/10.21741/9781644903452-2

Chapter 2

Mesoporous Materials for Adsorptive Removal of Dyes

Angel R. Hernandez-Martinez[1,2], A.L. Ramos-Jacques[3]

[1]Facultad de Química, Universidad Autónoma de Querétaro, Centro Universitario, Cerro de Las Campanas, Querétaro, Mexico

[2]Facultad de Química, Universidad Autónoma de Querétaro, Campus Pedro Escobedo, Pedro Escobedo, Querétaro, Mexico

[3]EditAcademy, Queretaro, Mexico

angel.ramon.hernandez@gmail.com

Abstract

Water pollution is an environmental challenge globally, particularly in relation to industrial effluents. Adsorption has been widely used for wastewater treatment. This chapter addresses the use of mesoporous materials for dyes adsorption in wastewater, presenting recent advances in their synthesis, the properties that make them suitable for this purpose, and examples of successful applications. The development of mesoporous materials as adsorbents is aligned with current trends towards the use of sustainable and inexpensive materials, and their structure allows for greater interaction between contaminants and the surface of these materials, improving the adsorption efficiency of contaminants.

Keywords

Wastewater Treatment, Materials Engineering, Synthesis, Biodegradable Adsorbents, Industrial Effluents

Contents

Materials Research Forum LLC
https://doi.org/10.21741/9781644903452-2

1. Introduction

Water pollution is one of the most pressing environmental challenges globally, particularly in relation to industrial effluents. Among the industries that contribute the most to water pollution is textile industry, responsible for approximately 20% of industrial pollution of water worldwide [1-5]. Synthetic dyes, such as reactive and azo dyes, are broadly used in dyeing processes, resulting in the release of toxic and non-biodegradable compounds into water sources [1, 5, 6]. This situation poses serious risks to both aquatic ecosystems and public health, given that many of these dyes contain aromatic compounds that can release potentially carcinogenic amines [4, 7].

Traditional methods for treating those wastewaters include physical, chemical and biological processes, each with advantages and limitations. However, adsorption has emerged as one of the most efficient and cost-effective techniques for removing pollutants, especially dyes, from

Materials Research Foundations 173 (2025) 40-59
https://doi.org/10.21741/9781644903452-2

wastewater [1, 4, 7]. Additionally, mesoporous materials, due to their high surface area and ability to host a variety of functional compounds, exhibit properties that could facilitate dye adsorption [6-8].

In this context, mesoporous materials, which feature structures with pores of size between of 2 to 50 nm, offer significant advantages. Their structure allows for greater interaction between contaminants and material surface, improving the adsorption efficiency of contaminants such as dyes [3, 8]. Furthermore, these materials can be chemically modified to increase their selectivity towards certain types of dyes, making them a flexible and efficient tool for wastewater treatment [3,4,9].

The development of mesoporous adsorbents is also aligned with current trends towards the use of sustainable and unexpensive materials, such as agro-industrial waste. These can be transformed into high-quality materials for adsorption, reducing both the costs and the environmental impact of industrial processes [2-4, 7].

This chapter addresses the use of mesoporous materials in the adsorption of dyes from wastewater, presenting recent advances in their synthesis, the properties that make them suitable for this purpose, and examples of successful applications.

2. Mesoporous Materials: Definition and Properties

Mesoporous materials are defined by the size of pores (diameters in nanometers), which gives them a considerable internal surface area and a remarkable ability to adsorb molecules, especially contaminants in liquid solutions. Simply put, mesopores function as extremely efficient sponges, where small molecules can penetrate and become trapped in their structure. However, it is not only about the size of the pores, but also the connectivity and internal arrangement of these pores, which largely determines their capacity to adsorb and transport substances.

As we delve deeper into the analysis of mesoporous materials, it is necessary to understand that their performance depends on the synergy between textural properties

(for example specific surface area and pore volume) and hierarchy in pore arrangement. These properties make mesoporous materials a promising option in applications such as contaminant removal, catalysis, and energy storage [7-11].

2.1 Definition and Structural Characteristics

Mesoporous materials are defined by the IUPAC classification, which states that pores in this type of materials present diameters between 2 and 50 nm, positioning them between micropores (<2 nm) and macropores (>50 nm) [Ref 4]. This intermediate pore diameters allows mesopores to offer a balance between the adsorption of smaller molecules and easier access for larger molecules, resulting in higher performance in processes where mass transport is crucial, such as in the adsorption of dyes [10].

In many cases, mesoporous materials present an ordered structural arrangement, which improves their performance in adsorption and catalysis processes [Fig. 1]. A typical example of ordered mesoporous materials is MCM-41, which presents a structure of cylindrical pores arranged in a hexagonal shape. This ordered structure maximizes the accessibility of the pores and the specific surface, providing efficient routes for the transport of adsorbed molecules [11]. On the other hand, there are also materials with disordered structures, such as some activated carbons, which although they do not have a regular arrangement, have a significant pore volume that makes them excellent adsorbents for the removal of contaminants. [7-9].

The presence of ordered pores, as in MCM-41 and SBA-15 materials, or disordered pores, as in biomass-derived activated carbons, increases interaction between the adsorbate molecules and the internal surfaces of the material, improving the efficiency of the adsorption processes. [7, 11].

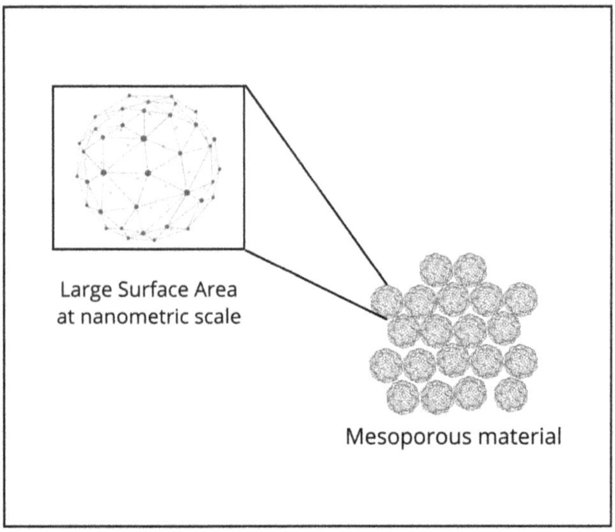

Figure 1. Structural arrangement of mesoporous materials could favor a large surface area for improving adsorption properties.

2.2 Textural Properties

Textural properties are key parameters determining the adsorption capacity of mesoporous materials, these properties could be pore volume and diameter and specific surface area. A high specific surface area, typically greater than 1000 m^2/g, favors the adsorption of a large amount of molecules. Furthermore, the total pore volume, which can reach values of up to 1 cm^3/g in certain materials, promotes the accumulation of large amounts of contaminants, such as industrial dyes. [7-14]. Evaluation of these properties is carried out using techniques such as nitrogen adsorption at low temperatures, using models such as Brunauer-Emmett-Teller (BET) to determine the specific surface area, and the BJH method (Barrett, Joyner and Halenda) to calculate the size and distribution of the pores[10,12,15].

2.3 Hierarchy and Connectivity of Pores

In addition to pore size, the behavior of the material is significantly influenced by the connectivity between pores. Well-connected porous systems allow for more efficient diffusion of molecules through the material, which improves access to all internal surface areas. This hierarchy concept refers to the coexistence of micro- and mesopores, where micropores provide additional adsorption areas, while mesopores allow for faster transport of molecules into the micropores. [10,14].

Recent studies have introduced the concept of a “hierarchy-connectivity factor” (HCF), which includes both the textural properties and the connectivity of the pore network, providing a more complete description of the behavior of these materials in specific applications such as adsorption and catalysis. [10 - 16].

3. Adsorption Mechanisms

Adsorption is a process by which molecules (adsorbate) adhere to the surface of a solid material (adsorbent). In the case of mesoporous materials, this phenomenon is facilitated by a high surface area and porous nature of the material, which provides multiple active sites for the capture of molecules. Adsorption mechanisms are essential to understand and optimize the design of effective adsorbents, such as those used for removing dyes in aqueous solutions.

Broadly speaking, adsorption mechanisms can be divided into these main categories; (i) physical adsorption (physisorption) and (ii) chemical adsorption (chemisorption). These mechanisms differ in the nature of the interactions between the adsorbate and the adsorbent, the energy involved in the process, and the reversibility of the adsorption. However, in mesoporous systems, both mechanisms can coexist and complement each other to maximize selectivity and adsorption capacity of the material [7,10,14].

3.2 Physical Adsorption (Physisorption)

Physical adsorption also known as physisorption is produced when adsorbate molecules adhere to the surface of the adsorbent through van der Waals forces, which are weak and non-specific interactions. In this scenario, the adsorption energy is relatively low (usually between 4 and 40 kJ/mol) and the process is typically reversible. This means that adsorbed molecules can be easily desorbed by decreasing the concentration or changing the system conditions, such as temperature or pressure [10 - 15].

In mesoporous materials, physisorption is favored by the high specific surface area and the presence of mesopores that allow better accessibility of the adsorbate molecules to the active sites. The nature of mesopores also facilitates the diffusion of the adsorbate through the material, which is particularly useful for large molecules such as industrial dyes. Furthermore, in some systems, the presence of interconnected micropores within the mesoporous network can act as molecule traps, increasing the total adsorption capacity [7, 10 - 16].

A relevant aspect of physisorption in mesoporous materials is the phenomenon of capillarity in the pores, which can increase the amount of adsorbate retained within the material. As the adsorbate concentration increases, the pores progressively fill, forming a liquid phase within the mesopores. This process is responsible for the adsorption profiles observed in typical isotherms such as type IV, where adsorption increases sharply at high relative pressures due to capillary condensation [6,7,10,14].

Materials Research Foundations 173 (2025) 40-59 https://doi.org/10.21741/9781644903452-2

3.3 Chemical Adsorption (Chemisorption)

Chemical adsorption (chemisorption) entails the chemical bonds formation between adsorbate and adsorbent surface. This mechanism involves much higher adsorption energies (in the range between 40 to 400 kJ/mol), meaning that the process is more selective and less reversible than physisorption [7 - 12]. In chemisorption, adsorbate molecules not only adhere to the adsorbent's surface, but can interact with specific functional groups present on the surface, leading to increased affinity and specificity towards certain types of contaminants.

In mesoporous materials, chemisorption can be promoted by functionalization of their surface, that means the introduction of active chemical groups on the inner surface of the material. These functional groups, such as amines, carboxylic acids, or thiol groups, can preferentially interact with charged dyes (anionic or cationic), improving both selectivity and adsorption efficiency [7-10]. This surface modification can also alter the hydrophilic or hydrophobic properties of the material, which is crucial for optimizing the adsorption of dyes in aqueous solutions.

A typical example of chemisorption on mesoporous materials is the adsorption of cationic dyes, such as methylene blue, on materials functionalized with anionic groups. In these systems, electrostatic interactions between the opposite charges of the adsorbate and the adsorbent play a crucial role in the adsorption mechanism, resulting in a high retention capacity [7-16].

3.4 Adsorption Kinetics

Adsorption kinetics describe the rate at which adsorbate molecules are adsorbed onto the surface of the adsorbent. This process is governed by several factors, including diffusion in the film, intra-particle diffusion, and adsorption at active sites. In the case of mesoporous materials, adsorption kinetics are usually controlled by intra-particle diffusion, which is the process by which adsorbate molecules move from the outside of the material to the inside of the pores [7-14].

In mesoporous systems, diffusion in mesopores is generally rapid because of their relatively large size, which facilitates the access of adsorbate molecules. However, when the system contains micropores, intra-particle diffusion can become the rate-limiting step, especially if the adsorbate molecules are large. This phenomenon is relevant when studying adsorption isotherms that present delays in equilibrium, indicating that the adsorbate is encountering resistance when penetrating the micropores [10].

Adsorption kinetics are commonly modeled using pseudo-first-order and pseudo-second-order kinetic models, which assume that the adsorption rate is proportional to the adsorbate concentration or the amount of adsorbate available at the active sites of the adsorbent. These models are particularly useful for predicting the adsorption capacity and behavior of mesoporous adsorbents under practical conditions. [7-9].

3.5 Adsorption Isotherm Models

Isotherm models are fundamental tools to describe how adsorbate molecules are distributed between liquid phase and adsorbent surface in equilibrium. The widely used models in the study of adsorption on mesoporous materials include the Langmuir and Freundlich isotherms.

- Langmuir isotherm: The assumption for this model is that adsorption takes place at homogeneous sites on the adsorbent surface and that the adsorption capacity is finite, i.e. only a monolayer of adsorbate can form on the surface. This model is useful to describe systems where chemical adsorption predominates, such as in functionalized materials [7,10].

- Freundlich isotherm: Unlike Langmuir, this model is empirical and used to describe systems where adsorption occurs on heterogeneous surfaces, which is more common in physical adsorption. This model is applicable in systems where adsorption is not limited to a monolayer formation and is useful for describing adsorption on mesoporous materials with disordered structures [9-14].

3.6 Factors Affecting Adsorption

Adsorption behavior on mesoporous materials is influenced by several operational and structural factors. The most relevant include:

1. pH: The pH of the solution can alter the charge properties of both the adsorbent and the adsorbate, affecting electrostatic interactions and adsorption capacity. For example, in acidic solutions, anionic functional groups on mesoporous materials can increase the adsorption of cationic dyes [7].

2. Temperature: Temperature affects both the adsorption energy and the diffusion of molecules within the pores. In general, physical adsorption tends to decrease at higher temperatures, while chemisorption could be favored by an increment in temperature[10].

3. Initial adsorbate concentration: The amount of adsorbate in the solution is also a determining factor. At low concentrations, adsorption sites are available and adsorption is rapid, but at higher concentrations, saturation of the active sites can slow down the process [7-14].

4. Contact time: This is the period of time required to reach adsorption equilibrium depends on the specific system and the pore structure of the material. Mesoporous materials, with their accessible pores, tend to reach equilibrium more quickly than microporous ones [7,8,12].

3.7 Energy Distribution at Adsorption Sites

A novel and relevant concept in the study of adsorption mechanisms is the energy distribution at adsorption sites, which addresses the heterogeneous nature of adsorbent materials, such as mesoporous ones. This approach allows a deeper understanding of the physicochemical interactions between adsorbed molecules and active sites on the surface of the adsorbent. Through the energy distribution of the sites, it is possible to assess the surface heterogeneity of the adsorbent, which is key to optimizing its performance in removing contaminants [17-19].

Not all adsorption sites have the same interaction energy with the adsorbate molecules, which directly affects the adsorption capacity and process efficiency. This approach allows identifying how molecules are distributed among the different sites on the adsorbent surface, optimizing performance under various operating conditions.

Energy distribution modeling can be performed using advanced models, for example, the Dubinin-Radushkevich (D-R) model, which has been adapted for liquid systems. This model provides an energy distribution function that describes how many active sites possess a certain adsorption energy, which facilitates the understanding of adsorption behavior and allows the design of more efficient and selective materials [17].

 Materials Research Forum LLC
https://doi.org/10.21741/9781644903452-2

This approach is particularly useful in dye removal, where the differential interaction of dyes with adsorption sites of different energies can be modeled and optimized. In summary, the energy distribution at adsorption sites is a powerful tool to improve the efficiency of mesoporous materials in industrial applications, such as the adsorption of contaminants in water treatment [17-19].

4. Synthesis of Mesoporous Materials for Dye Adsorption

The mesoporous materials' synthesis, specifically tailored for the adsorption of dyes, is a complex process that involves precise control over the material's structure, surface chemistry, and pore distribution. These materials must exhibit high surface areas, optimal pore sizes, and suitable chemical functionality to maximize adsorption capacity, particularly for large and structurally diverse dye molecules such as azo, anthraquinone, and methylene blue.

Table 1 summarizes the synthesis methods, types of mesoporous materials produced, their pore sizes, surface areas, and the corresponding advantages for dye adsorption.

Table 1: Synthesis Methods, Material Types, and Characteristics of Mesoporous Materials for Dye Adsorption.

Synthesis Method	Material Type	Pore Size (nm)	Surface Area (m^2/g)	Advantages	Reference
Template-Assisted	SBA	4-30	500-1000	Well-ordered structure, tunable pore size, high thermal stability	[26-30]
Template-Assisted	MCM-41	2-10	800-1200	Highly ordered hexagonal structure, large surface area	[7-12, 20-25, 31-33]
Sol-Gel	Functionalized Mesoporous Silica	2-50	600-800	Low-temperature synthesis, easy incorporation of functional groups	[34 - 37]
Hydrothermal	Metal-Organic Frameworks (MOFs)	1-2	>1000	High crystallinity, excellent adsorption capacity for dyes	[38 -41]
Microwave-Assisted Synthesis	Mesoporous Carbon	2-50	700-900	Fast synthesis, good control over pore size, high adsorption capacity	[42, 43]
Free-Radical Polymerization	Hydrogels (PHEMA, DMAa)	5-100	Variable	High water uptake, adaptable for pH or temperature-responsive systems	[24, 25, 44]

Materials Research Forum LLC
https://doi.org/10.21741/9781644903452-2

4.1 Template-Assisted Synthesis Methods

A common approach for synthesizing mesoporous materials is template-assisted synthesis, which employs surfactants or organic molecules as structure-directing agents. This method typically involves self-assembly, where micelles act as templates around which the precursor material condenses, creating an organized mesoporous framework. After removing the organic template via calcination or solvent extraction, a porous network remains [7-12, 20-25].

Silica-based materials, such as SBA-15 and MCM-41, are classic examples synthesized through this approach. For instance, in the synthesis of SBA-15, triblock copolymers are used as templates in acidic conditions to form mesostructures, an example is Pluronic P123. This process results in materials with pore sizes from 4 to 30 nm, making them suitable for adsorbing large dye molecules like methylene blue and reactive black [31-33].

4.2 Sol-Gel Synthesis

The sol-gel method is another popular technique for producing mesoporous materials. It involves the hydrolysis and polycondensation of metal alkoxides (such as tetraethyl orthosilicate, TEOS) to form a gel-like network. The advantage of the sol-gel method is the ability to synthesize materials at relatively low temperatures while allowing for the incorporation of functional groups directly into the silica framework. These functional groups enhance the material's selectivity toward specific dyes.

For example, by incorporating amino groups or thiol groups into the mesoporous silica, the material can selectively adsorb anionic or cationic dyes, respectively. This functionalization is key for improving dye removal efficiencies from aqueous solutions, particularly for dyes with specific charges [10, 34 - 37].

Recent advancements in sol-gel techniques also allow for the creation of hierarchical structures, which combine both mesopores and micropores. These hierarchical materials provide faster diffusion paths for dye molecules, making them highly efficient for large-scale water treatment applications [7].

4.3 Hydrothermal and Microwave-Assisted Synthesis

Hydrothermal synthesis is a method where precursors are dissolved in a solution and subjected to high temperatures and pressures in a sealed environment. This technique allows for the crystallization of materials with well-defined mesoporous structures. Hydrothermal synthesis is particularly useful for producing metal-organic frameworks (MOFs) and carbon-based mesoporous materials that are known for their exceptional adsorption capacities [10, 38 -41].

Microwave-assisted synthesis is a more recent advancement, offering rapid heating and shorter reaction times compared to conventional methods. This technique is highly efficient in producing mesoporous materials with controlled pore sizes and high surface areas. For instance, microwave-assisted synthesis has been successfully used to produce mesoporous carbon materials that demonstrate enhanced adsorption for organic dyes like Congo red and methyl orange [42, 43].

4.4 Polymeric Mesoporous Materials and Hydrogels

Polymeric mesoporous materials, such as hydrogels, have gained significant attention because of their high water uptake capacity and flexibility in modifying surface chemistry. Hydrogels synthesized from polymers like poly(2-hydroxyethyl methacrylate) (PHEMA) and N,N-dimethylacrylamide (DMAa) are highly effective for dye removal, particularly in pH-responsive systems. These materials can adsorb large quantities of water and dyes, making them highly efficient adsorbents for wastewater treatment [17, 24, 25, 44].

The synthesis of hydrogels typically involves free-radical polymerization, followed by cross-linking to create a three-dimensional porous network. Recent studies show that incorporating nanoparticles or metal ions into the hydrogel matrix can significantly improve its adsorption performance by creating additional active sites for dye molecules [24].

Hydrogels could also be designed to respond to external stimuli, such as pH or temperature changes, making them highly versatile for environmental applications. For example, temperature-sensitive hydrogels can shrink or swell depending on the surrounding environment, which can be leveraged to control the adsorption and release of dyes [17, 25].

4.5 Post-Synthesis Functionalization

While many mesoporous materials exhibit excellent structural properties, their surface chemistry often needs modification to improve their dye adsorption capabilities. Post-synthesis functionalization involves attaching specific functional groups to the internal surfaces of the mesopores. This can be achieved through silanation, ion exchange, or grafting techniques.

For example, mesoporous silica materials can be functionalized with sulfonic acid groups to enhance the adsorption of basic dyes or with amino groups to increase the adsorption of acidic dyes. These modifications create more active sites on the material surface, improving its affinity for specific dye molecules. The functionalization process also allows for the fine-tuning of the material's hydrophilic or hydrophobic properties, which can further optimize dye adsorption [7-10, 24].

Additionally, post-synthesis functionalization can improve the reusability of mesoporous materials by making them more resistant to fouling and easier to regenerate. This is crucial for developing sustainable water treatment technologies [17].

Modifications to the synthesis process, such as adjusting the concentration of surfactants or using dual templates, can further fine-tune pore sizes and distributions. These adjustments are critical for improving adsorption efficiency, as they promote interactions between the dye molecules and the adsorbent's surface [8, 9, 17,25].

Table 2 summarizes the functional groups used for post-synthesis modifications, the corresponding target dyes, adsorption capacities, pH ranges, and reusability performances across different mesoporous materials.

Table 2: Functionalization and Dye Adsorption Efficiency of Mesoporous Materials

Material Type	Functional Group	Target Dye	Adsorption Capacity (mg/g)	pH Range	Reusability	Reference
SBA-15	Amino (-NH_2)	Methylene Blue (Cationic)	150-200	4-7	High reusability for up to 5 cycles	[28 - 30, 45, 46]
MCM-41	Sulfonic Acid (-SO_3H)	Congo Red (Anionic)	100-180	2-6	Moderate, decreased after 3 cycles	[7-12, 20-25, 31-33]
Functionalized Mesoporous Silica	Thiol (-SH)	Reactive Black (Anionic)	120-220	3-7	Stable up to 4 cycles	[34-37,47,48]
Metal-Organic Frameworks (MOFs)	Carboxyl (-COOH)	Acid Orange (Anionic)	200-300	3-5	High reusability, stable for 6 cycles	[38 -41]
Mesoporous Carbon	Hydroxyl (-OH)	Methyl Orange (Anionic)	180-250	2-8	Excellent, up to 8 cycles	[42, 43]
Hydrogels (PHEMA, DMAa)	None (pH-Responsive)	Brilliant Blue (Anionic)	250-300	3-9	Highly reusable, >8 cycles	[24, 25, 44,49]

5. Specific Applications: Dye Adsorption

The practical use of mesoporous materials for the adsorption of dyes is a rapidly expanding field, driven by the growing need for efficient and sustainable methods to treat wastewater contaminated with industrial dyes. The unique structural characteristics of mesoporous materials, such as high surface areas, tunable pore sizes, and ease of functionalization, make them ideal candidates for this application. Various studies have explored the adsorption capacities of these materials for different types of dyes, including anionic, cationic, and non-ionic dyes, focusing on improving adsorption efficiency, selectivity, and reusability.

5.1 Cationic Dyes Adsorption

Cationic dyes, such as methylene blue and rhodamine B, are widely used in the textile industry and are known for their high toxicity and resistance to biodegradation. Mesoporous materials like SBA-15 and MCM-41, functionalized with amino groups (-NH_2), have demonstrated excellent adsorption capacities for these dyes. The electrostatic attraction between the negatively charged mesoporous surfaces and the positively charged dye molecules is the primary mechanism driving adsorption in these systems. For instance, SBA-15 functionalized with amines has shown adsorption capacities of up to 200 mg/g for methylene blue at an optimal pH range of 4-7.

Additionally, these materials exhibit good reusability, maintaining their performance over multiple adsorption-desorption cycles [7-30].

5.2 Anionic Dyes Adsorption

Anionic dyes, such as Congo red and methyl-orange, pose a different challenge due to their negative charge. Functionalizing mesoporous materials with positively charged groups, such as quaternary ammonium salts or sulfonic acid groups (-SO_3H), has been an effective strategy for adsorbing these dyes. Hernandez-Martinez and colleagues [24,25] have made significant contributions to this area by employing smart materials that respond to environmental stimuli, such as pH, to enhance dye adsorption. Their work on hydrogels, particularly those based on N,N-dimethylacrylamide (DMAa) and poly(2-hydroxyethyl methacrylate) (PHEMA), demonstrates how these materials can selectively adsorb anionic dyes like Acid Orange and Congo red at different pH levels [44].

The smart hydrogels developed by Hernandez-Martinez exhibit a swelling-shrinking behavior in response to pH changes, which significantly improves dye adsorption kinetics. At acidic pH values, the hydrogels swell, increasing the pore volume and surface area available for adsorption. When the pH increases, the hydrogels shrink, promoting the desorption of the adsorbed dye molecules. This pH-responsive behavior not only enhances the adsorption capacity but also facilitates the regeneration of the adsorbent, making these materials highly efficient and sustainable for multiple cycles of use. Hernandez-Martinez's work highlights the potential of these smart materials for large-scale applications in wastewater treatment, particularly for the removal of toxic anionic dyes [24,25,44].

5.3 Applications in Mixed Dye Solutions

In real-world wastewater scenarios, industrial effluents often contain mixtures of different dyes. The challenge is to develop adsorbents that can selectively remove multiple types of dyes in a single treatment process. Mesoporous silica materials functionalized with a combination of amino and sulfonic acid groups have been successfully applied to mixed dye solutions, where both cationic and anionic dyes coexist. The dual functionality of the material allows it to adsorb both types of dyes simultaneously, significantly improving the overall efficiency of the adsorption process. Studies have shown that these materials can remove up to 90% of mixed dyes, such as methylene blue and methyl orange, in a single cycle [34 - 37].

5.4 Environmental and Industrial Implications

The use of mesoporous materials for dye adsorption has important environmental and industrial implications. These materials offer a sustainable solution for the treatment of dye-contaminated wastewater, reducing the need for more energy-intensive and costly processes like chemical oxidation or biological degradation. Moreover, their high reusability makes them economically attractive for industrial applications. Development of smart, pH-responsive materials further enhances the sustainability and effectiveness of these systems, positioning mesoporous materials as leading candidates for future wastewater treatment technologies [44].

Materials Research Foundations 173 (2025) 40-59 https://doi.org/10.21741/9781644903452-2

6. Modeling and Simulation in the Adsorption of Dye

Modeling and simulation are essential tools in understanding and optimizing the adsorption processes in mesoporous materials. These techniques provide detailed insights into the behavior of dye molecules during adsorption, including interaction mechanisms, diffusion dynamics, and adsorption capacities. Several models have been developed to describe the adsorption equilibrium and kinetics of dyes on mesoporous materials, with the most widely used being the Langmuir and Freundlich isotherms, as well as kinetic models such as pseudo-first and second order models[50, 51].

6.1 Adsorption Isotherms

The Langmuir isotherm model assumes that the adsorption takes place on a homogeneous surface, with a finite number of identical sites, and that adsorption is limited to the formation of a monolayer. This model is particularly useful for describing the adsorption of dyes on mesoporous materials like SBA-15 and MCM-41, where the surface is relatively uniform and well-ordered. The Langmuir constant, derived from this model, can be used to predict the maximum adsorption capacity of the material. Studies on SBA-15 functionalized with amino groups for methylene blue adsorption have shown excellent agreement with the Langmuir isotherm, suggesting a strong monolayer adsorption mechanism [30 - 33].

In contrast, the Freundlich isotherm model applies to heterogeneous surfaces and multilayer adsorption, where the adsorption energy decreases as more molecules occupy the surface. This model is more applicable to materials like mesoporous carbon, where the surface is less uniform and may contain various types of adsorption sites. For example, mesoporous carbon materials functionalized with hydroxyl groups have shown better fit with the Freundlich model for the adsorption of methyl orange, reflecting the heterogeneous nature of the surface [24, 25, 49, 50].

6.2 Kinetic Models

Kinetic models are crucial for understanding the rate at which adsorption occurs. As mentioned before, the models more widely used are the pseudo-first-order and pseudo-second-order kinetic models for describing dye adsorption on mesoporous materials. The pseudo-first-order model assumes that the rate of occupancy of adsorption sites is directly proportional to the number of available sites, whereas the pseudo-second-order model posits that the adsorption rate depends on the square of the unoccupied sites. In most cases, the pseudo-second-order model has been found to provide a better fit, especially for functionalized mesoporous materials.

For instance, research on pH-responsive hydrogels demonstrated that the pseudo-second-order model accurately describes the adsorption kinetics of anionic dyes like Acid Orange, indicating that chemisorption is the dominant mechanism. This is consistent with the presence of strong electrostatic interactions between the dye molecules and the functional groups on the hydrogel surface [44].

6.3 Diffusion Models

Understanding the diffusion process within the mesopores is critical, especially for large dye molecules. The Weber-Morris intraparticle diffusion model is commonly applied to mesoporous systems to assess whether the rate of adsorption is limited by the diffusion of dye molecules into the pores. This model helps in identifying whether the adsorption process is controlled by external

Materials Research Foundations 173 (2025) 40-59 https://doi.org/10.21741/9781644903452-2

diffusion, intraparticle diffusion, or a combination of both. In systems where large mesoporous materials like SBA-15 are used, the diffusion of dye molecules through the pores often becomes the limiting step in the adsorption process [26-30].

6.4 Molecular Dynamics and Computational Models

Advances in computational techniques, such as molecular dynamics (MD) simulations and density functional theory (DFT), have allowed researchers to gain more detailed insights into the adsorption mechanisms at the atomic level. These simulations provide information about the interaction energies between dye molecules and the mesoporous surfaces, helping to predict adsorption behaviors and optimize the design of adsorbents [52] .

For example, DFT studies on functionalized mesoporous silica have shown how different functional groups influence the adsorption energy of various dyes, enabling the development of more efficient adsorbents tailored for specific dye types. MD simulations have also been used to study the diffusion of dye molecules within mesopores, providing valuable information on the optimal pore sizes for maximizing adsorption rates [53].

7. Future Perspectives

The field of mesoporous materials for dye adsorption is rapidly evolving, with numerous opportunities for further development. As environmental regulations become more stringent and industries seek more sustainable wastewater treatment solutions, the demand for efficient and reusable adsorbents will continue to grow. Several key areas present promising avenues for future research and application.

One of the most exciting developments in the field is the use of smart materials, which respond to external stimuli like pH, temperature, or ionic strength. These materials can adapt their structure to optimize adsorption under varying environmental conditions, significantly improving efficiency. Future work could focus on the development of multi-responsive materials capable of reacting to several stimuli simultaneously, enhancing their versatility for diverse wastewater treatment scenarios [24,25,44,49].

7.1 Sustainable and Biodegradable Adsorbents

While current mesoporous materials, particularly silica-based and carbon-based systems, exhibit high adsorption capacities, their production often involves energy-intensive processes and non-renewable resources. The future of mesoporous materials for dye adsorption lies in the development of more sustainable alternatives, such as biodegradable polymers and biosilica. These materials can be derived from renewable resources and designed to break down naturally after use, reducing the environmental impact of wastewater treatment technologies [44, 54].

7.3 Hybrid Materials for Improved Selectivity

Another promising direction is the development of hybrid materials, which combine mesoporous materials with other functional materials, such as nanocomposites or metal oxides, to improve selectivity and adsorption capacity. By incorporating additional active sites or catalytic components, these hybrid systems can simultaneously remove multiple contaminants, including

 Materials Research Forum LLC
https://doi.org/10.21741/9781644903452-2

dyes and heavy metals, from wastewater. This multifunctionality makes them highly attractive for industrial applications [20, 33].

7.4 Scaling Up for Industrial Applications

Despite the significant advances in mesoporous materials for dye adsorption, scaling these technologies for industrial applications remains a challenge. Future research should focus on addressing the economic and practical issues associated with large-scale production, including the development of cost-efficient synthesis methods and the optimization of material durability for long-term use. Also, the integration of these materials into existing wastewater treatment systems, such as membrane filtration or biological treatment units, could enhance the overall efficiency of the treatment process [43].

7.5 Circular Economy and Reusability

As sustainability becomes increasingly important, the concept of a circular economy—in which materials are reused and recycled—will play a pivotal role in the future of mesoporous adsorbents. Future research should explore methods for regenerating and reusing mesoporous materials without compromising their adsorption capacities. The development of easily regenerable adsorbents, such as those capable of pH-triggered desorption, will be key for lowering both the cost and the environmental impact of wastewater treatment processes [33, 43, 54].

Conclusion

The use of mesoporous materials for the adsorption of dyes presents a highly promising solution to the environmental challenges posed by industrial wastewater contamination. These materials, characterized by their high surface areas, tunable pore sizes, and ease of functionalization, offer excellent adsorption capacities for both cationic and anionic dyes. Through advancements in synthesis techniques—such as template-assisted, sol-gel, and hydrothermal methods—researchers have developed a wide range of materials, from silica-based systems to polymeric hydrogels, that can be tailored for specific adsorption applications.

The integration of functional groups, such as amino, carboxyl, and thiol groups, into the mesoporous frameworks has significantly enhanced dye selectivity and adsorption efficiency. Studies have demonstrated that these modifications not only increase the capacity for dye removal but also improve the reusability of the materials, composing them more sustainable for industrial applications. Furthermore, the development of smart materials, such as pH-responsive hydrogels, represents a breakthrough in dynamic adsorption systems, allowing for enhanced control over adsorption and desorption processes.

Modeling and simulation techniques, including isotherm and kinetic models, as well as molecular dynamics and density functional theory simulations, have provided deeper insights into the adsorption mechanisms. These tools have been invaluable for optimizing material design and predicting adsorption performance under various conditions. The combination of experimental data and computational models is crucial for advancing the field.

Looking forward, the future of mesoporous materials for dye adsorption lies in the development of more sustainable, biodegradable, and hybrid systems. The focus on smart and stimuli-responsive materials, coupled with the goal of scaling up for industrial applications, will drive innovation in wastewater treatment technologies. As industries move toward a circular economy, the reusability and regeneration of adsorbents will be key to minimizing the environmental footprint of these processes.

In conclusion, mesoporous materials are set to play a critical role in the future of dye adsorption, offering scalable, efficient, and environmentally friendly solutions to one of the most pressing industrial challenges of our time. Through continued research and technological advancements, these materials will undoubtedly become integral to sustainable wastewater treatment systems worldwide.

References

[1] G. Ravindiran, R.M. Jeyaraj, G. Nandipati, S. Seerangagounder, N.l Al-Zaqri, A. Boshaala, G. Hayder, Prevention of groundwater contamination from the pollutants released from dyeing industries using biochar produced from palm shell, Urban Climate 49 (2023) 101515. https://doi.org/10.1016/j.uclim.2023.101515

[2] J.S. Beck, J.C. Vartuli, W.J. Roth, M.E. Leonowicz, A new family of mesoporous molecular sieves prepared with liquid crystal templates, J. Am. Chem. Soc. 114 (1992) 10834–10843.

[3]. R.i Dariush, Pseudo-second-order kinetic equations for modeling adsorption systems for removal of lead ions using multi-walled carbon nanotube, J. nanostructure Chem. 3 (2013): 1-6. https://doi.org/10.1186/2193-8865-3-55

[4] C.T. Kresge, M.E. Leonowicz, W.J. Roth, J.C. Vartuli, J.S. Beck, Ordered mesoporous molecular sieves synthesized by a liquid-crystal template mechanism, Nature 359 (1992) 710–712.

[5] S. Dutta, S.l Adhikary, S. Bhattacharya, D. Roy, S. Chatterjee, A. Chakraborty, D. Banerjee, A. Ganguly, S. Nanda, P. Rajak, Contamination of textile dyes in aquatic environment: Adverse impacts on aquatic ecosystem and human health, and its management using bioremediation, J. Environ. Manage. 353 (2024), 120103. https://doi.org/10.1016/j.jenvman.2024.120103

[6] D. Khandelwal, I. Rana, V. Mishra ,K.R. Ranjan, P. Singh, Unveiling the impact of dyes on aquatic ecosystems through zebrafish – A comprehensive review, Environ. Res. 261 (2024) 119684. https://doi.org/10.1016/j.envres.2024.119684

[7] F. Benmahdi, S. Khetta, A. Khelali, M. Kolli, High-quality mesoporous adsorbent derived from pomegranate pomace for efficient removal of azo dye from textile wastewater: Influencing factors, reusability, and mechanism studies, J Indian Chem Soc. 101 (2024) 101320. https://doi.org/10.1016/j.jics.2024.101320

[8] K.K. Beltrame, A.L. Cazetta, P.S. De Souza, L. Spessato, T.L. Silva, V.C. Almeida, Adsorption of caffeine on mesoporous activated carbon fibers prepared from pineapple plant leaves, Ecotoxicol. Environ. Saf. 147 (2018) 64–71, https://doi.org/10.1016/j.ecoenv.2017.08.034.

[9] P. Zhang, M. He, W. Teng, F. Li, X. Qiu, K. Li, H. Wang, Ordered mesoporous materials for water pollution treatment: Adsorption and catalysis, Green Energy & Environment, 9(8) (2024), 1239-1256. https://doi.org/10.1016/j.gee.2023.11.001

[10] J. Castro-Gutiérrez, J. Jagiello, A. Celzard, V. Fierro, Introducing a novel Hierarchy-Connectivity factor for characterizing micro-mesoporous materials, Chemical Engineering Journal 497 (2024) 154938. https://doi.org/10.1016/j.cej.2024.154938

[11] F. Liebau, Ordered microporous and mesoporous materials with inorganic hosts: definitions of terms, formula notation, and systematic classification, Microporous and Mesoporous materials 58 (2003) 15-72. https://doi.org/10.1016/S1387-1811(02)00546-2

[12] A. Taguchi, F. Schüth, Ordered mesoporous materials in catalysis, Microporous Mesoporous Mater. 77 (2005) 1–45.

[13] L. Mercier, T.J. Pinnavaia, Direct synthesis of hybrid organic-inorganic nanoporous silica by a sol-gel process, Chem. Mater. 12 (2000) 188–196.

[14] J. Pérez-Ramírez, D. Verboekend, A. Bonilla, S. Abelló, Zeolite Catalysts with Tunable Hierarchy Factor by Pore-Growth Moderators, Adv. Funct. Mater. 19 (2009) 3972–3979, https://doi.org/10.1002/adfm.200901394.

[15] M.R. Benzigar, S.N. Talapaneni, S. Joseph, K. Ramadass, G. Singh, J. Scaranto, U. Ravon, K. Al-Bahily, A. Vinu, Recent advances in functionalized micro and mesoporous carbon materials: synthesis and applications, Chem. Soc. Rev. 47 (2018) 2680–2721, https://doi.org/10.1039/C7CS00787F.

[16] W. Xin, Y. Song, Mesoporous carbons: recent advances in synthesis and typical applications, RSC Adv. 5 (2015) 83239–83285, https://doi.org/10.1039/C5RA16864C,

[17] Q. Hu, R. Lan, L. He, H. Liu, X. Pei, A critical review of adsorption isotherm models for aqueous contaminants: Curve characteristics, site energy distribution and common controversies, Journal of Environmental Management 329 (2023) 117104. https://doi.org/10.1016/j.jenvman.2022.117104

[18] M.C. Carter, J.E. Kilduff, W.J. Weber, Site energy distribution analysis of preloaded adsorbents. Environ. Sci. Technol. 29, (1995) 1773–1780. https://doi.org/10.1021/acsagscitech.3c00341

[19] A. Sarkar, B. Paul, Evaluation of the performance of zirconia-multiwalled carbon nanotube nanoheterostructures in adsorbing As(III) from potable water from the perspective of physical chemistry and chemical physics with a special emphasis on approximate site energy distribution,Chemosphere 242 (2020) 125234. https://doi.org/10.1016/j.chemosphere.2019.125234

[20] H. Huan, L. Hou, F. Zhu, J. Li, M. Xu, Controllable thermal and pH responsive behavior of PEG based hydrogels and applications for dye adsorption and release, RSC Adv., 8 (2018) 9334-9343. https://doi: 10.1039/C8RA01018H

[21] L. Jianghong, W. Xiaohang, Y. Dandan, R. Jingwei, X. Jian, Recent Advances in Preparation and Application of Functionalized Mesoporous Materials.On Recent Innovations in Chemical Engineering (Formerly Recent Patents on Chemical Engineering) 12.3 (2019): 239-247. https://doi.org/10.2174/2405520412666190705125639

[22] Seifikar, Fatemeh, S. Azizian, M. Sillanpää, Microwave-assisted synthesis of carbon powder for rapid dye removal." Materials Chemistry and Physics 250 (2020) 123057. https://doi.org/10.1016/j.matchemphys.2020.123057

[23] G. Vishal, Md. Ahmaruzzaman, Smart materials for remediation of aqueous environmental contaminants, Journal of Environmental Chemical Engineering 9(6) (2021) 106486. https://doi.org/10.1016/j.jece.2021.106486

[24] A.R. Hernandez-Martínez, J.A. Lujan-Montelongo, C. Silva-Cuevas, J.D. Mota-Morales, M. Cortez-Valadez, Á.J. Ruíz-Baltazar, M. Cruze J. Herrera-Ordonez, Swelling and methylene blue adsorption of poly(N,N-dimethylacrylamide-co-2-hydroxyethyl methacrylate) hydrogel, Reactive and Functional Polymers 122 (2018) 75–84. https://doi.org/10.1016/j.reactfunctpolym.2017.11.008

[25] A.R. Hernández-Martínez, C. Silva-Cuevas, D. Rangel-Miranda, J.A. Lujan-Montelongo, Adsorption and swelling studies of 2-hydroxyethyl methacrylate- and N,N-dimethylacrylamide-based porous copolymers and their possible applications for QCM-sensors, Applied Surface Science 572 (2022) 151508. https://doi.org/10.1016/j.apsusc.2021.151508

[26] H. Shao, Y. Wang, H. Zeng, J. Zhang, Y. Wang, M. Sillanpää, X. Zhao, Enhanced photoelectrocatalytic degradation of bisphenol a by BiVO4 photoanode coupling with peroxymonosulfate, J. Hazard. Mater., 394 (2020), Article 121105, 10.1016/j.jhazmat.2019.121105

[27] S. Ben Hammouda, F. Zhao, Z. Safaei, D.L. Ramasamy, B. Doshi, M. Sillanpää, Sulfate radical-mediated degradation and mineralization of bisphenol F in neutral medium by the novel magnetic Sr2CoFeO6 double perovskite oxide catalyzed peroxymonosulfate: Influence of co-existing chemicals and UV irradiation, Appl. Catal. B Environ., 233 (2018), pp. 99-111, 10.1016/j.apcatb.2018.03.088

[28] M. Ghaedi, A. Ansari, F. Bahari, A.M. Ghaedi, A. Vafaei, A hybrid artificial neural network and particle swarm optimization for prediction of removal of hazardous dye brilliant green from aqueous solution using zinc sulfide nanoparticle loaded on activated carbon, Spectrochim. Acta Part A Mol. Biomol. Spectrosc., 137 (2015), pp. 1004-1015, 10.1016/j.saa.2014.08.011

[29] P. Assefi, M. Ghaedi, A. Ansari, M.H. Habibi, M.S. Momeni, Artificial neural network optimization for removal of hazardous dye Eosin Y from aqueous solution using Co2O3-NP-AC: isotherm and kinetics study, J. Ind. Eng. Chem., 20 (2014), pp. 2905-2913, 10.1016/j.jiec.2013.11.027

[30] Anbia, Mansoor, and Samira Salehi. "Removal of acid dyes from aqueous media by adsorption onto amino-functionalized nanoporous silica SBA-3." Dyes and Pigments 94.1 (2012): 1-9. https://doi.org/10.1016/j.dyepig.2011.10.016

[31] J.A.S. Costa, R.A. de Jesus, D.O. Santos, J.B. Neris, R.T. Figueiredo, C.M. Paranhos, Synthesis, functionalization, and environmental application of silica-based mesoporous materials of the M41S and SBA-n families: A review, Journal of Environmental Chemical Engineering, 9(3) (2021) 105259. https://doi.org/10.1016/j.jece.2021.105259

[32] R.K.Khaled, M.A. Wahba, M.D. Badry, M.F. Zawrah, E.A. Heikal, Highly ordered pure and indium-incorporated MCM-41 mesoporous adsorbents: synthesis, characterization and evaluation for dye removal. Journal of Materials Science, 57(7) (2022) 4504-4527. https://doi.org/10.1007/s10853-022-06877-7

[33] R.J. Kadhim, F.H. Al-Ani, Q.F. Alsalhy, MCM-41 mesoporous modified polyethersulfone nanofiltration membranes and their prospects for dyes removal, International Journal of Environmental Analytical Chemistry 103(4) (2023) 828-848. https://doi.org/10.1080/03067319.2020.1865326

[34] M.A. Tahir, H.N. Bhatti, I. Hussain, I.A. Bhatti, M. Asghar, Sol–gel synthesis of mesoporous silica–iron composite: Kinetics, equilibrium and thermodynamics studies for the adsorption of Turquoise-Blue X-GB dye. Zeitschrift für Physikalische Chemie, 234(2), (2020) 233-253. https://doi.org/10.1515/zpch-2019-1443

[35] Q. Lei, J. Guo, A. Noureddine, A. Wang, S., Wuttke, C.J. Brinker, W. Zhu, Sol–gel-based advanced porous silica materials for biomedical applications. Advanced Functional Materials, 30(41) (2020). 1909539. https://doi.org/10.1002/adfm.201909539

[36] L. Trupp, M.C. Marchi, B.C. Barja, Lanthanide–based luminescent hybrid silica materials prepared by sol-gel methodologies: a review. Journal of Sol-Gel Science and Technology, 102(1) (2022) 63-85. https://doi.org/10.1007/s10971-021-05678-9

[37] Y. Han, L. Zhang, W. Yang, Synthesis of Mesoporous Silica Using the Sol–Gel Approach: Adjusting Architecture and Composition for Novel Applications. Nanomaterials, 14(11) (2024) 903. https://doi.org/10.3390/nano14110903

[38] J. Liang, Y. Ge, Z. He, Q. Yun, G. Liu, S. Lu, L. Zhai, B. Huang, H. Zhang, Wet-chemical synthesis and applications of amorphous metal-containing nanomaterials, Nano Res (2021) 1-21. https://doi.org/10.1007/s12274-021-4007-6

[39] J. Zhu, P.-Z. Li, W. Guo, Y. Zhao, R. Zou, Titanium-based metal–organic frameworks for photocatalytic applications, Coord. Chem. Rev., 359 (2018) 80-101. https://doi.org/10.1016/j.ccr.2017.12.013

[40] Q. Du, P. Wu, Y. Sun, J. Zhang, H. He, Selective photodegradation of tetracycline by molecularly imprinted, ZnO@NH2-UiO-66 composites, Chem. Eng. J., 390 (2020) 124614. https://doi.org/10.1016/j.cej.2020.124614

[41] A. Hayat, S. Rauf, B. Al Alwan, A. El Jery, N.Almuqati, N., Melhi, S.,Melhi, A.A. Mohammed, A. Yas, S. Muhammad, O. Yasin, W. Lv, Recent advance in MOFs and MOF-based composites: synthesis, properties, and applications. Materials Today Energy, (41) (2024) 101542. https://doi.org/10.1016/j.mtener.2024.101542

[42] G. Durán-Jiménez, J. Rodriguez, L. Stevens, E.T. Kostas, C. Dodds, Microwave pyrolysis of waste biomass and synthesis of micro-mesoporous activated carbons: The role of textural properties for CO2 and textile dye adsorption. Chemical Engineering Journal, 488 (2024) 150926. https://doi.org/10.1016/j.cej.2024.150926

[43] Kopac, T., and S. D. Lin, A review on the characterization of microwave-induced biowaste-derived activated carbons for dye adsorption, International Journal of Environmental Science and Technology 21 (2024), 8717–8748. https://doi.org/10.1007/s13762-024-05583-y

[44] A.S. Umesh, Y.M. Puttaiahgowda, S. Thottathil, Enhanced adsorption: reviewing the potential of reinforcing polymers and hydrogels with nanomaterials for methylene blue dye removal. Surfaces and Interfaces, 51 (2024), 104670. https://doi.org/10.1016/j.surfin.2024.104670

[45] K. Shi, Z. Wang, J. Luo, G. Qian, Y. Luo, M. Li, H. Yao, Construction of novel NH2-Al-SBA-15 based mesoporous molecular sieve and application in the adsorption of methylene blue. Journal of Porous Materials, (2024) 1-13. https://doi.org/10.1007/s10934-024-01647-z

[46] P.N. Diagboya, J. Junck, S.O. Akpotu, R.A. Düring, Isolation of aqueous pesticides on surface-functionalized SBA-15: glyphosate kinetics and detailed empirical insights for atrazine. Environmental Science: Processes & Impacts, 26(2) (2024) 323-333. https://doi.org/10.1039/D3EM00425B

[47] R. Wang, S. Gu, C. Zhai, L. Deng, R. Li, X. Wang, Y. He, One-Step Synthesis of Thiol-Functional Mesoporous Silica Nanospheres and Selective Removal of Cationic Dyes. ChemistrySelect, 9(19) (2024) e202401349. https://doi.org/10.1002/slct.202401349

[48] R.P. Dhavale, V.G. Parale, H. Choi, T. Kim, K.Y. Lee, V.D. Phadtare, H.H. Park, Epoxy-thiol crosslinking for enhanced mechanical strength in silica aerogels and highly efficient dye adsorption, Applied Surface Science, 642 (2024) 158619. https://doi.org/10.1016/j.apsusc.2023.158619

[49] A.R.Hernandez-Martinez, Poly(2-Hydroxyethyl methacrylate-co-N,N-dimethylacrylamide)-Coated Quartz Crystal Microbalance Sensor: Membrane Characterization and Proof of Concept, Gels 7(4) (2021) 151. doi.org/10.3390/gels7040151

[50] J.P. Lima, G. Alvarenga, A.C. Goszczynski, G.R. Rosa, T.J. Lopes, Batch adsorption of methylene blue dye using Enterolobium contortisiliquum as bioadsorbent: experimental, mathematical modeling and simulation. Journal of Industrial and Engineering Chemistry, 91 (2020) 362-371. https://doi.org/10.1016/j.jiec.2020.08.029

[51] Q. Ibrahim, L. Creedon, S. Gharbia, A literature review of modelling and experimental studies of water treatment by adsorption processes on nanomaterials. Membranes, 12(4) (2022) 360. https://doi.org/10.3390/membranes12040360

[52] R. Pelalak, R. Soltani, Z. Heidari, R.E. Malekshah, M. Aallaei, A.Marjani, S.Shirazian, Molecular dynamics simulation of novel diamino-functionalized hollow mesosilica spheres for adsorption of dyes from synthetic wastewater. Journal of Molecular Liquids, 322 (2021) 114812. https://doi.org/10.1016/j.molliq.2020.114812

[53] H. Hajjaoui, M. Khnifira, A. Soufi, M. Abdennouri, S. Kaya, R. Akkaya, N. Barka, Experimental, DFT and MD simulation studies of Mordant Black 11 dye adsorption onto polyaniline in aqueous solution. Journal of Molecular Liquids, 364 (2022) 120045. https://doi.org/10.1016/j.molliq.2022.120045

[54] H.M. Munjur, M.N. Hasan, M.R. Awual, M.M. Islam, M.A. Shenashen,J. Iqbal, Biodegradable natural carbohydrate polymeric sustainable adsorbents for efficient toxic dye removal from wastewater. Journal of Molecular Liquids, 319 (2020) 114356. https://doi.org/10.1016/j.molliq.2020.114356

Materials Research Forum LLC
https://doi.org/10.21741/9781644903452-3

Chapter 3

Magnetic Mesoporous Materials for the Removal of Pollutants

Tatheer Fatima[1], Nadia Akram[1]*, Khalid Mahmood Zia[1], Zumaira Siddique[1], Fozia Anjum[1]

[1]Department of Chemistry, Government College University Faisalabad, Faisalabad-38000, Pakistan

nadiaakram@gcuf.edu.pk

Abstract

The special combination of large surface area, variable porosity, and magnetic recoverability has made, magnetic mesoporous materials extremely effective and adaptable adsorbents for the removal of pollutants. These materials have improved adsorption capabilities for a variety of contaminants, such as heavy metals, organic dyes, pharmaceuticals, and emerging pollutants. They are usually made of mesoporous silica, carbon or with other embedded magnetic nanoparticles. The magnetic characteristics make regeneration and separation simple, which lowers secondary pollutants and operating expenses. Selectivity and adsorption efficiency are further enhanced by functionalization with different chemical groups. With an emphasis on their potential for effective and sustainable water and air purification applications, this chapter reviews the synthesis techniques, adsorption processes, and current developments in the use of magnetic mesoporous materials for environmental remediation.

Keywords

Mesoporous Materials, Functionalization, Water Treatment, Soil Remediation, Future Perspectives

Contents

Materials Research Forum LLC
https://doi.org/10.21741/9781644903452-3

Materials Research Forum LLC
https://doi.org/10.21741/9781644903452-3

1. Introduction

Environmental pollution is one of the biggest socio-environmental problems in the contemporary world and has dire impacts on the health of global ecosystems, human beings, and the economy. Environmental problems such as air pollution, water pollution, and soil pollution remain rampant due to industrial activities, increased urbanization, and farming practices. As stated by the WHO, ambient air pollution causes the death of approximately 4.2 million deaths per year which is associated with heart disease, stroke, Chronic Obstructive Pulmonary Disease, lung cancer and acute lower respiratory infections in children [1]. For the former, according to [2] more than 2.2 billion people have no access to safely managed drinking water services and it highlights the importance of proper water purification techniques.

The conventional remediation strategies including chemical, physical or biological treatment techniques have certain limitations concerning effectiveness, costs, and environmental consequences [3]. Secondary pollutants may be generated during chemical treatment, several physical treatments are energy-consuming and may not be effective on some pollutants while biological treatments are slow and may depend on certain environmental factors. Therefore, the creation of new remediation methods that will be effective, cheap in implementation, and non-hazardous for the external environment is a critical requirement.

A new class of materials with ordered pore systems in the range from 2 to 50 nanometers in size is mesoporous materials and they are considered as candidates for environmental applications [4]. These materials are characterized by adjustable pore size and large surfaces that increase their adsorption/elimination capabilities concerning pollutant strains.

For example, Mesoporous silica has been investigated profoundly in the process of water purification. As expected, its large SSA makes it an effective adsorbent for pollutants, and, in addition, due to the possibility of controlling pore size in the mesoporous structure, it is possible to select the target substance [5]. Likewise, mesoporous carbons do not degrade rapidly and exhibit thermal stability therefore can be used in water as well as air treatment [5].

The peculiarities of the structure of the mesoporous material also enable its functionalization using adding functional groups that would also bolster pollutant elimination. For instance, by grafting amino groups on mesoporous silica the adsorption capacity for heavy metals, such as lead and cadmium can be enhanced [6]. Such change is processed to allow the development of special materials that can constructively solve given environmental issues.

The incorporation of magnetic properties into mesoporous materials greatly improves the applicability of the materials in pollutant removal. Advanced uses of magnetic mesoporous materials have upgraded mesoporous structures with high surface area and port size tunability with the nanoparticles property of magnetism thus giving better performance and ease to handle.

Materials Research Forum LLC
https://doi.org/10.21741/9781644903452-3

Iron oxide (Fe_3O_4) is preferred most of the time because of its high magnetic value and its dispersion in the biological system is innocuous [7]. Due to their specific properties, these nanoparticles help to simplify the process of separating adsorbent from the treated medium in the case when they are incorporated into mesoporous materials: the usage of an external magnetic field makes it possible to separate the nanoparticles from the medium during several minutes, thus decreasing the cost of the operation [8]. This magnetic separation technique is very useful in water treatment ranging from the separation of the designed adsorbent to water containing no traces of adsorbents after the pollutant capture as pointed out by [9].

Additionally, it is noted that magnetic mesoporous materials possess higher adsorption capacity for pollutants as compared to non-magnetic mesoporous materials because of the combined effect of the mesoporous structure of the material and the magnetic filler added to it. For example, magnetic mesoporous silica has enhanced uptake capability for organic dyes and heavy metals than non-magnetic mesoporous silica, this was due to the available active place caused by the magnetic nanoparticles [9].

In air purification, magnetic mesoporous materials can easily filter out particulate matter and also the volatile organic compounds (VOCs) in the polluted air streams [10]. The magnetic properties help to avoid the problem of adsorbent regeneration by the magnetic induction heating of the fixed-bed and accomplish the desorption and decomposition of the adsorbed pollutants; thus the life span and effectiveness of the material are enhanced [11].

The usage to magnetic mesoporous materials in soil remediation is also of significant emphasis. These materials can be applied for the extraction of organic and inorganic pollutants from the soil, thus they serve as useful equipment in environmental remediation. For example, mesoporous carbon containing embedded iron oxide nanoparticles was demonstrated to remove PAHs Summing up, the introduction of the magnetic properties into the mesoporous materials can be viewed as a step forward in utilizing the advanced materials for environmental applications. These materials not only possess high capacities of adsorption but they are also easy to be recovered as well as their performances are improved; thus these materials are potential candidates in dealing with the existing pollution problems in the world today. Further studies and investigations in this regard are required to enhance the characteristics of these materials when applied in large structures, as well as possibilities to use them in other environmental conditions. From contaminated soils [12].

2. Fundamentals of Mesoporous Materials

Mesoporous materials are a category of nanomaterials that have the features of pores with a size between 2 and 50 nanometers (nm). These materials have a very well-defined mesoporous structure about which the large surface area and pore size can be conveniently adjusted. Mesoporous materials usually have a BET surface area that varies between 500 and 1500 m^2/g, which when combined with high adsorption, make these materials suitable to be applied in several areas such as catalysis, the medical field particularly in drug delivery techniques, and environmental fields [13]. According to the type of pore size and its distribution, surface area, and stability, mesoporous materials can be described. Since the pore size in the mesoporous range is controllable, the molecules of pollutants according to their size and shape can selectively be adsorbed on the surface of the material. The large surface area enhances the desorption and catalysis of the material expanding its usage in the environmental field. Moreover, mesoporous

Materials Research Foundations 173 (2025) 60-92
https://doi.org/10.21741/9781644903452-3

materials can show great thermal and mechanical stability, which is essential for their proper functioning in changing conditions [14].

2.1 Types of Mesoporous Structures

Mesoporous materials can be categorized according to their type as silica, carbon, or metal. Each type has different characteristics and is applied with effectiveness in certain cases.

2.1.1 Mesoporous Silicas

Among all mesoporous materials, mesoporous silicas occupy a special place and are investigated most frequently. It is prepared using surfactant templates which are arranged in micelles that are used by silica precursors to form condensation. The surfactant template is then removed and this results in a mesoporous structure [15]. Some of them include MCM-41 and SBA-15 because of their large surface area and long narrow pore sizes. The benefits that mesoporous silicas contain are thermal stability, chemical stability, and modification possibilities. That is why these mesoporous silicas can be used as catalysts, adsorbents, and carriers for drug delivery. The functionalized mesoporous silicas are useful in the adsorption of heavy metals and small organic pollutants and thus can be used in water purification[15].

2.1.2 Mesoporous Carbons

There are several methods of preparing mesoporous carbons among which a templating technique consists in the impregnation of the template structure with a carbon source followed by carbonization. The exotemplate is then separated leaving behind a mesoporous carbon structure [16]. Some of the great characteristics of mesoporous carbons include; electrical conductivity, thermal stability and large surface area. These properties make the mesoporous carbons suitable for energy storage applications, catalysis and environmental treatment. Organic polluting agents, heavy metals as well as gases can be readily adsorbed by mesoporous carbons because of the huge available surface area and porosity [16].

2.1.3 Mesoporous Metals

Mesoporous metals consist of oxides and alloys which are prepared with the help of soft and hard templating methods and metal precursors are incorporated into the templates that are later on removed. These materials possess such characteristics as electronic, catalytic, and optical properties that are due to the nanostructure of these materials [17]. The categorization of mesoporous metals has the following benefits; high surface region, controllable pore diameter and functionality as a catalyst. Such properties make the materials suitable for catalytic processes, sensing and pollutant removal. Organic pollutants can be degraded by Mesoporous metals and heavy metals can be adsorbed from water thus improving the application of Mesoporous metals in environmental remediation [18].

Materials Research Foundations 173 (2025) 60-92
https://doi.org/10.21741/9781644903452-3

2.2 Synthesis Methods for Mesoporous Materials

Several methods are typically used in synthesizing mesoporous materials, and each of them allows the properties such as the pore size and surface area, as well as the stability of the formed structure to be controlled. Some of these synthesis techniques include the sol-gel method, templating techniques, and self-assembly techniques.

2.2.1 Sol-Gel Method

Sol-gel involves a process of converting a system originally in a liquid state containing mostly colloids known as the sol to a gel that is in a solid state. It is one of the most extensively used techniques for synthesizing mesoporous silicas and involves hydrolyzing and condensing silica precursors while using surfactants to provide a template for the mesoporous structure [19]. The benefits of the sol-gel method include; flexibility and tunability of the desired materials, porosity and surface area. Tem is employed for the synthesis of mesoporous materials such as MCM-41 and SBA-15, which are used in the catalytic and adsorption activities [13].

2.2.2 Templating Methods

Templating methods employ the usage of templates in developing the mesoporous architecture. There are two main types of templating methods: These are the two types of templating, namely the soft templating and the hard templating. Soft templating employs surfactant micelles as solvents for assembly of the precursor material its structure being based on the micelles. Following the creation of the mesoporous structure, the surfactant is eliminated and the pore remains[15]. Hard templating involves using solid templates that act as a mould for the formation of the porous structure wherein the template is impregnated with the precursor material. SBA-15 template is then evaporated by a chemical or thermal process, which organizes itself into a mesoporous structure[16]. The above templating methods make it possible to accurately control the size and pattern of the pore leading to the synthesis of well-ordered mesoporous materials.

2.3 Self-Assembly

Self-assembly can be defined as the process through which molecules arrange themselves in a specific structure without the help of a template. In the application of self-assembly in mesoporous materials, the interaction forces involved are hydrogen bonding, van der Waals and electrostatic forces. This method is widely employed in the preparation of mesoporous silicas, and carbons [14]. Some of the advantages of self-assembled structures are mainly the fact that the pores within the formed structures are of equal size. Self-assembly is utilized for generating different mesoporous materials for use in catalysis, adsorption as well as drug delivery. In conclusion, the structure and features of mesoporous materials, such as high surface area, provide superior conditions and an efficient framework for the usage of the materials in environmental remediation. Therefore, the knowledge of the sequence of synthesis method for mesoporous structures and the types of mesoporous structures is very essential for enhancing their efficiency in polluting agents removal. Further investigations in this research area are imperative to reach the optimum utilization of mesoporous materials to combat the growing pollution problems all over the world. Therefore, the prospect of mesoporous materials in environmental applications is bright and holds good promise in the future due to the continuous improvement in synthesis methods and the functionalization of the mesoporous materials to achieve better performance and environmental compatibility.

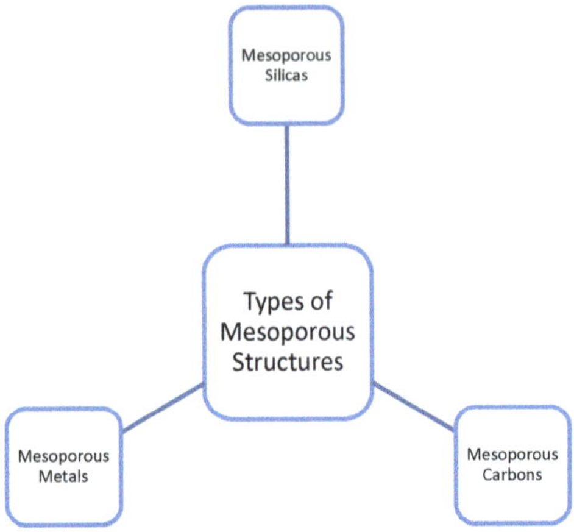

Figure 1: Types of Mesoporous Structures

3. Magnetic Functionalization of Mesoporous Materials

3.1 Magnetic Nanoparticles and Their Properties

MNPs have been attracting attention because of their unusual magnetic behaviour and usability in wide-ranging applications such as biomedical, data storage, and environmental areas. These nanoparticles generally can be in the size of between 1 to 100 nanometers and include a unique property known as super paramagnetism which refers to particles that are magnetized only when placed in a magnetic field and which are not magnetized when the field is withdrawn [7]. This property is especially useful in environmental end uses because it affords the possibility of simple and rapid collection of MNPs by magnetic force, thus reducing the quantity of secondary pollution. Thus, the magnetic properties of nanoparticles are controlled by their size, articles, and the type of dispersed phase. Typical magnetic materials are iron oxides, for example, magnetite Fe_3O_4 and maghemite gamma Fe_2O_3, cobalt, nickel, and combinations of them. Among these, iron oxide nanoparticles are most explored because they pose less toxicity risks to the human body are chemically stable and are inexpensive [20]. In environmental remediation, MNPs are used for the aspect of adsorption and degradation of pollutants and hence can be applied in water treatment, soil decontamination, and air treatment.

3.2 Methods for Incorporating Magnetic Components into Mesoporous Structures

The combination of the magnetic functionalized components with mesoporous structures comes as a unique characteristic in which the mesoporous has a high surface area and tunable size with pore size in addition to the magnetic nature of the nanoparticles to enhance the pollutant-removing capability of the final hybrid material. There are several methods through which, the magnetic

Materials Research Foundations 173 (2025) 60-92 https://doi.org/10.21741/9781644903452-3

components are introduced into mesoporous matrices, such as impregnation, co-precipitation and in-situ synthesis.

3.2.1 Impregnation

Impregnation involves the use of a solution of the magnetic precursor where the mesoporous material is soaked and then dried then sintered to convert the precursor into magnetic nanoparticles. This method is very simple and can enable the establishment of much control over the quantitative proportion of the magnetic component to be incorporated. For example, mesoporous silica can be attempted with an iron nitrate solution and then by a calcination process to obtain iron oxide nanoparticles in the silica structure [21]. The major overplus of impregnation is its ease and versatility but a drawback is the nanoparticle's inability to evenly deposit in the pores.

3.2.2 Co-precipitation

Co-precipitation is common for obtaining magnetic mesoporous materials, for instance, synthesizing iron salts precipitated in the large-pore matrix with the help of a base. This process usually takes place at ambient conditions and entails the deposition of two or more metals, which results in the formation of mixed oxide nanoparticles. For example, the co-precipitation of iron salts in the presence of ammonia in a mesoporous silica matrix can synthesis iron oxide nanoparticles within the pores [22]. It is a highly effective method in the aspect of large-scale production since it provides a uniform distribution of the nanoparticles.

3.2.3 In-situ Synthesis

One technique of producing magnetic nanoparticles involves the preparation of the mesoporous structure and at the same process synthesizing the magnetic nanoparticles within the structure. This method would provide a high extent of nanoparticle dispersion and good contact between the magnetic and mesoporous phases. An in-situ synthesis example is the hydrothermal synthesis of a reaction mixture of the silica precursor, surfactant, and iron salts in obtaining mesoporous silica with embedded iron oxide nanoparticles [23]. The process of in-situ synthesis is beneficial when creating a uniform dispersion of the nanoparticles and obtaining a firm bond between the nanoparticles and the mesoporous structure, however, using this method might need a complicated synthesis environment.

3.3 Characterization Techniques for Magnetic Mesoporous Materials

It is important to describe the magnetic mesoporous materials so that one can fully know their structural, morphological and magnetic properties that affect their efficiency in environmental applications. Several characterization techniques are used for these materials such as X-ray diffraction (XRD), scanning electron microscopic (SEM) techniques, transmission electron microscopic (TEM), and vibrating sample magnetometer (VSM).

3.3.1 X-ray Diffraction (XRD)

XRD technique is employed in establishing the crystal structure and phase distribution of magnetic mesoporous materials. From the diffraction patterns, one would be able to determine the type and degree of crystallinity of the magnetic nanoparticles that are embedded in the mesoporous matrix.

For instance, XRD can prove the synthesis of magnetite (Fe_3O_4) or maghemite (γ-Fe_2O_3) nanoparticles in mesoporous silica [24]. XRD also gives information about the size of the crystalline domains and this is very relevant when it comes to the determination of the magnetic properties of the material.

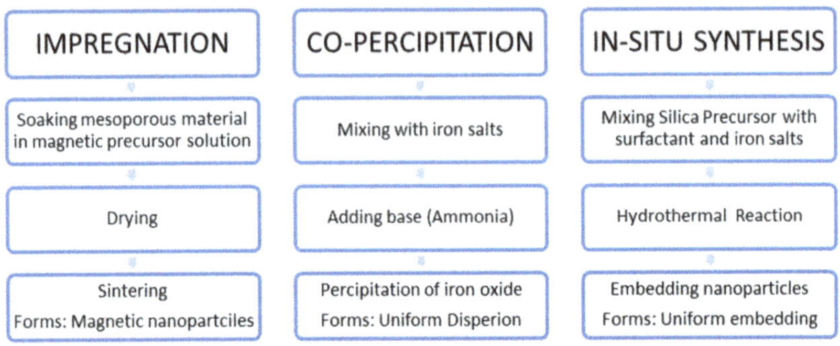

Figure 2 Incorporating Magnetic Components into Mesoporous Structures

3.2.2 Scanning Electron Microscopy (SEM)

SEM offers good surface morphology and particle size of magnetic mesoporous materials. For investigating the distribution and deposition of MNPs on the surface and within the pores of mesoporous structures it is especially efficient. The transmission electron microscope can help in identifying the homogeneous distribution of the nanoparticles and any porosity or surface roughness for that matter [25]. Also, energy dispersive spectroscopy (EDS) in seam with SEM can determine elemental composition proving the existence of magnetic elements, for example, iron.

3.2.3 Transmission Electron Microscopy (TEM)

The TEM has the capability of showing higher surface detail and internal pore structure of magnetic mesoporous materials. This offers a direct visualization of these magnetic nanoparticles in the mesoporous matrix plus information on the size shape and distribution of the particles. Besides, the application of TEM enables to determine the thickness of the mesoporous walls and the size of the pores – characteristics that influence the adsorption properties of the material [24]. An examination of the nanoparticles can be made through the use of high-resolution transmission electron microscopy (HRTEM) in an attempt to decipher the crystallographic structure of the material in question.

3.2.4 Vibrating Sample Magnetometry (VSM)

VSM is employed to determine magnetic characteristics of magnetic mesoporous materials like magnetization, coercivity, and remanence. This technique involves the use of a magnetic field that is applied to the sample through vibration and finding the magnetic moment. VSM offers information on the superparamagnetic character of the nanoparticles as well as the kind of response they offer to an external magnetic field [26]. Familiarity with these magnetic attributes is crucial for the enhancement of absolute performance in uses such as magnetic separation and pollutant elimination in specific regions.

Therefore, the magnetic functionalization of mesoporous materials has been established as a huge progress in environmental cleanup processes. Using magnetic nanoparticles in conjunction with mesoporous structures allows researchers to produce a new composite material with a higher adsorption capability, recoverability and overall higher effectiveness of pollutant removal. The method of magnet incorporation including impregnation, co-precipitation, and in-situ synthesis provides different benefits and issues that have an impact on the material's overall performance. The properties of these materials such as the structural and magnetic can be explored through characterization techniques like XRD, SEM, TEM, and VSM to improve their synthesis and use in environmental cleanup. Further studies in improvements of this field could be made to enhance the application of magnetic mesoporous materials in the mitigation of pollution around the globe.

4. Pollutants Targeted for Removal

Pollution of the environment is prevalent in the world and concerns such aspects as air, water, and soil. Such pollutants comprise heavy metals, organic chemicals, and particulate materials which are dangerous to human health and the environment.

Heavy Metals

Lead (Pb), cadmium (Cd) mercury (Hg) and arsenic (As) are called where Heavy Metals due to their risky abilities to upset the equilibrium of living organisms and the environment and have hazardous impacts. This is from several sources, among them being industrial effluent, mining and also from leaching that occurs in agricultural practice [27]. Heavy metals are formulated as dangerous to human health when taken within water bodies significantly affect the human body causing health problems such as neurological diseases and kidney problems. Another disadvantage of heavy metals in the soils is that they are toxic to the plant which hinders vigor and also gets into the food chain posing a threat to the fauna/ benefiting the flora. Some of the magnetic mesoporous structures based on doped have been proven to have high efficiency in the removal of heavy metals from water. For instance, the synthesis of amino-functionalized magnetic mesoporous silica for the selective adsorption of Pb (II) and Cd (II) ions in aqueous solutions. As pointed out by Zhu et al in regards to the earlier research done they established that functionalized mesoporous silica has very high adsorption capability of Pb(II) and Cd(II) ions; this infers that it can be used as a water purging material.

Organic Compounds

Heterocyclic compounds, pesticides and herbicides, pharmaceuticals, dyes, and volatile organic compounds (VOC) are some of the organic pollutants that are found in the environment as a result of industrial activities, use of chemicals in agriculture and poor disposal of products. Among these chemicals identified as POPs, there are PCBs and PAHs which pose a high risk to human health, more specifically, they are persistent, accumulate in the tissues, and may cause cancer and endocrine disruption [28]. These compounds are capable of polluting air, water, and land and thus they have unlawful effects on the environment as well as health in the long run. A great deal of research has been conducted to remove organic pollutants like dyes and pharmaceuticals using magnetic mesoporous materials. For example, magnetic mesoporous carbon has been used to adsorb and remove methylene blue from the wastewater which is a commonly used dye. The large surface of the material and its magnetic characteristic enabled rapid adsorption and convenience in separation by a magnetic field [28]. Besides, magnetic mesoporous silica has been used to adsorb pharmaceuticals including ibuprofen and diclofenac from water proving its effectiveness in the removal of different organic pollutants [29].

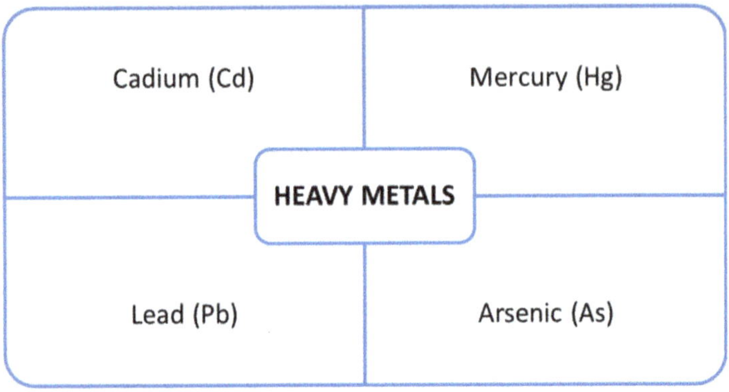

Figure 3 Heavy Metals

Particulate Matter

Some of the organic pollutants include Heterocyclic compounds, pesticides and herbicides, pharmaceuticals, dyes, and volatile organic compounds (VOC). They include PCBs and PAHs out of the identified chemicals that are classified as POPs and are said to be carcinogenic and cause endocrine disruption; they also accumulate in tissues. These compounds are a friend of air, water and soil pollution and hence unlawful in so far as their effects on the environment, as well as health with time, are concerned. Recent studies have been carried out to eliminate organic pollutants such as dyes and pharmaceuticals through the magnetic mesoporous material. For instance, magnetic mesoporous carbon is used to adsorb and eliminate methylene blue from the wastewater which is often used as dye. The large surface of the material, and its magnetic characteristic allow for a quicker adsorption and ease in which they can be separated using a magnetic field [5]. In addition, magnetic mesoporous silica has been applied to adsorb pharmaceuticals such as ibuprofen as well as diclofenac from water establishing its efficiency in eradicating various organic contaminants [29].

4.1 Adsorption Mechanisms of Pollutants on Mesoporous Surfaces

Due to the applied mesoporous material, the pollutants adhere to its surface through general forces such as Van der Waals force, electrostatic force, and chemisorption. Knowledge of these mechanisms is vital in the further enhancement of the removal of pollutants in magnetic mesoporous materials.

4.1.1 Van der Waals Forces

They are weak, attractive intermolecular forces that influence the adsorption of non-polar pollutants on porous high-surface area media. These forces are very important when it comes to the adsorption of organic species including VOCs and PAHs on the hydrophobic mesoporous media such as mesoporous carbon. Because of the relatively large surface area and pore volume, the Van der Waals forces are favored thus allowing great adsorption of nonpolar pollutants [30].

4.1.2 Electrostatic Interactions

Coulombic forces are those experienced by the charged pollutant molecules and a surface containing functional groups of the opposite charge to mesoporous materials. They are critical for the adsorption of metals and other ionizable heavy metals. For example, magnetic mesoporous silica containing amino groups (positively charged) can adsorb heavy negative metal ions such as Pb(II) and Cd(II) by electrostatic forces[6]. Likewise, sulfonic acid functionalized mesoporous silica (Negatively charged) adsorbs positively charged organic pollutants like methylene blue through electrostatic intercalation [31].

4.1.3 Chemical Bonding

The formation of covalent and coordination chemical bonds plays a major role in increasing the adsorption of particular pollutants on mesoporous surfaces. These strong interactions usually originate from an active functional group on the mesoporous material that reacts with the pollutant molecules. For instance, thiol-functionalized mesoporous silica can chemisorb with Hg Hg(II) and possess high capacity and selectivity for Hg(II) ion adsorption [32]. The coordination bonds between metal ions and functional groups, carboxyl or hydroxyl groups for instance are also believed to facilitate the adsorption of heavy metals on mesoporous materials [5].

5. Applications in Water Treatment

Magnetic mesoporous materials have been considered for drinking water treatment because of their excellent properties such as a large BET surface area, adjustable pore size, and easy separation by applying a magnetic field. Several examples and case studies here demonstrate the implementation of these materials in water treatment and indeed success of the use of these materials. One example is magnetic mesoporous silica for metal ions removal from water possibly for application in wastewater treatments [32] prepared magnetic mesoporous silica with amino groups by using anycetyltrimethyl ammonium bromide and evaluated the efficiency of the prepared adsorbent to remove lead (Pb) and cadmium (Cd) ions from the aqueous solutions. The study established that the material held large adsorption capacities of 300 mg/g for Pb(II) and 150 mg/g for Cd(II) which proved to be considerably more than those of the other basic adsorbents. The magnetic characteristic enabled the efficient and convenient recovery of the material after treatment from the water, which depicted its effectiveness in the removal of heavy metals as well as its applicability.

Another successful application is the elimination of organic pollutants through magnetic mesoporous carbon [33] studied the application of magnetic mesoporous carbon in the adsorption of methylene blue which is one of the most used dyes in industries and is present in wastewater. The effectiveness of the adsorbent in the present study was determined to be 450 mg/g, which is much higher than that of the activated carbon. The magnetic properties of this adsorbent made it possible to regenerate and reuse the adsorbent hence was environmentally friendly in the treatment of wastewater.

The magnetic mesoporous silica was efficient in the adsorption of ibuprofen and diclofenac from the solution. That produced a removal efficiency of more than 90% according to the study, this they attributed to the large surface area and the functionality presented by the mesoporous structure of the material. The possibility for the easy separation and regeneration of the adsorbent by the action of a magnetic pull, long the other key advantage of employing magnetic mesoporous materials in water purification.

5.1 Comparison with Traditional Water Treatment Methods

Magnetic mesoporous materials have been discussed for water treatment for drinking water due to factors like larger BET surface area, tunable pore size and ease of magnetically separating the adsorbent from the water again. The following examples and case studies shown here illustrate the use of these materials in water treatment and the success of the use of these materials.

For instance, magnetic mesoporous silica for the removal of metal ions from water for probably in wastewater treatments [24] synthesized magnetic mesoporous silica with amino groups using anycetyltrimethyl ammonium bromide and examined the possibility of the synthesized adsorbent for the removal of Pb and Cd ions from the aqueous solutions. The study found that the material was able to contain the large adsorption capacities that were measured at 300mg/g for Pb (II) and 150mg/g for Cd (II) which were way above those of the other basic adsorbents. The magnetic characteristic allowed for easy and convenient collection of the material after it has undergone the treatment from water thus illustrating its functionality in the removal of the abovementioned heavy metals and its practicality.

Another successful application is the removal of organic pollutants with the help of magnetic mesoporous carbon. The works by Wang and his group pose remarkable research on the efficiency of magnetic mesoporous carbon in the adsorption of methylene blue which is one of the widely used dyes by industries and is existing in wastewater. The values of adsorbent efficiency of the present study were obtained as 450 mg/g, which marked greater than the activated carbon. This adsorbent possessed magnetic properties that enabled them to re-use the adsorbent after treatment hence environmentally friendly in the treatment of wastewater.

In his study, Boehm et al. (2012) showed that Magnetic mesoporous silica was efficient in the adsorption of ibuprofen and diclofenac from solution. This according to the study resulted in a removal efficiency of more than 90% of the pollutant, which they attributed to the large surface area, and functionality presented by the mesoporous structure of the material. Opportunity to separate and reuse the adsorbent by applying the magnetic field, along with the other major beneficial factor of using magnetic mesoporous materials in the water purification process.

5.1.1 Filtration

Filtration is a widely used method of purifying water from suspended impurities and other solids and particulate material. It is, however, less effective in removing dissolved materials such as heavy metals as well as organic pollutants. Magnetic mesoporous materials have a large specific surface area and adjustable pore size to remove a large amount of pollutants especially those that can hardly be filtered [6].

Materials Research Foundations 173 (2025) 60-92
https://doi.org/10.21741/9781644903452-3

5.1.2 Coagulation

Coagulation includes the addition of chemicals to water that will encourage the compression and sedimentation of solids and certain dissolved substances. This process may be helpful but it mostly involves the use of many chemicals and results in the formation of a thick slug which has to undergo treatment and dumping. Another kind of mesoporous material in water treatment is magnetic materials which are friendly to the environment since they can directly adsorb contaminants without employing any other chemicals and can be easily separated from water with a magnetic field, minimizing the production of wastes [23].

5.1.3 Activated Carbon Adsorption

Many organic pollutants are adsorbed using activated carbon since it has a very large surface area and pores. However, it has limitations in terms of regeneration and reuse meaning that after using the data once, the ability to use it again decreases. This diver is a combination of the features of magnetic mesoporous materials in comparison with other types of mesoporous materials that have either the same or higher adsorption capacity due to ease of regeneration through magnetic separation. This cuts the chances of operational costs and environmental pollution related to the frequent replacement and disposal of activated carbon [34].

5.2 Challenges and Opportunities

Thus, as promising as magnetic mesoporous materials may seem for future water treatment plants, several problems have to be overcome.

5.2.1 Cost of Production

Therefore, among the main obstacles that can be singled out, costs are the biggest problem. The process of fabrication of the magnetic mesoporous materials is very complex and consists of two general procedures, the synthesis of mesoporous and the loading of the magnetic nanoparticles. This can be more expensive as compared with other more conventional materials such as activated carbon. On the same note, it is wise to explain that like in any other field, progress in the synthesis techniques used, including improved synthesis methods, costs are likely to come down [22].

5.2.1 Material Stability

Their long-term magnetic mesoporous structure stability in different water treatment-related environments is another issue. Fluency of pH, temperature and effects of the other ions that may be present in a system also alters the properties of the material. It is crucial to precipitate the research for the creation of new strong and durable admixtures that are capable of performing under these circumstances with minimum losses [20].

5.2.3 Regeneration and Reuse

There is also the long-term magnetic mesoporous structure stability in the different water treatment-related environments is also a big question. Regularity of the pH, temperature and effects of the other ions that may exist in a system also change the properties of the material. Thus, it is necessary to advance the research for the development of new strong and durable admixtures for the situation described above with minimum losses [20].

Materials Research Forum LLC
https://doi.org/10.21741/9781644903452-3

5.2.4 Environmental Impact

Another concern is long-term magnetic mesoporous structure stability in the various other conditions that exist in water treatment-related fields. Flow of pH, temperate and influence of the other ions that might be in a system also change the characteristics of the material. Thus, it is important to precipitate the research for the creation of new strong and durable admixtures performing under such conditions with minimum losses [20].

5.3 Opportunities for Advancement

Still, enormous prospects can be pointed in the direction of improvements in magnetic mesoporous materials in water treatment. Further studies to improve the existing materials or the development of new materials can result in higher capacities, better stability, or ease of regeneration, among others. Interactions with developers and policymakers may help in the development of the utilisation of such materials in large-scale water treatment processes.

In addition, the combination of magnetic mesoporous materials with other efficient treatment technologies like membrane process and photocatalytic process results in the development of an integrated system that results in an improvement of the overall efficiency of the process. For instance, when using magnetic mesoporous materials together with photocatalytic materials, it proposes a solution for both adsorption and photodegradation of pollutants under light exposure [20].

Thus, MMMOs can be considered as a breakthrough innovation in the water treatment field. Due to their high adsorption capacities, ease of regeneration and compatibility with versatile applications, they have been reported as preferable over the traditional approaches[29]. These challenges entail cost, stability, and environmental hardness, which becomes a barrier to their full application in large water treatment plants. Hence, with further studies and developments, magnetic mesoporous materials can contribute to offering clean and safe water for people globally.

6. Applications in Air Filtration

6.1 Role of Magnetic Mesoporous Materials in Air Pollutant Capture

Major sources of air pollution include industries, automobiles, and combustion processes that are all emitting fumes that are risky to human health and the surrounding environment. Magnetic mesoporous materials have proven very effective as adsorbents of gaseous pollutants such as PM and VOCs because of their high surface area, adjustable pore size, and magnetic properties.

6.1.1 Particulate Matter (PM)

Airborne particles mainly dust and particulate matter of which 10 microns in diameter and below are considered fine particles. Such particles of sizes 5 micrometres (PM2. 5) for example, can enter the respiratory system and cause respiratory and cardiovascular diseases [35]. These fine particles can be adsorbed by magnetic mesoporous materials because of the infinity of their surfaces as well as their porosity. For example, [36] confirmed that magnetic mesoporous silica can be used for PM2.5. It was identified that the extensive porosity of the material gave immense area for particle adsorption and the magnetic character enabled ease in retrieving the material by applying a magnetic force.

Materials Research Forum LLC
https://doi.org/10.21741/9781644903452-3

6.1.2 Volatile Organic Compounds (VOCs)

VOCs are a series of organic compounds that are easy to evaporate at room temperature, which becomes the source of indoor as well as outdoor air pollution. Some of the primary contributors to VOCs are industrial operations, cars and other vehicles, and home and office items. Magnetic mesoporous materials can adsorb VOCs from the air because of their large surface area and prospects of modifying their surface chemistry to improve adsorption properties. [34] conducted a study in which Mesoporous Carbon with magnetic properties was used to adsorb benzene; a prototype of VOCs. The material demonstrated a high surface area for benzene capture and hence this work points to the use of the material to capture volatile organic compounds in air filtering systems.

6.2 Efficiency and Selectivity of Magnetic Mesoporous Materials in Air Filtration

6.2.1 Capture Efficiency

Mesoporous materials have high capture efficiencies because of their large surface area and because of the magnetic mesoporous materials' potential to eliminate airborne pollutants. Adsorption efficiency is the product's capacity to catch pollutants from the air as announced on the label. Magnetic pore size. For instance, while discussing the findings of the study by [36], revealed a capture efficiency of above 90% for PM2. 5 using magnetic mesoporous silica. In the same manner, magnetic mesoporous carbon had a capture efficiency of 85% for benzene as analyzed by Wang et al. in their study.

6.2.2 Selectivity

Selectivity is the specific property of the material that can fix one sort of pollutant more effectively than the other. This is because the functionalization of Magnetic mesoporous materials can be conducted for specific pollution. For example, magnetic mesoporous silica with amino groups introduced can selectively adsorb formaldehyde, a representative indoor VOC, because of the good compatibility between the amino group and formaldehyde [31]. This kind of adsorption is useful in certain specific air treatment processes, where only certain contaminants are to be eliminated.

6.3 Integration of Magnetic Mesoporous Materials into Air Purification Systems

The enhancement of the air purification systems which are already in use with magnetic mesoporous materials can make them more efficient and productive. Such materials can be integrated into different air purification systems such as filtration equipment, catalytic converters, and adsorption systems.

6.3.1 Filtration unit

Air purifiers are useful devices that assist in the removal of particulate matter from the next possible surroundings. Another strategy for increasing the proficiency of the filtration process in situations whereby small particles are involved subsequent in the trapping and removal of such particles is by the incorporation of magnetic mesoporous materials in the creation of the filter material. For example, magnetic mesoporous silica can be introduced in HEPA filters and this makes the filters to be more efficient in trapping PM2. Other fine particles and number 5. Another related characteristic is that this material has field capability and high regenerative properties of this material making it be reused constantly thus reducing its maintenance costs and at the same time being friendly to the environment as noted by [37].

6.3.2 Catalytic Converters

The catalytic converters are applied in industries and the automobile industry particularly to minimize emissions. In such systems, magnetic mesoporous materials may act as catalysts or support materials to boost the efficiency of VOCs and other gaseous pollutants' abatement. For example, magnetic mesoporous carbon can be applied to enhance the physical deposition of catalytic metals, such as platinum or palladium, for the oxidation of VOCs at lower temperatures and, consequently, enhancing the efficiency of catalytic converters [38].

6.3.3 Adsorption Systems

Adsorption technologies are commonly used to address the issue of VOC removal from indoor and outdoor air. Magnetic mesoporous materials can be used for the present systems because of their large adsorption capacities and receiving special pollutants. The fact that regeneration can be easily induced through a magnetic field makes these materials more advantageous in adsorption systems. For example, [39] confirmed that magnetic mesoporous silica can effectively adsorb toluene, one of the VOCs. This material demonstrated high absorption and could be simply regenerated using a magnetic field, thus it can be recommended for usage in air purification systems to eliminate VOC.

6.3.4 Potential Advantages

The integration of magnetic mesoporous materials into air purification systems offers several advantages over traditional materials:

Selectivity: Functionalization of the mesoporous surfaces enhances the selectivity of these materials for specific pollutants, enabling targeted air purification.

Ease of Regeneration: The magnetic properties of these materials facilitate easy regeneration and reuse using a magnetic field, reducing operational costs and environmental impact.

Versatility: Magnetic mesoporous materials can be used in various air purification technologies, including filtration units, catalytic converters, and adsorption systems, making them versatile options for different applications.

Environmental Sustainability: The ability to regenerate and reuse magnetic mesoporous materials reduces the need for frequent replacement and disposal, contributing to more sustainable air purification solutions.

In conclusion, magnetic mesoporous materials hold significant potential for enhancing air purification systems by improving the capture efficiency and selectivity for airborne pollutants. Their integration into existing technologies can offer practical advantages, including ease of regeneration and environmental sustainability. Continued research and development in this field are essential to optimize these materials for large-scale applications and to fully harness their potential in providing cleaner air.

7. Applications in Soil Remediation

7.1 Techniques for Using Magnetic Mesoporous Materials in Soil Cleanup

Magnetic mesoporous materials applied for soil remediation involve these techniques; in-situ and ex-situ have their pros as well as cons. These methods make use of factors such as inherent large surface area, controlled pore size distribution and magnetic behaviour of mesoporous materials for the removal of contaminants like OCs and inorganics from the soil.

7.1.1 In-Situ Remediation

In-situ bioremediation deals with the treatment of polluted soil without having to dig it up. This reduces interruptions to the environmental experience and the overall cost demanded in the removal and transport of soil. Magnetic mesoporous materials entail practical application at the contaminated site by directly injecting them; they interact with the pollutants in question. One of the in-situ methods is soil washing where a washing solution is disposed of in the soil containing magnetic mesoporous materials. The materials collect and purify pollutants, and a magnetic field is used to get back the adsorbents with the pollutants. For instance, [23] used magnetic mesoporous silica to remediate soil contaminated with heavy metals as a quasi-real-time process. It successfully removed lead (Pb) and cadmium (Cd) and could be easily magnetically separated thus having possibilities of application on a large scale.

7.1.2 Ex-Situ Remediation

Ex-situ treatment implies that polluted soil must be dug up and treated at another location or in a specific area. This method affords an opportunity to control the treatment conditions and is furnished with the hope of utilizing the treated soil again. It can be treated using ex-situ processes like magnetic mesoporous materials applicable in soil washing chemical extraction and bioremediation. In ex-situ soil washing, magnetic mesoporous materials are introduced to the contaminated soil in a solution. The material accumulates the pollutants, and they are subsequently collected using a magnetic field. It has been established that this technique is very useful in eliminating a greater variety of contaminants. For instance, an ex-situ study which was conducted by [39] employed magnetic mesoporous carbon for the removal of polycyclic aromatic hydrocarbons (PAHs) in contaminated soil. The high adsorption capacity of the material and its magnetism ensured the effective removal of pollutants and adsorbent regain.

7.2 Case Studies on the Removal of Organic and Inorganic Contaminants from Soil

Some of the illustrations explain how magnetic mesoporous materials can be used in realistic scenarios of soil management to remove organic and inorganic pollutants.

7.2.1 Organic Contaminants

Being on the list of priority substances, organic contaminants, including PAHs and pesticides, are highly resistant to degradation and present serious threats to people and the environment. These mesoporous materials have been postulated to be effective in adsorbing the aforementioned pollutants because of their large surface area and tunable surfaces. The sorption of persistent polycyclic aromatic hydrocarbons (PAHs) in soils has been studied in detail; [38] magnetic mesoporous carbon in cleaning the contaminated soil. This research concluded that the material significantly adsorbed the PAHs in the soil thus lowering their concentration. The magnetic properties enabled simple and efficient separation and reuse of the adsorbent; thus, the material's applicability can be evaluated on a commercial scale in removing toxic metals from contaminated soil [40] also conducted another study related to the concern of this research work; in this study, magnetic mesoporous silica was employed as the adsorbent for the removal of pesticides from agricultural soil. The silica surface modified in this functionalized exhibited notably high adsorption capacity for some of these common pesticides like atrazine and chlorpyrifos. The practical utilization of the adsorbent was made possible through the regeneration of the same by applying a magnetic field; this aspect made the remediation process more sustainable.

7.2.2 Inorganic Contaminants

Considering PAHs and pesticides as the list of priority substances, it should be mentioned that they are characterized by high non-biodegradability and pose potential threats to people and the environment. These mesoporous materials have been proposed to be useful in adsorbing the aforementioned pollutants due to the availability of large surfaces and surface chemistries can be tailored. Techniques in the removal of sorbed persistent polycyclic aromatic hydrocarbons (PAHs) in soils have been reported [34] demonstrating the use of magnetic mesoporous carbon in cleaning the contaminated soil. This research pointed out that the material fully absorbed the concentration of the PAHs in the soil hence reducing their concentration. The magnetic properties allow for easy and convenient recovery and reusability of the adsorbent; hence, the material's feasibility can be assessed on a commercial level for treating contaminated soil with toxic metals. [40] also carried out another study in the same line as the concern of this research work; in this study, magnetic mesoporous silica was used as the adsorbent for the degradation of pesticides from agricultural land. For a good number of these common pesticides such as atrazine and chlorpyrifos, the surface of this silica on functionalization was observed to have a highly enhanced adsorption capability. The actual application of the adsorbent was by regenerating the same through the application of a magnetic field; such a feature contributed to the sustainability of the remediation process.

7.3 Environmental Impact and Sustainability Considerations

The application of magnetic mesoporous materials in the removal of soil pollutants has some benefits to the environmental populace including; conservation of chemical utilization, formation of unwanted products and possibly the reactivation of adsorbents. However, these products raise ecological impacts, which have to be considered in reducing the effects of such types of material on the environment.

7.3.1 Environmental Benefits

Reducing the chemical additives used in cleaning is also considered one of the significant environmental rewards of using magnetic mesoporous materials as they are characterized by a very high adsorption capacity. This means that the chances of secondary pollution and the general impacts of pollution on the environment are eliminated to a very large extent. Magnetic properties of these materials make it easier to retrieve and recycle them hence minimizing the quantity of waste produced during the remediation exercise. For instance, due to the more efficient adsorption of magnetic mesoporous carbon used to clean PAHs-contaminated soil, the used adsorbent material could be regenerated and reused at least several times in the method developed by [34].

Materials Research Forum LLC
https://doi.org/10.21741/9781644903452-3

7.3.2 Ecological Impacts

Thus, the utilization of magnetic mesoporous materials has its benefits, however, it is crucial to examine any possible ecological consequences related to their application. This manufacturing process sometimes requires the use of chemicals and energy hence an impact on the environment is possible. Thus, there is a need to draw attention to the need to use eco-friendly methods of synthesis that will have a lesser effect on the environment.

Another area of concern is the possible release of nanoparticles into the environment when carrying out the remediation exercise. Even though magnetic mesoporous materials are created to be stable, their activity in the environment needs to be controlled and threats to ecosystems should be avoided. More research about the stability of such materials in the long run, and the possible toxicity of the products if they are released into the environment is needed.

7.3.3 Sustainability Considerations

To extend the sustainability of magnetic mesoporous materials for soil rectification the following measures can be implemented. The employment of green synthetic methods with environment-friendly precursors and energy-conserving methods can go a long way to minimize the disadvantageous effects of material production on the environment. However, the use of these materials in combination with other types of sustainable remedial methods like the use of plants in remediating the contaminated areas, or biological remediation can improve the rate of remediation. For instance, the incorporation of magnetic mesoporous materials with phytoremediation improves the bioavailability of contaminants in plants and has a method of regenerating the adsorption material. The integration of this concept can enhance the efficiency and sustainability of amelioration processes of soil [18]. Thus, it may be concluded that magnetic mesoporous materials have a vast capability of soil remediation since they have a high adsorption rate, can be easily recovered and have the ability to adsorb both organic and inorganic materials. The real-life examples prove them to be useful in combating soil pollution whereas, the economic and environmental concerns such as environmental effects and sustainability show the necessity of devising eco-friendly synthesis techniques and studying their effects on the environment. Further investigation and development work prove that magnetic mesoporous materials will be vital in the proper management of soil remediation in the future.

8. Mechanisms and Kinetics of Pollutant Removal

8.1 Adsorption/Desorption Kinetics of Pollutants on Magnetic Mesoporous Materials

The rate constants of adsorption and desorption processes are important for Mm-MOFs and their effectiveness in eliminating pollutants. Adsorption kinetics relates to the rate at which pollutants are removed from the water phase and deposited on the surface of the adsorbent while desorption kinetics is the rate at which pollutants are desorbed from the adsorbent surface and returned to the water phase.

8.1.1 Rate of Adsorption and Desorption

The rate of adsorption depends on several factors among which are pollutants' concentration, the type and size of the adsorbent's surface and pores, temperature, and competing ions or molecules. The high porosity, large specific surface area of magnetic mesoporous materials along with the approval of appropriate pore sizes offers many adsorption sites for pollutants, thus resulting in high uptake rates. For example, [41] have studied the adsorption of lead (Pb) ions on magnetic mesoporous silica and have noted that the adsorption was rather fast, and most of the process was completed in the initial half an hour; however, the final equilibrium was achieved gradually. The desorption kinetics, however, depends on the effectiveness of the interaction between the pollutant and the adsorbent surface. Molecules with a capability of chemical reactions show slower diffusion rates than the ones with weak interactions like van der Waals forces. Knowledge of these kinetics is vital in the regeneration and recycling of the adsorbent material. For instance, magnetic mesoporous carbon was discovered to have the ability to be regenerated by magnetic separation alongside heat treatment thus giving consecutive cycles of adsorption and desorption without great reduction of effectiveness [34].

8.1.2 Factors Affecting Kinetics

Several factors affect the adsorption/desorption kinetics of pollutants on magnetic mesoporous materials:

Table 1 Factors affecting the Kinetics of pollutants on magnetic mesoporous materials:

Factor	Description
Surface Area and Pore Size	Higher surface areas and optimal pore sizes enhance adsorption sites and facilitate faster diffusion.
Temperature	Increased temperatures enhance adsorption rates, but may affect desorption rates and equilibrium.
Concentration Gradient	Higher initial concentrations create a steeper gradient, driving faster adsorption until equilibrium.
pH and Ionic Strength	pH and ionic strength influence the charge and speciation of pollutants and the adsorbent surface.

8.1.3 Models Used to Describe Adsorption Kinetics

Various models are used to describe the kinetics of adsorption processes, including:

Pseudo-First-Order Mode: According to this model, it is presumed that the rate of occupancy of the adsorption sites depends on the availability of these sites. Most commonly, it is applied to those systems, for which adsorption happens on the uniform surface [41].

Pseudo-Second-Order Model: The data being reduced using this model is based on the principle that the adsorption of samples occurs through second-order kinetics, which often signifies chemisorption as the limiting step [42].

Intra-Particle Diffusion Model: Regarding the kinetics, this model views diffusion inside the pores of the adsorbent as being rate controlling, this is so especially for materials of mesoporousity with well-defined pore structures [43].

8.2 Thermodynamic Aspects of Pollutant Adsorption

The process is supported by thermodynamics where feasibility, spontaneity or otherwise and the nature of the interaction between the pollutants and the adsorbent are analyzed. Concerning the kinetics, this model considers diffusion within the pores of the adsorbent as controlled by rate; it is especially so for materials that mesoporousity having highly defined structures[43]. Has with enthalpy (ΔH), entropy (ΔS), and Gibbs free energy (ΔG) changes.

8.2.1 Enthalpy (ΔH)

The sign of enthalpy change of adsorption will help to know whether the process of adsorption is accompanied by the release of energy or requires energy. When the change in enthalpy is negative, that is when enthalpy is released during the adsorption process ($\Delta H < 0$), it is referred to as exothermic and is usually associated with physisorption which is caused by van der Waals forces. The exothermic processes ($\Delta H < 0$) are those that release heat and may be chemisorption that includes stronger interactions like coordinate covalent etc. For instance, the process of heavy metals adsorption on magnetic mesoporous silica exhibits exothermic tendencies, typical for physisorption [6].

8.2.2 Entropy (ΔS)

The change of entropy in the process of adsorption can show the state of the solid-liquid interface. Entropy changes in the system are positive ($\Delta S > 0$), this points to an increase in randomness in terms of water molecules desorbed from the adsorbent surface or the pollutant molecules in the system may exhibit more disorder upon the adsorption process. Phase II and III are characterized by an exothermic process that causes negative entropy changes ($\Delta S < 0$) attributing to the formation of more ordered structural layers on the adsorbent surface.

8.2.3 Free Energy (ΔG)

The extent of the process is characterized by the free energy change of adsorption (ΔG). If $\Delta G < 0$ it would denote that the adsorption is spontaneous, otherwise, if $\Delta G > 0$ the process will follow non-spontaneous behaviour. ΔG of a reaction will be less than 0 if the reaction is spontaneous, that is, the reactants will be converted to products more often than the products to reactants. The No Passing Rule tells us that ΔG of a reaction will be less than zero if the reaction is spontaneous as the reactants will tend to where TTT is the temperature in Kelvin differing from the basic unit. For example, the ΔG of the adsorption of VOCs on magnetic mesoporous carbon is negative suggesting spontaneous adsorption [34].

8.3 Modeling Approaches to Predict Pollutant Removal Efficiencies

Computer-aided tools and simulations are used to estimate the efficiency of magnetic mesoporous materials in the eliminative purification of pollutants. Such models are useful for evaluating the material design and elaborating adsorption processes and removal effects in different situations.

8.3.1 Molecular Dynamics (MD) Simulations

MD simulations offer molecular-level information on the interactions and diffusive behaviour of pollutants on the adsorbent. Here, MD is capable of explaining the function of certain components such as functional groups, interfacial area, pore size, and the roles of temperature and pH on the adsorption outcomes. For instance, MD simulations have been applied in the adsorption of metal ions on mesoporous silica with functional groups to reveal the interaction mechanisms and the most favourable functional group strategies [39].

8.3.2 Density Functional Theory (DFT) Calculations

Quantum mechanistic results on adsorbent and pollutant electronic structures can be made with DFT calculations. These calculations can forecast the adsorption energy, the active sites and the chemical nature of the interaction of adsorbent-pollutant species. The relevance of the present study to the work reported in the literature is explained by the DFT study on the adsorption of organic pollutants on magnetic mesoporous carbon indicating that π-π interactions and hydrogen bonding play an important role in the adsorption process [44].

8.3.3 Isotherm Models

Isotherm models give the state at which there is a balance of distribution of the concentration of the pollutants in the liquid phase and its adsorption in the solid phase. Commonly used isotherm models include:

Table 2 Isothermal Models

Isotherm Model	**Description**	**Reference**
Langmuir Isotherm	Assumes monolayer adsorption on a homogeneous surface with a finite number of identical sites.	[45]
Freundlich Isotherm	Empirical model describing adsorption on heterogeneous surfaces with sites of varying affinities.	[46]
Temkin Isotherm	Considers the effects of indirect adsorbate/adsorbate interactions, with adsorption heat decreasing linearly.	[47]

Consequently, for assessment and possibly improvement of the magnetic mesoporous material performance in environmental applications, these mechanisms and kinetics of pollutant degradation are significantly important to be determined. Kinetically, there are studies of adsorption/desorption kinetics, and the effectiveness is governed by thermodynamics principles, while the computational data give insight into the efficiency of the affecting materials. Future studies in this direction will improve the design and utilization of magnetic mesoporous materials for the more efficient removal of pollution, hence making it more sustainable.

Materials Research Forum LLC
https://doi.org/10.21741/9781644903452-3

9. Future Perspectives and Challenges

9.1 Emerging Trends in Magnetic Mesoporous Materials for Pollutant Removal

Several developing trends for the field of magnetic mesoporous materials used for pollutant removal have been identified towards improving their performances and expanding their usage. A new trend is the design of the materials that serve both the adsorption and the catalytic degradation of the pollutant. Apart from trapping pollutants, these materials also promote the depollution process of pollutants into less hazardous forms making them useful in the enhancement of total remediation performance, as postulated by [45].

Yet another great trend is in the application of green synthesis methods. Science is shifting concern towards using green routes for the fabrication of MM materials using natural templates and green reagents. This shift seeks to minimize a material production's environment Furthermore, recently, magnetic mesoporous materials have coupled with other superior technologies, including photocatalysis and electrocatalysis. This can be especially useful for characterising a hybrid system since such systems can benefit from the features of many technologies for high levels of pollutant removal capability. For example, incorporation with magnetic mesoporous materials improves the photocatalyst properties, such as the ability to degrade the organic pollutants under light irradiation using titanium dioxide (TiO2) impact thus making it more sustainable [46].

9.2 Innovations in Material Design and Synthesis

The enhancement of the properties of these materials and the extension of their uses depends on the advances in the design and synthesis of magnetic mesoporous materials. Any prospects involve approaches in designing hierarchical structures that integrate both mesoporous and microporous characteristics. Such hierarchical materials can possess multilayered porosity, which can further improve the adsorption capacity and selectivity of the pollutants [39].

One more new approach is biomimetic synthesis when the processes and materials of the biosphere are the basis for the creation of synthetic adsorbents. For instance, employing templating, which uses plant fibres or animal shells, can go a long way in developing mesoporous structures that are as efficient as the existing natural systems[47]. Besides, this approach of preparing adsorbents creates highly effective adsorbents while encouraging the use of renewable resources.

Nanotechnology is also employed in increasing the synthesis of magnetic mesoporous materials. The use of nanotechnology allows the creation of materials with desired characteristics, i.e.. , specific size of the pores, certain functionalities at the surface, and magnetic properties. The nano deposition techniques such as ALD and MLD enable newcomers and small additions to precise control at the atomic layer level and improve efficiency for civilization's pollutant removal[48].

9.3 Challenges in Practical Implementation and Commercialization

The chapters have provided a promising scenario of how magnetic mesoporous materials can be used despite the limitations with the following challenges threading when moving towards the application and commercialization of the materials.

Materials Research Foundations 173 (2025) 60-92
https://doi.org/10.21741/9781644903452-3

9.3.1 Regulatory Challenges

The first among material regulations is the concern of safety and environmental requirements of the materials. The substances that have to penetrate the market have to undergo some tests and prove that they are safe for human use and or the environment. Magnetic mesoporous material may be demanded by the regulating agencies to produce lengthy data regarding their toxicity, biodegradable nature, and other potential long-term impacts on the environment. Compliance with these regulations may take much time and may at times prove to be very costly[49].

9.3.2 Economic Challenges

Currently, the cost of production is one major factor that hampers the use of magnetic mesoporous materials. The synthesis processes often involve expensive precursors and both laboratory to industrial scale have economic issues linked to the large-scale manufacturing facilities and the capital investments required [50]. As a result, the researchers are looking for ways to reduce costs, for instance, by employing cheaper reagents and precursors, enhancing the yields of the synthesis reactions, and employing easy-scalable processes. It will also be much easier to realize commercialization since industry partners will offer some of the needed resources such as large-scale production knowledge.

9.3.3 Technical Challenges

Four technical issues remain crucial within the context of improving the application of magnetic mesoporous materials in practice. Thus, one of the great challenges for technical progress is the choice of the material's stability and its adaption to different conditions. Effects like changing the pH or the temperature or the addition of competing ions will add to the effects stated above or the deterioration of the material. It was also emphasized that more research should be conducted regarding concrete technology for the production of materials that could serve these conditions without serious degradation [51]. Regeneration of the used materials and their reuse is another technical consideration that requires initial elaboration. Comparatively, using magnetic mesoporous materials supplies the advantage of simplicity in separation due to the utilization of a magnetic field, the regeneration process should also be considered to maximize the adsorption capacity at each cycle. This has called for the development of efficient regeneration methods to reduce energy requirements and loss of materials [52].

9.3.4 Integration into Existing Systems

Another challenge is the incorporation of magnetic mesoporous materials in the current treatments of pollutants. Such materials have to be compatible with contemporary technologies and physical plants to be integrated. This may call for future change in the current systems or even the invention of new mixed systems that consist of magnetic mesoporous materials with normal technologies [44].

9.3.5 Market Acceptance

There is, however, a need to have it accepted by the market entailing some commercialization of magnetic mesoporous materials. This requires persuading key audiences such as industries, policymakers, and society on the advantages and the stability of these materials. However by focusing on the pilot and case projects, it is possible to show that the magnetic mesoporous materials are cost-effective and more efficient while being environmentally friendly [39].

9.3.6 Future Directions

To overcome these challenges and fully realize the potential of magnetic mesoporous materials for pollutant removal, future research should focus on several key areas:

Sustainable Synthesis: Green synthesis that concerns the discovery and application of sustainable carbon feedstocks and techniques with lower pollution rates.

Material Optimization: Nanotechnology and biomimicry of advanced mesoporous magnetic materials

Regeneration Techniques: Developing improved practical technologies for the regeneration of used adsorbents and reapplication of adsorbent materials.

Integration Strategies: The creation of product designs that allow the magnetic mesoporous materials to fit into current technologies with relative ease.

Economic Viability: Take for instance, flexible electronics and the efforts made to look for ways of reducing costs and expanding the production of the materials in an attempt to make them affordable for diverse uses in the market

Thus, magnetic mesoporous materials seem to have a great potential for pollutant elimination but several problems have to be solved to pave the way for their real application and utilization on an industrial scale. Further work in the development of structure-property relationships, synthesis of new and improved materials, and practical implementation will be vital in overcoming these challenges and growing the discipline. The researcher's input, industries, and regulatory stakeholders shall also be instrumental in promoting the successful application of these materials for a sustainable advanced environment treatment programme.

Conclusion

This chapter has shown that magnetic mesoporous materials are very useful in the decontamination of air, water, and soil of numerous pollutants. These materials are specific as compared to the conventional remediation techniques because of their vast surface area, tunable pore size and magnetic nature. Magnetic mesoporous materials possess greater adsorption capacities of different Pollutants such as heavy metals, Organic compounds, and Particulate matter. The fact that PSMs can be easily and quickly separated and recovered using an external magnetic charge makes them convenient and preferable. The incorporation of functional groups into the meso-porous structure enhances their selectivity towards particular contaminant and their efficiency. This paper has reviewed the adsorption and desorption mechanisms and kinetics by identifying the factors affecting the adsorption and desorption rate and the models that have been adopted in describing the processes. Thermodynamics is used to devise the enthalpy, entropy and free energy changes which give information about the adsorption process and its possibility, Further, computational modelling is applied to estimate the efficiency of removal and the performance of the adsorptive material. Some actual applications of magnetic mesoporous materials have been described in some of the following case studies. For example, in water treatment, the incorporation of magnetic mesoporous silica and carbon has demonstrated impressive removal of heavy metals and organic

Materials Research Forum LLC
https://doi.org/10.21741/9781644903452-3

compounds. Along the same lines, air filtration has also seen these materials prove capable of collecting particulate material and volatile organic compounds (VOCs) and thus improving the quality of air. They have success in adsorbing both the organic and inorganic pollution in soil remediation strategies, which makes them versatile to any application. However, there exist studies showing positive results; the conversion and transfer of the basic research from the laboratory to an operational environment and eventually the market requires various considerations. The factors that have major impacts include the regulatory, economic, and technical challenges that should be overcome to optimize the application of these materials. Strengthening and prolonging the action of the materials concerning environmental factors, searching for the best way to regenerate, and finally, making their production a reasonable price are the crucial steps to the practical implementation of the created materials. To advance the field of magnetic mesoporous materials for pollutant removal, future research should focus on several key areas: To further advance the field of magnetic mesoporous materials for pollutant removal, future research should focus on several key areas: It is crucial to employ synthesis methods that do not cause any negative impacts on the environment are sustainable and that use renewable feedstock. This embraces the use of natural templates, benign chemicals, energy reduction and minimizing wastes produced within the company. Sustainable synthesis addresses not only the effects that it has on the environment but also the principles of green chemistry and sustainable development. The other aspect is to upgrade the properties of magnetic mesoporous materials by using innovative designs along with synthesis strategies that help to boost performance. Recent ideas like the hierarchy of structures, biomimetic synthesis and nanotechnology have resulted in having materials with better adsorption abilities, selective nature and long life. Furthermore, combining various functionalities into one material, i.e.. , catalytic and photocatalytic activity, seems to produce materials with the ability to remove pollutants on the spot. Regeneration and reuse of the magnetic mesoporous materials have to be efficient to cut costs and ensure an environmentally friendly process. Regeneration studies should be directed toward the creation of effective regeneration techniques that would not show significant adsorption capacity decay in sequential adsorption-regeneration cycles, that would require low energy input and that would not affect unfavourably the properties of the adsorbent material. Instead, further investigations on new regeneration methods, for example, microwave or ultrasound-assisted desorption methods, can improve the applicability of these materials. Some of the studies have also shown that combining magnetic mesoporous materials with other methods of pollutant removal can result in a combination that is greater than the sum of the individual values. For instance, the use of these materials in conjunction with membrane filtration, catalytic converters or AOPs can enhance the removal of these complicated contaminants. The hybrid systems that will incorporate the strong characteristics of several technologies to tackle various environmental issues will be another important development in the field. I believe that to prove the functionality and application of magnetic mesoporous materials, several field tests and pilot projects should be carried out. Such studies are essential for evaluating the stability of the materials and potential problems that may arise in the actual environmental conditions and defining the most efficient schemes of treatment. Elimination and minimization strategies include involving industry partners and regulators to help in piloting the projects and subsequent commercialization. To acquire essential information about the performance of magnetic mesoporous materials compared to methods applied conventionally, overall economic evaluations have to be performed. This comes under assessing the manufacturing costs and SO cost of installation, running, and maintenance and possible savings resulting from decreased waste disposal and enhanced efficiency of pollutant removal. Finding out the methods of saving costs and studying the possibilities of the mass production of these materials will be critical for their affordability. By successfully incorporating magnetic mesoporous materials in environmental systems, such a possibility is of immense value. These materials present a durable and effective answer to clean the air, water, and soil from many types of pollutants, helping to face some of the most important environmental issues. Due to their high adsorption rate, good regenerative characteristics and the possibility of designing them for particular contaminants, they are valuable in numerous applications. Magnetic mesoporous materials in air purification reduce the concentration of aerosols and VOCs in the air reducing health disorders due to air pollution. Such use of materials can improve the efficiency of existing technologies incorporated in filtration systems, and catalytic converters, offering total solutions to air pollution issues. In water treatment, these materials can efficiently remove metals, organics, and new-generation emerging contaminants such as PPCPs. Due to their large adsorption rates and efficient magnetic regeneration tactics, they can suitably serve industrial water treatment plants and municipalities. The creation of composite materials with catalytic capabilities can also improve the process of water purification since most of the contaminants can be degraded locally. In soil remediation, magnetic mesoporous materials present the way forward, for decontaminating areas contaminated with pollutants. Along with the adsorption of organic-inorganic pollutants, ease in the recovery and reuse of these materials make them useful in any kind of soil contamination problem. In this respect, they aid in the remediation of ecosystems and the protection of human health through the elimination of toxic concentrations of pollutants in the soil. In general, applying magnetic mesoporous materials in environmental remediation can be considered sustainable by following green chemistry principles. Thus, being able to provide efficient and eco-friendly ways of cleaning up pollutants in the air, water, and soil, these materials can help in the global mission of attaining environmentally healthy communities and ecosystems.

References

[1] Hoffmann, B., et al., WHO air quality guidelines 2021-aiming for healthier air for all: a joint statement by medical, public health, scientific societies and patient representative organisations. International journal of public health, 2021. 66: p. 1604465. https://doi.org/10.3389/ijph.2021.1604465

[2] Fund, U.N.C.s. and W.H. Organization, Progress on household drinking water, sanitation and hygiene 2000-2022: Special focus on gender. 2024: World Health Organization.

[3] Marcu, D., et al., Contaminants of emerging concern (CECs) and male reproductive health: challenging the future with a double-edged sword. Toxics, 2023. 11(4): p. 330. https://doi.org/10.3390/toxics11040330

[4] Taguchi, A. and F. Schüth, Ordered mesoporous materials in catalysis. Microporous and mesoporous materials, 2005. 77(1): p. 1-45. https://doi.org/10.1016/j.micromeso.2004.06.030

[5] Li, C.M., et al., Functionalized porous silica-based nano/micro particles for environmental remediation of hazard ions. Nanomaterials, 2019. 9(2): p. 247. https://doi.org/10.3390/nano9020247

[6] Li, Y. and J. Shi, Hollow-structured mesoporous materials: chemical synthesis, functionalization and applications. Advanced Materials, 2014. 26(20): p. 3176-3205. https://doi.org/10.1002/adma.201305319

[7] Laurent, S., et al., Magnetic iron oxide nanoparticles: synthesis, stabilization, vectorization, physicochemical characterizations, and biological applications. Chemical reviews, 2008. 108(6): p. 2064-2110. https://doi.org/10.1021/cr068445e

[8] Zhao, X., et al., Synthesis of magnetic metal-organic framework (MOF) for efficient removal of organic dyes from water. Scientific reports, 2015. 5(1): p. 11849. https://doi.org/10.1038/srep11849

[9] Acharya, G.R., Dynamics of Confined Liquids at Multiple Time Scales. 2022, Wayne State University.

[10] Guo, G., et al., Hydroconversion of Kraft lignin for biofuels production using bifunctional rhenium-molybdenum supported zeolitic imidazolate framework nanocatalyst. Bioresource Technology, 2021. 321: p. 124443. https://doi.org/10.1016/j.biortech.2020.124443

[11] Hu, M., et al., Direct carbonization of Al-based porous coordination polymer for synthesis of nanoporous carbon. Journal of the American Chemical Society, 2012. 134(6): p. 2864-2867. https://doi.org/10.1021/ja208940u

[12] Sevilla, M. and A.B. Fuertes, The production of carbon materials by hydrothermal carbonization of cellulose. Carbon, 2009. 47(9): p. 2281-2289. https://doi.org/10.1016/j.carbon.2009.04.026

[13] Beck, J.S., et al., A new family of mesoporous molecular sieves prepared with liquid crystal templates. Journal of the American Chemical Society, 1992. 114(27): p. 10834-10843. https://doi.org/10.1021/ja00053a020

[14] Zhao, D., et al., Nonionic triblock and star diblock copolymer and oligomeric surfactant syntheses of highly ordered, hydrothermally stable, mesoporous silica structures. Journal of the American Chemical Society, 1998. 120(24): p. 6024-6036. https://doi.org/10.1021/ja974025i

[15] Kresge, a.C., et al., Ordered mesoporous molecular sieves synthesized by a liquid-crystal template mechanism. nature, 1992. 359(6397): p. 710-712. https://doi.org/10.1038/359710a0

[16] Ryoo, R., et al., Ordered mesoporous carbons. Advanced Materials, 2001. 13(9): p. 677-681. https://doi.org/10.1002/1521-4095(200105)13:9<677::AID-ADMA677>3.0.CO;2-C

[17] Stein, A., B.J. Melde, and R.C. Schroden, Hybrid inorganic-organic mesoporous silicates-nanoscopic reactors coming of age. Advanced materials, 2000. 12(19): p. 1403-1419. https://doi.org/10.1002/1521-4095(200010)12:19<1403::AID-ADMA1403>3.0.CO;2-X

[18] Sun, W., et al., Overexpression of a Freesia hybrida flavonoid 3-O-glycosyltransferase gene, Fh3GT1, enhances transcription of key anthocyanin genes and accumulation of anthocyanin and flavonol in transgenic petunia (Petunia hybrida). In Vitro Cellular & Developmental Biology-Plant, 2017. 53: p. 478-488. https://doi.org/10.1007/s11627-017-9836-3

[19] Brinker, C.J. and G.W. Scherer, Sol-gel science: the physics and chemistry of sol-gel processing. 2013: Academic press.

[20] Gupta, A.K. and M. Gupta, Synthesis and surface engineering of iron oxide nanoparticles for biomedical applications. biomaterials, 2005. 26(18): p. 3995-4021. https://doi.org/10.1016/j.biomaterials.2004.10.012

[21] Yang, H., M. Zhu, and Y. Li, Sol-gel research in China: a brief history and recent research trends in synthesis of sol-gel derived materials and their applications. Journal of Sol-Gel Science and Technology, 2023. 106(2): p. 406-421. https://doi.org/10.1007/s10971-022-05750-y

[22] Sun, S., et al., Monodisperse mfe2o4 (m= fe, co, mn) nanoparticles. Journal of the American chemical society, 2004. 126(1): p. 273-279. https://doi.org/10.1021/ja0380852

[23] Xin, W. and Y. Song, Mesoporous carbons: recent advances in synthesis and typical applications. Rsc Advances, 2015. 5(101): p. 83239-83285. https://doi.org/10.1039/C5RA16864C

[24] Alphandéry, E., Biodistribution and targeting properties of iron oxide nanoparticles for treatments of cancer and iron anemia 2.

[25] Jiao, Y., et al., The Efficient Removal for Cr (Vi) of Magnetic Adsorbent with Mesoporous Structure of Walnut Shell Filled by Fe3o4. Available at SSRN 4128533.

[26] Lu, A.H., E.e.L. Salabas, and F. Schüth, Magnetic nanoparticles: synthesis, protection, functionalization, and application. Angewandte chemie international edition, 2007. 46(8): p. 1222-1244. https://doi.org/10.1002/anie.200602866

[27] Jaishankar, M., et al., Toxicity, mechanism and health effects of some heavy metals. Interdisciplinary toxicology, 2014. 7(2): p. 60-72. https://doi.org/10.2478/intox-2014-0009

[28] Jones, K.C. and P. De Voogt, Persistent organic pollutants (POPs): state of the science. Environmental pollution, 1999. 100(1-3): p. 209-221. https://doi.org/10.1016/S0269-7491(99)00098-6

[29] Liu, C., et al., Effect of aluminium dust on secondary organic aerosol formation in m-xylene/NO x photo-oxidation. Science China Earth Sciences, 2015. 58: p. 245-254. https://doi.org/10.1007/s11430-014-5023-0

[30] Ruthven, D.M., Principles of adsorption and adsorption processes. 1984: John Wiley & Sons.

[31] Zhang, Z., et al., Efficient removal of methylene blue using the mesoporous activated carbon obtained from mangosteen peel wastes: Kinetic, equilibrium, and thermodynamic studies. Microporous and Mesoporous Materials, 2021. 315: p. 110904. https://doi.org/10.1016/j.micromeso.2021.110904

[32] Shan, H., et al., Thiol-Functionalized Nitrogen-Rich Covalent Organic Framework For Fluorescence Detection and Removal of Hg (Ii) from Water. Available at SSRN 4875131.

[33] Li, Q., et al., Preparation, characterization, and highly effective mercury adsorption of L-cysteine-functionalized mesoporous silica. New Journal of Chemistry, 2014. 38(1): p. 248-254. https://doi.org/10.1039/C3NJ00799E

[34] Karim, A.H., Synthesis of Carbon Nanotubes-mesostructured Silica Nanoparticles Composites for Adsorption of Methylene Blue. 2015, Universiti Teknologi Malaysia.

[35] Pope III, C.A. and D.W. Dockery, Health effects of fine particulate air pollution: lines that connect. Journal of the air & waste management association, 2006. 56(6): p. 709-742. https://doi.org/10.1080/10473289.2006.10464485

[36] Liu, J., et al., The rise of two-dimensional-material-based filters for airborne particulate matter removal. Advanced Fiber Materials, 2023. 5(2): p. 461-483. https://doi.org/10.1007/s42765-022-00242-8

[37] Tian, Z., et al., Efficient capture of airborne PM by nanotubular conjugated microporous polymers based filters under harsh conditions. Journal of Hazardous Materials, 2022. 423: p. 127047. https://doi.org/10.1016/j.jhazmat.2021.127047

[38] Liu, B., et al., Graded Synthesis of Hierarchical Activated Carbon and Mesoporous Silica Microspheres from High Carbon Coal Gasification Fine Slag for Ultra-High Capacity Adsorption of Methyl Blue. Available at SSRN 4175025.

[39] Adireddy, S.R., High Yield Solvothermal Synthesis of Hexaniobate Based Nanocomposites via the Capture of Preformed Nanoparticles in Scrolled Nanosheets. 2013, University of New Orleans. https://doi.org/10.1021/cm402352k

[40] Sarkar, S. and B. Bandyopadhyay, Aldehyde as a potential source of aminol in troposphere: Influence of water and formic acid catalysis on ammonolysis of formaldehyde. Atmospheric environment, 2019. 213: p. 223-230. https://doi.org/10.1016/j.atmosenv.2019.05.069

[41] Karpinska, K., Generation and functionalization of biodegradable nanoparticles for biomedical applications. 2020: The University of Manchester (United Kingdom).

[42] Ho, Y.-S. and G. McKay, Pseudo-second order model for sorption processes. Process biochemistry, 1999. 34(5): p. 451-465. https://doi.org/10.1016/S0032-9592(98)00112-5

[43] Weber Jr, W.J. and J.C. Morris, Kinetics of adsorption on carbon from solution. Journal of the sanitary engineering division, 1963. 89(2): p. 31-59. https://doi.org/10.1061/JSEDAI.0000430

[44] Yang, H., et al., A water-insoluble and visible light induced polyoxometalate-based photocatalyst. Chemical communications, 2010. 46(14): p. 2429-2431. https://doi.org/10.1039/b919868g

[45] Dzumbira, W., et al., Separation and remediation of environmental pollutants using metal-organic framework-based tailored materials. Environmental Science and Pollution Research, 2021: p. 1-21.

[46] Shahbazi, K., M. Marzi, and H. Rezaei, Heavy metal concentration in the agricultural soils under the different climatic regions: a case study of Iran. Environmental earth sciences, 2020. 79(13): p. 324. https://doi.org/10.1007/s12665-020-09072-6

[47] Liu, N., et al., Hierarchical architectures of Ag clusters deposited biomimetic membrane: synthesis, emulsion separation, catalytic and antibacterial performance. Separation and Purification Technology, 2020. 241: p. 116733. https://doi.org/10.1016/j.seppur.2020.116733

[48] Pal, N., J.-H. Lee, and E.-B. Cho, Recent trends in morphology-controlled synthesis and application of mesoporous silica nanoparticles. Nanomaterials, 2020. 10(11): p. 2122. https://doi.org/10.3390/nano10112122

[49] Williams, A.J., et al., Sourcing data on chemical properties and hazard data from the US-EPA CompTox Chemicals Dashboard: A practical guide for human risk assessment.

Materials Research Forum LLC
https://doi.org/10.21741/9781644903452-3

Environment international, 2021. 154: p. 106566. https://doi.org/10.1016/j.envint.2021.106566

[50] Srivastava, A.K., et al., Coupled catalytic dephosphorylation and complex phosphate ion-exchange in networked hierarchical lanthanum carbonate grafted asymmetric bio-composite membrane. Journal of Colloid and Interface Science, 2022. 606: p. 2024-2037. https://doi.org/10.1016/j.jcis.2021.09.173

[51] Fizir, M., et al., Design approaches, functionalization, and environmental and analytical applications of magnetic halloysite nanotubes: A review. Clays and Clay Minerals, 2022. 70(5): p. 660-694. https://doi.org/10.1007/s42860-022-00210-8

[52] Peres, R.M., J.S. Forero, and R.J. Corrêa, Tuning the graphene oxide chemistry by excimer formation: The stabilization of electron/hole pair on sp2 surface. Journal of Photochemistry and Photobiology A: Chemistry, 2021. 414: p. 113266. https://doi.org/10.1016/j.jphotochem.2021.113266

Materials Research Foundations 173 (2025) 93-123 | https://doi.org/10.21741/9781644903452-4

Chapter 4

Mesoporous Materials for Liquid Chromatographic Separation

Vipul D. Prajapati[1*], Princy Shrivastav[2]

[1]Department of Pharmaceutics, SSR College of Pharmacy (Permanently Affiliated to Savitribai Phule Pune University), Sayli, Silvassa, U.T. of D.N.H. & D.D., India.

[2]Department of Pharmaceutical Chemistry, SSR College of Pharmacy (Permanently Affiliated to Savitribai Phule Pune University), Sayli, Silvassa, U.T. of D.N.H. & D.D., India.

vippra2000@yahoo.com

Abstract

This chapter explores the creative application of mesoporous materials to improve liquid chromatography's efficacy and efficiency. Mesoporous materials have special benefits in chromatographic applications, such as enhanced loading capacity, stronger selectivity, and better separation performance. These materials are distinguished by their highly organized pore architectures and huge surface areas. The production and functionalization of mesoporous materials are covered in detail in this chapter, along with an emphasis on their favourable physicochemical characteristics as chromatographic media. The chapter also examines the different uses of mesoporous materials in liquid chromatography, including preparative and process-scale procedures as well as analytical separations. The useful advantages and future possibilities of these materials in chromatography are demonstrated through case studies and a discussion of current developments in the field. Through this exploration, the chapter aims to present mesoporous materials as a versatile and powerful tool in the ongoing development of chromatographic techniques.

Keywords

Mesoporous Materials, Liquid Chromatography, Chromatographic Separation, Pore Structure, Surface Area, Functionalization, Stationary Phase

Contents

1. Introduction

The last few decades have seen remarkable growth in the field of liquid chromatography (LC) due to the ongoing search for high-capacity, highly selective, and efficient separation methods. Among the numerous advancements, the emergence of mesoporous materials has been identified as a crucial breakthrough that has completely changed the field of chromatographic science.

Mesoporous materials (MMs) have shown remarkable promise in improving the efficiency of liquid chromatographic separations. They are distinguished by their huge surface areas and highly organized pore architectures.

Due to large surface area, variable pores size, and capacity to functionalize the surfaces of mesoporous substances with different chemical groups, MMs have several advantages. These features help chromatographic procedures achieve more selectivity, greater loading capacity, and improved mass transfer [1,2]. Varieties of MMs, such as mesoporous silica, mesoporous alumina, and hybrid organic-inorganic frameworks, have been explored for their unique properties and applications in LC [3,4].

This chapter explores the special qualities and benefits of using mesoporous materials in LC. We examine the synthesis, characterisation, and functionalization of these materials, emphasizing how they can be customized to fulfil particular chromatographic requirements. Our goal is to present a thorough understanding of the role those different mesoporous materials, including as silicas, aluminas, and hybrid organic-inorganic frameworks, play in enhancing analyte retention, resolution, and separation efficiency [3].

Additionally, we go over the real-world uses for mesoporous materials in several chromatographic modes, including size-exclusion, normal-phase, and reversed-phase chromatography. Insights into the practical advantages and constraints of these materials are provided by case studies and real-world examples that demonstrate their successful integration into chromatographic systems. Presenting the most recent developments and potential paths for the application of mesoporous materials, this chapter is a useful tool for scientists, researchers, and practitioners working in the field of LC. In examining the relationship between material science and chromatographic methods, we highlight how mesoporous materials have revolutionized analytical separations and have the potential to spur further advancements.

While mesoporous materials offer significant advantages, there are still challenges to be addressed, such as improving mechanical stability, achieving precise pore size control, and developing cost-effective synthesis methods [5]. Future research may focus on integrating mesoporous materials with advanced detection techniques, developing hybrid materials, and exploring the use of "smart" mesoporous materials that are sensitive to environmental changes.

2. Fundamentals of Liquid Chromatographic Separation

The International Union of Pure and Applied Chemistry (IUPAC) defines LC as a technique for physically separating components in mixtures by dividing the components between two phases. LC was initially developed at the turn of the 20th century. The second of these phases is stable, but one of them is moving and permeates it. Liquids are the mobile or moving phase in LC. Since then, the method has expanded greatly to encompass preparative LC, microbore LC, capillary LC, and high-pressurized liquid chromatography (HPLC) [6].

It is a mostly preferred analytical methods for separating and analysing complex mixtures of biomolecules and other compounds. Liquid type chromatographic technique separates compounds according to their differential partitioning between two phases namely a stationary as well as a mobile. Varieties of the components interact with the stationary phase at different degrees as the sample mixture is added to the system, which causes the components to separate as they elute at different times. LC may be of different types, including HPLC, which is a widely used technique that employs high level pressure to push the mobilized phase through the columns, allowing for faster separations and higher resolution; Size Exclusion type of chromatography (SEC) that separates biomolecules based on size; Affinity type of chromatography which utilizes specific attraction between a target biomolecule and a ligand immobilized on the stagnant immobile phase for selective separation; and Ion Exchange type of chromatography (IEC) that differentiates biomolecules according to their charge, using a charged stagnant phase to attract oppositely charged molecules of the substance [7].

2.1 Principles of LC separation

The principles of LC separation are based on the interaction between a sample mixture and a stationary phase, facilitated by a mobile phase. Biomolecules or other chemicals are divided in LC according to how differently they partition into the stagnant and the mobile phase. As the blend flows through the column, the various components' interactions with the stagnant phase result in differing retention times. In normal phase types of LC, separation occurs based on the adsorption of analytes onto the stagnant phase. The stronger the attraction with the stationary phase, the longer the retention time. The choice of mobile phase affects the separation efficiency, resolution, and the overall behavior of the analytes. Common mobile phases include water, organic solvents, or mixtures thereof [6,7].

2.2 Types of interactions in LC

In LC, various types of interactions play the potential role in the separation of biomolecules and other compounds. These interactions determine how analytes interact with the stagnant and the movable phase during the separation phenomena. The primary types of interactions involved in LC are adsorption, partition, ion-exchange, hydrophobic, hydrophilic, electrostatic, and size exclusion interactions.

2.2.1 Adsorption

In adsorption chromatography, analytes adhere to the surface layer of the stagnant phase. The strength of this interaction affects the retention time of the analytes. This interaction can be governed by the polarity of the stagnant phase and the analytes. Nonpolar analytes tend to be retained longer on nonpolar immobile phases due to robust hydrophobic interactions. It relies on the differential adsorption of various components in a mixture of evaluation onto a stationary solid phase [8].

The principle behind adsorption chromatography involves the interaction between the adsorbate (the substances being separated) and the adsorbent (the stationary phase). When a mixture is passed through a column containing the adsorbent, different components interact with the solid external surface at various level according to their chemical features. Stronger interactions lead to longer retention times, while weaker interactions allow for quicker elution. Such differential retention resulting in the separation of the components as they migrate through the columns [8,9].

2.2.2 Partitioning

This type of chromatographic technique in general involving the distribution of components of the mixture between two immiscible liquid phases: a stagnant and a mobile phase. This method is

 Materials Research Forum LLC
https://doi.org/10.21741/9781644903452-4

particularly effective for separating compounds according to their differing affinities towards of chromatographic phases [10].

The partitioning of solutes between two liquid phases forms the foundation of partition chromatography. The ratio of a solute's strength or proportion in the immobile phase (Cs) to its strength or proportion in the mobile phase (Cm) at steady equilibrium is known as the partition coefficient (K), which is a crucial factor:

$$K = \frac{Cs}{Cm} \quad (1)$$

This coefficient determines how a solute will distribute itself between the two phases, influencing its retention time and separation efficiency. When the mobile phase flows, components of the mixture interact differently with the immobile phase, resulting to their separation as they moving through the chromatographic system [8,10].

The stationary phase is prepared by coating a solid support with a liquified solvent that is immiscible with the movable phase. The sample mixture is allowed to apply to the stationary phase, allowing the components to interact with both phases. A suitable mobile phase is chosen according to the nature of the analytes and the desired separation. The movable phase is subjected to move through the stagnant phase, facilitating the separation of components based on their partition coefficients. After the separation is complete, the chromatogram is analysed, often using visualization techniques to identify the separated components [11].

2.2.3 Ion Exchanging Chromatography (IEC)

It is a powerful technique of differentiating components that utilizes the electrostatic attractions between charged the molecules and oppositely charged immobile phases to achieve separation. This method is mostly used in biochemistry and analytical chemistry for the purification and estimation of nucleic acids, polypeptides, and other charged biomolecules [12].

IEC operates on the phenomena of reversible exchanging of the ions presented between the target in a sample solution and on an ion exchanger. It is of two types [13];

1. Cation-Exchange Chromatography: In this method, the stagnant immobile phase is negatively charged, attracting positively charged ions (cations). This is effective when the pH of the buffer is below the isoelectric point (pI) of the protein or molecule of interest, ensuring that it carries a positive charge.

2. Anion-Exchange Chromatography: Here, the immobile phase is positively charged, attracting oppositely charged ions (anions). This method is suitable when its pH is above the pI, allowing the target molecules to carry a negative charge.

The separation process occurs as the sample passes through a column packed with the ion exchangers, where components interact with the stationary phase based on their charge. Molecules with a higher affinity for the stagnant phase will be retained longer, leading to separation as they elute at different times [13,14].

Materials Research Forum LLC
https://doi.org/10.21741/9781644903452-4

2.2.4 Hydrophobic Interactions

These play a crucial role in hydrophobic interactions type of Chromatographies (HIC), a LC method used primarily for the separation and purification of biomolecules, particularly proteins. This method capitalizes on the differences in surface hydrophobicity among proteins, allowing for effective separation based on their hydrophobic characteristics [15].

HIC operates on the principle that proteins have both hydrophobic and hydrophilic regions. When proteins are exposed to a high-salt buffer, the salt ions shield the hydrophilic regions and promote the aggregation of hydrophobic areas, facilitating interactions with hydrophobic ligands attached to the stationary phase of the chromatography column [15]. The key steps in this process include equilibration of chromatography column with a high-salt buffer, then the sample containing proteins is applied to the column, where hydrophobic proteins bind to the stationary phase through hydrophobic interactions. Proteins with greater hydrophobicity will bind more strongly. Next, the salt concentration in the buffer is gradually decreased. As the ionic strength lowers, the hydrophobic interactions are weakened, leading to the elution of proteins. Typically, the least hydrophobic proteins elute first, followed by more hydrophobic ones as the salt concentration continues to decrease [15,16].

HIC typically operates under non-denaturing conditions, preserving protein functionality and activity. The method offers excellent resolution and selectivity for separating proteins based on their hydrophobic characteristics. It can be utilized to the broad range of biomolecules, including various class of proteins, peptides, and nucleic acids [17].

2.2.5 Hydrophilic Interactions

Hydrophilic interaction type of LC (HILIC) is a powerful separation method used for the evaluation of polar and hydrophilic compounds that are critical to retain on standard reversible phase columns. The moving phase in HILIC is made up of a large proportion of organic solvent and a small amount of water, while the stationary phase is hydrophilic [18,19].

Partitioning and adsorption work together to form the foundation of the HILIC separation mechanism. The organic-rich moving phase and the hydrated-enriched layer on the stagnant phase's surface are separated by the polar analytes. In addition, polar molecules are retained due to H-bonding, ionic interactions, and dipole-dipole interactions [19]. HILIC has been successfully utilized to the analysis of the broad range of polar compounds, including carbohydrates, amino acids, peptides, nucleosides, and ionic compounds. It is especially essential for the evaluation of compounds that are not well retained on reversed-phase columns [18].

However, HILIC requires careful method development to optimize the separation, as the retention mechanism is governed and affected by varieties of the factors, such as the immobile phase chemistry, movable phase composition, and the fundamental nature of the analytes. Proper conditioning and equilibration of the HILIC column are also crucial for reproducible results. Its versatility and sensitivity make it an essential tool in various fields, including metabolomics, and glycomics [18,19].

2.2.6 Size Exclusion type Chromatography (SEC)

It is a type of LC method that separating the molecules of the examination according to their size, rather than their chemical properties. The mechanism includes the need of a porous stagnant phase, typically formed from the materials like dextran, agarose, or polyacrylamide, which allows molecules to penetrate into the pores based on their size [20].

The chromatography column is packed with fine, porous beads. These beads have defined pore sizes that allow molecules to enter and be retained within the pores. Then, the sample containing the molecules to be separated is injected into the column. Molecules are sorted according to size as the sample flows through the column. Small size molecules can move through the pores and stay in longer than larger molecules, which cannot pass through and elute fast. Depending on their size, the molecules elute from the column in a particular order. First to elute are larger molecules, then smaller molecules [20].

The dimension of the pores in the immovable phase determines the range of molecular weights that can be separated. Larger pores allow for the separation of larger molecules, while smaller pores are suitable for smaller molecules. The molecules' retention time is also significantly influenced by their molecular weight. Smaller molecules retain information for longer periods of time than larger molecules do. The shape and flexibility of the molecules determine their hydrodynamic volume, which in turn affects how long they retain information [20].

3. Mesoporous Materials: Definition and Characteristics

Mesoporous materials are defined as materials characterized by pores with dimensions in the range between 2 to 50 nm. These compounds exhibit unique structural features that make them highly versatile for numerous functions, particularly in the fields of catalysis, drug deliveries, and biomolecules separations. Mesoporous materials possess ordered arrangements of pore networks that allow for significant surface areas and density. The high surface area (typically between 400 and 1000 m^2/g) and tunable pore sizes enable these materials to interact effectively with a wide range of molecules, making them suitable for applications in adsorption, separation, and catalysis [21]. The factors which are considered while selecting a mesoporous material in LC is depicted in Figure 1.

The ability to control the pore size during synthesis enables the optimization of these compounds for desired applications, such as size exclusion chromatography or drug delivery systems. The large surface area facilitates enhanced interaction with biomolecules, leading to improved separation efficiency and capacity. Mesoporous materials, especially those based on silica, exhibit excellent chemical stability under various conditions, making them needful for the broad range of functionalities. The external surfaces of mesoporous materials can be easily changed with specific functional moieties. This functionalization enhances selectivity for target molecules and improves the performance of separation techniques. Also, many mesoporous silica materials are biocompatible, making them unique for biomedical usage, including desired drug deliveries and diagnostics. In summary, mesoporous materials are characterized by their unique pore structures, high surface areas, and potential for functionalization, making them valuable in numerous scientific and industrial applications. Their versatility and effectiveness in biomolecule separation, catalysis, and drug delivery highlight their significance in contemporary materials science [22,23]. Table 1 gives a comparison of mesoporous materials with traditional stationary phases used in LC.

Materials Research Foundations 173 (2025) 93-123 https://doi.org/10.21741/9781644903452-4

***Table 1**. Comparative features of mesoporous materials with traditional stationary phases for the purpose of use in liquid chromatography.*

Features	Mesoporous Materials	Traditional Stationary Phases
Pore Size [2, 3]	Ranging from 2-50 nm, allowing for selective separation based on size.	Larger pores of varied sizes (60-100 Å), which may limit resolution for smaller molecules.
Surface Area [5]	High specific surface area (>300 m^2/g), enhancing interaction with analytes.	Comparatively lower surface area, leading to reduced interaction capacity.
Separation Efficiency [7, 8]	High efficiency, especially for large molecules	Good for smaller molecules; reduced efficiency for large molecules
Retention Mechanism [10]	Highly tuneable surface chemistry, customizable selectivity	Based on adsorption, partitioning, or ion exchange
Functionalization [3, 21]	Easily functionalized with various chemical groups to tailor selectivity and retention.	Limited functionalization options, often restricted to standard alkyl chains.
Stability [21]	Chemically stable, generally robust but can vary based on synthesis method, careful handling required, high thermal stability.	Established mechanical stability but can be prone to packing issues in high-pressure applications, silica-based phases degrade under extreme pH and temperature.
Selectivity [12, 20]	Enhanced selectivity due to tuneable pore structure and surface chemistry.	Selectivity often limited to established methods and materials, less adaptable to novel applications.
Reproducibility [3]	High reproducibility in synthesis and chromatographic performance when properly controlled.	Established reproducibility, but variations can occur based on batch production methods.
Cost & Availability [3]	More expensive, less commercially available	Cost-effective, widely available
Applications [1-3]	Bio-separations, pharmaceutical analysis, environmental monitoring	Widely used in pharmaceuticals, food safety, environmental analysis
Ease of Use & Implementation [1]	Requires more optimization and specialized knowledge	User-friendly, well-established protocols

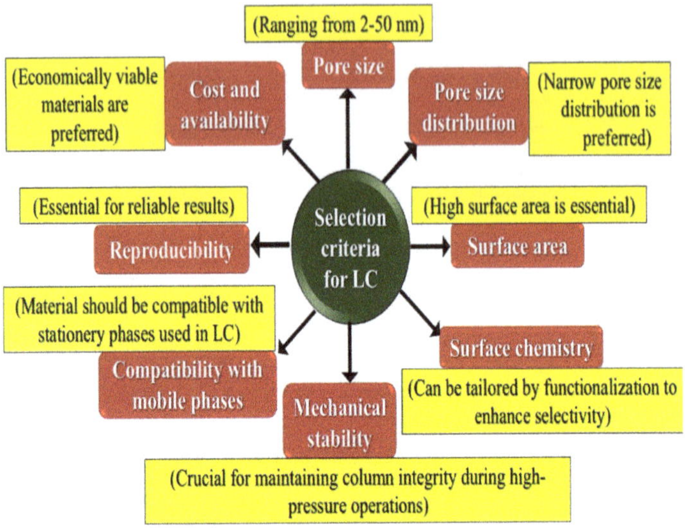

Figure 1. An outlook diagram showing the selection criteria of mesoporous materials for suitable application in liquid chromatography.

4. Synthesis and Characterization of Mesoporous Materials for LC

4.1 Methods for synthesizing mesoporous materials suitable for LC

There are several methods for synthesizing mesoporous materials that are suitable for LC applications. Some key synthesis techniques are template method, sol-gel processing, microwave-assisted synthesis, etc.

4.1.1 Template Method

The template method for synthesizing mesoporous materials is used broadly that allows for the controlled formation of materials with specific pore structures and sizes. This method requires the use of templates, either hard or soft, to guide the synthesis of the desired material, followed by the elimination of such template to create the final mesoporous backbone [24].

The first step of this method involves selecting and preparing the appropriate template. Templates can be classified as hard templates which are typically rigid structures and provide a defined shape and size for the mesoporous material, and soft templates which include surfactants or biomolecules that can self-assemble to create a mesostructured environment. The next step is the synthesis of the target material around the template. This often involves sol-gel processes, co-precipitation, or other chemical methods where the precursor materials interact with the template to form a solid structure. After the synthesis, the template is removed to reveal the mesoporous structure. This can be completed using variety of techniques, including chemical etching, and thermal decomposition [25].

For successful template synthesis, several requirements must be met. The template's surface properties should match those of the raw materials to ensure effective replication. The template should have the unique structure that can be easily replicated in the final product. The template must be removable without damaging the mesoporous structure. It should be strong enough to maintain its shape during the synthesis process [26].

The template method is extensively used in various fields. Mesoporous materials synthesized via this method can serve as catalysts with large surfaces and tunable pore dimensions. These materials are effective for adsorbing pollutants or other substances due to their large area of surfaces and the porosity. Mesoporous materials can be designed for controlled release of drugs in biomedical applications using this method [24].

4.1.2 Sol-Gel Processing

One popular technique for creating mesoporous materials is the sol-gel procedure. It entails the creation of a network (gel) through the gelation of a colloidal suspension (sol). Precursor solution preparation is the initial stage in this process. Precursor solutions are usually made of metal salts or alkoxides dissolved in the proper solvent, like alcohol or water. The required composition of the mesoporous material determines the choice of precursor. After that, the precursor solution experiences condensation and hydrolysis reactions with an acid or basic catalyst present. Alkoxide groups (-OR) are changed to hydroxyl groups (-OH) during hydrolysis. After that, condensation takes place, producing alcohol or water as byproducts and creating metal-oxygen-metal (M-O-M) connections.

A structure-directing agent, such as a surfactant, is added to the sol in the next step. The surfactant molecules self-assemble into micelles, which behaves as the templates for the development of mesopores. The surfactant can be cationic, anionic, and non-ionic. The sol is aged, allowing the condensation reactions to continue and the network to grow. During this stage, the micelles and the inorganic network interact, subjected to the development of a gel with the unique mesoporous structure. The gel is dried, either by evaporation (xerogel) or by supercritical drying (aerogel), to remove the solvent and surfactant. The dried material is then calcined at higher level of heat to remove the remaining organic components and to further condense the inorganic network, resulting in the final mesoporous material [27,28].

The sol-gel process allows for excellent command over the composition, pore size, and topology of the mesoporous materials by maintaining the parameters such as the type and proportion of precursors, catalysts, surfactants, and reaction conditions. Some examples of mesoporous materials synthesized via the sol-gel process include mesoporous silica, mesoporous metallic oxides, mesoporous mixed metallic oxides, and mesoporous carbon materials [29].

4.1.3 Microwave-Assisted Synthesis (MAS)

This type of synthesis of mesoporous materials is an innovative approach that leverages microwave energy to enhance the efficiency and speed of the synthesis process. This type of technique offers numerous benefits over conventional methodologies, such as improved uniformity, reduced reaction times, and the ability to control the synthesis environment more precisely [30].

Microwave radiation provides uniform heating throughout the reaction mixture, which leads to more consistent and reproducible results. This contrasts with traditional heating methods, where temperature gradients can occur. The application of microwave energy significantly accelerates chemical reactions. For instance, synthesis processes that might typically take hours can often be completed in minutes, enhancing productivity and reducing energy consumption [30,31]. The involvement of microwaves can also responsible to the formation of mesoporous materials with improved structural integrity and surface characteristics. This is particularly beneficial for applications in catalysis, drug delivery, and adsorption [32,33]. Recent advancements have demonstrated the feasibility of scaling up microwave-assisted synthesis methods. For example, flow MAS has been developed, allowing for the continuous production of mesoporous compounds in larger quantities while maintaining uniform properties [30].

The general process for microwave-assisted synthesis of mesoporous materials typically involves several steps. A solution containing the necessary precursors (such as metal alkoxides or silicates) is prepared. Surfactants or structure-directing components may also be included to facilitate the formation of mesopores. The precursor solution is subjected to microwave irradiation, which promotes rapid hydrolysis and condensation reactions. This step is crucial for forming the mesoporous structure. The mixture is allowed to gel and age, during which the structure begins to stabilize and develop porosity. If surfactants or templates are used, they are typically removed through calcination or solvent extraction, resulting in the final mesoporous material. The synthesized materials are characterized using standard technologies such as transmission type electron microscopy (TEM), X-ray diffraction (XRD), and nitrogen adsorption-desorption isotherms to evaluate their structural and porosity properties [32,34].

4.1.4 Emulsion-Oriented Assembly

Emulsion-oriented assembly is the advanced new approach for development of mesoporous materials, leveraging the unique properties of emulsions to create structured porous architectures. Such technique typically involving the development of oil-in-water (O/W) emulsions, where surfactants stabilize the emulsion and facilitate the organization of precursors into mesoporous structures [27].

The process begins with the preparation of an oil-in-water emulsion. This emulsion consists of an organic phase (oil) containing a silica precursor, such as tetraethyl orthosilicate (TEOS), and an aqueous phase containing surfactants that stabilize the emulsion. The surfactants exhibit a critical role in controlling the dimension and distribution of droplets within the emulsion. During the emulsion formation, the surfactants cluster between the oil-water layers, creating a structured environment where the silica precursor can condense. This self-assembly process subjects to the development of a matrices that defines the mesoporous structure. The emulsion is allowed to age, promoting further condensation of the silica precursor and the development of a gel-like structural composition. This stage is crucial for developing the desired porosity and stability of the mesoporous material. After gelation, the organic components of the emulsion are removed, typically through calcination or solvent extraction, to yield the final mesoporous material. The removal process must be carefully controlled to preserve the integrity of the mesoporous structure.

The emulsion method allows for potential regulation over the pore dimension and arrangement of the mesoporous material by altering the composition and proportion of the surfactants and precursors. Mesoporous materials synthesized via this method often exhibit high surface areas, making them suitable for utilities in catalysis, adsorption, and drug deliveries. This approach can be adapted to synthesize varieties of mesoporous compounds, including silica, metal oxides, and hybrid organic-inorganic materials [27,29].

4.2 Surface modification techniques to enhance chromatographic performance

Such techniques are essential for enhancing the chromatographic performance of mesoporous materials, particularly in applications such as LC. These modifications can improve selectivity, retention, and overall separation efficiency. Some key surface modification techniques are given in Figure 2.

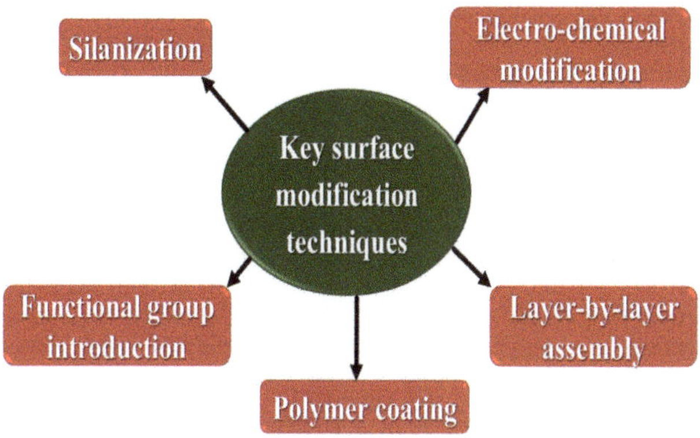

Figure 2. An outlook chart showing key surface modification techniques.

4.2.1 Silanization

Silanization is a widely used method for the surface level changing of mesoporous materials, particularly mesoporous silica nanoparticles (MSNs). This process includes the covalent attachment of organosilane compounds to the exterior surface of the mesoporous material, introducing various functional groups that can be further utilized for various applications [35].

In the silanization process, the first step typically involves activation of the mesoporous surface. It is activated by exposing it to hydroxyl groups (-OH), which are typically generated through treatment with acids, bases, or oxidizing agents. This step creates reactive sites for the subsequent silanization reaction. The activated mesoporous material is then reacted with organosilane compounds, such as alkyltrialkoxysilanes or aryltrialkoxysilanes. These compounds contain a silicon atom bonded to three alkoxy groups (e.g., methoxy or ethoxy) and an organic functional group (e.g., amino, carboxyl, or thiol). During the reaction, the alkoxy groups undergo hydrolysis and condensation, forming covalent Si-O-Si bonds with the surface hydroxyl groups, while the organic type functional moiety remains exposed on the surface. After the silanization reaction, the modified mesoporous material is typically purified to remove any unreacted silanes or byproducts. This can be done through filtration, centrifugation, or washing with appropriate solvents. The purified material is then dried to remove any remaining solvent [36,37].

This process is advantageous as a wide range of organosilane compounds are available, allowing for the introduction of various functional groups, such as amino, carboxyl, thiol, or epoxy groups, depending on the desired application [38]. The formation of covalent Si-O-Si bonds between the organosilane and the mesoporous surface ensures strong and stable surface modification, preventing the loss of functional groups during subsequent applications. Also, this process can be carried out without significantly altering the porous framework of the mesoporous compound, maintaining its large surface and accessibility [39].

4.2.2 Functional Group Introduction

This process involves the covalent attachment of organosilane compounds containing specific functional groups to the surface of the mesoporous compound. The introduction of these functional moieties can significantly alter the surface properties and enhance the performance of mesoporous materials in various applications. Certain functional moieties, including hydroxyl (-OH) as well as amino ($-NH_2$) moieties, can enhance the biocompatibility of mesoporous materials, making them more useful for biomedical usage like drug deliveries and tissue engineering. The introduction of hydrophobic (e.g., alkyl groups) or hydrophilic (e.g., carboxyl (-COOH) groups) functional moieties can modify the wettability of the mesoporous surface, influencing the adsorption and release behavior of molecules [40]. Functional group modification can promote the attachment and growth of electroactive bacteria on the anode surface of bio-electrochemical systems (BES), leading to improved biofilm development and enhanced electron transfer. The introduction of antibacterial functional moieties on air cathodes in BES can help suppress the formation of aerobic biofilms, improving the efficiency of oxygen reduction reactions [41].

Common functional groups used are amino group, carboxyl group, thiol group, and epoxy group. Amino ($-NH_2$) groups introduce positive charges and enhance the adsorption of negatively charged molecules. They can also be used for further functionalization [36,42]. Carboxyl (-COOH) groups introduce negative charges and improve the adsorption of positively charged molecules. They can also be used for covalent attachment of biomolecules. Thiol (-SH) groups provide reactive sites for the immobilization of metal nanoparticles or biomolecules. They can also enhance the adsorption of heavy metals. Epoxy groups serve as reactive sites for further functionalization and can be used for the covalent attachment of biomolecules [42].

Several techniques, such as grafting, co-condensation, and post-synthesis modification, are commonly used for the introduction of functional groups on mesoporous compounds. In grafting, the mesoporous material is reacted with organosilane compounds containing the desired functional groups, leading to the generation of covalent Si-O-Si bonds. In co-condensation, which is the other technique, the organosilane compound is added during the synthesis of the mesoporous compound, allowing for the incorporation of functional groups within the porous structure [36]. Post-synthesis modification involves treatment of mesoporous material with plasma or it is subjected to chemical reactions to introduce functional groups on the surface [41].

4.2.3 Polymer Coating

Polymer coating is a powerful versatile and efficient approach for the surface modification of mesoporous compounds. This technique enhances the characteristics of mesoporous compounds, making them suitable for various applications [39]. The first step involves selecting an appropriate polymer for coating. Common choices include poly(ethylene glycol) (PEG), polystyrene,

Materials Research Forum LLC
https://doi.org/10.21741/9781644903452-4

polyvinyl alcohols (PVA), and polyacrylic acid. The choice of polymer according to the desired characteristics of the final compound, such as hydrophilicity, biocompatibility, and functionality. The external surface of the mesoporous compound is often activated to enhance the adhesion of the polymer. This can be achieved through methods such as plasma treatment, chemical etching, or silanization to introduce functional groups that promote bonding with the polymer. The polymer is applied to the surface of the mesoporous material using various techniques, including dip coating, spray coating, and grafting. After the coating is applied, the material is cured (if necessary) to ensure proper bonding and to remove any solvent. This step may involve heating or exposure to UV light, depending on the polymer used [37].

Polymer coatings can improve the mechanical and thermal stability of mesoporous materials, making them more robust for various applications. Coated mesoporous materials can be designed for controlled drug release, as the polymer can act as a barrier that regulates the diffusion of drugs. The introduction of biocompatible polymers can enhance the compatibility of mesoporous materials for biomedical applications, reducing toxicity and improving cellular uptake. Polymer coatings can introduce specific functional groups that enhance the interaction of the mesoporous compounds with target moeties, improving their performance in applications such as adsorption and catalysis [37].

4.2.4 Layer-by-Layer (LbL) Assembly

This type of assembly is the potential method for the desired changes on the surface of mesoporous compounds, allowing for the critical control of film properties and the introduction of functional groups. This method covers the ordered deposition of alternating layers of materials, typically polyelectrolytes, onto the desired surface of the mesoporous substrate, allowing for precise regulations over the thickness, composition, and functional properties of the resulting films [43].

The first step involves selecting suitable materials for the assembly. Common choices include polyelectrolytes, nanoparticles, and biomolecules. The materials should have complementary charges to facilitate electrostatic interactions during the layering process. This process consists of repeated cycles of adsorption and rinsing. In each cycle, a layer of one compound is deposited onto the mesoporous surface, followed by rinsing to remove unbound material. This is then followed by the layering of the next layer, creating a multilayer structure. The assembly continues until the needful number of layers is achieved. The thickness and properties of the multilayer film can be regulated by maintaining the number of deposition cycles and the concentration of the solutions used. After the desired layers are deposited, the films may undergo curing or crosslinking to enhance their stability and adhesion to the mesoporous substrate. This step can involve heat treatment or chemical crosslinking agents [44,45].

LbL assembly allows for the incorporation of a wide variety of materials, including polymers, nanoparticles, and biomolecules, enabling the customization of surface properties. The technique provides precise regulations over the thickness of the developed films, allowing for the creation of multilayer structures with specific functionalities. The multilayer films formed through LbL assembly often exhibit improved mechanical and chemical stability compared to single-layer coatings. LbL assembly can introduce specific functional moieties on the surface of mesoporous compounds, enhancing their performance in various utilizations such as drug deliveries, catalysis, and sensing [46].

Materials Research Foundations 173 (2025) 93-123
https://doi.org/10.21741/9781644903452-4

4.2.5 Electrochemical Modification

Electrochemical modification is an innovative technique for the required changes on the surface of mesoporous compounds, offering advantages in terms of control, environmental impact, and stability. This method utilizes electrochemical processes to introduce functional groups or alter the surface features of the compounds, enhancing their performance in various applications [47]. The electrochemical modification process typically involves an electrochemical cell where the mesoporous material acts as an electrode. The setup includes a working electrode (the mesoporous material), a counter electrode, and a reference electrode, along with an electrolyte solution containing the desired functionalizing agents. During the electrochemical process, the application of an electric current induces the deposition of functional moieties or materials onto the surface of the mesoporous material. This can be achieved either through electrochemical reduction or electrochemical oxidation. The electrochemical modification can lead to the generation of covalent bonds between the active functional moieties and the surface of the mesoporous material, resulting in a stable and well-defined surface modification [37].

The electrochemical process allows for precise regulations over the amount and type of functional moieties introduced onto the surface, enabling tailored modifications for specific applications. These methods often require milder conditions and can reduce the need for hazardous chemicals, making them more environmentally friendly compared to traditional chemical modification methods. The covalent bonding formed during this process often results in enhanced stability of the functional groups, preventing leaching during subsequent applications [39,48].

4.3 Methods for characterizing mesoporous materials suitable for LC

Characterizing mesoporous materials suitable for LC involves several analytical techniques that assess their structural, textural, and chemical properties. The primary methods used for this purpose include electron microscopy techniques, XRD, thermogravimetric analysis, etc. The advantages, limitations, and applications of all these techniques is summarised in Table 2.

4.3.1 Nitrogen Adsorption-Desorption Isotherms

These are a fundamental technique for characterizing mesoporous materials, providing critical information about their surface area, pore size distribution, and porosity. This method is widely used in the evaluation of materials intended for applications in LC and other fields [49].

Nitrogen adsorption isotherms are classified according to the IUPAC into several types, each indicating different adsorption behaviours. Type I isotherms are characteristic of microporous materials, where adsorption occurs rapidly at low relative pressures (P/P_0). Type II isotherms are typical for non-porous or macroporous compounds, showing a continuous increase in adsorption with pressure. Type IV isotherms are commonly observed in mesoporous materials, characterized by a hysteresis loop, indicating capillary condensation within the pores [50].

The Brunauer-Emmett-Teller (BET) method is employed to calculate the specific surface area of mesoporous materials from the nitrogen adsorption data. The BET equation relates the amount of nitrogen adsorbed at different pressures to the surface area, allowing for the estimation of the total surface area available for adsorption. The pore size distribution can be derived from the adsorption isotherm using models such as the Barrett-Joyner-Halenda (BJH) method. This analysis provides insights into the size and distribution of mesopores, which is critical for understanding the material's behavior in LC. The total pore volume and the micropore volume can also be calculated

Materials Research Foundations 173 (2025) 93-123
https://doi.org/10.21741/9781644903452-4

from the adsorption isotherms. These parameters are essential for evaluating the capacity of mesoporous materials to hold liquids or gases, which is particularly relevant in applications like LC [51].

The characterization of mesoporous materials using nitrogen adsorption isotherms helps in selecting appropriate materials for packing LC columns, ensuring optimal separation efficiency. Understanding the surface area and pore size distribution allows researchers to predict how effectively a mesoporous material can adsorb target analytes, which is crucial for method development in LC. By tailoring the pore network fram of mesoporous materials, researchers can enhance the selectivity and resolution of separations in LC [51,52].

Table 2. *Characterization techniques for mesoporous materials with respect to their pros, cons, and applications for suitability in liquid chromatography.*

Characterization Technique	Advantages	Limitations	Applications
Nitrogen Adsorption-Desorption Isotherms [50-52]	• Versatile • Standardized procedures available • Carried out at cryogenic temperature • High sensitivity and resolution • Can be combined with other characterization techniques like XRD, SEM, and TEM. • Cost-effective.	• Equilibration issues • Relies on several assumptions, which may not be true for real materials. • Temperature-dependent • Complex calculations	• Characterization of porous materials • Surface area determination • Pore size distribution analysis • Understanding adsorption mechanisms • Catalyst development
Transmission Electron Microscopy [53-56]	• Can achieve high magnification • Exceptional resolution • Versatile imaging modes • Detailed internal structure visualization	• Extensive sample preparation needed • Radiation damage to sensitive samples • Limited field of view • Operates under high vacuum conditions • Requires samples to be static • Expensive and complex • Generates two-dimensional images only	• Crystal structure analysis of semiconductors, metals, and ceramics • Characterization of nanoparticles and nanostructures • Visualise cellular structures, viruses, and organelles

Materials Research Foundations 173 (2025) 93-123
https://doi.org/10.21741/9781644903452-4

Scanning Electron Microscopy [57]	• High resolution • Large depth of field • Wide range of materials can be analyzed • Rapid imaging • Three-dimensional visualization of samples • Less complex sample preparation • Non-destructive	• High cot • Extensive training required • Limited internal structure information • Requires a high-vacuum environment • Radiation damage to sensitive samples • Colourless image • Time-consuming	• Analyzing the microstructure, defects, and grain boundaries in materials like semiconductors, metals, and ceramics • Visualizing the surface morphology of cells, tissues, and microorganisms
X-ray Diffraction [49, 57]	• Detailed crystal structure analysis • Identify and quantify different phases in a sample • Non-destructive testing • High sensitivity • Rapid data collection	• Inability to analyse amorphous materials • Sample must be homogenous and preferably single-phase • Sample preparation needed • Limited depth profiling • Analysing non-isometric crystalline structures can be complex • No real-space imaging	• Phase identification of unknown crystalline materials • Analysis of phase composition in mixtures • Microstructural characterization • Analysis of material behavior in different environments
FTIR Spectroscopy [58, 59]	• Non-destructive • No external calibration is required, giving accurate results • Both organic and inorganic compounds can be identified • Multiple scans in a shorter time, hence faster data collection	• Interference from strong signals • Overlapping absorption bands • Analysis limited to sample surface • Requires a flat surface for analysis	• Analysis of surface functionalization • Identify and quantify surface hydroxyl groups in mesoporous materials • Investigate mechanisms of chemical reactions • Identification of adsorbed species

		• Spectra can be affected by factors like crystal temperature, pressure, and contact forces	• Monitoring chemical reactions
Thermogravimetric Analysis [60, 61]	• Require small sample size • Both solid and liquid samples can be analysed • Minimal sample preparation • Can be coupled with other techniques like FTIR, MS, or GC/MS	• Inability to Distinguish Between Processes • Some materials may decompose at high temperatures • Limited to Mass Changes • Sensitive to Environmental Conditions • Complex Sample Preparation	• Characterization of material stability • Monitoring synthesis processes • Functional group analysis • Assessment of pore structure • Evaluation of additives and modifiers
Dynamic Light Scattering [62-65]	• Rapid analysis • High sensitivity • Non-destructive • Real-time monitoring	• Sensitivity to Large Particles • May not be able to detect smaller particles in the presence of larger ones • Not well-suited for analysing highly polydisperse samples	• Determine the hydrodynamic size of mesoporous nanoparticles in colloidal suspensions • Assessment of the stability of mesoporous nanoparticles
Mercury Intrusion Porosimetry [66-69]	• Accurately measures wide pore size range • High pressure capability • Comprehensive pore structure analysis • Rapid measurement	• Destructive technique • Hysteresis effect can complicate data interpretation • Not effective for very small pore (<3 nm)	• Pore size distribution analysis • Total pore volume measurement • Characterization of structural properties • Assessment of material stability

4.3.2 Transmission Type of Electron Microscopy (TEM)

This is the most potential method for the bulk characterization of mesoporous substances, providing detailed insights into their structural and morphological properties [53]. TEM allows for imaging at the nanoscale, enabling the observation of the bulk structure of mesoporous substances

[54]. It can also provide information on the morphology of mesoporous substances, revealing the arrangement of pores and the overall particle shape. This morphological characterization is vital for understanding how the structure affects the material's functionality. This methodology can also be utilised to assess the crystallinity of mesoporous substances. High-Resolution TEM (HRTEM) can identify the crystalline phases present and provide information on defects and disorder within the material. Apart from that, Selected Area Electron Diffraction (SAED) patterns obtained from TEM can provide information about the crystallographic features of the mesoporous substances. This helps in identifying the symmetry and phase of the materials, which is important for understanding their properties [55]. TEM can be coupled with different spectroscopic methods namely Energy Dispersive X-ray (EDS) and Electron Energy Loss Spectroscopy (EELS) to provide compositional information. This is particularly useful for analysing the elemental composition of mesoporous materials and understanding the interactions between different components [56].

4.3.3 Scanning Type of Electron Microscopy (SEM)

It is a widely used methodology for characterizing the surface topography and particle size distribution of mesoporous compounds. SEM provides high-resolution images that reveal the external features and overall structure of these materials.

SEM can generate the detailed features about the surface topography of mesoporous substances, such as the presence of pores, cracks, or irregularities. SEM allows for the estimation of particle size and the assessment of size distribution. This is particularly important for mesoporous materials, as the particle size can affect their performance in applications like LC. SEM can also be coupled with EDS to provide information about the elemental composition of mesoporous materials. This technique is useful for identifying the presence and distribution of different elements within the material [54].

There are also certain challenges associated with this method of characterization of mesoporous materials. Mesoporous materials, especially those with insulating frameworks, can experience charging effects under the electron beam, which can lead to image distortion. To overcome this, samples may need to be coated with a conductive material, such as gold or carbon. Another issue is that the higher energy electron beams mostly applied in SEM can potentially cause structural damage or collapse in mesoporous materials, especially those with low mechanical stability. Careful sample preparation and optimization of imaging conditions are necessary to minimize this issue [57].

4.3.4 X-ray Diffraction (XRD)

It is a crucial methodology for characterizing mesoporous substances, particularly in understanding their crystallographic structure and phase purity. XRD is primarily used to identify the crystalline phases present in mesoporous materials. By analysing the diffraction patterns, scientists can identify the symmetry and arrangement of atoms within the material. It helps assess the degree of crystallinity of mesoporous materials. Crystalline materials produce sharp diffraction peaks, while amorphous materials yield broad peaks or halos. This information is essential for evaluating the material's potential applications, as crystallinity can influence properties like stability and reactivity. Although mesoporous materials often exhibit a degree of disorder, XRD can provide insights into the average pore structure. The main positions and the intensities of the

Materials Research Foundations 173 (2025) 93-123
https://doi.org/10.21741/9781644903452-4

diffraction peaks can indicate the occurrence of ordered mesopores and their arrangement. XRD can be used for quantitative phase analysis, allowing scientists to generate the relative amounts of different crystalline phases present in a sample. This is in general useful when working with mixed-phase materials [57].

Mesoporous materials often exhibit diffraction peaks at low angles due to their large pore sizes. These peaks can overlap, making it challenging to resolve individual contributions from different phases. The short-range order in mesoporous materials may complicate the interpretation of XRD patterns, as the peaks may not be as distinct as in highly crystalline materials. Despite these challenges associated with low-angle diffraction peaks and short-range order, XRD remains an essential technique for researchers working with mesoporous materials in various applications, including catalysis, adsorption, and material development [49].

4.3.5 Fourier Transform Type of Infrared Spectroscopy (FTIR)

For the purpose of characterizing mesoporous compounds, FTIR is a highly effective technique that yields important details regarding the chemistry of their surface and the existence of the functionality groups. The functionality groups added during the surface modification of mesoporous materials can be easily identified using this technique.

It can confirm the successful attachment of functionality groups onto the exterior surface of mesoporous materials during silanization or other modification processes. It can detect the presence of organic components, such as surfactants or templates, within the porous structure. This information is crucial for evaluating the effectiveness of template removal during the development of mesoporous compounds. This technique can also provide insights into the chemical interactions between mesoporous materials and target molecules, such as drugs or catalysts, by detecting changes in the functional group signatures. FTIR can be used for both qualitative and quantitative estimation of mesoporous materials. The technique generates a "fingerprint" spectrum that can be compared to reference spectra for identification of compounds frame.

The technique offers several advantages. FTIR analysis is relatively faster and does not require extensive sample development, making it essential for routine characterization. It can detect functional groups present at low concentrations on the surface of mesoporous materials. This can also be applied to the number of mesoporous compounds, including silica, metal oxides, and carbon-based materials. Apart from these advantages, FTIR also has certain limitations. Interpreting FTIR spectra can be challenging, especially for complex mixtures or materials with overlapping functional group signatures. Effective identification of unknown materials using FTIR requires a comprehensive reference library of spectra.

Overall, FTIR Spectroscopy is a valuable tool for characterizing mesoporous materials, providing insights into their surface chemistry and the success of surface modification processes. By identifying the presence and nature of functional groups, FTIR contributes to the formation and optimization of mesoporous compounds for various purpose of usage [58,59].

4.3.6 Thermogravimetric Analysis (TGA)

It is a type of thermal analysis technique that determines the differentiation in weight of a sample as it is subjected to controlled temperature variations. This method is particularly valuable for characterizing mesoporous materials, providing insights into their thermal stability, composition, and decomposition behavior.

This method is used to evaluate the thermal stability of mesoporous compounds by monitoring weight changes over a specified temperature range. A stable material will show minimal weight loss, while significant mass loss indicates decomposition or degradation. This information is crucial for determining the upper temperature limits for the application of these materials. TGA can help identify the composition of mesoporous materials by analysing the weight loss associated with different components. For instance, the loss of water, organic solvents, or other volatile substances can be quantified, providing insights into the material's purity and the presence of residual solvents or templates used during synthesis. The TGA curve plots weight loss as a function of temperature, allowing researchers to identify specific temperature ranges where decomposition occurs. This data can reveal the thermal degradation mechanisms of the material, which is essential for the usage such as catalysis and drug deliveries.

TGA can also be used to study the kinetics of thermal decomposition processes. By analysing the rate of weight loss at different temperatures, scientists can gain insights into the reaction mechanisms and the activation energy required for decomposition. The presence of specific functional moieties can be inferred from the TGA data, particularly when combined with other techniques like Differential Thermal Analysis (DTA) or FTIR. This combination allows for a more comprehensive understanding of the material's thermal properties and chemical interactions [60,61].

4.3.7 Dynamic Light Scattering (DLS) Method

It is also a widely used methodology for characterizing mesoporous compounds, particularly in assessing their size, stability, and dispersion in colloidal systems. DLS determines the time-dependent changes in the intensity of scattered light caused by the brownian motion of particles in a suspension. This data is used to calculate the hydrodynamic diameter of the particles, providing information on their size distribution. This is particularly important for mesoporous materials, as particle size can significantly affect their performance in usage such as drug deliveries and catalysis [62].

DLS can also be employed to evaluate the stability of mesoporous nanoparticles in suspension. By analysing the observation of the possible changes in size over time, scientists can determine whether the particles are aggregating or remaining stable, which is critical for ensuring consistent performance in various applications. It can provide insights into the inner structure of mesoporous materials. The high inner volume of mesoporous matrices allows for the encapsulation of active molecules, and DLS can help assess how these materials behave in a colloidal state, which is relevant for applications like drug delivery and stabilization of sensitive molecules [63,64]. The technique is often used in conjunction with other characterization methods, such as TEM and nitrogen sorption measurements, to provide a comprehensive understanding of mesoporous materials. This combination can help overcome some limitations of DLS, such as its inability to provide detailed structural information [62,65].

Particle size and stability can be tracked in real time using DLS, a rapid and non-destructive technology that doesn't change the sample. It is appropriate for assessing small mesoporous nanoparticles because it can detect size changes in the nanometer range. Numerous materials, such as polymers, various nanostructured materials, and silica-based mesoporous nanoparticles, can be treated with this approach [64].

Besides, in concentrated suspensions, multiple scattering can affect the accuracy of dimension measurements, leading to overestimation of particle size using this technique. Also, DLS assumes that particles are spherical, which may not be the case for all mesoporous materials. This can result into inaccuracies in dimension determination if the particles have irregular shapes. Another issue is that DLS can be sensitive to particle aggregation, which may complicate the interpretation of results, especially in systems where stability is a concern [65].

4.3.8 Mercury Intrusion Porosimetry (MIP)

It is a widely utilized technique for characterizing the pore framework of mesoporous substances. This method provides valuable insights into pore size distribution, pore volume, and overall porosity, which are critical for understanding the performance of these materials.

MIP operates on the principle that mercury, a non-wetting liquid, can be forced into the pores of a substance under high pressure [66]. The Washburn equation, which provides the link between pressure and pore diameter, can be used to calculate the pore size distribution since the pressure needed to force mercury into the pores is related to the dimension of the pores:

$$P = -\frac{2\gamma cos\theta}{r} \tag{2}$$

Where, r means the pore radius, γ means the surface relevant tension of mercury, θ means the contact angle (typically close to 140° for mercury), and P means the applied pressure. This equation allows for the determination of pore sizes based on the pressure at which mercury intrudes. MIP provides a direct measurement of the total pore volume of the mesoporous compound by calculating the volume of mercury that penetrates the pores at various pressures [67].

This technique can measure a broad distribution of pore sizes, from macropores down to mesopores, making it essential for various types of porous materials. The method provides quantitative data on pore volume and size distribution, which are essential for evaluating the material's performance in applications. MIP is relatively quick compared to other characterization techniques, allowing for efficient analysis of multiple samples [68].

Although, the technique requires high pressures (up to 415,000 kPa), which can potentially deform or crush soft materials, affecting the accuracy of the measurements. This is particularly relevant for materials like paper or membranes that may not withstand such pressures [66]. Since mercury does not wet most materials, the method may not accurately reflect the behavior of wetting fluids in real applications. This can lead to discrepancies when interpreting pore size distributions for hydrophilic materials. Samples must be dry before analysis, as the presence of liquids in the pores can prevent accurate measurements. This requirement can be a disadvantage when characterizing hydrophilic or moisture-sensitive materials [69].

5. Applications of Mesoporous Materials in Liquid Chromatographic Separation

Mesoporous materials, particularly mesoporous silica, have found significant applications in liquid chromatographic separation due to their unique features, such as larger surface area, tunable pores size, and thermal stabilities. The larger surface area of mesoporous silica enhances the resolution of molecules, making it suitable for preparative chromatography applications. Mesoporous spheres (MSs) have been prepared and used as stationary phases in LC [5].

HPLC can employ mesoporous silica as a stationary phase in both the normal phase and the reverse phase. Molecule resolution is improved by mesoporous silica's significantly larger surface area than that of commercially available chromatography grade silica. It has also been used to separate enantiomers in chiral HPLC. Chiral compounds can be effectively separated thanks to mesoporous materials' wide surface area and adjustable pore size. Mesoporous compounds are also appropriate for adsorption applications due to their high porosity and surface area. They have the potential to be used in sensor applications since they can be employed as adsorbents to remove gasses or **contaminants [5]. Table 3 depicts a brief summary of few types of commercially available** mesoporous materials that can be used in liquid chromatography.

Table 3*. Types of mesoporous materials used in commercial scale for liquid chromatography.*

Commercial Example	Grades Available	Manufactured / Developed by	Key Characteristics	Reference
Mesoporous Silica				
MCM-41	MSE PRO MCM-41 Nano powder	Mobil Oil Corporation	Hexagonal array of uniform mesopores, amorphous, can be functionalized using organic groups or metal ions.	[70]
MCM-48	-		Cubical structure, good thermal stability, high specific surface area.	[71]
MCM-50	-		Layered structure, unlike MCM-41 and MCM-48.	[72]
SBA-11	-	University of California, Santa Barbara	Three-dimensional cubic structure, open 3D cage-type pores	[73]
SBA-12	-		Hexagonal arrangement of pores, larger pore diameters, high surface area.	
SBA-15	Five grades available from SBA-1 to SBA-5		Hexagonal arrangement of cylindrical mesopores, high specific surface area, thermally stable	
SBA-16	-		Three-dimensional cubic structure, which allows for interconnected pores.	

6. Challenges and Future Perspectives

While mesoporous compounds, particularly mesoporous silica, have shown great potential for liquid chromatographic separation, there are still several challenges and future perspectives to consider. Achieving precise control over the pore size distribution of mesoporous compounds remains a challenge. Variations in pore size can affect the separation efficiency and resolution. The particle size of mesoporous materials needs to be optimized for efficient packing in chromatographic columns. Smaller particles can lead to high backpressures, while larger particles may compromise separation performance. These materials can suffer from mechanical instability under high pressures, which is common in LC. Improving the mechanical stability of these materials is crucial for their long-term performance and durability in chromatographic applications. Surface functionalization of mesoporous materials is often necessary to enhance their selectivity and separation capabilities. However, the introduction of functional moieties can

Materials Research Foundations 173 (2025) 93-123 https://doi.org/10.21741/9781644903452-4

sometimes lead to pore blockage or reduced surface area, which may compromise the material's performance [5].

Continued research into the synthesis and design of mesoporous compounds with precisely controlled pore sizes and morphologies will enable the development of highly efficient stationary phases for LC. The production of monodisperse mesoporous particles with uniform size and shape is a promising direction for improving column packing efficiency and separation performance. The combination of mesoporous materials with other functional components, such as polymers or nanoparticles, can lead to the formation of hybrid materials with enhanced selectivity, stability, and separation capabilities. The miniaturization of mesoporous-based chromatographic systems and their integration with microfluidic devices can enable the development of portable, high-throughput, and cost-effective separation platforms for various applications, including environmental monitoring, food analysis, and pharmaceutical quality control. Advances in computational modelling and simulation techniques can help in the rational design and optimization of mesoporous compounds for specific chromatographic applications, reducing the need for extensive experimental trials [5,88].

Conclusion

In conclusion, the integration of mesoporous materials into liquid chromatographic separation processes represents a significant advancement in the area of analytical chemistry. The unique structural attributes of mesoporous materials, including their large surface area, tunable pore sizes, and high stability, provide unparalleled benefits in terms of separation efficiency, selectivity, and capacity. This chapter has demonstrated the diverse applications and versatile nature of mesoporous materials, from enhancing analytical separations to supporting large-scale preparative processes. Through detailed case studies and the latest research developments, it is evident that mesoporous materials hold immense potential to address current challenges in chromatographic separation, paving the way for more efficient and precise analytical techniques. Future research and development will likely focus on further optimizing these materials and exploring new functionalization strategies to expand their applicability and performance in various chromatographic contexts. Thus, mesoporous materials are poised to play a crucial role in the evolution of chromatographic technologies, driving innovation and improving outcomes in numerous scientific and industrial applications.

References

[1] D. Zhao, Q. Huo, J. Feng, B.F. Chmelka, G.D. Stucky, Nonionic Triblock and Star Diblock Copolymer and Oligomeric Surfactant Syntheses of Highly Ordered, Hydrothermally Stable, Mesoporous Silica Structures, J. Am. Chem. Soc. 120 (1998) 6024-6036. https://doi.org/10.1021/ja974025i

[2] J.S. Beck, J.C. Vartuli, W.J. Roth, M.E. Leonowicz, C.T. Kresge, K.D. Schmitt, C.T.W. Chu, D.H. Olson, E.W. Sheppard, S.B. McCullen, J.B. Higgins, J.L. Schlenker, A new family of mesoporous molecular sieves prepared with liquid crystal templates, J. Am. Chem. Soc. 114 (1992) 10834-10843. https://doi.org/10.1021/ja00053a020

 Materials Research Forum LLC
https://doi.org/10.21741/9781644903452-4

[3] L. Zhao, H. Qin, R. Wu, H. Zou, Recent advances of mesoporous materials in sample preparation, J. Chromatogr. A 1228 (2012) 193-204. https://doi.org/10.1016/j.chroma.2011.09.051

[4] R. Eivazzadeh-Keihan, K.K. Chenab, R. Taheri-Ledari, J. Mosafer, S.M. Hashemi, A. Mokhtarzadeh, A. Maleki, M.R. Hamblin, Recent advances in the application of mesoporous silica-based nanomaterials for bone tissue engineering, Mater. Sci. Eng. C Mater. Biol. Appl. 107 (2020) 110267. https://doi.org/10.1016/j.msec.2019.110267

[5] K.W. Gallis, A.G. Eklund, S.T. Jull, J.T. Araujo, J.G. Moore, C.C. Landry, The Use of Mesoporous Silica in Liquid Chromatography, in: A. Sayari, M. Jaroniec (Eds), Studies in Surface Science and Catalysis, Elsevier Inc., United Kingdom, 2000, pp. 747-755. https://doi.org/10.1016/S0167-2991(00)80279-7

[6] A. Soliven, S. Kayillo, R.A. Shalliker, Liquid Chromatography Reversed Phase, in: P. Worsfold, C. Poole, A. Townshend, M. Miró (Eds.), Reference Module in Chemistry, Molecular Sciences and Chemical Engineering, Encyclopedia of Analytical Science, Elsevier Academic Press, United Kingdom, 2013, pp. 238-243. https://doi.org/10.1016/B978-0-12-409547-2.00305-X

[7] I. François, K. Sandra, P. Sandra, Comprehensive liquid chromatography: Fundamental aspects and practical considerations-A review, Anal. Chim. Acta. 641 (2009) 14-31. https://doi.org/10.1016/j.aca.2009.03.041

[8] K. Robards, P.R. Haddad, P.E. Jackson, High-performance Liquid Chromatography-Separations, in: Principles and Practice of Modern Chromatographic Methods, Elsevier Academic Press, United Kingdom, 2004, pp. 305-380. https://doi.org/10.1016/B978-0-08-057178-2.50009-1

[9] D.S. Hage, Chromatography, in: N. Rifai, A.R. Horvath, C.T. Wittwer (Eds.), Principles and Applications of Clinical Mass Spectrometry, Elsevier Inc., United Kingdom, 2018, pp. 1-32. https://doi.org/10.1016/B978-0-12-816063-3.00001-3

[10] C.F. Poole, S.K. Poole, Foundations of retention in partition chromatography, J. Chromatogr. A. 1216(10) (2009) 1530-1550. https://doi.org/10.1016/j.chroma.2008.10.092

[11] R. Consden, Partition chromatography on paper, its scope and application, Nature. 162(4114) (1948) 359-361. https://doi.org/10.1038/162359a0

[12] M.K. Gupta, P.K. Biswas, Chromatography: Basic principle, types, and applications, in: A.K. Bhatt, R.K. Bhatia, T.C. Bhalla (Eds.), Basic Biotechniques for Bioprocess and Bioentrepreneurship, Elsevier Academic Press, United Kingdom, 2023, pp. 173-182. https://doi.org/10.1016/B978-0-12-816109-8.00010-6

[13] K.Z. Masoodi, S.M. Lone, R.S. Rasool, Ion-exchange chromatography, in: K.Z. Masoodi, S.M. Lone, R.S. Rasool (Eds.), Advanced Methods in Molecular Biology and Biotechnology, Elsevier Academic Press, United Kingdom, 2021, pp. 151-154. https://doi.org/10.1016/B978-0-12-824449-4.00027-X

[14] W. Wang, Z. Liu, Y. Liu, Z. Su, Y. Liu, Plant polypeptides: A review on extraction, isolation, bioactivities and prospects, Int. J. Biol. Macromol. 207 (2022) 169-178. https://doi.org/10.1016/j.ijbiomac.2022.03.009

[15] J.A. Queiroz, C.T. Tomaz, J.M. Cabral, Hydrophobic interaction chromatography of proteins. J. Biotechnol. 87 (2001) 143-159. https://doi.org/10.1016/S0168-1656(01)00237-1

[16] N. Galeotti, E. Hackemann, F. Jirasek, H. Hasse, Prediction of the elution profiles of proteins in mixed salt systems in hydrophobic interaction chromatography, Sep. Purif. Technol. 233 (2020) 116006. https://doi.org/10.1016/j.seppur.2019.116006

[17] J.L. Ochoa, Hydrophobic (interaction) chromatography, Biochimie. 60 (1978) 1-15. https://doi.org/10.1016/S0300-9084(78)80193-X

[18] S. Moldoveanu, V. David, Hydrophilic interaction liquid chromatography, in: Essentials in Modern HPLC Separations, Elsevier Inc., United Kingdom, 2022, pp. 447-477. https://doi.org/10.1016/B978-0-323-91177-1.00017-X

[19] A. Cavazzini, M. Catani, A. Felinger, Hydrophilic interaction liquid chromatography, in: S. Fanali, B. Chankvetadze, P.R. Haddad, C.F. Poole, M.L. Riekkola (Eds.), Liquid Chromatography Fundamental and Instrumentation, Elsevier Inc., United Kingdom, 2023, pp. 227-249. https://doi.org/10.1016/B978-0-323-99968-7.00030-8

[20] D. Schrag, M. Corbier, S. Raimondi, Size exclusion-high-performance liquid chromatography (SEC-HPLC), Methods Mol. Biol. 1131 (2014) 507-512. https://doi.org/10.1007/978-1-62703-992-5_31

[21] V. Cauda, G. Canavese, Mesoporous Materials for Drug Delivery and Theranostics, Pharmaceutics. 12(1108) (2020) 1-3. https://doi.org/10.3390/pharmaceutics12111108

[22] J. Zhao, F. Gao, Y. Fu, W. Jin, P. Yang, D. Zhao, Biomolecule separation using large pore mesoporous SBA-15 as a substrate in high performance liquid chromatography, Chem. Commun. (2002) 752-753. https://doi.org/10.1039/b110637f

[23] M. Faraji, Y. Yamini, F. Noormohammadi, M. Adeli, Application of magnetic nanomaterials in environmental monitoring, in: M. Ahmadi, A. Afkhami, T. Madrakian (Eds.), Magnetic Nanomaterials in Analytical Chemistry, Elsevier Inc., United Kingdom, 2021, pp. 155-189. https://doi.org/10.1016/B978-0-12-822131-0.00005-4

[24] H. Li, L. Wang, Y. Wei, W. Yan, J. Feng, Preparation of Templated Materials and Their Application to Typical Pollutants in Wastewater: A Review, Front. Chem. 10 (882876) (2022) 1-20. https://doi.org/10.3389/fchem.2022.882876

[25] H. Ghaedi, M. Zhao, Review on Template Removal Techniques for Synthesis of Mesoporous Silica Materials, Energy & Fuels. 36(5) (2022) 2424-2446. https://doi.org/10.1021/acs.energyfuels.1c04435

[26] Y. Liu, J. Goebl, Y. Yin, Templated synthesis of nanostructured materials. Chem. Soc. Rev. 42 (2013) 2610-2653. https://doi.org/10.1039/C2CS35369E

[27] T. Zhao, A. Elzatahry, X. Li, D. Zhao, Single-micelle-directed synthesis of mesoporous materials, Nat. Rev. Mater. 4 (2019) 775-791. https://doi.org/10.1038/s41578-019-0144-x

[28] Z. Li, L. Liu, Z. Wang, P. Gao, G.K. Li, Synthesis and Application of Mesoporous Materials: Process Status, Technical Problems, and Development Prospects: A Mini-Review, Energy Fuels, 37 (2023) 3413-3427. https://doi.org/10.1021/acs.energyfuels.2c03882

[29] J.A. Cecilia, R. Moreno, M. Retuerto Millán, Mesoporous Materials: From Synthesis to Applications, Int. J. Mol. Sci. 20(3) 3213 (2019) 1-4. https://doi.org/10.3390/ijms20133213

[30] M. Benítez, C. Rodríguez-Carrillo, S. Sánchez-Artero, J. El Haskouri, P. Amorós, J.V. Ros-Lis, Scaled-up microwave-assisted batch and flow synthesis and life cycle assessment of a silica mesoporous material: UVM-7, Green Chem. 26 (2024) 785-793. https://doi.org/10.1039/D3GC02875E

[31] J. Quinchia, J. Sánchez, J.C. Poveda, A. Moreno, Microwave-assisted synthesis of mesostructured aluminosilicates (MA) for the preparation of nano-MoS2/MA catalysts active in decalin hydrocracking. Microporous Mesoporous Mater. 301 (2020) 110226. https://doi.org/10.1016/j.micromeso.2020.110226

[32] N.H.N. Kamarudin, A.A. Jalil, S. Triwahyono, S.N. Timmiati, Microwave-assisted synthesis of mesoporous silica nanoparticles as a drug delivery vehicle, Malaysian J. Anal. Sci. 20(6) (2016) 1382-1389. https://doi.org/10.17576/mjas-2016-2006-17

[33] Y. Cao, H. Wei, Z. Xia, Advances in microwave assisted synthesis of ordered mesoporous materials. Trans. Nonferrous Met. Soc. China. 19 (2009) s656-s664. https://doi.org/10.1016/S1003-6326(10)60127-6

[34] L. López-Pérez, M.A. López-Martínez, K. Djanashvili, K. Góra-Marek, K.A. Tarach, M.E. Borges, M.C. Ignacio, Process Intensification of Mesoporous Material's Synthesis by Microwave-Assisted Surfactant Removal, ACS Sustainable Chem. Eng. 45(8) (2020) 16814-16822. https://doi.org/10.1021/acssuschemeng.0c05438

[35] A. Shiekh, A. Mushtaq, U. Jabeen, F. Bashir, M. Zahra, F. Behlil, N. Hina, I. Hafeez, Surface Modification of Mesoporous Silica Nanoparticles with Hexamethyl Disilazane as Smart Carriers for Tocopherol Acetate, Nano Biomed. Eng. 14(3) (2022) 216-224. https://doi.org/10.5101/nbe.v14i3.p216-224

[36] E. Ahmadi, N. Dehghannejad, S. Hashemikia, M. Ghasemnejad, H. Tabebordbar, Synthesis and surface modification of mesoporous silica nanoparticles and its application as carriers for sustained drug delivery, Drug Deliv. 21(3) (2014) 164-172. https://doi.org/10.3109/10717544.2013.838715

[37] S. Porrang, N. Rahemi, S. Davaran, M. Mahdavi, B. Hassanzadeh, A.M. Gholipour, Direct surface modification of mesoporous silica nanoparticles by DBD plasma as a green approach to prepare dual-responsive drug delivery system, J. Taiwan Inst. Chem. Eng. 123 (2021) 47-58. https://doi.org/10.1016/j.jtice.2021.05.024

[38] Y.L. Khung, D. Narducci, Surface modification strategies on mesoporous silica nanoparticles for anti-biofouling zwitterionic film grafting, Adv. Colloid Interface Sci. 226(Pt B) (2015) 166-186. https://doi.org/10.1016/j.cis.2015.10.009

[39] S. Kovtareva, L. Kusepova, G. Tazhkenova, T. Mashan, K. Bazarbaeva, E. Kopishev, Surface Modification of Mesoporous Silica Nanoparticles for Application in Targeted Delivery Systems of Antitumour Drugs, Polymers (Basel). 16(8) (2024) 1105. https://doi.org/10.3390/polym16081105

[40] C. Li, S. Cheng, Functional group surface modifications for enhancing the formation and performance of exoelectrogenic biofilms on the anode of a bioelectrochemical system, Crit. Rev. Biotechnol. 39(8) (2019) 1015-1030. https://doi.org/10.1080/07388551.2019.1662367

[41] E.M. Conners, K. Rengasamy, A. Bose, Electroactive biofilms: how microbial electron transfer enables bioelectrochemical applications, J. Ind. Microbiol. Biotechnol. 49(4) (2022) kuac012 1-12. https://doi.org/10.1093/jimb/kuac012

[42] J. Deng, Y. Song, Z. Lan, Z. Xu, Y. Chen, B. Yang, H. Hao, The surface modification effect on the interfacial properties of glass fiber-reinforced epoxy: A molecular dynamics study, Nanotechnol. Rev. 11(1) (2022) 1143-1157. https://doi.org/10.1515/ntrev-2022-0068

[43] Z. Zhang, J. Zeng, J. Groll, M. Matsusaki, Layer-by-layer assembly methods and their biomedical applications, Biomater. Sci 10(15) (2022) 4077-4094. https://doi.org/10.1039/D2BM00497F

[44] S. Zhao, F. Caruso, L. Dähne, G. Decher, B.G. De Geest, J. Fan, N. Feliu, Y. Gogotsi, P.T. Hammond, M.C. Hersam, A. Khademhosseini, N. Kotov, S. Leporatti, Y. Li, F. Lisdat, L.M. Liz-Marzán, S. Moya, P. Mulvaney, A.L. Rogach, S. Roy, D.G. Shchukin, A.G. Skirtach, M.M. Stevens, G.B. Sukhorukov, P.S. Weiss, Z. Yue, D. Zhu, W.J. Parak, The Future of Layer-by-Layer Assembly: A Tribute to ACS Nano Associate Editor Helmuth Möhwald, ACS Nano. 13(6) (2019) 6151-6169. https://doi.org/10.1021/acsnano.9b03326

[45] J.J. Richardson, J. Cui, M. Björnmalm, J.A. Braunger, H. Ejima, F. Caruso, Innovation in Layer-by-Layer Assembly, Chem. Rev. 116 (2016) 14828-14867. https://doi.org/10.1021/acs.chemrev.6b00627

[46] C. Wang, M.J. Park, H. Yu, H. Matsuyama, E. Drioli, H.K. Shon, Recent advances of nanocomposite membranes using layer-by-layer assembly, J. Memb. Sci. 661 (2022) 120926 1-30. https://doi.org/10.1016/j.memsci.2022.120926

[47] R. Guillet-Nicolas, L. Marcoux, F. Kleitz, Insights into pore surface modification of mesoporous polymer-silica composites: introduction of reactive amines, New J. Chem. 34 (2010) 355-366. https://doi.org/10.1039/b9nj00478e

[48] A. Fakhar, M. Sadeghi, M. Dinari, Stepwise surface modification of mesoporous silica and its use in poly(urethane-urea) composite films, Polym. Int. 71 (2022) 107-116. https://doi.org/10.1002/pi.6290

[49] L. Ludescher, R. Morak, S. Braxmeier, F. Putz, N. Hüsing, G. Reichenauer, O. Paris, Hierarchically organized materials with ordered mesopores: adsorption isotherm and adsorption-induced deformation from small-angle scattering, Phys. Chem. Chem. Phys. 22 (2020) 12713-12723. https://doi.org/10.1039/D0CP01026J

[50] Z.A. ALOthman, A Review: Fundamental Aspects of Silicate Mesoporous Materials, Mater. 5 (2012) 2874-2902. https://doi.org/10.3390/ma5122874

[51] J. Goscianska, A. Olejnik, R. Pietrzak, Comparison of ordered mesoporous materials sorption properties towards amino acids, Adsorption. 19 (2013) 581-588. https://doi.org/10.1007/s10450-013-9481-z

[52] N.S. Ali, H.N. Harharah, I.K. Salih, N.M. Cata Saady, S. Zendehboudi, T.M. Albayati, Applying MCM-48 mesoporous material, equilibrium, isotherm, and mechanism for the effective adsorption of 4-nitroaniline from wastewater, Sci. Rep. 13 (2023) 9837. https://doi.org/10.1038/s41598-023-37090-4

[53] T. Ohsuna, Z. Liu, S. Che, O. Terasaki, Characterization of Chiral Mesoporous Materials by Transmission Electron Microscopy, Small. 1(2) (2005) 233-237. https://doi.org/10.1002/smll.200400048

[54] S.M. Shawky, A.A. Abo-AlHassan, H. Lill, D. Bald, S.F. EL-Khamisy, M.E. El-Zeiny, Efficient Loading and Encapsulation of Anti-Tuberculosis Drugs using Multifunctional Mesoporous Silicate Nanoparticles, J. Nanosci. Curr. Res. 1(1) (2016) 1-9.

[55] S. Hudson, D.A. Tanner, W. Redington, E. Magner, K. Hodnett, S. Nakahara, Quantitative TEM analysis of a hexagonal mesoporous silicate structure, Phys. Chem. Chem. Phys. 8(29) (2006) 3467-3474. https://doi.org/10.1039/b605581h

[56] M.L. Ruiz-González, A. Torres-Pardo, J.M. González-Calbet, The Role of Transmission Electron Microscopy in the Early Development of Mesoporous Materials for Tissue Regeneration and Drug Delivery Applications, Pharmaceutics 13(12) (2021) 2200. https://doi.org/10.3390/pharmaceutics13122200

[57] W. Wei, L. Ziwei, H. Jiajie, Z. Yi, L. Yongsheng, Microstructure Characterization of Mesoporous Materials by FE-SEM, J. Test. Eval. 40(3) (2012) 496-500. https://doi.org/10.1520/JTE104285

[58] G.A. Eimer, M.B. Gómez Costa, L.B. Pierella, O.A. Anunziata, Thermal and FTIR spectroscopic analysis of the interactions of aniline adsorbed on to MCM-41 mesoporous material, J. Colloid Interface Sci. 263(2) (2003) 400-407. https://doi.org/10.1016/S0021-9797(03)00038-9

[59] S. Kumar, A. Sharma, D. Gautam, S. Hooda, Characterization of Mesoporous Materials, in: A. Uthaman, S. Thomas, T. Li, H. Maria (Eds.) Advanced Functional Porous Materials. Engineering Materials, Springer Cham., Switzerland AG, 2022, pp. 175-204. https://doi.org/10.1007/978-3-030-85397-6_6

[60] P. Basu, Analytical Techniques, in: P. Basu (Ed.), Biomass Gasification, Pyrolysis and Torrefaction, Academic Press, Elsevier Inc., United Kingdom, 2018, pp. 479-495. https://doi.org/10.1016/B978-0-12-812992-0.00023-6

[61] D. Camuffo, Measuring Time of Wetness and Moisture in Materials, in: Microclimate for Cultural Heritage, Elsevier B.V., United Kingdom, 2019, pp. 459-482. https://doi.org/10.1016/B978-0-444-64106-9.00019-5

[62] M. Kaasalainen, V. Aseyev, E. von Haartman, D.S. Karaman, E. Mäkilä, H. Tenhu, J. Rosenholm, J. Salonen, Size, Stability, and Porosity of Mesoporous Nanoparticles Characterized with Light Scattering, Nanoscale Res. Lett. 12(1) (2017) 74. https://doi.org/10.1186/s11671-017-1853-y

[63] F. Babick, Dynamic light scattering (DLS), in: V.D. Hadoroaba, W.E.S. Unger, A.G. Shard (Eds.), Characterization of Nanoparticles, Elsevier Inc., United Kingdom, 2020, pp. 137-172. https://doi.org/10.1016/B978-0-12-814182-3.00010-9

[64] Z. Jia, J. Li, L. Gao, D. Yang, A. Kanaev, Dynamic Light Scattering: A Powerful Tool for In Situ Nanoparticle Sizing. Colloids and Interfaces, 7(15) (2023) 1-18. https://doi.org/10.3390/colloids7010015

[65] S.K. Filippov, R. Khusnutdinov, A. Murmiliuk, W. Inam, L.Y. Zakharova, H. Zhang, V.V. Khutoryanskiy, Dynamic light scattering and transmission electron microscopy in drug

delivery: a roadmap for correct characterization of nanoparticles and interpretation of results, Mater. Horiz. 10 (2023) 5354-5370. https://doi.org/10.1039/D3MH00717K

[66] R. Guillet-Nicolas, R. Ahmad, K.A. Cychosz, F. Kleitz, M. Thommes, Insights into the pore structure of KIT-6 and SBA-15 ordered mesoporous silica - recent advances by combining physical adsorption with mercury porosimetry. New J. Chem. 40 (2016) 4351-4360. https://doi.org/10.1039/C5NJ03466C

[67] S.P. Rigby, New methodologies in mercury porosimetry, Stud. Surf. Sci. Catal. 144 (2002) 185-192. https://doi.org/10.1016/S0167-2991(02)80133-1

[68] C. Schlumberger, M. Thommes, Characterization of Hierarchically Ordered Porous Materials by Physisorption and Mercury Porosimetry-A Tutorial Review, Adv. Mater. Interfaces. 8(4) (2021) 1-25. https://doi.org/10.1002/admi.202002181

[69] D. Majda, M. Zimowska, K. Tarach, K. Góra-Marek, B.D. Napruszewska, A. Michalik-Zym, Water thermoporosimetry as a tool of characterization of the textural parameters of mesoporous materials, J. Therm. Anal. Calorim. 127 (2017) 207-220. https://doi.org/10.1007/s10973-016-5400-3

[70] D. Khan, Shaily. Synthesis and catalytic applications of organo-functionalized MCM-41 catalyst: A review, Appl. Organomet. Chem. 37(3) (2023) 1-9. https://doi.org/10.1002/aoc.7007

[71] N.S. Ali, H.N. Harharah, I.K. Salih, N.M. Cata Saady, S. Zendehboudi, T.M. Albayati, Applying MCM-48 mesoporous material, equilibrium, isotherm, and mechanism for the effective adsorption of 4-nitroaniline from wastewater, Sci. Rep. 13 (2023) 9837. https://doi.org/10.1038/s41598-023-37090-4

[72] F.M. Bobonich, A.S. Kovalenko, Yu.G. Voloshina, A.S. Korchev, V.N. Solomakha, A.P. Philippov, Y.G. Il'in, Adsorptive Properties of Template-Containing Silica-Based MCM-41 and MCM-50 Materials, Adsorp. Sci. Technol. 20(6) (2002) 595-605. https://doi.org/10.1260/026361702321039528

[73] D. Zhao, Q. Huo, J. Feng, B.F. Chmelka, G.D. Stucky, Nonionic Triblock and Star Diblock Copolymer and Oligomeric Surfactant Syntheses of Highly Ordered, Hydrothermally Stable, Mesoporous Silica Structures, J. Am. Chem. Soc. 120(24) (1998) 6024-6036. https://doi.org/10.1021/ja974025i

[74] T.V. Sagar, B. Abhishek, Production techniques of functional solid catalysts, in: C.M. Hussain, P. Sudarsanam (Eds.), Advanced Functional Solid Catalysts for Biomass Valorization, Elsevier Inc., United Kingdom, 2020, pp. 39-75. https://doi.org/10.1016/B978-0-12-820236-4.00002-7

[75] C. Jo, K. Kim, R. Ryoo, Syntheses of high quality KIT-6 and SBA-15 mesoporous silicas using low-cost water glass, through rapid quenching of silicate structure in acidic solution, Microporous and Mesoporous Mater. 124(1-3) (2009) 45-51. https://doi.org/10.1016/j.micromeso.2009.04.037

[76] H. Aghavandi, A. Ghorbani-Choghamarani, FDU-12@AGA-Pd: A green, novel, recyclable, and highly versatile mesoporous catalyst for C-C coupling reaction and synthesis of tetrazoles, J. Organomet. Chem. 1015 (2024) 123209. https://doi.org/10.1016/j.jorganchem.2024.123209

[77] L.M. Henning, G.J. Smales, M.G. Colmenares, M.F. Bekheet, U. Simon, A. Gurlo, Synthesis and properties of COK-12 large-pore mesocellular silica foam. Nano Select 4(3) (2023) 202-212. https://doi.org/10.1002/nano.202200223

[78] C.G. Visconti, Alumina: A Key-Component of Structured Catalysts for Process Intensification, T. Indian Ceram. Soc. 71 (2012) 123-136. https://doi.org/10.1080/0371750X.2012.738481

[79] G. Busca, The surface of transitional aluminas: A critical review, Catal. Today. 226 (2014) 2-13. https://doi.org/10.1016/j.cattod.2013.08.003

[80] Y. Jeong, M. Cui, J. Choi, Y. Lee, J. Kim, Y. Son, J. Khim, Development of modified mesoporous carbon (CMK-3) for improved adsorption of bisphenol-A, Chemosphere. 238 (2020) 124559. https://doi.org/10.1016/j.chemosphere.2019.124559

[81] M.M. Rahman, M.G. Ara, M.A. Alim, M.S. Uddin, A. Najda, G.M. Albadrani, A.A. Sayed, S.A. Mousa, M.M. Abdel-Daim, Mesoporous Carbon: A Versatile Material for Scientific Applications, Int. J. Mol. Sci. 22(9) (2021) 4498. https://doi.org/10.3390/ijms22094498

[82] P. Van Der Voort, D. Esquivel, E. De Canck, F. Goethals, I. Van Driessche, F.J. Romero-Salguero, Periodic Mesoporous Organosilicas: from simple to complex bridges; a comprehensive overview of functions, morphologies and applications, Chem. Soc. Rev. 42 (2013) 3913-3955. https://doi.org/10.1039/C2CS35222B

[83] H. Dai, X. Yuan, L. Jiang, H. Wang, Z. Jin, Z. Jingjing, T. Xiong, Recent advances on ZIF-8 composites for adsorption and photocatalytic wastewater pollutant removal: Fabrication, applications and perspective, Coord. Chem. Rev. 441 (2021) 213985. https://doi.org/10.1016/j.ccr.2021.213985

[84] P. Shen, B. Zhang, Y. Wang, X. Liu, C. Yu, T. Xu, S.S. Mofarah, Y. Yu, Y. Liu, H. Sun, H. Arandiyan, Nanoscale niobium oxides anode for electrochemical lithium and sodium storage: a review of recent improvements. J Nanostructure Chem. 11 (2021) 33-68. https://doi.org/10.1007/s40097-020-00367-5

[85] Y. Jiao, X. Zhang, H. Yang, H. Ma, J. Zou, Mesoporous tantalum oxide nanomaterials induced cardiovascular endothelial cell apoptosis via mitochondrial-endoplasmic reticulum stress apoptotic pathway, Drug Deliv. 30 (2023) 108-120. https://doi.org/10.1080/10717544.2022.2147251

[86] Z. Yu, S. Zhu, L. Zhang, S. Watanabe, Mesoporous single crystal titanium oxide microparticles for enhanced visible light photodegradation, Opt. Mater. (Amst) 127 (2022) 112297. https://doi.org/10.1016/j.optmat.2022.112297

[87] B. Leonetti, A. Perin, E.K. Ambrosi, G. Sponchia, P. Sgarbossa, A. Castellin, P. Riello, A. Scarso, Mesoporous zirconia nanoparticles as drug delivery systems: Drug loading, stability and release, J. Drug Deliv. Sci. Technol. 61 (2021) 102189. https://doi.org/10.1016/j.jddst.2020.102189

[88] Z. Guo, Y. Chen, S. Wang, J. Pang, Z. Liu, Dendritic Mesoporous Silica Nanospheres: Toward the Ultimate Minimum Particle Size for Ultraefficient Liquid Chromatographic Separation, ACS Appl. Mater. Interfaces. 13 (2021) 22970-22977. https://doi.org/10.1021/acsami.1c03985

Materials Research Forum LLC
https://doi.org/10.21741/9781644903452-5

Chapter 5

Bentonite Clay and its Potential Applications

Sara Efigênia Dantas de Mendonça y Araújo, Adrian Lima Roberto, Jayane Ellen Pereira Farias, João Augusto Oshiro Junior*

Pharmaceutical Sciences Postgraduate Program, Center for Biological and Health Sciences, State University of Paraíba – UEPB, Campina Grande, Paraíba, Brazil

* joaooshiro@yahoo.com.br

Abstract

Bentonite is the commercial name given to a group of natural clays made up of a mixture of minerals belonging to the smectites. Its composition varies according to the geographical region of origin and the way the bentonite is extracted, which can vary the material's properties. The independent negative charge, the presence of easily exchangeable cations, and swelling are fundamental properties that can vary in intensity depending on the composition of these clays. Montmorillonite (MMT) is the most abundant component of Bentonite and has an excellent ability to adsorb positive charges from toxins when they come into contact with water. In addition, this clay can be exposed to additional treatments, which can modify its structure and increase its properties. As a result, this clay can be used in a wide range of areas, from pharmaceuticals and biomedicine to agriculture and engineering. Therefore, this chapter will provide the reader with a comprehensive overview of bentonite's various applications, focusing on its use in the production of pharmaceutical formulations, in cosmetic products, in the controlled release of drugs, in detoxification pharmacological activity, cholesterol reduction, anti-inflammatory action, as well as its potential in the treatment of bone tissue and cancer, among other areas. By exploring its unique properties and versatility, the chapter highlights the fundamental role of this clay in sectors such as health, beauty, and materials technology.

Keywords

Smectite, Montmorillonite, Environment-Friendly, Phyllosilicates, Low Cost

Contents

Materials Research Foundations 173 (2025) 124-144
https://doi.org/10.21741/9781644903452-5

1. Introduction

Clays and clay minerals are hydrated alumino-silicates that belong to the phyllosilicate family. [1]. Bentonite is the commercial name given to a group of natural clays composed of a mixture of minerals belonging to the smectites. Although it has no fixed chemical composition, bentonite consists mainly of SiO2 and Al2O3 plus various metal oxides. In PUBCHEM it is found under the CID (compound identification) record 517273 and the molecular formula Al2H2O6Si. Its composition varies according to the geographical region of origin and how the bentonite is extracted [2, 3]. However, it is common for Bentonite clay to have several layers. The space between these layers is filled with interchangeable cations. [4].

Smectites are phyllosilicates of the 2:1 type, so they are made up of two tetrahedral layers surrounding an octahedral layer, which means that these clays have a negative charge in the elemental layer and are pH charge independent. This group differs from kaolinites, which are formed by a 1:1 structure, consisting of a tetrahedral layer with a central silicon and an octahedral layer with aluminum, in addition to having a neutral charge. Another interesting feature of smectites is that the structure of this group is somewhere between crystalline and amorphous. The structural distribution of smectite and the presence of cations within it allows them to be easily exchanged and allows them to hydrate easily, which leads to their ability to swell in the presence of water, as shown in Fig. 1. These properties allow clays of this class to be used in a variety of applications. [5].

Among the components that make up Bentonite, the most common cations are sodium and calcium. Sodium Bentonite (Na-Be) has a greater swelling capacity than calcium Bentonite because it has monovalent cations and therefore requires less energy to hydrate the cations, but Calcium Bentonite (Ca-Be) is the most prevalent in the world. The most abundant component of this clay is Montmorillonite (MMT), which is predominantly negatively charged due to isomorphic substitution and is directly related to water absorption and swelling capacity. In addition, the ability to adsorb positive charges from toxins comes from the electronegative charge behavior of clays when they come into contact with water [6].

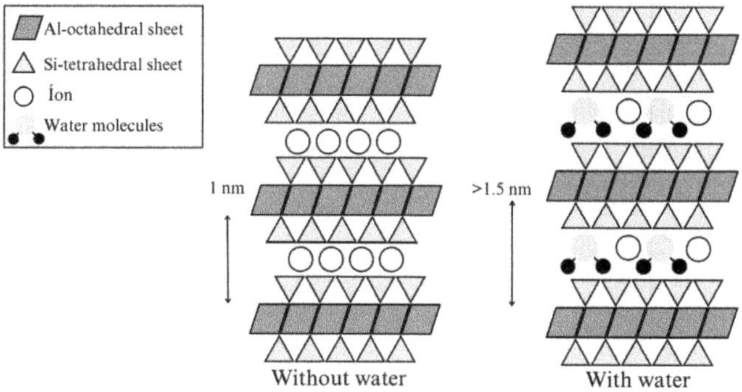

Figure 1. Influence of the presence of water in the space between the Smectite layers

The presence of MMT in Bentonite clay was discovered in 1890 in the United States of America (USA). MMT is on PUBCHEM under the CID 71586775, where its molecular formula is Al2H2O12Si4. The most commonly found MMT is sodium MMT (Na-MMT), containing Na between the clay layers [2]. Bentonite MMT is a smectite that has properties of swelling, adsorption, agglutination, as well as particle capture, as when in contact with water this component attracts the smallest particles and decants to the bottom. These properties allow this component to be used in various areas such as engineering, cosmetics, and pharmaceuticals. However, one of the disadvantages of MMT for use in humans is its inability to pass through the mucous membranes of the gastrointestinal tract, limiting its use to external areas [7].

In this way, clays can be widely used in the engineering, pharmaceutical, and food industries. One of the most common applications in the pharmaceutical industry is its use as an excipient and as an active pharmaceutical ingredient (API). As an API it is used in anti-inflammatory treatments, cancer treatment, and dermatological purposes, among others. As for its use as an excipient, it is widely used in various pharmaceutical forms liquids, solids, and semi-solids. The latter group benefits mainly from this clay's thixotropic capacity when in contact with water [2, 8].

The great industrial value of this product is due to its ability to be applied in various areas. Bentonite clay is widely exploited throughout the world. The distribution of global extraction, shown in Fig. 2, indicates that the USA is the largest extractor of bentonite clay, accounting for around 25% of global production, followed by India and China, which together account for almost 27% of production. [9, 10]. Regardless of where it is extracted, this clay usually has a high swelling capacity, low permeability, and adsorption capacity [3].

Humans have been exploring the properties of this clay for around 3,000 years. The first records range from its use as an excipient to its medicinal use. Its use for medicinal purposes began after the practice of geophagy by animals was observed. This was because this clay can provide essential minerals to the organisms of these beings, as well as having a high capacity to eliminate toxins, due to the presence of negatively charged sites that react with the positively charged sites present in toxins when in contact with the acidic environment of the stomach [8, 11].

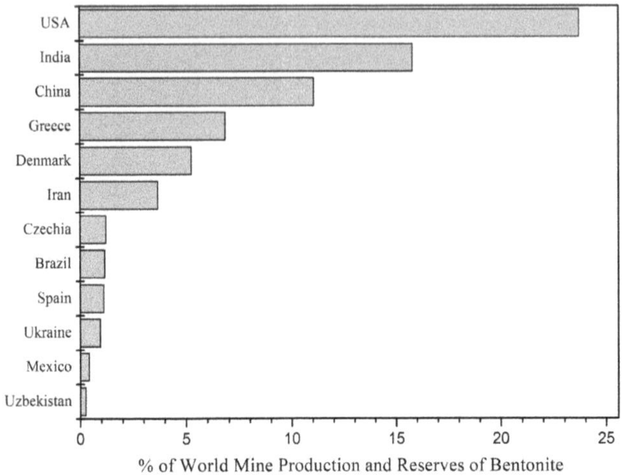

Figure 2. Distribution of the main countries responsible for Bentonite exploitation

To improve the properties of bentonite, this clay can be exposed to additional treatments that can modify its structure and increase its textural properties. Among the treatments used are chemical activation, mechanochemical activation, intercalation, thermochemical activation, ultrasonic or microwave irradiation, and acid and basic treatments. In acid treatment, the most commonly used agent is H2SO4, which provides a greater surface area than pure clay. In addition, it is a reliable technique that causes controlled dissolution and the promotion of active sites, which can lead to a promising catalytic application. [4].

Given all these unique characteristics of Bentonite and its main properties, such as its ability to swell, adsorb, and interact with fillers, this chapter will explore in detail its applications in the pharmaceutical field. Bentonite will be used both as an excipient in various formulations and as an API, highlighting its functions in controlled release systems, its anti-inflammatory, detoxifying, and cholesterol-reducing action, as well as its therapeutic potential in cancer treatments and bone tissue regeneration. The following sections will detail how these properties can be used to develop solutions for the medical sciences.

2. Pharmaceutical Formulations

2.1 Excipients for Pharmaceutical and Cosmetic Products

Clays and their minerals are used in the pharmaceutical and cosmetics industry as excipients, disintegrants, binders, diluents, emulsifiers, thickeners, and anti-caking agents. [12]. Evaluating the potential use of Bentonite as excipients involves carrying out tests to assess the mineralogical composition, such as X-ray diffraction (XRD), energy dispersive X-ray (EDX), Fourier transform infrared (FTIR) and morphology by scanning electron microscopy (SEM). Other essential tests are the ones from pharmacopeia, mainly regulated by the European and American pharmacopeia, such as the evaluation of cation exchange capacity (CEC), specific surface area, pH, powder flow characteristics, swelling capacity, sedimentation volume, rheological properties, and microbiological tests. These tests make it possible to assess both the potential of Bentonite and parameters that can be adjusted in large-scale production. Depending on the purpose of the product in which this clay will be used, additional tests can be applied to investigate more specific parameters, such as CEC [13, 14, 15, 16].

In addition, the potential of clays can vary depending on where they are exploited. Algerian bentonite from the Maghnia region, for example, has constituents that are not ideal for use in the pharmaceutical and cosmetics fields. However, after the application of treatments such as Na+ ion exchange and sedimentation, as carried out by Babahoum and Hamou [14], in addition to the complete removal of quartz and feldspar, its properties were also enhanced. The purified Bentonite had a more porous surface compared to the raw one, which may have improved the CEC, a drastic increase in the percentage of fine particle size, and an increase in acidity, which enhances the clay's adsorbent capacity. The USP's rheological properties and hygienic and pharmacopoeial quality tests showed results within the specifications of the pharmacopeia and regulatory institutions. Thus, purified Algerian Bentonite is qualified for use in cosmetic products such as powders, creams, and emulsions, and has a high capacity for use as a pharmaceutical adsorbent.

In the case of Iranian Bentonites, even without a purification process, they demonstrate good properties for use as a pharmaceutical material, as seen in Modabberi et al. [15] in Iran, these clays are associated with volcanic rocks of the stratified type and include thin layers of gypsum. This alteration is irregular and is more intense in some deposits which show generalized calcification. In an analysis of different mines, those from the Zagros (ZBe) and Chah-Golestan (CGBe) regions stood out for meeting the requirements of the American and European pharmacopeias in terms of mineralogical composition and microbiological study. The SCC and specific surface area showed that the Bentonite obtained from ZBe has the highest values and is therefore the best material among those analyzed for pharmaceutical uses in which the specific surface area is relevant. In addition, all the dispersions produce flocculated systems when dispersed and thixotropy, which makes it possible to use them in liquid and semi-solid products. Thus, the samples demonstrated their ability to be used in topical dosage forms as rheological additives.

One of the current concerns is the reuse of industrial waste, and the green extraction of polyphenols from Bentonite waste is an example of this practice. Bentonite can be pre-treated and extracted for use in excipients enriched with functional polyphenols and characterized to produce functional antioxidant excipients for cosmetic and pharmaceutical purposes, as in Di Prima et al. [16]. The addition of the PEG 200 polymer enhanced the extraction of polyphenols and phenolics and showed greater antioxidant power by DPPH, with 20 μL of extract being as powerful as 30 μg of standard gallic acid, and it also proved to be more cytocompatible than extracts with propylene glycol. In addition, the antioxidant trend proved to be constant, rather than a rapid consumption of the initial radicals. A stability test of the extracts proved that storing them for 6 months at 4°C allowed the extracts to remain antioxidantly effective. These findings have brought closer the possibility of using residual Bentonite as a new starting material for obtaining polyphenol-enriched raw materials for use in pharmaceutical and cosmetic products.

One of the most commonly used functions for Bentonite as an excipient is to alter the release of the drug. This alteration usually leads to a more sustained release, which can be caused by the inherent characteristics of this clay when in contact with the media. However, in some cases, interactions can occur between the drug and Bentonite or one of its components, which can result in an immediate release profile. In both cases, Bentonite is capable of altering the release profile of drugs and can help deliver them according to what is desired and expected in the treatment. It is therefore important to analyze the characteristics of formulations prepared with this clay to correctly map the profile of the product developed. These analyses can be carried out employing dissolution profile studies at different pH levels. [13, 17].

Bentonite can be used as an excipient in tablet production to influence the release profile. This is due to the swelling capacity offered by this clay. This fact was proven in a study in which three formulations were produced by direct compression, containing three drugs with different solubilities, paracetamol (PA), diclofenac sodium (SD), and metformin HCl (MET). To analyze these formulations, tests according to USP were used, in which all parameters were in compliance with USP 34, but the three formulations had low and very low fluidity. A dissolution profile study was carried out, at pHs of 1.2, 4.5, and 6.8, in which it was observed that at all pHs PA and MET at pH 6.8 showed sustained and immediate release, while SD at pH 6.8 showed an extended-release profile. The PA and SD profiles were obtained by the tablets swelling in the medium, while MET probably had an intensification of disintegration due to the interaction between Si-OH and H2O. [13].

Hydrogels are another example of a formulation in which Bentonite can be applied. The analysis of three hydrogels that contained diclofenac acid nanocrystals (DCF) was based on two clays and a 50:50 ratio of both in the third formulation: Bentonite and Palygorskite (PAL), Be-DCF, PAL-DCF, and BD: PAL-DCF. The rheological analysis of the hydrogels showed pseudoplastic non-Newtonian flow behavior, where PAL-DCF had a viscosity seven times higher than Be-DCF. In a structural assessment, Be-DCF was more porous. EDX suggested the presence of silicon and calcium and small amounts of chlorine related to DCF. Thermal analysis suggested that Bentonite acted as a thermal protector to DCF, by delaying the thermal event concerning free DCF. At 8h, Bentonite allowed a constant release of DCF, which may have been faster than in PAL-DCF due to the lower viscosity which facilitated the drug's exit from the hydrogel, as well as providing good permeation into the skin. Bentonite in hydrogels therefore has great potential as an alternative strategy for increasing the topical bioavailability of DCF and improving the drug's solubility and dissolution rate. [17].

2.2 Delivery System

Bentonite is an excellent carrier and has several benefits, such as cost, stability, and adsorption capacity. In addition, this clay can improve the mechanical properties of nanocomposites [18]. MMT, one of the main constituents of Bentonite, has a compact structure that promotes a highly ordered network arrangement and a flat, regular surface. The crystalline structure of MMT allows for greater surface area for chemical activation and CEC, making it suitable for transporting organic cationic compounds, such as many antimicrobial agents. Bentonite and MMT are natural nano-clays that have a large internal surface area which allows them to act as drug delivery systems. When incorporated, the drugs are found between the layers of the clays, and release normally occurs by the mechanism of ionic charge exchange. Their use in oral administration has attracted increasing attention, since the drug molecules on the surface of the clay layers, when in an aqueous medium, due to their ability to swell, lead to the gradual release of the drug into the physiological environment [19, 20].

 Materials Research Forum LLC
https://doi.org/10.21741/9781644903452-5

The use of bentonite in nanocomposites was evaluated by Behera et al. [21], in which the antimicrobial activity of bentonite nanocomposites against *Escherichia coli* was assessed. The nanocomposite was obtained by the precipitation method, obtaining an average nanoparticle size of 450 nm, with an elongated shape. Ge et al. [20], evaluated the antimicrobial activity of silver nanoparticles (AgNP) associated with an MMT modified with lectin (AgNPs@Mt-LEC) against the strains of *Staphylococcus aureus* (ATCC 25923) and *E. coli* (ATCC 25922). It was observed that after preparing the nanoparticles, the spacing between the MMT layers was increased, which suggests that the AgNPs were accommodated in this space. The nanoparticles formed were spherical and interspersed between the MMT layers, ranging from 10 to 30 nm in diameter. Thus, the AgNPs were uniformly introduced into the MMT, and the nanomaterials formed.

3. Pharmacological Activities

3.1 Detoxification properties

Exposure to xenobiotics is constant throughout a human's life, present in products and foods. For this reason, the human body has developed strategies to deal with these toxins and reduce their impact, called xenobiotic detoxification or xenobiotic metabolism. This detoxification is not a reaction, but a process involving multiple reactions and multiple participants. In addition, mycotoxins such as Aflatoxins (AF), which are extremely lethal secondary fungal metabolites, can be found in food crops all over the world, which can result in the emergence of serious diseases such as cancer in which AFs have a grade 1 risk, immunological and neurological disorders, and liver diseases. One of the detoxification strategies is adsorption by Bentonite, due to its high capacity to adsorb materials such as bacteria, viruses, heavy metal ions, and other toxins in humans, as well as its low cost, non-toxicity, and ease of modification. As the detoxification method takes place through adsorption, tests such as pH are extremely important for evaluating the behavior of Bentonite. [22, 23].

The modification of Bentonites can optimize the surface properties of this material, such as adsorption, which directly influences the detoxifying activity of the material. Saeed et al. [23] this study analyzed organo-bentonites modified with benzyl-tri-n-butyl ammonium bromide (BTB), benzethonium chloride (BTC), and dioctyl sodium sulfoccinate (DSS) for mycotoxin adsorption. To optimize the adsorption process, some parameters were analyzed in isolation. The pH analysis showed that the maximum removal of mycotoxins occurred at pH 5.0 due to the effective interaction of mycotoxins with the surface of the protonated modified clay, while the ideal contact time was 30 minutes. The Langmuir adsorption isotherm model was the best fit, suggesting that adsorption occurs on a homogeneous monolayer surface without any interaction with adjacent adsorbed molecules, and the values of the dimensionless constant RL were <1, demonstrating favorable adsorption. Maximum adsorption occurred at low temperatures, since at high temperatures the Van der Waal interaction is weakened, and the values observed suggest spontaneous adsorption at some temperatures.

The ingestion of toxic iron components is quite common, with around 5,000 cases occurring annually in the USA alone. Bentonite can be an alternative measure in acute treatment, mainly because it can prevent iron from entering the bloodstream. In Moosavi et al. [24] the efficacy of MMT in acute iron poisoning in rats was evaluated, using MMT obtained from Bentonite from the city of Ghaen in South Korea and purified. The doses tested 0.5 g/kg, 1.0 g/kg, and 1.5 g/kg were able to adsorb and remove Fe, but 1.5 was not ideal due to the reduction in the rats' activity. The study suggested two probable mechanisms, the first related to Fe's ability to bind to Na-saturated MMT particles, reducing Fe bioavailability. The other mechanism would be due to the decreased absorption of Fe as the clay particles mix with the intake, creating a coating that reduces the availability of the element. In addition, the most common side effect of activated charcoal administration, constipation, was not present in the tested doses of the formulation.

The presence of FA is a more frequent concern in agriculture, whether in animal husbandry such as livestock and poultry or the production of food for humans, as the microorganisms that produce this mycotoxin proliferate in cereals. These mycotoxins can cause a reduction in the growth rate of animals, mortality, and consequently economic losses. Type B1 FA (AFB1) is the most toxic of the FAs and can cause acute toxicity, teratogenicity, mutagenicity, and carcinogenicity, although types B2, G1, and G2 can also be highlighted. One of the methods currently used to eliminate traces of toxins without reducing the nutritional value of the food is the addition of clay. [25, 26, 27].

Activations in the clays can be factors that enhance their predominant properties, for example, a highly activated Na-Be (Na-BeAt) can present a different adsorption than ordinary bentonites. In Wang et al. [26] a Na-BeAt was evaluated for its aflatoxin sorption capacity, comparing it to Ca-Be. Which, the absolute enthalpy values in the interaction between AFB1 and the Na-BeAt indicated that AFB1 is chemically firmly adsorbed to the clay surfaces and after heating, a large amount of AFB1 binds within the negatively charged interlayer of these clays and small amounts bind at the edges and basal surfaces. Based on the dosimetry results, the predicted adsorbent concentration for Ca-Be and Na-BeAt was 2.4 g and 0.3 g of clay/day, respectively, of a safe daily intake. To indicate toxicity, the hydra bioassay was used, and 100% mortality was achieved in 92 h with 20 mg/kg of AFB1 in the medium. Thus, Na-BeAt is highly effective for binding AFB1 and may be applicable for formulations aimed at removing these aflatoxins from the body.

The presence of FA in animal feed is a frequent concern among poultry farmers. Evaluating the impact of the presence of Bentonite in the treatment of affected animals makes it possible to gauge the potential for reducing the damage caused by FA in feed. In Mgbeahuruike et al. [25] the detoxification of poultry feed contaminated with FAs was analyzed by comparing Bentonite with activated carbon (AC) and Fuller's earth (FuE) for birds contaminated with FAs in Nigeria. Among these adsorbents, Bentonite had the highest detoxification capacity for AFG1 and AFB1. In addition, Bentonite was more effective in preventing the histopathological effects caused by aflatoxicosis, and both adsorbents were able to increase the prey of the chickens by 63-100% of their weight when compared to the control group that was fed feed contaminated with AFs.

In agreement with these results, the study by Mahrose et al. [27] In which Japanese quails were analyzed, found that the birds in which 1% Bentonite clay was added to their feed obtained an improvement in growth performance and a reduction in the effect of AFB1 when compared to the groups without added clay. Another positive effect observed was weight gain, which was related to the clay's ability to increase food retention time in the chicks' intestines or to increase digestibility due to the presence of Bentonite. Serum concentrations of biochemical components such as total protein, albumin, total cholesterol, creatinine, and uric acid were also observed, making it possible to state that the addition of Bentonite reduced the effects caused by AFB1.

3.2 Cholesterol-Lowering

The prevalence of overweight and obesity in the world has caused increasing concern about the influence of these conditions on the health of the population. Eating fatty foods changes parameters such as cholesterol over time, which can lead to cardiovascular disease. Cholesterol homeostasis in the body is maintained by the balance of endogenous synthesis, intestinal absorption, and secretion of bile acid and cholesterol. As it is an insoluble molecule, its intestinal absorption requires several processes, so the mass of cholesterol excreted and the efficiency of intestinal absorption of cholesterol are routes for new therapies aimed at reducing cholesterol levels, such as intestinal absorption inhibitors, which act on the absorption of dietary cholesterol. Bentonite and its main component, MMT, have a large internal surface area, high cation exchange capacity, adsorption, swelling, and protection of the gastrointestinal mucosa, so they can act to adsorb excess lipids. [22, 28].

In the study by Turgut et al. [22] analyzed the development of cholesterol-lowering and detoxification formulations using bentonite and herbal ingredients, investigating the effects of potentially effective herbal ingredients and combinations of food-grade bentonite clay on detoxification and normalization of blood cholesterol levels to prevent cardiovascular diseases and develop effective formulations. In this study, flaxseed, psyllium seed, turmeric, and grape seed were evaluated. Bentonite was food-grade and supplied by the Alya Mineral Company in Turkey. Mice with this diet-induced condition were used to analyze its activity against hypercholesterolemia. It was determined that serum LDL and triglyceride levels were significantly reduced in the groups containing Bentonite in the blood samples compared to the group of hypercholesterolemic mice. In addition, Bentonite increases the fecal excretion of lipids by immobilizing lipids in the gastrointestinal system, which demonstrates that the presence of this component in a formulation can potentiate the decrease in LDL levels.

Adsorption capacity is one of the main properties observed in Bentonite clay when it comes to reducing cholesterol absorption. In García-Guzman et al. [28] this capacity was evaluated among five Bentonite and MMT clays, namely MMT nano-clay (NMMT), MMT K10, MMT K30, and two Bentonites treated with H2SO4 and lysine. Among the clays tested, the MMT clays showed the best profiles in terms of cation exchange capacity, surface area, and porosity. Thus, when the three clays were tested for cholesterol adsorption, although both showed adsorption, the MMTK10 clay showed enhanced double adsorption, even though cholesterol is a neutral molecule and the attraction to the surface is weak, the clay is soon adsorbed by hydrophobic interactions. In addition, the data obtained showed that both clays conformed to the Freundlich model, where the adsorption process occurs in multilayers with interactions between the adsorbed molecules, the model suggesting that the adsorption energy decreases exponentially as the adsorption sites are completed.

In addition to the adsorption capacity, it is necessary to understand the profile of the clay concerning a high-fat diet (HFD). This investigation was carried out by Xu et al. [29] Where MMT was investigated for its potential to act as a preventative agent for obesity and hyperlipidemia due to its ability to adsorb lipids from the diet and increase their excretion in the digestive tract. The rats tested had diet-induced hyperlipidemia, so serum lipid concentrations of total cholesterol, triglycerides, and LDL-C were lower in the HFD+MMT-treated group than in the HFD-fed group on the 15th and 25th days of treatment, suggesting that MMT is capable of reducing the extent of HFD-induced hypercholesterolemia. Fecal water content was regulated with the HFD-MMT group

Materials Research Forum LLC
https://doi.org/10.21741/9781644903452-5

and triglyceride and total cholesterol contents were higher in the HFD+MMT with a greater difference in triglycerides. The HFD+MMT also showed the ability to reduce weight gain when compared to the control groups. Thus, MMT could be an excellent nutraceutical for adsorbing excess lipids during the consumption of a fatty diet.

3.3 Anti-inflammatory activity

Anti-inflammatory activity is one of the main actions of Bentonite and may be associated with the clay's structural factors. The swelling of this clay is a property that is related to its expansive capacity, which can result in physical occlusion and cause a decrease in the migration of leukocyte cells to the site of inflammation, reducing the inflammatory process and its signs and symptoms. In addition, one of the main components of Bentonite, MMT, can interact with signaling mediated by COX-2 and interleukin 10, which is an anti-inflammatory cytokine, thus playing a role in direct action. [30, 31].

In Silva et al. [31], the anti-inflammatory action of bentonites collected in India, Argentina, Hungary, and Indonesia in response to edema in the ears of mice was investigated. To this end, six mice were used, in which 12-O-tetradecanoliforbol-13-acetate (TPA) was applied under anesthesia to induce an inflammatory response. A dispersion consisting of 1 mg of bentonite diluted in 20 μL of a 1:1 solution of water and acetone was prepared and 10 μL of this dispersion was applied to the right ear of the mice. The result was a suppression of edema formation, possibly associated with physical occlusion, making it difficult for leukocyte cells to move into the inflamed area. Although all the bentonites showed anti-inflammatory activity, the one obtained in Argentina showed an anti-inflammatory response 3.7 times greater than the one from Hungary, due to the greater expandability of the Argentine bentonite. This shows that the parameter that most influences this activity is the clay's capacity for expansion and not the place of extraction.

Meanwhile, in the study by Lee et al. [30], explored the anti-inflammatory activity of bentonite *in vivo* and *in vitro* studies. As a result, they found that both bentonite complexes reduced the expression of inflammatory markers such as interleukin 1 beta and cyclooxygenase 2 (COX-2). In addition, there was a reduction in the synthesis of prostaglandin E2 and an increase in the expression of interleukin 10, an anti-inflammatory cytokine. In the *in vitro* test, 3D4/2 cells showed a decrease in the expression of interleukin 1 beta, COX-2, and prostaglandins, as well as a small reduction in the expression of interleukin 10. On the other hand, in the HaCaT lines there was also a decrease in the expression of interleukin 1 beta, COX-2, and prostaglandin, but, unlike the 3D4/2 cells, there was a considerable increase in the expression of interleukin 10. Based on the results obtained, the authors pointed out that, due to bentonite's properties, it could be used as a basis for the preparation of medical and cosmetic products.

3.4 Bones treatment

MMT, especially in the form of Ca-MMT, shows positive activity in bone formation, promoting osteoblast differentiation and showing good alkaline phosphatase (ALP) mining, which is a biomarker of osteoblastic phenotypes. [32, 33]. When MMT is incorporated into polymer matrix materials, it facilitates the biomineralization and osteogenic differentiation of mesenchymal stem cells, which are necessary for bone formation. This effect is attributed to the bioactive compounds present in its composition, such as Ca, Mg, and Si, which stimulate the deposition of hydroxyapatite resulting in the differentiation of stem cells into osteogenic lineages and regulating

 Materials Research Forum LLC
https://doi.org/10.21741/9781644903452-5

cell behavior through the process of biomineralization. The bone-like mineral layers help to integrate the implants and bone tissue, which provides a favorable environment for stem cell adhesion, growth, and differentiation. [34].

Kim et al. [32], the presence of osteoblastogenesis and osteoclastogenesis was investigated using *in vivo* and *in vitro* tests, using clay collected in the Gyeongiu region of South Korea, which was subjected to a process of purification and isolation of montmorillonite (MMT). Cultures of osteoblast-like cells (MG-63) and osteoclasts (RAW 264.7) were carried out. To determine alkaline phosphatase (ALP) activity, MG-63 cells were seeded in 96-well plates. For 7 days, MMT was added at the indicated concentrations. The *in vivo* test used female Wistar rats that had undergone ovariectomy (OVX). It was observed that when treated with MMT, the group showed an increase in ALP activity. There was a 350% increase in mining in the MMT-treated group compared to the control. In addition, the tests showed a decrease in 5osteoclastogenesis and greater differentiation of osteoblasts in the cells treated with MMT. In general, an increase in the expression of markers related to the process of bone regeneration, formation, and metabolism was observed. Due to its significant anti-osteoporosis activity, MMT plays an important role in the treatment of bone metabolism disorders.

Due to the properties contained in MMT, Wu et al. [34] determined the scaffolding activity of nanocomposites formed with hydroxyethyl cellulose (HEC) and soy protein isolate (SPI), with the addition of MMT nanosheets. The study showed that the nanocomposites had good viability and cytocompatibility. *In vitro* activity showed an increase in biomineralization and the deposition of crystallized minerals. It was observed that ALP activity increased significantly at concentrations of 2% and 3% MMT. When stained with alizarin red S, these concentrations showed better osteogenic performance, as well as greater calcium deposition. *In vivo*, when treating rat calvarial defects with the nanocomposites, there was greater new bone formation in the defective regions with MMT compared to the same nanocomposites without MMT. These materials therefore appear to be a promising alternative for bone tissue engineering applications.

The incorporation of MMT into biomaterials can favor bone formation. In Singh et al. [33] this activity was evaluated by adding hydrophilic nano-montmorillonite to a solution of benzoyl peroxide with diethylene glycol dimethacrylate, followed by xanthan gum. This material was heated in microwaves, followed by cooling to synthesize the biomaterial. Biocompatibility was tested with osteosarcoma (SaOS-2) and osteoblast-like (MG-63) cells, and alkaline phosphatase activity was assessed using SaOS-2 cells. *In vitro*, the biomaterial showed greater ALP activity than the control, which may favor osteogenesis. In addition, its biomining showed high calcium deposition. In the future, the material could be used to prepare 3D scaffolds for bone reconstruction.

3.5 Cancer treatment

Currently, some clay minerals such as MMT have been tested in systemic treatment. One of the factors that favor its use against cancer is its size, which ranges from 1 to 30 nm to several micrometers, so MMT has been successfully injected into tumors or tumor resection sites, resulting in increased efficacy and reduced toxicity of other therapeutic agents. In addition, MMT is a promising agent in the treatment of various types of cancer, due to its ability to inhibit the proliferation and survival of tumor cells. This property is attributed to its intrinsic activity, with antiproliferative action against various cancer cell lines, such as HepG2, HT-29, MRC-5, MDA-

MB-231, MCF-7, and SK-Mel-28. MMT's activity is related to its specificity for cells with a high proliferation rate [35, 36]

Studies such as Sabzeravi et al. [35] investigated the activity of MMT against breast tumor cells. MMT was isolated from bentonite by adding 6 g/L bentonite in water, followed by sonication, stirring, and purification with NaCl. The antiproliferative effect was assessed using 3-(4,5-dimethylthiazol-2-yl)-2,5-diphenyltetrazolium bromide (MTT). MRC-5 cell lines and breast tumor-specific cells MDA-MB-231 and MCF-7 were used. The results showed that, at low concentrations, MMT can induce cell growth factors, while at higher doses it inhibits cell growth such as HepG2, and HT-29 by inducing cytotoxicity in the MDA-MB-231, MCF-7, and MRC-5 cell lines. MTT showed a more aggressive action on the MDA-MB-231 strain due to its high proliferation rate. Thus, in the parameters evaluated, MMT appears to be a promising alternative for the development of anti-tumor drugs for breast cancer.

In the work of Abduljauwad et al. [36], evaluated the ability of Na-MMT and palygorschite nano-clays in the treatment of melanoma. For the research, Na-MMT and palygorschite were purified, ground, and centrifuged to obtain nanoparticles. A suspension was then prepared containing 20 mg of each nano-clay in 10 mL of buffered saline solution. The *in vitro* activity of the nano-clays was determined against the SK-Mel-28 melanoma tumor cell line. For the *in vivo* test, Nude mice were used, to which the tumor cells were added. A significant reduction in viable cells was seen after 3 days, with the application of different concentrations of the nano-clays. The higher the concentration of both Na-MMT and palygorschite, the lower the viability of the SK-Mel-28 cells. On the other hand, there was no damage to human epidermal melanocytes. It is therefore compatible with human cells and can be used to treat other types of cancer.

The efficiency of the association of cytochrome c (cytC) with MMT was investigated by Hristova and Zhivkov [37], where the process for obtaining the suspension was carried out by adding the MMT k10 powder to water, followed by adding different concentrations of the MMT solution to an aqueous solution of ferric equine heart with cytC. A cytC-MMt suspension was thus obtained. Metastatic colorectal adenocarcinoma cells (ATCC CCL-227) were used to determine activity. It was observed that cytC-MMT at a concentration of 10:3 mg/mg showed a cytotoxic effect on cancer cells of 95% after 94 hours, the higher the concentration, the greater the cytotoxic effect of cytC-MMT. In addition, this study showed that cytC and MMT had no activity when free, only when they were complexed, which indicates a synergistic activity of the two compounds.

3.6 Hemostatic activity

Hemostasis is one of the stages in the wound healing process, in which uncontrollable bleeding and bacterial infection are crucial challenges. Hemostatic materials are used clinically to stop bleeding and prevent infection, such as conventional dressings, microporous aluminosilicates or polymer-based materials such as sponges and films. The problem with current materials is that in the case of conventional dressings, they can lead to lengthy treatments, microporous aluminosilicates can cause serious exothermic reactions, inflammation or thrombosis and polymer-based materials can lead to infections by promoting a moist environment. MMT, the component with the highest proportion in Bentonite, is considered the most effective hemostatic among natural phyllosilicates, mainly due to its water absorption properties and ability to adsorb bacteria and bacterial enterotoxins. In addition, studies show that MMT can stimulate the activation of coagulation factors in the blood, promoting the formation of blood clots and stopping bleeding.

However, the use of free MMT is not recommended, as it can cause problems such as thrombosis, which is why the FDA has limited the use of hemostatic products made from smectites since 2007 [38, 39] .

In order to reduce the risk of side effects, MMT can be complexed with other compounds, as analyzed by Li et al. [39] where a sponge composed of MMT and graphene (GMCS) was evaluated for the treatment of hemostasis. The *in vivo* model of hemostatic performance showed that after compression with the GMCS with the bleeding site there was complete hemostasis in 85 s. The dynamic coagulation test showed that when in contact with fresh arterial blood, the absorbance value of hemoglobin decreased sharply within 2 min, making coagulation 2x more effective than the control group. This activity is probably due to the rapid absorption of plasma and enrichment of blood cells on the surface of the sponge and the absorbed plasma coming into contact with the MMT. The thromboembolic risk showed that no thrombus or blood clot was formed in the vessels, especially near the injured arteries treated with GMCS, and the presence of MMT improved the hemostatic mechanism compared to free GS. Hence, this graphene composite sponge can act as a platform for introducing different hemostatic mechanisms.

Another innovative hemostatic material is the complex of silver MMT (Ag-MMT) and agarose (AGA) in a three-dimensional mesh gel (AMA) studied by Zhang et al. [38], where the association with agarose made it possible to reduce the risk of side effects and make the compound more stable and biocompatible. AMA efficiently promoted blood coagulation and prevented blood loss, while MMT in the system improved hemostatic performance by 15%. The antibacterial action occurred through silver, but the presence of this component in the MMT interlayers promoted a slow and lasting release. *In vitro*, AMA showed negligible toxicity even at 500 μg/mL and had a lower hemolysis rate than AGA, indicating that it is a safe and efficient material. In the *in vivo* mouse liver model, the hemostasis time and blood loss for wounds treated with AMA were lower than that of conventional dressings. In addition, histological analysis showed that AMA provided excellent wound healing with a faster epithelial regeneration process and was superior to the groups of isolated components.

4 Skin Applications

4.1 Pelotherapics

Pelotherapy is becoming increasingly popular and, with this progress, it is necessary to understand the use of peloids, that is a mixture of fine-grained natural materials of geological and/or biological origin, mineral or sea water and, usually, organic compounds with biological metabolic activity. As it is made up of a combination of solid and liquid phases, a period of maturation is necessary to promote interaction between the components. To prepare these products, it is essential to understand their thermal behavior, defined by parameters such as specific heat, thermal capacity, and cooling kinetics. [40, 41]

The work of Armijo et al. [40], carried out the characterization of eight samples rich in smectite (M1 and M2 activated with Na^+, M5, M6, and M7 containing mainly divalent cations in the interlayer), palygorskite (M3), sepiolite (M4) and kaolin (M8). It was observed that the palygorschite sample had the lowest specific heat capacity and the sepiolite had the highest, confirming the values already reported in the literature. Another parameter analyzed was the ability to release heat slowly based on Newton's law of cooling, which is proven when the temperature between the hot and cold object is not very high, as is the case with peloids or pastes. M1, M2, M3, and M6 reached a Tr of approximately 9 minutes with 70% water, and M1 and M6 followed by M5 and M4 were the samples that obtained the highest Tr for this water content, releasing heat more slowly. In addition, the water retention capacity of clays and the presence of fine-grained clay minerals are essential for understanding thermal behavior. Thus, bentonite activated by Na^+ showed better properties for pelotherapy purposes.

These mineralogical and thermal properties, such as the specific heat capacity and heat dissipation of the peloids, were also investigated by Karakaya et al. [42], in 20 samples of peloids from spas in Turkey, given their therapeutic importance. Among the samples analyzed, the most abundant mineral was Ca-montmorillonite. Dissipation and specific capacity were determined by checking the temperature drop time and comparing it with the average time taken to apply the dressing to the target site of 15 to 20 minutes, proving that the samples are suitable for pelotherapy. In this process, the recommended temperature for the peloid is above body temperature by at least 5°C. Thus, it was noted that sepiolite has a long dissipation period, while illite and kaolinite have a short period, indicating that sepiolite and mixtures of Na-MMT and Ca-MMT are more suitable for therapeutic use, including treatment of musculoskeletal and neurological problems.

4.2 Cosmetics

Widely used in the cosmetics industry, clays contain a large number of minerals that can greatly improve the health and appearance of the skin. These include calcium and potassium, which improve circulation and strengthen tissues, zinc and magnesium, which rejuvenate and enliven the skin, iron, which acts as an antiseptic and promotes cell renewal, and silicon, which promotes hydration and a calming effect. Titanium, a prominent mineral in the field of cosmetology, has been used due to its great ability to reflect UV radiation depending on the size of the particles and has therefore been used in photoprotection formulations. Thus, smectite clays are used in the formulation of photoprotective cosmetic emulsions, given their great capacity for absorbing substances, which is broad protection against the effects of the sun, as well as being able to be applied to facial rejuvenation and improving the skin's barrier function. Another extremely important point is their wide availability and good cost-benefit ratio. [43, 44, 45].

In the work of Duarte et al. [45] prepared photoprotective cosmetic emulsions with Brazilian smectite clays from natural clays - Branca (BRA) and Verde Inferior (VI) - and commercial clays - Brasgel (BRG) of the bentonite type and Cloisite (CL) of the Na-MMT type. When checking the viscosity at different speeds, it was observed that the viscous behavior of the emulsions decreases as the speed of rotation increases, characterizing them as non-Newtonian pseudoplastic fluids, ideal for topical cosmetic formulations, given their better application and spreadability. He also pointed out that the pH was adequate, close to 7.0 and the differences in values between the formulations were between 6.6 and 7.5. In the analysis of the *in vitro* SPF, which were generally low but within compliance, it was observed that the BRA and CL cosmetic emulsions, prepared from emulsions without chemical UV filters, were the lowest, indicating that these clays do not have the potential to absorb UV radiation. Thus, it is clear that the natural Brazilian commercial clays of Bentonite and MMT can be used to prepare cosmetic emulsions and expand the cosmetics industry by offering natural and economical solutions.

MMT has also been used to reduce the dermal absorption of benzene, toluene, and xylene (BTX), which are volatile organic compounds that often appear in wastewater due to various industrial activities and accidents, and can cause various health problems. Prolonged exposure to contaminated water can compromise the skin barrier, facilitating the absorption of BTX and other chemicals. Thus, Wang and Philips [43] investigated the addition of MMT calcium sodium clays enriched with chlorophyll (at 5%). To characterize the MMT sorbents, the apparent density and moisture percentage were checked, and it was observed that the addition of lipophilic chlorophyll increased hydrophobicity and, consequently, the attraction of hydrophobic compounds. From the XDR patterns before and after chlorophyll, it was noted that it acts to support the layer-network structure, increasing the basal spacing. Through characterization, it was observed that the formulations have characteristics suitable for topical application with a uniform shear on the skin, which are desired points for cosmetics.

Hence Wang et al. [44] examined the performance of Ca-MMT and acid-processed MMT (APM) clays in the adsorption and detoxification of mixtures of glyphosate (GLP) and aminomethylphosphonic acid (AMPA) in contaminated water. *In vitro* thermodynamic and isothermal adsorption and desorption analyses were carried out, as well as in silico computational modeling, which showed that GLP and AMPA were absorbed by the clays through electrostatic bonds and hydrogen bonds, based on the correlation of the chemical bond graphs with the Langmuir model used to determine the capacity and affinity values, thus showing that MMT clays can be useful for the remediation of mixtures of LPG and AMPA in water and soils and contribute to the protection of the species.

5. Other Applications

5.1 Catalyst agent

Smectites, the family to which bentonite belongs, have been used as catalysts for many years. The acidity of these clays is related to the presence of small amounts of substitutions between aluminum and silicon in the structural network, giving rise to AlO4-, which requires a compensator for the charge generated, thus forming Bronsted-type acid sites when the compensator is hydrogen. Bentonite is widely used as a heterogeneous catalyst in organic transformations such as condensation reactions between carbonyl compounds and amines. Natural clay can have two routes of origin for Bronsted acidity: from compensating cations and from specific sites such as the edges of the clay layers. The modulation of acidity can be due to the nature of the clay itself, modification processes such as pillarization or delamination, and the insertion of chemical species. The delamination process is used for high molecular weight reactions. This process can be carried out using microwaves, ultrasound, cold drying, or the incorporation of large polyhydroxycations. [46, 47].

A 100% increase in the acidity of natural Bentonite due to modulation processes enhances the catalytic capacity of this material. In Amaya et al. [46], delamination with polyvinyl acid (PVA) showed improved catalytic action in 2-propanol dehydration-dehydrogenation reactions, compared to natural Bentonite (NB), where delaminated Bentonite (DB) effectively produced propene and acetone, due to the material's intrinsic acid sites. In a decane hydroconversion reaction, this reaction normally follows a bifunctional mechanism, requiring an acidic site and a metallic site for the reaction mechanism, when the metallic function is omitted, there is a low conversion of 5% by NB and an increase of 10% by DB, which shows the increased ability to access the characteristic sites of the natural mineral after this modification process. It can therefore be said that the catalytic behavior allows the important effect of the mineral and the nature of the incorporated cation to be established, serving as a tool for the synthesis of catalytic supports.

In the production of biofuels, Bentonite can be used as a catalyst in oil reactions, as analyzed by Rabie et al. [48] where a chemically modified Bentonite reacted with hydrochloric acid to act as a heterogeneous catalyst in the catalytic cracking of two inedible oils. The reactivity of the catalyst was assessed by the variation in total acid sites: the natural Bentonite obtained 0.333 mmol/g of acid sites and the modified Bentonite obtained 1.2 mmol/g of acid sites, demonstrating high reactivity. The biofuels obtained after the reaction of castor and jatropha oils with natural Bentonite showed no product at the temperature applied, while the physical and combustible properties of the modified Bentonite were compared with ASTM (American Society for Testing and Materials) specifications, in terms of density, viscosity, pour point, cloud points, flash and fire points, sulfur content, ash and carbon residue, all parameters were following ASTM, demonstrating the effectiveness of Bentonite in developing biofuels.

5.2 Engineering

Due to its high viscoelasticity, bentonite can be used as a construction material, especially for roads, stabilizing the unbound layers of sidewalks [49]. In addition, due to its low permeability and high swelling capacity, it can be used as a backfill material for closing cavities and can be used in the disposal of radioactive waste. [50]. In the work of Barbieri et al. [50] determined the sealing capacity of bentonite in empty spaces, using Ca-Be and Na-Be. The tests were conducted under pressure, evaluating the filling and homogenization capacity in fixed-volume cavities over 100 days. The results showed that the swelling of bentonite occurs through the plastic flow of the clay and is directly related to the proportion of water and clay. This process allowed the bentonite to fill the voids, reduce the cohesion of the material, and present a measurable swelling pressure and could be an option for the landfill of radioactive materials.

While in the study by Harrington et al [49] a study was carried out on the use of calcium-based bentonite (Ca-Be) and sodium-based bentonite (Na-Be) as stabilizing agents for the lower layers of roads, also known as unbound layers. The bentonites were mixed with crushed rock aggregates and natural gravel and then subjected to tests to determine the liquidity limit, plasticity, and waviness. The study revealed that the addition of bentonite improved the rigidity and stability of the material. Of the two, Ca-Be achieved greater stability compared to Na-Be, both with crushed rock aggregate and natural gravel. Thus, the use of bentonite appears to be a viable and environmentally friendly option as a stabilizing agent for unpaved roads.

5.3 Soil Quality

Soil can suffer problems that damage crops and consequently the food industry. One of the causes is the long-term application of mineral fertilizers to soils with a low percentage of organic compounds, which accelerates the mineralization of soil organic matter (SOM) and interrupts the metabolic processes of the organisms present in the soil, consequently leading to soil erosion processes that directly affect the productivity of the land. Environmental factors also affect soil quality; it is estimated that by 2050 climate change could lead to a 9% global agricultural decline. Bentonite can be used as a soil additive to reclaim polluted land and remediate soil contaminated with heavy metals. This clay can improve soil nutrient retention and help restore bare land due to its high cation exchange capacity. To identify soil health, some indicators are evaluated, mainly microbial biomass carbon (MIC), dehydrogenase activity (DHA), and abundance of the soil microbial community (ACMS) [51, 52].

The work of Datta et al. [51] investigated the impact of bentonite applied alone and in combination with manure on different soil microbial properties, as they believed that the combination of these components would result in the modification of properties and possible synergism in improving soil quality in the Czech Republic. The lowest DHA value was found when the application of bentonite increased soil moisture and improved enzyme activity when complexed with manure. In addition, the application of bentonite without or in combination with manure was able to increase Cmic compared to the control groups. The immobilization of microorganisms on the surface of bentonite causes a change in the properties of bentonite that facilitates the bioconversion of adsorbed polluting compounds into decomposition products that are non-toxic to the soil and the environment. These results prove the environmental decontamination potential of the manure-bentonite combination.

The process of photosynthesis can be affected from the structure of the leaf to the properties of the soil. The use of super absorbents such as Bentonite in the soil can retain water and nutrients that are slowly released as needed by crop growth, increasing soil productivity. In Mi et al. [52] studied the effect of Bentonite on field water retention capacity, millet photosynthesis, and grain quality for three years. Bentonite was mixed with sandy soil, where it extended deep into the 0-40 cm layer, increasing soil water retention and available soil water by up to 12% and 10% respectively, improving cultivation in semi-arid regions. The presence of ions in the clay also contributes to increased crop growth and improved photosynthesis, the parameters of photosynthesis had a significant increase, resulting in improved soil grain quality, with a higher protein, fat, and grain fiber content of 28.10 and 20% respectively compared to the control without Bentonite in all three years. Thus, the application of Bentonite can be an effective strategy for improving millet production in semi-arid regions.

Conclusion

Clays have been widely used in everything from pharmaceutical and biomedical applications to engineering and agriculture. Bentonite and its main constituent, montmorillonite, are low-cost, non-toxic materials that have a high swelling capacity, cation exchange capacity, and the presence of ions in their interlayers, allowing these materials, in addition to acting as an agent, to carry, act as an excipient or modify the release of other agents. In addition, the structure and therefore the properties of this clay can be different depending on where it is extracted, sometimes requiring purification, which can reduce some of the clay's limitations, and modification methods can also be carried out which can enhance the specific actions of this material. The unique properties of this clay and its wide availability are also factors that favor the wide range of research carried out into its applications. Thus, the performance of the materials in the applications observed in this work proved to be effective, in Clinical trials [53] by searching for the terms Bentonite and Montmorillonite, it is possible to find current studies investigating the action of these materials in terms of bone regeneration and protein adsorption. In this way, it can be concluded that although these materials have been used for many years by mankind, they can act in various areas, and, therefore, their applications have the great capacity to be expanded over the years.

Materials Research Forum LLC
https://doi.org/10.21741/9781644903452-5

References

[1] A. B. Mapossa, A. H. da Silva Júnior, C. R. S. de Oliveira, and W. Mhike, "Thermal, Morphological and Mechanical Properties of Multifunctional Composites Based on Biodegradable Polymers/Bentonite Clay: A Review," Aug. 01, 2023, Multidisciplinary Digital Publishing Institute (MDPI). https://doi.org/10.3390/polym15163443.

[2] F. Uddin, "Montmorillonite: An Introduction to Properties and Utilization," in Current Topics in the Utilization of Clay in Industrial and Medical Applications, InTech, 2018. https://doi.org/10.5772/intechopen.77987.

[3] L. D. Maxim, R. Niebo, and E. E. McConnell, "Bentonite toxicology and epidemiology – a review," Nov. 09, 2016, Taylor and Francis Ltd. https://doi.org/10.1080/08958378.2016.1240727.

[4] D. Gandhi, R. Bandyopadhyay, and B. Soni, "Naturally occurring bentonite clay: Structural augmentation, characterization, and application as a catalyst," Mater Today Proc, vol. 57, pp. 194–201, Jan. 2022. https://doi.org/10.1016/j.matpr.2022.02.346.

[5] M. Belghazdis and E.-K. Hachem, "Clay and Clay Minerals: A Detailed Review," International Journal of Recent Technology and Applied Science, vol. 4, no. 2, pp. 54–75, Sep. 2022. https://doi.org/10.36079/lamintang.ijortas-0402.367.

[6] F. E. Jalal, Y. Xu, X. Li, B. Jamhiri, and M. Iqbal, "Fractal approach in expansive clay-based materials with special focus on compacted GMZ bentonite in nuclear waste disposal: a systematic review," Environmental Science and Pollution Research, vol. 28, pp. 43287–43314, 2021. https://doi.org/10.1007/s11356-021-14707-7/Published.

[7] N. Yaghmaeiyan, M. Mirzaei, and R. Delghavi, "Montmorillonite clay: Introduction and evaluation of its applications in different organic syntheses as catalyst: A review," Results Chem, vol. 4, Jan. 2022. https://doi.org/10.1016/j.rechem.2022.100549.

[8] E. Srasra and I. Bekri-Abbes, "Bentonite Clays for Therapeutic Purposes and Biomaterial Design," Curr Pharm Des, vol. 26, no. 6, pp. 642–649, Feb. 2020. https://doi.org/10.2174/1381612826666200203144034.

[9] BRITISH GEOLOGICAL SURVEY, WORLD MINERAL PRODUCTION 2018-2022. Keyworth: British Geological Survey, 2024. [Online]. Available: www.geologyshop.com

[10] United States Geological Survey, "Mineral commodity summaries 2023," Feb. 31, 2023, United State Geological Survey, Reston. https://doi.org/10.3133/mcs2023.

[11] D. Borah, H. Nath, and H. Saikia, "Modification of bentonite clay & its applications: a review," Sep. 01, 2022, De Gruyter Open Ltd. https://doi.org/10.1515/revic-2021-0030.

[12] C. Nomicisio et al., "Natural and Synthetic Clay Minerals in the Pharmaceutical and Biomedical Fields," May 01, 2023, MDPI. https://doi.org/10.3390/pharmaceutics15051368.

[13] J. A. Alkrad, R. Abu Shmeis, I. Alshwabkeh, H. Abazid, and M. A. Mohammad, "Investigation of the potential application of sodium bentonite as an excipient in formulation of sustained release tablets," Asian J Pharm Sci, vol. 12, no. 3, pp. 259–265, May 2017. https://doi.org/10.1016/j.ajps.2017.01.004.

[14] N. Babahoum and M. Ould Hamou, "Characterization and purification of Algerian natural bentonite for pharmaceutical and cosmetic applications," BMC Chem, vol. 15, no. 1, Dec. 2021. https://doi.org/10.1186/s13065-021-00776-9.

Materials Research Forum LLC
https://doi.org/10.21741/9781644903452-5

[15] S. Modabberi, A. Namayandeh, A. López-Galindo, C. Viseras, M. Setti, and M. Ranjbaran, "Characterization of Iranian bentonites to be used as pharmaceutical materials," Appl Clay Sci, vol. 116–117, pp. 193–201, Nov. 2015. https://doi.org/10.1016/j.clay.2015.03.013.

[16] G. Di Prima et al., "Green Extraction of Polyphenols from Waste Bentonite to Produce Functional Antioxidant Excipients for Cosmetic and Pharmaceutical Purposes: A Waste-to-Market Approach," Antioxidants, vol. 11, no. 12, Dec. 2022. https://doi.org/10.3390/antiox11122493.

[17] M. Ruggeri et al., "Bentonite- and Palygorskite-Based Gels for Topical Drug Delivery Applications," Pharmaceutics, vol. 15, no. 4, Apr. 2023. https://doi.org/10.3390/pharmaceutics15041253.

[18] B. Subash, R. Sasikala, K. Jayamoorthy, and P. Magesan, "Heterojunction of Bentonite Clay supported Bi2O3/ZnO composite for the detoxification of azo dyes under UV-A light illumination," Chemical Physics Impact, vol. 6, Jun. 2023. https://doi.org/10.1016/j.chphi.2022.100152.

[19] S. ul Haque, A. Nasar, and Inamuddin, "Montmorillonite clay nanocomposites for drug delivery," in Applications of Nanocomposite Materials in Drug Delivery, Elsevier, 2018, pp. 633–648. https://doi.org/10.1016/B978-0-12-813741-3.00028-5.

[20] M. Ge, J. Li, S. Song, N. Meng, and N. Zhou, "Development and antibacterial performance of silver nanoparticles-lecithin modified montmorillonite nanoparticle hybrid," Appl Clay Sci, vol. 183, Dec. 2019. https://doi.org/10.1016/j.clay.2019.105334.

[21] S. K. Behera et al., "Antibacterial Efficacy of ZnO/Bentonite (Clay) Nanocomposites against Multidrug-Resistant Escherichia coli," ACS Omega, vol. 9, no. 2, pp. 2783–2794, Jan. 2024. https://doi.org/10.1021/acsomega.3c07950.

[22] R. Turgut, M. Kartal, E. K. Akkol, İ. Demirbolat, and H. Taştan, "Development of Cholesterol-Lowering and Detox Formulations Using Bentonite and Herbal Ingredients," Front Pharmacol, vol. 12, Dec. 2021. https://doi.org/10.3389/fphar.2021.775789.

[23] M. Saeed, K. Farooq, M. Nafees, M. Arshad, M. S. Akhter, and A. Waseem, "Green and Eco-Friendly Removal of Mycotoxins with Organo-Bentonites; Isothermal, Kinetic, and Thermodynamic Studies," Clean (Weinh), vol. 48, no. 9, Sep. 2020. https://doi.org/10.1002/clen.201900427.

[24] P. Mirhoseini Moosavi et al., "Efficacy of orally administered montmorillonite for acute iron poisoning detoxification in rat," Appl Clay Sci, vol. 103, pp. 62–66, Jan. 2015. https://doi.org/10.1016/j.clay.2014.11.010.

[25] A. C. Mgbeahuruike, T. E. Ejioffor, O. C. Christian, V. C. Shoyinka, M. Karlsson, and E. Nordkvist, "Detoxification of Aflatoxin-Contaminated Poultry Feeds by 3 Adsorbents, Bentonite, Activated Charcoal, and Fuller's Earth," Journal of Applied Poultry Research, vol. 27, no. 4, pp. 461–471, Dec. 2018. https://doi.org/10.3382/japr/pfy054.

[26] M. Wang, S. E. Hearon, and T. D. Phillips, "A high-capacity bentonite clay for the sorption of aflatoxins," Food Addit Contam Part A Chem Anal Control Expo Risk Assess, vol. 37, no. 2, pp. 332–341, Feb. 2020. https://doi.org/10.1080/19440049.2019.1662493.

[27] K. M. Mahrose et al., "Role of clay in detoxification of aflatoxin B1 in growing Japanese quail with reference to gender," Vet Res Commun, vol. 45, no. 4, pp. 363–371, Dec. 2021. https://doi.org/10.1007/s11259-021-09817-z.

[28] P. García-Guzmán, L. Medina-Torres, M. J. Bernad-Bernad, F. Calderas, and O. Manero, "Study of the cholesterol adsorption and characterization of montmorillonite and bentonite clay." [Online]. Available: https://doi.org/./j.mtcomm.

[29] P. Xu et al., "Preventive obesity agent montmorillonite adsorbs dietary lipids and enhances lipid excretion from the digestive tract," Sci Rep, vol. 6, Feb. 2016. https://doi.org/10.1038/srep19659.

[30] J. Y. Lee, H. N. Suh, K. Y. Choi, C. W. Song, and J. H. Hwang, "Regenerative and anti-inflammatory effect of a novel bentonite complex on burn wounds," Vet Med Sci, vol. 8, no. 6, pp. 2422–2433, Nov. 2022. https://doi.org/10.1002/vms3.908.

[31] J. Cervini-Silva, Antonio-Nieto-Camacho, S. Kaufhold, K. Ufer, and E. R. de Jesús, "The anti-inflammatory activity of bentonites," Appl Clay Sci, vol. 118, pp. 56–60, Dec. 2015. https://doi.org/10.1016/j.clay.2015.08.039.

[32] G. J. Kim et al., "Effect of nano-montmorillonite on osteoblast differentiation, mineral density, and osteoclast differentiation in bone formation," Nanomaterials, vol. 10, no. 2, Feb. 2020. https://doi.org/10.3390/nano10020230.

[33] A. Singh, S. Kumar, C. Goswami, and L. Goswami, "Montmorillonite-reinforced xanthan gum-based biomaterial enhances biomineralization suitable for bone tissue engineering," Mater Today Chem, vol. 38, Jun. 2024. https://doi.org/10.1016/j.mtchem.2024.102067.

[34] M. Wu et al., "Nanoclay mineral-reinforced macroporous nanocomposite scaffolds for in situ bone regeneration: In vitro and in vivo studies," Mater Des, vol. 205, Jul. 2021. https://doi.org/10.1016/j.matdes.2021.109734.

[35] A. G. Sabzevari, H. Sabahi, M. Nikbakht, M. Azizi, H. Dianat-Moghadam, and Z. Amoozgar, "Exploring the Potential of Montmorillonite as an Antiproliferative Nanoagent against MDA-MB-231 and MCF-7 Human Breast Cancer Cells," Cells, vol. 13, no. 2, Jan. 2024. https://doi.org/10.3390/cells13020200.

[36] S. N. Abduljauwad, H. ur R. Ahmed, and V. T. Moy, "Melanoma treatment via non-specific adhesion of cancer cells using charged nano-clays in pre-clinical studies," Sci Rep, vol. 11, no. 1, Dec. 2021. https://doi.org/10.1038/s41598-021-82441-8.

[37] S. H. Hristova and A. M. Zhivkov, "Montmorillonite colloid plates with adsorbed cytochrome c: in vitro cytotoxic effect on colon cancer cell culture," Cancer Nanotechnol, vol. 12, no. 1, Dec. 2021. https://doi.org/10.1186/s12645-021-00095-7.

[38] J. Zhang et al., "Montmorillonite/Agarose Three-Dimensional Network Gel Sponge for Wound Healing with Hemostatic and Durable Antibacterial Properties," ACS Appl Nano Mater, vol. 6, no. 18, pp. 17263–17275, Sep. 2023. https://doi.org/10.1021/acsanm.3c03708.

[39] G. Li et al., "Graphene-Montmorillonite Composite Sponge for Safe and Effective Hemostasis," ACS Appl Mater Interfaces, vol. 8, no. 51, pp. 35071–35080, Dec. 2016. https://doi.org/10.1021/acsami.6b13302.

[40] F. Armijo et al., "Thermal behaviour of clays and clay-water mixtures for pelotherapy," Appl Clay Sci, vol. 126, pp. 50–56, Jun. 2016. https://doi.org/10.1016/j.clay.2016.02.020.

[41] M. V. Fernández-González, M. I. Carretero, J. M. Martín-García, A. Molinero-García, and R. Delgado, "Peloids prepared with three mineral-medicinal waters from spas in Granada. Their suitability for use in pelotherapy," Appl Clay Sci, vol. 202, Mar. 2021. https://doi.org/10.1016/j.clay.2020.105969.

[42] M. Ç. Karakaya, N. Karakaya, and H. C. Vural, "Thermal Properties of Some Turkish Peloids and Clay Minerals for Their Use in Pelotherapy," Geomaterials, vol. 06, no. 04, pp. 79–90, 2016. https://doi.org/10.4236/gm.2016.64007.

[43] M. Wang and T. D. Phillips, "Green-Engineered Barrier Creams with Montmorillonite-Chlorophyll Clays as Adsorbents for Benzene, Toluene, and Xylene," Separations, vol. 10, no. 4, Apr. 2023. https://doi.org/10.3390/separations10040237.

[44] M. Wang, K. J. Rivenbark, and T. D. Phillips, "Adsorption and detoxification of glyphosate and aminomethylphosphonic acid by montmorillonite clays," Environmental Science and Pollution Research, vol. 30, no. 5, pp. 11417–11430, Jan. 2023. https://doi.org/10.1007/s11356-022-22927-8.

[45] R. R. de A. Duarte, T. P. Galdino, A. M. Rodrigues, M. V. L. Fook, G. de A. Neves, and S. M. de L. Silva, "Photoprotective Cosmetic Emulsions Based on Brazilian Smectite Clays," Sustainability (Switzerland), vol. 15, no. 19, Oct. 2023. https://doi.org/10.3390/su151914563.

[46] J. Amaya, L. Bobadilla, L. Azancot, M. Centeno, S. Moreno, and R. Molina, "Potentialization of bentonite properties as support in acid catalysts," Mater Res Bull, vol. 123, Mar. 2020. https://doi.org/10.1016/j.materresbull.2019.110728.

[47] V. A. Cardozo, R. Sánchez-Obregón, H. Salgado-Zamora, and R. Jiménez-Juárez, "Bentonite clay: An efficient catalyst for the synthesis of 2-substituted benzimidazoles," Monatsh Chem, vol. 146, no. 8, pp. 1335–1337, Aug. 2015. https://doi.org/10.1007/s00706-015-1423-x.

[48] A. M. Rabie, E. A. Mohammed, and N. A. Negm, "Feasibility of modified bentonite as acidic heterogeneous catalyst in low temperature catalytic cracking process of biofuel production from nonedible vegetable oils," J Mol Liq, vol. 254, pp. 260–266, Mar. 2018. https://doi.org/10.1016/j.molliq.2018.01.110.

[49] D. M. Barbieri et al., "Calcium bentonite and sodium bentonite as stabilizers for roads unbound," Clean Eng Technol, vol. 6, Feb. 2022. https://doi.org/10.1016/j.clet.2021.100372.

[50] J. F. Harrington, K. A. Daniels, A. C. Wiseall, and P. Sellin, "Bentonite homogenisation during the closure of void spaces," International Journal of Rock Mechanics and Mining Sciences, vol. 136, Dec. 2020. https://doi.org/10.1016/j.ijrmms.2020.104535.

[51] R. Datta et al., "Bentonite-based organic amendment enriches microbial activity in agricultural soils," Land (Basel), vol. 9, no. 8, Aug. 2020. https://doi.org/10.3390/LAND9080258.

[52] J. Mi, E. G. Gregorich, S. Xu, N. B. McLaughlin, and J. Liu, "Effect of bentonite as a soil amendment on field water-holding capacity, and millet photosynthesis and grain quality," Sci Rep, vol. 10, no. 1, Dec. 2020. https://doi.org/10.1038/s41598-020-75350-9.

[53] National Library of Medicine, "Clinical Trials." Available at: https://clinicaltrials.gov/. Accessed on: Sept. 2024.

Materials Research Foundations 173 (2025) 145-159 https://doi.org/10.21741/9781644903452-6

Chapter 6

Sustainable Mesoporous Material for Environmental Remediation and Advanced Medicinal Applications

Hojjat Tayefi[1], Masoud Salehipour[2], Shahla Rezaei[1], Mehdi Mogharabi-Manzari[1,3*]

[1]Pharmaceutical Sciences Research Center, Hemoglobinopathy Institute, Mazandaran University of Medical Sciences, P.O. Box 48175–861 Sari 4847193698, Iran

[2]Department of Biology, Faculty of Biological Sciences, Parand Branch of Islamic Azad University, P.O. Box 37613–96361, Parand, Tehran, Iran

[3]Thalassemia Research Center, Hemoglobinopathy Institute, Mazandaran University of Medical Sciences, Sari, Iran

*m.mogharabi@mazums.ac.ir

Abstract

Mesoporous materials have gained significant attention in recent years due to their unique structural properties, such as high surface area, tunable pore sizes, and robust chemical stability. This paper explores the development of sustainable mesoporous materials and their dual applications in environmental remediation and advanced medicine. In environmental contexts, these materials are demonstrated to effectively adsorb pollutants, degrade hazardous substances, and facilitate catalytic processes for water and air purification. In medicinal applications, their biocompatibility, controlled drug delivery capabilities, and potential for diagnostic imaging offer promising advancements in targeted therapies and regenerative medicine. By utilizing eco-friendly synthesis methods and renewable resources, this research highlights the potential of mesoporous materials as sustainable solutions to global environmental and healthcare challenges. Through a multidisciplinary approach, the study underscores the transformative role of mesoporous materials in fostering sustainable development across diverse scientific and industrial sectors.

Keywords

Mesoporous Materials, Environmental Remediation, Drug Delivery, Nanoparticles, Controlled Release

Contents

Materials Research Foundations 173 (2025) 145-159
https://doi.org/10.21741/9781644903452-6

1. Introduction

Mesoporous materials are characterized by pore diameters ranging from 2 to 50 nanometers, and their classification is based on pore size. These materials are categorized into three groups including microporous materials, with pore sizes less than 2 nm, mesoporous materials, with pore sizes between 2 and 50 nm, and macroporous materials, with pore sizes exceeding 50 nm [1]. The defining features of mesoporous materials include their pore structure, surface area, and chemical composition. The history of these materials goes far back from today in the early 1970s decade when the first mesoporous silica was produced still this early work wasn't notable for others until later advancements were made [2]. In 1990 a crucial breakthrough occurred when Japanese researchers synthesized the first mesoporous silica nanoparticles (MSNs) Two years later, in 1992 there was a major milestone in the history of these materials and that major milestone was the discovery of MCM-41 by Kresge et al [3]. They used a surfactant templating method to create highly mesoporous silica. This pivotal moment was the beginning of tons of research in the area of ordered mesoporous material (OMM) that happened years later to develop and rectify their characteristics for example in 1998 SBA-15 was made by researchers, another magnificent type of mesoporous silica, SBA-15 had a larger pore size and also was better in thermal stability in compare to MCM-41 [4].

Regarding pore structure, these materials can be classified as either ordered or disordered. Ordered mesoporous materials, such as MCM-41, exhibit a highly regular pore arrangement, which often enhances their performance in various applications compared to disordered materials with irregular pore structures [5]. These structural characteristics, in combination with other properties, make mesoporous materials valuable for a wide range of applications in catalysis, adsorption, and drug delivery systems. Ordered mesoporous materials have a highly regular, well-defined pore architecture, which leads to improved diffusion, stability, and accessibility of the active sites, contributing to superior performance in applications like catalysis and separation processes [6]. In contrast, disordered mesoporous materials exhibit irregular pore structures, which may still be useful but often offer less control over molecular interactions. Another key characteristic of mesoporous materials is their tunable chemical composition [7]. They are typically synthesized from silica, alumina, or other metal oxides, and can be functionalized with organic or inorganic groups to modify their surface chemistry. This allows for the tailoring of hydrophobicity, hydrophilicity, and catalytic properties. The wall thickness of mesoporous materials also plays a role in their stability, particularly at high temperatures and under harsh chemical conditions [8]. Additionally, their pore size distribution can be tuned during synthesis, allowing for the selective adsorption of specific molecules or ions. These versatile properties collectively enable mesoporous materials to be highly customizable, making them excellent candidates for a wide range of scientific and industrial applications. Mesoporous materials have a wide range of applications because of their spectacular properties such as catalysis, drug delivery, gas separation and adsorption, and also energy storage. They have been utilized in industrial and scientific fields for decades since their discovery in the late 20th century [9].

In this chapter, we conducted a comprehensive exploration of the unique characteristics of mesoporous materials, along with their potential applications in biomedical and environmental remediation.

2. Mesoporous silica

Mesoporous silica nanoparticles also known as MSNs have a variety of materials and common types of them are: MCM-41, MCM-48, SBA-15, and SBA-16. Mcm-41 synthesized in the early 1990s shows a hexagonal arrangement of uniform pores but SBA-16 synthesized later has a better characteristic such as larger pore size and also more complex structure [10]. In the process of synthesis of these materials usually use tetraethyl orthosilicate (TEOS) that combines with a surfactant or template that helps the material to form its structure. Having a tunable pore size, high surface area, and also strong chemical stability make them promising for drug delivery systems [11]. They can easily encapsulate chemical compounds and deliver them to the targeted organs and tumors which makes them very competitive with other nanomaterials in the area of chemotherapy and targeted drug delivery. These mechanical properties that are mentioned insinuate that these materials are capable of enhancing the solubility and permeability of loaded molecules. Mesoporous silica materials are synthesized through a sol-gel process in the presence of a structure-directing agent, typically surfactants such as cetyltrimethylammonium bromide (CTAB) [12]. During synthesis, the surfactant molecules self-assemble into micelles, which act as templates for the formation of the silica framework. Silica precursors, such as tetraethyl orthosilicate (TEOS), are then hydrolyzed and condensed around these micelles [13,14]. The subsequent removal of the surfactant, either by calcination or solvent extraction, results in a highly porous silica network with a well-defined mesoporous structure. The pore size, surface area, and pore volume can be tuned by adjusting synthesis conditions, such as pH, temperature, and surfactant concentration.

The regular pore structure provides excellent accessibility and diffusion properties, which are beneficial for applications such as drug delivery, catalysis, and adsorption. Additionally, mesoporous silica exhibits good thermal and chemical stability, making it suitable for use in harsh environments [15]. The surface of mesoporous silica can also be functionalized with organic or inorganic groups, enabling the tailoring of its surface chemistry for specific applications, such as improving biocompatibility for biomedical uses or enhancing selectivity in catalytic reactions. These versatile properties make mesoporous silica materials highly valuable in a wide range of fields, from environmental remediation to nanomedicine [16,17].

3. Mesoporous carbon

Mesoporous carbons are promising materials for a vast range of applications like catalysis, environmental remediation, drug delivery systems, and especially energy storage. There are several ways to produce these materials for example template carbonization which evolves utilizing silica and other temples to form an ordered structure and self-assembly techniques that employ copolymers to form mesoporous materials during polymerization and also pyrolysis of organic resources that involve thermal degradation of materials like sucrose or biomass to create porous material [18]. The high surface area of these materials which exceeds over 2500 m^2/g makes them suitable for catalysis and adsorption processes. Possessing flexible pore sizes that can be controlled during the synthesis process allows them to be characterized based on specific needs. Another important feature of these materials is their high thermal steadiness makes them appropriate for high-thermal applications. Mesoporous carbon has a variety types of materials such as ordered mesoporous carbon (OMC) which has a well-defined structure that often originated from silica templates [19,20]. another type is disordered mesoporous carbon which lacks a regular structure and its synthesis method is usually pyrolysis. Another group is functionalized mesoporous carbon, modified and enhanced with various functional groups, and also magnetic mesoporous carbon that is designed to have magnetic properties to deliver drugs and chemical compounds to a specific organ or tissue this group also has a beneficial property in separation processes [21].

4. Mesoporous metal oxides

Mesoporous metallic materials are synthesized using various templating methods, with soft and hard templates being the most common. In soft templating, surfactants or block copolymers self-assemble into micelles that serve as templates for the metal precursors, which are subsequently reduced to form a mesoporous metallic structure [22]. The removal of the surfactant via chemical treatment or calcination results in a highly porous metal network. Hard templating, on the other hand, involves the use of preformed mesoporous silica or carbon frameworks. Metal precursors infiltrate these hard templates and are then reduced, followed by template removal through dissolution or chemical etching, leaving behind a mesoporous metallic material. Metals such as gold, platinum, palladium, and nickel are commonly synthesized in mesoporous forms through these methods [23].

Mesoporous metallic materials exhibit several distinctive characteristics and properties. They have large surface areas and highly ordered pore structures, typically with pore sizes ranging from 2 to 50 nanometers [24]. These materials possess excellent electrical and thermal conductivity, making them ideal for use in applications such as catalysis, sensing, and energy storage. Their mesoporosity enhances mass transport and diffusion, improving the performance of catalytic processes by providing increased access to active sites. Additionally, mesoporous metals exhibit remarkable mechanical stability and can be functionalized to enhance their chemical reactivity. The high surface area, coupled with their tunable pore size and structure, makes mesoporous metallic materials versatile and highly efficient for use in heterogeneous catalysis, electrochemical energy conversion, and biomedical applications such as drug delivery and bio-sensing. [25]

5. Metal-Organic frameworks

Mesoporous metal-organic frameworks (MOFs) are synthesized through a variety of methods, including solvothermal, microwave-assisted, and template-assisted synthesis [26]. The solvothermal method is the most common, where metal ions and organic linkers are mixed in a solvent and heated to promote self-assembly into a highly porous crystalline structure. To create mesopores, templating strategies are often employed, using surfactants or block copolymers that act as pore-forming agents. Once the MOF structure is formed, the templates are removed by calcination or solvent extraction. Another approach to creating mesoporous MOFs involves post-synthetic modification, where additional porosity is introduced into the framework through etching or linker removal [27]. Mesoporous MOFs possess several unique characteristics and properties that distinguish them from traditional MOFs. They combine the high surface area and tunable pore sizes of conventional MOFs with the enhanced mass transport and accessibility provided by their mesoporous structure. The regular arrangement of mesopores within the crystalline framework allows for efficient diffusion of larger molecules, making them ideal for applications in catalysis, gas storage, and drug delivery [28]. Mesoporous MOFs also exhibit high thermal and chemical stability, particularly when composed of robust metal-ligand bonds. Additionally, the chemical functionality of MOFs can be easily tailored by modifying the organic linkers or introducing functional groups onto the framework, allowing for precise control over adsorption properties and catalytic activity. The combination of large surface area, tunable porosity, and customizable chemistry makes mesoporous MOFs highly versatile materials with potential in energy storage, environmental remediation, and biomedical applications [29].

Furthermore, mesoporous MOFs can be engineered to exhibit selective adsorption and catalytic activity, which is particularly advantageous for targeted applications such as carbon capture, hydrogen storage, and pollutant degradation. The tunability of the pore size and functionality allows for the incorporation of specific guest molecules, enhancing their role in applications like drug delivery, where controlled release is crucial [30]. The hierarchical porosity of mesoporous MOFs also facilitates enhanced diffusion pathways, which can significantly improve the kinetics of reactions, making them attractive for use in catalytic converters and chemical reactors.

Moreover, the integration of mesoporous MOFs with other materials, such as polymers or nanoparticles, can further enhance their properties and expand their functionality. For instance, composite materials combining mesoporous MOFs with conductive polymers can improve charge transport for energy-related applications, such as batteries and supercapacitors [31]. The development of hybrid systems, where mesoporous MOFs are used alongside metal nanoparticles, can lead to synergistic effects that enhance catalytic performance. As research continues to explore the potential of mesoporous MOFs, their versatility and customizable nature position them as promising candidates for addressing critical challenges in areas such as environmental sustainability, energy conversion, and healthcare. The ongoing advancements in synthesis techniques and functionalization strategies are likely to pave the way for innovative applications, underscoring the significance of mesoporous MOFs in modern materials science.

6. Covalent organic frameworks

Covalent organic frameworks (COFs) are synthesized through the reticular chemistry approach, utilizing the covalent bonding of organic building blocks to form extended two- or three-dimensional structures [32]. The synthesis often involves a dynamic covalent bond formation, where the choice of linkers and reaction conditions, such as solvent, temperature, and pH, can be precisely controlled to yield materials with specific pore sizes and morphologies. The incorporation of rigid organic linkers, such as boronic acids, imines, or aldehydes, promotes the formation of stable networks with well-defined mesoporosity. Post-synthetic modifications can also enhance the porosity of COFs or introduce functional groups to tailor their properties further. The unique characteristics and properties of mesoporous COFs make them particularly appealing for various applications [33]. They possess exceptionally high surface areas, often exceeding 1000 m^2/g, and exhibit tunable pore sizes and shapes, allowing for selective adsorption and storage of small molecules, gases, or biomolecules. The stability of COFs is attributed to the strong covalent bonds within their frameworks, which enables them to maintain structural integrity under diverse conditions, including heat and humidity. Furthermore, the organic nature of COFs allows for extensive functionalization, enabling modifications that can impart specific chemical properties, such as hydrophobicity or catalytic activity [34]. The ordered pore structure facilitates efficient diffusion and transport of guest molecules, enhancing their potential in applications such as gas separation, catalysis, drug delivery, and sensing. As research progresses, the design and synthesis of mesoporous COFs continue to evolve, promising innovative solutions to challenges in areas such as environmental remediation, energy storage, and pharmaceuticals [35].

7. Environmental application of mesoporous material

One of the most prominent uses of porous materials is in the field of gas adsorption and separation. For instance, MOFs and COFs can be engineered to selectively capture greenhouse gases, such as carbon dioxide (CO_2) and methane (CH_4), from industrial emissions or the atmosphere [36]. Their tunable pore sizes and high surface areas facilitate the adsorption of these gases, contributing to efforts in mitigating climate change. Additionally, these materials can be employed in the capture and storage of volatile organic compounds (VOCs), which are prevalent pollutants from industrial processes and vehicle emissions [37]. The ability of porous materials to selectively adsorb specific gases makes them vital in the development of efficient air purification systems and carbon capture technologies. In addition to gas and water treatment, porous materials are increasingly utilized in soil remediation efforts. The high adsorption capacity of these materials allows for the immobilization and removal of hazardous substances, such as pesticides, heavy metals, and hydrocarbons, from contaminated soils [38]. For example, activated carbon, a widely used porous material, can effectively adsorb organic pollutants, reducing their bioavailability and preventing further contamination of groundwater. Similarly, functionalized MOFs and COFs can be employed to selectively capture and degrade pollutants in soil, contributing to the restoration of contaminated sites. The versatility of porous materials in environmental applications not only aids in pollution control but also promotes sustainable practices by facilitating the recycling of waste materials and reducing the environmental footprint of various industrial processes [39]. As research continues to advance in the design and synthesis of novel porous materials, their potential for addressing pressing environmental challenges is expected to grow, paving the way for innovative solutions in pollution mitigation and resource recovery.

Moreover, porous materials are making significant strides in the development of sustainable energy technologies, particularly in the context of environmental applications. For instance, mesoporous materials and MOFs are increasingly being explored as efficient platforms for hydrogen storage, which is essential for the advancement of hydrogen fuel cell technology [40,41]. Their high surface areas and tunable pore structures allow for the encapsulation of hydrogen

Materials Research Foundations 173 (2025) 145-159 https://doi.org/10.21741/9781644903452-6

molecules at lower pressures and temperatures, making them safer and more efficient storage solutions. This capability is critical for the transition to clean energy systems, as hydrogen fuel cells produce only water as a byproduct, offering a pollution-free energy source. Additionally, porous materials can play a vital role in battery technology, where their structures can facilitate ion transport, thereby enhancing the efficiency and longevity of energy storage devices [42].

Furthermore, the integration of porous materials into environmental sensing applications is an area of growing interest. By functionalizing porous substrates with specific receptors or chemical probes, researchers can develop highly sensitive and selective sensors for detecting environmental pollutants, including heavy metals, pesticides, and pathogens. The large surface area of porous materials enables the immobilization of a significant amount of sensing elements, which enhances the sensitivity and detection limits of the sensors. This capability is crucial for real-time monitoring of environmental quality and the early detection of contaminants, providing valuable data for regulatory compliance and public health protection [43–45]. The use of porous materials in catalysis offers promising avenues for environmental remediation. Many environmental pollutants are challenging to degrade or detoxify, necessitating efficient catalytic processes. Porous catalysts, such as those derived from MOFs or COFs, can facilitate the breakdown of organic pollutants through advanced oxidation processes or photodegradation. Their porous structure enhances mass transfer and reaction kinetics, leading to improved catalytic efficiency. Moreover, these materials can be tailored for specific catalytic reactions, allowing for the development of highly selective processes that minimize by-products and enhance the overall sustainability of chemical transformations.

The versatile properties of porous materials position them as crucial components in a wide array of environmental applications, from pollution control and resource recovery to sustainable energy and sensing technologies [46]. As ongoing research continues to unlock new functionalities and improve the performance of these materials, their role in addressing global environmental challenges will undoubtedly expand, contributing to a more sustainable and cleaner future.

8. Biomedical applications of mesoporous material

Mesoporous materials have several applications that position them at the forefront of biomedical research. Their application of these materials in biosensors, drug delivery, medical imaging, and other biomedical applications makes them promising to revolutionize treatment strategies in the future [47].

8.1 Biosensors

Enzymatic biosensors are usually used in the immobilization of enzymes for example mesoporous silica has been used to immobilize glucose oxidase for glucose detection[48]. The noticeable point is that the performance of these materials in this application is mostly affected by many factors like pore size and surface chemistry [49]. Non-enzymatic biosensors are usually mesoporous materials that have electrochemical characteristics that allow them to imitate enzyme behavior for example mesoporous carbon materials have been used for detecting biomolecules like glucose and hydrogen peroxide [50].

Porous nanoparticles have gained significant attention in the development of advanced biosensors due to their unique structural and functional properties. These nanoparticles are characterized by a high surface area and a well-defined porous architecture, allowing for increased immobilization of bioreceptors, enhanced mass transport, and improved signal amplification. The pores in these nanoparticles provide large surface-to-volume ratios, enabling efficient interaction between the analytes and the sensor's recognition elements. This makes porous nanoparticles highly effective in detecting trace amounts of biological and chemical substances [51]. Porous materials such as mesoporous silica nanoparticles, porous gold nanoparticles, and MOFs are commonly used in biosensor development. Their ability to encapsulate biomolecules, facilitate the diffusion of analytes, and amplify detection signals has led to their widespread application in fields such as medical diagnostics, environmental monitoring, and food safety.

Mesoporous silica nanoparticles possess a highly ordered porous structure with tunable pore sizes, making them ideal for the encapsulation of biomolecules like enzymes, antibodies, and DNA. Mesoporous silica nanoparticles improve the stability and loading capacity of biosensors by protecting biomolecules from degradation while allowing analyte diffusion through the pores [52]. This makes them particularly effective in enzyme-based biosensors, where the porous network enables efficient interaction between the enzyme and substrate. they have been used in glucose and cholesterol biosensors, improving sensitivity and response time by enhancing enzyme activity and facilitating electron transfer [53]. Porous metallic nanoparticles combine the benefits of metals excellent conductivity and biocompatibility with a highly porous structure that increases the surface area for biomolecule attachment. These nanoparticles are particularly useful in electrochemical biosensors, where they enhance electron transfer between the bioreceptor and electrode [54]. For example, the porous nature of gold nanoparticles allows for a higher loading of enzymes, antibodies, or DNA, making them ideal for applications requiring ultrasensitive detection, such as cancer biomarkers or pathogenic bacteria. Their surface plasmon resonance properties can also be leveraged in optical biosensors, where changes in light absorption or scattering are used to detect the presence of target molecules.

8.2 Drug delivery

Mesoporous materials have attracted significant attention as potential drug delivery vehicles due to their unique structural features. These materials offer an exceptionally high surface area and large pore volumes. Such properties allow them to encapsulate a wide range of therapeutic agents, including small-molecule drugs, proteins, and even nucleic acids [55]. One of the key advantages of mesoporous materials in drug delivery is their tunable pore size, which facilitates precise control over the loading and release of drugs. Additionally, the surface of these materials can be functionalized with various chemical groups or targeting ligands, enabling site-specific drug delivery and enhancing therapeutic efficacy. The applications of mesoporous materials in drug delivery are diverse [56]. One major application is controlled drug release. The highly porous structure of these materials allows for the sustained release of drugs over time, which can improve patient compliance by reducing the need for frequent dosing. For instance, mesoporous silica nanoparticles have been used to deliver anticancer drugs in a controlled manner, minimizing drug degradation and maintaining therapeutic levels in the bloodstream for extended periods [57]. Another important application is targeted drug delivery. By modifying the surface of mesoporous carriers with targeting molecules such as antibodies or peptides, these materials can be designed to selectively bind to specific cells, such as cancer cells, thereby delivering drugs directly to the diseased tissue and minimizing side effects on healthy cells [58]. Furthermore, mesoporous materials can be engineered to release their drug payload in response to specific external or internal stimuli, such as changes in pH, temperature, or the presence of certain enzymes. This stimuli-responsive behavior enables the release of drugs at the desired site of action, further enhancing their therapeutic potential [59].

Despite their promising applications, mesoporous materials have several advantages and disadvantages that must be carefully considered. Among their most notable advantages is their high drug-loading capacity. Owing to their large surface area and pore volume, mesoporous materials can encapsulate substantial amounts of therapeutic agents, making them particularly useful for delivering poorly soluble drugs. Additionally, the ability to control drug release kinetics through the adjustment of pore size and surface chemistry is a significant advantage, as it enables prolonged drug release and reduces the likelihood of burst release [60]. This controlled release profile is beneficial in achieving a sustained therapeutic effect while minimizing the risk of toxicity. Another advantage is the versatility of mesoporous materials. Their surfaces can be functionalized with a wide range of chemical groups or biological molecules, which enhances their ability to target specific tissues, respond to environmental stimuli, and interact with biological systems in a controlled manner. Importantly, many mesoporous materials, such as MSNs, exhibit good biocompatibility [61]. MSNs degrade into non-toxic byproducts, such as silicic acid, which can be safely excreted by the body, reducing concerns related to long-term accumulation and toxicity. However, mesoporous materials also have some limitations that need to be addressed. One of the primary concerns is potential toxicity. While silica-based mesoporous materials are generally considered biocompatible, their degradation products and the materials themselves may cause toxicity if they accumulate in certain organs, such as the liver or kidneys. Inflammation or immune responses may also occur if the materials are not adequately cleared from the body. Another challenge is the complexity and cost of synthesizing mesoporous materials with precise control over pore size, drug loading, and surface functionalization [62]. The synthesis processes often involve multiple steps, making large-scale production difficult and expensive. Moreover, the stability of mesoporous materials in biological environments is a critical issue. They must remain intact and retain their drug load in the bloodstream or other biological fluids until they reach the target site [63]. Premature release or leakage of the drug can reduce therapeutic efficacy and cause unwanted side effects. Lastly, although mesoporous materials have shown great promise in preclinical studies, their clinical translation remains limited due to the aforementioned challenges, as well as regulatory hurdles and the need for reproducibility in manufacturing [64].

Mesoporous materials hold significant potential for advancing drug delivery systems. Their high drug-loading capacity, controlled release profiles, and ability to target specific tissues make them attractive candidates for a wide range of therapeutic applications. However, issues related to potential toxicity, stability in biological environments, and the complexity of synthesis must be addressed to fully realize their clinical potential. Future research is likely to focus on improving the safety and scalability of these materials, optimizing their drug delivery capabilities, and overcoming the barriers that have hindered their broader use in clinical settings.

8.3 Imaging

Mesoporous materials, particularly mesoporous silica nanoparticles have emerged as significant tools in the field of biomedical imaging. Their unique structural properties, such as high specific surface area, tunable pore size, and excellent biocompatibility, make them suitable for various imaging modalities [65]. Mesoporous silica nanoparticles represent a promising platform for advancing biomedical imaging techniques. Their customizable properties and multifunctional capabilities position them at the forefront of research aimed at improving diagnostic accuracy and

 Materials Research Forum LLC
https://doi.org/10.21741/9781644903452-6

therapeutic outcomes in clinical settings. As research continues to evolve, MSNs are expected to play a critical role in non-invasive imaging technologies and targeted therapy strategies [66]. Mesoporous materials can be used in a wide range of medical imaging applications like optical imaging, magnetic resonance imaging (MRI), multimodal imaging, Computed Tomography (CT) and Positron Emission Tomography (PET).

Recent advancements have explored the use of MSNs in CT and PET imaging by integrating radiolabeled compounds within their porous structure, providing enhanced imaging capabilities for disease diagnosis. The mesoporous structure allows water molecules to access the internal matrix of MSNs, enhancing contrast in MRI applications. They can be combined with superparamagnetic iron oxide nanoparticles to improve imaging quality [67].

They can be functionalized with fluorescent dyes to create effective optical imaging agents. These nanoparticles exhibit strong fluorescence that is stable against environmental changes, making them suitable for in vivo tracking.

Conclusions

The exploration of mesoporous materials presents a promising avenue for addressing critical challenges in environmental remediation and advanced medicinal applications. By leveraging their unique structural properties and eco-friendly synthesis methods, these materials demonstrate exceptional capabilities in adsorbing and degrading pollutants, thus contributing to cleaner ecosystems. Furthermore, their biocompatibility and controlled drug delivery systems signify a transformative potential in medical therapies, paving the way for enhanced treatment efficacy and patient outcomes. The integration of mesoporous materials into various industrial and scientific sectors not only promotes sustainable practices but also facilitates innovations in healthcare and environmental management. Future research should focus on optimizing the synthesis and functionalization of these materials to expand their applicability and effectiveness, ensuring their role as essential components in the pursuit of a sustainable and healthier world.

Reference

[1]K. Ariga, A. Vinu, Y. Yamauchi, Q. Ji, J. Hill, Nanoarchitectonics for Mesoporous Materials, Bull. Chem. Soc. Jpn. 85 (2012) 1–32. https://doi.org/10.1246/bcsj.20110162

[2]M. Nazari, F. Zadehahmadi, M.M. Sadiq, A.L. Sutton, H. Mahdavi, M.R. Hill, Challenges and solutions to the scale-up of porous materials, Commun. Mater. 5 (2024) 170. https://doi.org/10.1038/s43246-024-00608-y

[3]C.T. Kresge, M.E. Leonowicz, W.J. Roth, J.C. Vartuli, J.S. Beck, Ordered mesoporous molecular sieves synthesized by a liquid-crystal template mechanism, Nature 359 (1992) 710–712. https://doi.org/10.1038/359710a0

[4]D. Zhao, J. Feng, Q. Huo, N. Melosh, G.H. Fredrickson, B.F. Chmelka, G.D. Stucky, Triblock Copolymer Syntheses of Mesoporous Silica with Periodic 50 to 300 Angstrom Pores, Science 279 (1998) 548–552. https://doi.org/10.1126/science.279.5350.548

[5]X.S. Zhao, G.Q. (Max) Lu, G.J. Millar, Advances in Mesoporous Molecular Sieve MCM-41, Ind. Eng. Chem. Res. 35 (1996) 2075–2090. https://doi.org/10.1021/ie950702a

[6]W. Li, D. Zhao, An overview of the synthesis of ordered mesoporous materials, Chem. Commun. 49 (2013) 943–946. https://doi.org/10.1039/C2CC36964H

[7]Z. Yang, Y. Lu, Z. Yang, Mesoporous materials: tunable structure, morphology and composition, Chem. Commun. (2009) 2270–2277. https://doi.org/10.1039/B820539F

[8]K. Lan, D. Zhao, Functional Ordered Mesoporous Materials: Present and Future, Nano Lett. 22 (2022) 3177–3179. https://doi.org/10.1021/acs.nanolett.2c00902

[9]S.L. Suib, A Review of Recent Developments of Mesoporous Materials, Chem. Rec. 17 (2017) 1169–1183. https://doi.org/10.1002/tcr.201700025

[10] M. Mogharabi-Manzari, M. Amini, M. Abdollahi, M. Khoobi, G. Bagherzadeh, M.A. Faramarzi, Co-immobilization of Laccase and TEMPO in the Compartments of Mesoporous Silica for a Green and One-Pot Cascade Synthesis of Coumarins by Knoevenagel Condensation, ChemCatChem 10 (2018) 1542–1546. https://doi.org/10.1002/cctc.201701527

[11] S.-H. Wu, C.-Y. Mou, H.-P. Lin, Synthesis of mesoporous silica nanoparticles, Chem. Soc. Rev. 42 (2013) 3862–3875. https://doi.org/10.1039/C3CS35405A

[12] J. Hwang, J.H. Lee, J. Chun, Facile approach for the synthesis of spherical mesoporous silica nanoparticles from sodium silicate, Mater. Lett. 283 (2021) 128765. https://doi.org/10.1016/j.matlet.2020.128765

[13] H. Qiu, S. Che, Chiral mesoporous silica: Chiral construction and imprinting via cooperative self-assembly of amphiphiles and silica precursors, Chem. Soc. Rev. 40 (2011) 1259–1268. https://doi.org/10.1039/C0CS00002G

[14] A.-M. Putz, K. Wang, A. Len, J. Plocek, P. Bezdicka, G.P. Kopitsa, T.V. Khamova, C. Ianăşi, L. Săcărescu, Z. Mitróová, C. Savii, M. Yan, L. Almásy, Mesoporous silica obtained with methyltriethoxysilane as co-precursor in alkaline medium, 11th Int. Conf. Phys. Adv. Mater. 424 (2017) 275–281. https://doi.org/10.1016/j.apsusc.2017.04.121

[15] L. Liu, X. Shan, X. Hu, W. Lv, J. Wang, Superhydrophobic Silica Aerogels and Their Layer-by-Layer Structure for Thermal Management in Harsh Cold and Hot Environments, ACS Nano 15 (2021) 19771–19782. https://doi.org/10.1021/acsnano.1c07184

[16] R.I. Ripon, Z.A. Begum, R. Miyazawa, K. Leonard, M. Ogata, B. Ahmmad, I.M.M. Rahman, Development and application of a new macrocyclic ligand-functionalized mesoporous silica sorbent for selective separation of radiocesium from environmental wastewater, Microchem. J. 207 (2024) 111649. https://doi.org/10.1016/j.microc.2024.111649

[17] T.M. Albayati, S.M. Alardhi, A.H. Khalbas, Z.J. Humdi, N.S. Ali, I.K. Salih, N.M.C. Saady, S. Zendehboudi, M.A. Abdulrahman, Comprehensive Review of Mesoporous Silica Nanoparticles: Drug Loading, Release, and Applications as Hemostatic Agents, ChemistrySelect 9 (2024) e202400450. https://doi.org/10.1002/slct.202400450

[18] X. Zhao, P. Gao, B. Shen, X. Wang, T. Yue, Z. Han, Recent advances in lignin-derived mesoporous carbon based-on template methods, Renew. Sustain. Energy Rev. 188 (2023) 113808. https://doi.org/10.1016/j.rser.2023.113808

[19] D. Gang, Z. Uddin Ahmad, Q. Lian, L. Yao, M.E. Zappi, A review of adsorptive remediation of environmental pollutants from aqueous phase by ordered mesoporous carbon, Chem. Eng. J. 403 (2021) 126286. https://doi.org/10.1016/j.cej.2020.126286

[20] J. Zhang, N. Zhang, F.M.G. Tack, S. Sato, D.S. Alessi, P. Oleszczuk, H. Wang, X. Wang, S. Wang, Modification of ordered mesoporous carbon for removal of environmental contaminants from aqueous phase: A review, J. Hazard. Mater. 418 (2021) 126266. https://doi.org/10.1016/j.jhazmat.2021.126266

[21] M.R. Benzigar, S.N. Talapaneni, S. Joseph, K. Ramadass, G. Singh, J. Scaranto, U. Ravon, K. Al-Bahily, A. Vinu, Recent advances in functionalized micro and mesoporous carbon materials: synthesis and applications, Chem. Soc. Rev. 47 (2018) 2680–2721. https://doi.org/10.1039/C7CS00787

[22] Y. Ren, Z. Ma, P.G. Bruce, Ordered mesoporous metal oxides: synthesis and applications, Chem. Soc. Rev. 41 (2012) 4909. https://doi.org/10.1039/c2cs35086f

[23] X. Deng, K. Chen, H. Tüysüz, Protocol for the Nanocasting Method: Preparation of Ordered Mesoporous Metal Oxides, Chem. Mater. 29 (2017) 40–52. https://doi.org/10.1021/acs.chemmater.6b02645

[24] Y. Zou, X. Zhou, J. Ma, X. Yang, Y. Deng, Recent advances in amphiphilic block copolymer templated mesoporous metal-based materials: assembly engineering and applications, Chem. Soc. Rev. 49 (2020) 1173–1208. https://doi.org/10.1039/C9CS00334G

[25] Y. Ren, Z. Ma, P.G. Bruce, Ordered mesoporous metal oxides: synthesis and applications, Chem. Soc. Rev. 41 (2012) 4909. https://doi.org/10.1039/c2cs35086f

[26] M. Salehipour, S. Rezaei, H.F. Asadi Khalili, A. Motaharian, M. Mogharabi-Manzari, Nanoarchitectonics of Enzyme/Metal–Organic Framework Composites for Wastewater Treatment, J. Inorg. Organomet. Polym. Mater. 32 (2022) 3321–3338. https://doi.org/10.1007/s10904-022-02390-1

[27] H. Furukawa, K.E. Cordova, M. O'Keeffe, O.M. Yaghi, The Chemistry and Applications of Metal-Organic Frameworks, Science 341 (2013) 1230444. https://doi.org/10.1126/science.1230444

[28] T. Limongi, F. Susa, M. Allione, E. Di Fabrizio, Drug Delivery Applications of Three-Dimensional Printed (3DP) Mesoporous Scaffolds, Pharmaceutics 12 (2020) 851. https://doi.org/10.3390/pharmaceutics12090851

[29] S. Nikpour, M. Salehipour, S. Rezaei, M. Mogharabi-Manzari, Design and fabrication of magnetic cross-linked laccase aggregate using superparamagnetic metal-organic frameworks for phenol removal, Biochem. Eng. J. 209 (2024) 109385. https://doi.org/10.1016/j.bej.2024.109385

[30] Y. Feng, Q. Ma, Z. Wang, Q. Zhang, L. Zhao, J. Cui, Y. Du, S. Jia, Multivariate mesoporous MOFs with regulatable hydrophilic/hydrophobic surfaces as a versatile platform for enzyme immobilization, Chin. J. Catal. 60 (2024) 386–398.

[31] X. Gao, Y. Dong, S. Li, J. Zhou, L. Wang, B. Wang, MOFs and COFs for Batteries and Supercapacitors, Electrochem. Energy Rev. 3 (2020) 81–126. https://doi.org/10.1007/s41918-019-00055-1

[32] K.T. Tan, S. Ghosh, Z. Wang, F. Wen, D. Rodríguez-San-Miguel, J. Feng, N. Huang, W. Wang, F. Zamora, X. Feng, A. Thomas, D. Jiang, Covalent organic frameworks, Nat. Rev. Methods Primer 3 (2023) 1. https://doi.org/10.1038/s43586-022-00181-z

[33] S. Daliran, M. Blanco, A. Dhakshinamoorthy, A.R. Oveisi, J. Alemán, H. García, Defects and Disorder in Covalent Organic Frameworks for Advanced Applications, Adv. Funct. Mater. 34 (2024) 2312912. https://doi.org/10.1002/adfm.202312912

[34] J. Wang, S. Zhuang, Covalent organic frameworks (COFs) for environmental applications, Coord. Chem. Rev. 400 (2019) 213046. https://doi.org/10.1016/j.ccr.2019.213046

[35] N. Zhang, A. Ishag, Y. Li, H. Wang, H. Guo, P. Mei, Q. Meng, Y. Sun, Recent investigations and progress in environmental remediation by using covalent organic framework-based adsorption method: A review, J. Clean. Prod. 277 (2020) 123360. https://doi.org/10.1016/j.jclepro.2020.123360

[36] J.-R. Li, R.J. Kuppler, H.-C. Zhou, Selective gas adsorption and separation in metal–organic frameworks, Chem. Soc. Rev. 38 (2009) 1477–1504. https://doi.org/10.1039/B802426J

[37] M. Alhamami, H. Doan, C.-H. Cheng, A Review on Breathing Behaviors of Metal-Organic-Frameworks (MOFs) for Gas Adsorption, Materials 7 (2014) 3198–3250. https://doi.org/10.3390/ma7043198

[38] P.A. Kobielska, A.J. Howarth, O.K. Farha, S. Nayak, Metal–organic frameworks for heavy metal removal from water, Coord. Chem. Rev. 358 (2018) 92–107. https://doi.org/10.1016/j.ccr.2017.12.010

[39] G. Lin, B. Zeng, J. Li, Z. Wang, S. Wang, T. Hu, L. Zhang, A systematic review of metal organic frameworks materials for heavy metal removal: Synthesis, applications and mechanism, Chem. Eng. J. 460 (2023) 141710. https://doi.org/10.1016/j.cej.2023.141710

[40] M. Hirscher, B. Panella, B. Schmitz, Metal-organic frameworks for hydrogen storage, Sel. Pap. 1st Int. Conf. Met. Org. Framew. Open Framew. Compd. MOF 2008 Oct. 8–10 2008 129 (2010) 335–339. https://doi.org/10.1016/j.micromeso.2009.06.005

[41] H.W. Langmi, J. Ren, B. North, M. Mathe, D. Bessarabov, Hydrogen Storage in Metal-Organic Frameworks: A Review, Adv. Electrochem. Mater. Manuf. 128 (2014) 368–392. https://doi.org/10.1016/j.electacta.2013.10.190

[42] Y. He, F. Chen, B. Li, G. Qian, W. Zhou, B. Chen, Porous metal–organic frameworks for fuel storage, Coord. Chem. Energy 373 (2018) 167–198. https://doi.org/10.1016/j.ccr.2017.10.002

[43] M. Aghaee, M. Salehipour, S. Rezaei, M. Mogharabi-Manzari, Bioremediation of organic pollutants by laccase-metal–organic framework composites: A review of current knowledge and future perspective, Bioresour. Technol. (2024) 131072. https://doi.org/10.1016/j.biortech.2024.131072

[44] M. Salehipour, S. Rezaei, J. Mosafer, Z. Pakdin-Parizi, A. Motaharian, M. Mogharabi-Manzari, Recent advances in polymer-coated iron oxide nanoparticles as magnetic resonance

imaging contrast agents, J. Nanoparticle Res. 23 (2021). https://doi.org/10.1007/s11051-021-05156-x

[45] M. Salehipour, S. Nikpour, S. Rezaei, S. Mohammadi, M. Rezaei, D. Ilbeygi, A. Hosseini-Chegeni, M. Mogharabi-Manzari, Safety of metal–organic framework nanoparticles for biomedical applications: An in vitro toxicity assessment, Inorg. Chem. Commun. 152 (2023) 110655. https://doi.org/10.1016/j.inoche.2023.110655

[46] M. Mogharabi-Manzari, M. Heydari, S. Sadeghian-Abadi, M. Yousefi-Mokri, M.A. Faramarzi, Enzymatic dimerization of phenylacetylene by laccase immobilized on magnetic nanoparticles via click chemistry, Biocatal. Biotransformation 37 (2019) 455–465. https://doi.org/10.1080/10242422.2019.1611788

[47] S. Rezaei, R. Yazdian-Robati, M. Mogharabi-Manzari, M. Salehipour, Editorial: Medical and Pharmaceutical Applications of Nanomaterials: From Diagnosis to Treatment, Front. Nanotechnol. 6 (2024). https://www.frontiersin.org/journals/nanotechnology/articles/10.3389/fnano.2024.1386250

[48] X. Yang, P. Qiu, J. Yang, Y. Fan, L. Wang, W. Jiang, X. Cheng, Y. Deng, W. Luo, Mesoporous Materials–Based Electrochemical Biosensors from Enzymatic to Nonenzymatic, Small 17 (2021) 1904022. https://doi.org/10.1002/smll.201904022

[49] M. Etienne, L. Zhang, N. Vilà, A. Walcarius, Mesoporous Materials-Based Electrochemical Enzymatic Biosensors, Electroanalysis 27 (2015) 2028–2054. https://doi.org/10.1002/elan.201500172

[50] K.A. Kilian, T. Böcking, J.J. Gooding, The importance of surface chemistry in mesoporous materials: lessons from porous silicon biosensors, Chem. Commun. (2009) 630–640. https://doi.org/10.1039/B815449J

[51] Y. Esmaeili, M. Khavani, A. Bigham, A. Sanati, E. Bidram, L. Shariati, A. Zarrabi, N.A. Jolfaie, M. Rafienia, Mesoporous silica@chitosan@gold nanoparticles as "on/off" optical biosensor and pH-sensitive theranostic platform against cancer, Int. J. Biol. Macromol. 202 (2022) 241–255. https://doi.org/10.1016/j.ijbiomac.2022.01.063

[52] M. Hasanzadeh, N. Shadjou, M. de la Guardia, M. Eskandani, P. Sheikhzadeh, Mesoporous silica-based materials for use in biosensors, TrAC Trends Anal. Chem. 33 (2012) 117–129. https://doi.org/10.1016/j.trac.2011.10.011

[53] J.L. Olloqui-Sariego, J.J. Calvente, R. Andreu, Immobilizing redox enzymes at mesoporous and nanostructured electrodes, Curr. Opin. Electrochem. 26 (2021) 100658. https://doi.org/10.1016/j.coelec.2020.100658

[54] Y. Feng, Y. Xu, S. Liu, D. Wu, Z. Su, G. Chen, J. Liu, G. Li, Recent advances in enzyme immobilization based on novel porous framework materials and its applications in biosensing, Coord. Chem. Rev. 459 (2022) 214414. https://doi.org/10.1016/j.ccr.2022.214414

[55] I. Izquierdo-Barba, Á. Martinez, A.L. Doadrio, J. Pérez-Pariente, M. Vallet-Regí, Release evaluation of drugs from ordered three-dimensional silica structures, Eur. J. Pharm. Sci. 26 (2005) 365–373. https://doi.org/10.1016/j.ejps.2005.06.009

[56] M. Manzano, M. Vallet-Regí, Mesoporous Silica Nanoparticles for Drug Delivery, Adv. Funct. Mater. 30 (2020) 1902634. https://doi.org/10.1002/adfm.201902634

[57] D. Shao, Q. Gao, Y. Sheng, S. Li, Y. Kong, Construction of a dual-responsive dual-drug delivery platform based on the hybrids of mesoporous silica, sodium hyaluronate, chitosan and oxidized sodium carboxymethyl cellulose, Int. J. Biol. Macromol. 202 (2022) 37–45. https://doi.org/10.1016/j.ijbiomac.2022.01.033

[58] M. Vallet-Regí, F. Schüth, D. Lozano, M. Colilla, M. Manzano, Engineering mesoporous silica nanoparticles for drug delivery: where are we after two decades?, Chem. Soc. Rev. 51 (2022) 5365–5451. https://doi.org/10.1039/D1CS00659B

[59] E. Vélez-Peña, R. Morales, C. Reyes-Escobar, C.C. Torres, M. Avello, K.P. Marrugo, J. Manzo-Merino, J.B. Alderete, C.H. Campos, Mesoporous mixed oxides prepared by hard template methodology as novel drug delivery carriers for methotrexate, J. Drug Deliv. Sci. Technol. 73 (2022) 103483. https://doi.org/10.1016/j.jddst.2022.103483

[60] J. Kim, H.S. Kim, N. Lee, T. Kim, H. Kim, T. Yu, I.C. Song, W.K. Moon, T. Hyeon, Multifunctional Uniform Nanoparticles Composed of a Magnetite Nanocrystal Core and a Mesoporous Silica Shell for Magnetic Resonance and Fluorescence Imaging and for Drug Delivery, Angew. Chem. Int. Ed. 47 (2008) 8438–8441. https://doi.org/10.1002/anie.200802469

[61] S. Kwon, R.K. Singh, R.A. Perez, E.A. Abou Neel, H.-W. Kim, W. Chrzanowski, Silica-based mesoporous nanoparticles for controlled drug delivery, J. Tissue Eng. 4 (2013) 2041731413503357. https://doi.org/10.1177/2041731413503357

[62] M. Mogharabi, M. Abdollahi, M.A. Faramarzi, Toxicity of nanomaterials; An undermined issue, DARU J. Pharm. Sci. 22 (2014). https://doi.org/10.1186/s40199-014-0059-4

[63] H. Fatima, S. Shukrullah, H. Hussain, H. Aslam, M.Y. Naz, Chapter 11 - Utility of various drug delivery systems and their advantages and disadvantages, in: R. Pratap Singh, K. RB Singh, J. Singh, C.O. Adetunji (Eds.), Nanotechnol. Drug Deliv. Pharm., Academic Press, 2023: pp. 235–258. https://doi.org/10.1016/B978-0-323-95325-2.00015-8

[64] K.B. Seljak, P. Kocbek, M. Gašperlin, Mesoporous silica nanoparticles as delivery carriers: An overview of drug loading techniques, J. Drug Deliv. Sci. Technol. 59 (2020) 101906. https://doi.org/10.1016/j.jddst.2020.101906

[65] B. Sun, X. Zhen, X. Jiang, Development of mesoporous silica-based nanoprobes for optical bioimaging applications, Biomater. Sci. 9 (2021) 3603–3620. https://doi.org/10.1039/D1BM00204J

[66] F.W. Pratiwi, C.W. Kuo, S.-H. Wu, Y.-P. Chen, C.Y. Mou, P. Chen, Chapter Six - The Bioimaging Applications of Mesoporous Silica Nanoparticles, in: F. Tamanoi (Ed.), The Enzymes, Academic Press, 2018: pp. 123–153. https://doi.org/10.1016/bs.enz.2018.07.006

[67] X. Du, X. Li, L. Xiong, X. Zhang, F. Kleitz, S.Z. Qiao, Mesoporous silica nanoparticles with organo-bridged silsesquioxane framework as innovative platforms for bioimaging and therapeutic agent delivery, Biomaterials 91 (2016) 90–127. https://doi.org/10.1016/j.biomaterials.2016.03.019

Materials Research Foundations 173 (2025) 160-202 https://doi.org/10.21741/9781644903452-7

Chapter 7

Mesoporous Materials for Photocatalytic Degradation of Pollutants

A.M. Gutiérrez-Peralta, C.E. Pérez-García, E. Mas-Hernández, E. Quiroz-Pérez, J.A. de Lira-Flores*

Universidad Autónoma de Querétaro, Facultad de Química, Cerro de las Campanas s/n Col. Las Campanas, Querétaro, Querétaro, 76010, México

*julio.lira@uaq.mx

Abstract

This chapter explores the use of mesoporous materials as photocatalysts for contaminant degradation. It explains the principles of photocatalysis, where materials such as TiO_2 and ZnO generate reactive species under light to oxidize contaminants. Factors influencing efficiency, such as light intensity, pH, and photocatalyst dosage, are discussed. Different synthesis methods of mesoporous materials, such as sol-gel, are also reviewed. Finally, a patent analysis is presented, highlighting the growth in the use of these materials in water and air purification technologies.

Keywords

Mesoporous Materials, Photocatalysis, Environmental Remediation, Contaminant Degradation. Semiconductor Oxides

Contents

1. Introduction

Many human activities emit pollutants that can cause diseases in humans, fauna, and flora. For this reason, different methods have been proposed for the elimination of pollutants, among which the photocatalytic degradation of pollutants using mesoporous semiconductors stands out since these have a high surface area and the mesoporous structure can facilitate the surface diffusion of reagents towards the active site of the catalytic surface. One use has been water purification by removing organic pollutants at low concentrations and eliminating pathogenic microorganisms without leaving transient residual chlorine in wastewater.

Photocatalysis is a promising tool for contaminant degradation in diverse environments. This chapter addresses the fundamental principles of photocatalysis, which are necessary to understand the processes that govern light-activated chemical reactions. In addition, the synthesis and characterization of mesoporous materials, which play a key role in improving photocatalytic efficiency due to their high surface area and adsorption capacity, will be explored.

 Materials Research Forum LLC
https://doi.org/10.21741/9781644903452-7

A comprehensive patent review will also be presented, allowing the identification of the most recent applications and technological innovations in the field of photocatalysis, including its use in water and air purification. Finally, the mechanisms of contaminant degradation will be discussed, highlighting the kinetic models used to describe and predict the behavior of the photocatalytic process under different conditions. This chapter will provide a comprehensive view of how these elements interrelate to advance the development of more effective technologies in environmental remediation.

2. Principles of Photocatalysis

2.1 Fundamentals of Photocatalysis

Photocatalysis is a process in which a material (semiconductor) is pulsed by electromagnetic radiation from the ultraviolet (UV) or visible region to accelerate a chemical reaction. The energy of the incident light must succeed in exciting the semiconductor's electrons from the valence band to the conduction band, creating electron-hole pairs. The gap between the valence band and the conduction band in materials is called the band gap, and it is the minimum amount of energy that the photon of light incident on the semiconductor should have to generate a jump from the valence band to the conduction band and the generation of the electron-hole, Fig. 1. The electron-hole pair formation allows simultaneous oxidation and reduction between the semiconductor and molecules [1-3].

2.1.1 Mechanisms of Action of Photocatalysts

Photocatalysts facilitate the generation of reactive species capable of oxidizing or reducing compounds adsorbed on their surface, enabling the degradation of pollutants or the production of valuable chemical products. The mechanism begins when a photocatalyst is activated by light, triggering a series of interconnected events that ultimately lead to chemical transformations on its surface. The principal events are:

- Light absorption: when a photocatalyst is irradiated with light, the photons must have energy greater than or equal to the forbidden band of the material, the electrons in the valence band are excited to the conduction band, and a free electron is generated in the conduction band and a hole in the valence band both migrate to the surface of the material. For example, for titanium dioxide (TiO_2), the value of its forbidden band is about 3.2 eV, which means that light in the ultraviolet range ($\lambda < 387$ nm) is needed to be excited.
- Charge migration and separation: The generated electrons (negative charge) and holes (positive charge) can move to the surface of the photocatalyst. However, there is the possibility of recombination of the electron-hole pairs back to their basal state, resulting in the release of energy in the form of heat or radiation. Recombination reduces the efficiency of the photocatalysis process.
- Redox reactions: electrons in the conduction band can reduce acceptor molecules such as oxygen (O_2) to form reactive species (superoxide O_2^-) or reduce metal ions, while holes in the valence band oxidize donor molecules such as water (H_2O) or OH^- ions, producing hydroxyl radicals (OH^-), which are highly reactive [4, 5].

- Degradation: The radicals formed (OH and $O_2^{\cdot}$) together with other molecules such as hydrogen peroxide (H_2O_2) can decompose organic or inorganic contaminants, eliminate bacteria or viruses, or even decompose toxic gases.

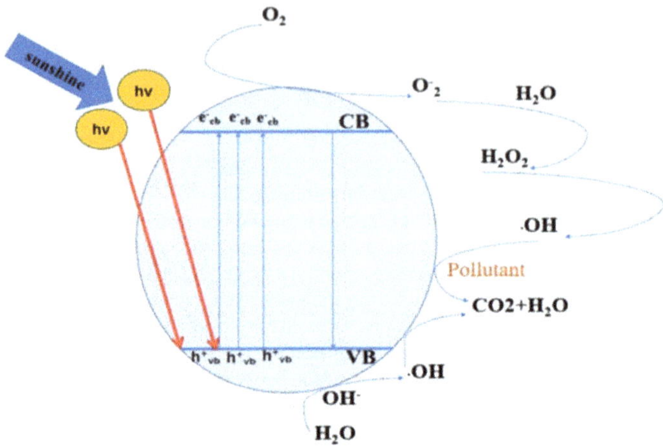

Figure 1. Photocatalytic reaction processes. Reproduced with permission from ref.[6]

2.1.2 Factors Affecting Photocatalysis

The efficiency of the photocatalysis process is influenced by several factors that affect the generation of reactive species and the kinetics of the redox reactions. The following paragraphs present some factors.

The intensity and wavelength of the light to which the photocatalyst material is exposed are fundamental because of the energy and amount it will promote the generation of reactive species (radicals). For this reason, the light intensity must be controlled to have the necessary number of photons to improve the reaction rate without reaching saturation or excess of light because if it is excessive, it can induce the recombination of the electron-hole pairs and the efficiency of the photocatalytic process decreases.

The second factor is the pH, which of the medium influences the surface charge of the photocatalyst as well as the nature of the species present in the solution. If the pH of the medium is acidic, there is a higher number of protonated species that can affect the redox behavior and interfere with the adsorption of contaminants. In contrast, in an alkaline medium, an excess of hydroxyl ions facilitates the formation of hydroxyl radicals, the main oxidizing species in the process.

Another factor initial concentration. The substrate or pollutant concentration impacts the efficiency of the photocatalysis when the substrate has low concentrations, the active sites of the photocatalyst are underutilized, and the degradation rate has a linear behavior concerning the

Materials Research Foundations 173 (2025) 160-202 https://doi.org/10.21741/9781644903452-7

concentration. An opposite effect can observed when concentrations are high; the system can become saturated, reducing the efficiency of the process due to the competition between molecules for the active sites on the semiconductor and the possible inhibition of the incident light by absorption of the species [2].

Temperature is a critical factor in photocatalysis because it can modify the reaction kinetics and the mobility of the molecules. Increasing the temperature favors the mobility and diffusion of species towards the active sites. However, higher temperatures can facilitate the recombination of electron-hole pairs and reduce the number of active sites, decreasing the efficiency of photocatalysis. It is of utmost importance to determine the appropriate temperature for the photocatalysis process depending on the material and substrate to reduce the recombination of electron-hole pairs and improve the efficiency of the process.

Finally, the dosage of photocatalyst is crucial as the amount of photocatalyst increases, so does the number of active sites on the surface, but there is a limiting concentration when the amount of photocatalyst is exceeded, i.e., the efficiency of the process does not increase due to the decrease in light transmission (shadow effect). Low dosages of the photocatalyst generate few available active sites, which limits the reaction rate, and an excess of photocatalysts can cause the agglomeration of particles, which reduces the surface area and decreases efficiency [5, 7].

In summary, factor optimization is crucial to improve the efficiency of photocatalytic processes, and it is essential to find a balance between them depending on the specific conditions of the system.

2.2 Common Photocatalytic Materials

Common photocatalytic materials are titanium dioxide (TiO_2), zinc oxide (ZnO), and cadmium sulfide (CdS). These materials are often selected based on their ability to generate reactive species under specific wavelengths, playing a vital role in pollutant degradation, water splitting, and air purification.

2.2.1 Titanium Dioxide

TiO_2 is a semiconducting oxide that can be presented in anatase, rutile, and brookite phases, being the anatase phase the most common with a band gap of approximately 3.2 eV that absorbs UV light with a wavelength of less than 385 nm. In addition to the fact that TiO_2 is a material with thermal and chemical stability, non-toxic, and accessible cost, it makes it attractive for diverse applications, highlighting its application in photocatalysis. In addition, the surface of TiO_2 has a high capacity to generate reactive species (hydroxyl radicals and superoxide radicals), which means that it can used as a photocatalyst.

However, it has the disadvantage of only absorbing UV light, representing approximately 5% of the solar spectrum. For this reason, ways are being developed to activate it with the energy of visible light radiation. The surface modification is achieved by doping with some metal to change the bandwidth value. In addition, it sought to decrease the recombination of pairs (electron/hole) by making TiO_2 heterostructure or in heterojunctions with other semiconductors, Fig. 2 [8-10].

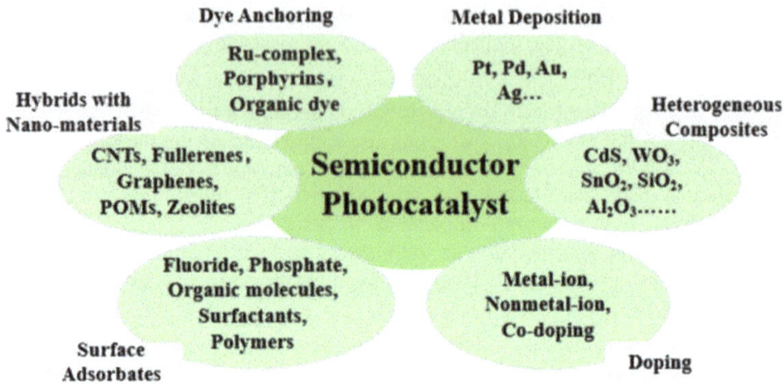

Figure 2. Modification methods of a semiconductor photocatalyst. Reproduced with permission from ref. [6]

TiO_2 is widely investigated and used in photocatalysis for degrading organic pollutants and dyes and for hydrogen generation in solar cells.

2.2.2 Zinc Oxide

ZnO is an abundant material in nature, and there are several synthesis methods of ZnO synthesis to obtain and produce low-cost material. In addition, ZnO possesses attractive properties such as thermal and chemical stability, an exciton energy of approximately 60 meV, non-toxic, biocompatible, and a band gap of approximately 3.37 eV, which classifies it as a transparent semiconducting oxide.

ZnO generally presents the hexagonal wurtzite structure to obtain a high degree of crystallinity in the material, which favors the mobility of the charge carriers. On the other hand, by synthesizing ZnO by different methods, different shapes and sizes (thin films, nanoparticles, nanotubes) can be obtained, increasing its surface area and making it a candidate used as a photocatalyst.

ZnO as a photocatalyst also has restrictions, such as the energy that activates it is visible light, which is a small fraction of the electromagnetic spectrum, and high recombination of the pairs (electron/hole). After an extended period, it loses chemical stability, which decreases photocatalytic efficiency. Several strategies must be explored to modify and improve the photocatalytic behavior through doping, increasing the surface area, heterojunctions, or heterostructures of ZnO with other metals or nonmetals [11-13].

The current use of ZnO as a photocatalyst has allowed the removal and degradation of pollutants such as dyes, drugs, and pesticides. It is also used for air purification and has been observed to have bactericidal properties in cosmetic products.

TiO_2 and ZnO are the most studied and used photocatalyst semiconductor oxides with similar forbidden bands and physicochemical properties. However, ZnO has advantages such as a higher electron-hole pair generation rate and a more versatile synthesis. Nevertheless, both semiconductors face similar challenges in limited absorption to UV light and high charge carrier recombination [14].

Other metallic and non-metallic materials ($SrTiO_3$, Ag_3PO_4, CdS, MoS_2, TaON) have demonstrated adequate photocatalytic properties, and their physical and chemical properties are currently being investigated to increase and modify their efficiency in the photocatalytic process. In addition, it is intended to increase their surface area to make photocatalytic materials with different morphologies (nanotubes, nanofibers, and nanoparticles) at the nanometric level [15].

3. Synthesis of Mesoporous Materials for Photocatalysis

Natural porous materials such as clays or zeolites have outstanding properties (adsorption, high specific surface area, chemical and thermal stability). However, they are limited by the heterogeneity of pore size and arrangement, but this has generated interest in reproducing materials with a large number of pores with defined and tunable structures (size and shape) in a controlled manner, thus resulting in materials with a large surface area, a desirable property (Fig. 3) that will improve performance in applications such as catalysis, photocatalysis, adsorption, separation (chromatography), energy storage and drug release [16, 17]. According to IUPAC, porous materials are classified as microporous with pore diameters $\leq$ 2 nm, mesoporous with pore diameters between 02-50 nm, and macroporous with pore diameters $\geq$ 50 nm [18, 19].

In the 1990s, Mobil Oil Corporation patented the synthetic porous material MCM-41 with an engineered and modifiable porous structure that exhibited better properties than natural porous materials, and from this developed other families of synthetic porous materials (SBA, FDU, KIT, MSU) based on silica and aluminosilicates [20]. This material promoted the creation of a new line of research worldwide due to the potential applications in different sectors and many publications on porous materials, from synthesis techniques to pore hierarchization, surface functionalization, and applications [21, 22].

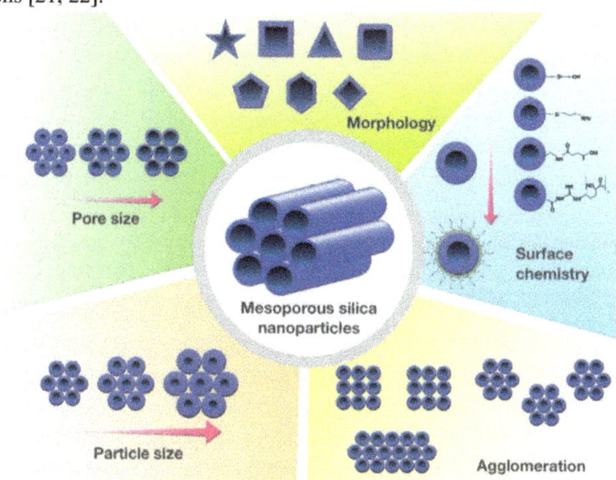

Figure 3. Tunable physicochemical properties of mesopore materials, Adapted from [17]

 Materials Research Forum LLC
https://doi.org/10.21741/9781644903452-7

The importance of mesoporous materials lies in the size of their pores; it allows larger interaction (diffusion) of molecules with the walls of the material. The pores can act with active sites that can host specific molecules (dyes, drugs, inhibitors, catalysts, additives, among others) and are released when exposed to a stimulus such as a change in pH, temperature, reducing or oxidizing agent, exposure to light and magnetic field; the capability to regulate the stimuli allows a controlled and specific release that is required in biomedical applications, sensors or catalysis. In addition, mesoporous materials have high regularity in shape, size, pore arrangement, and large pore volume, forming an open structure (skeleton of the solid) that allows better transfer of molecules. The structural skeleton is amorphous and have long porous channels arranged with different types of arrangements that, although it has an orderly and well-defined pattern, are not long-term, which is the reason for the lack of crystallinity. The arrangement determines which family of mesoporous materials it belongs to MCM, SBA, FDU, KIT, or COK (Fig. 4) [23, 24].

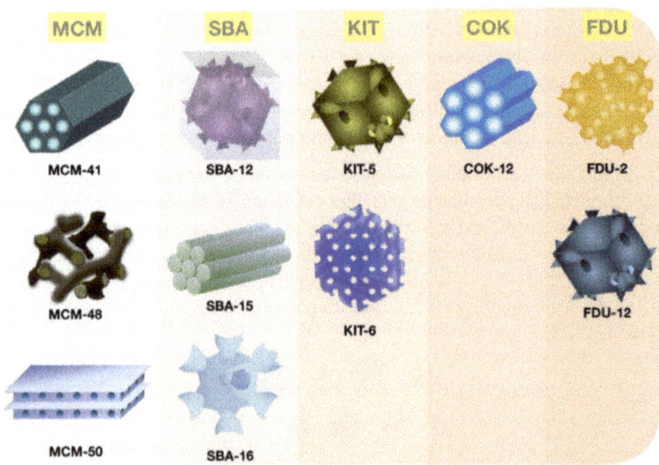

Figure 4. Families of traditional mesoporous materials. Adapted from [17]

The first synthetic mesoporous materials were based on clays and ceramic oxides (silica, alumina); however, the research for other options to obtain a solid porous structure based on polymeric materials (polyurethanes, polystyrene, polyesters), noble metals (Au and Pt), metal oxides (niobium oxide, titanium oxide, zirconium oxide) and carbon (activated carbon and ordered mesoporous carbon) has expanded, which present different properties and cover current needs such as energy demand (solar cells, capacitors), biomedical (drug release, bone or tissue regeneration), chemical (catalysis, supports, absorbents) and environmental (adsorbent, photocatalysis) [20, 25, 26].

The study of synthesis and surface functionalization of mesoporous materials spans more than three decades and is still ongoing. The choice of synthesis technique depends on the nature, type of mesoporous structure required, properties, and application for the material. In addition, the synthesis methods must allow control of diameter, shape, structure, uniform pore distribution, and pore volume. Several synthesis techniques include hard and soft templating methods, sol-gel, phase separation, template impregnation, and self-assembly-induced evaporation [27-29].

3.1 Mesoporous Materials Synthesis Methods

The most common and used way to synthesize synthetic mesoporous materials is generated by chemical reactions by deposition on substrate (sol-gel, spray coating), where an organic molecule (surfactant) consisting of a polar group and apolar group is the pattern or template to make the porous material and is the one that determines the pore size. There is also a solution prepared with the surfactant and a solvent to the minimum concentration necessary to generate micelles that can be spherical, hexagonal and lamellar (Fig. 5): The precursors of the material are added to the solution where it reacts with the micelles and starts the self-assembly (hydrolysis, condensation) of the material around the micelles (templates), later the templates must be eliminated (calcination, dissolution). Finally, the amorphous solid with a porous structure with an ordered arrangement is obtained [24, 28].

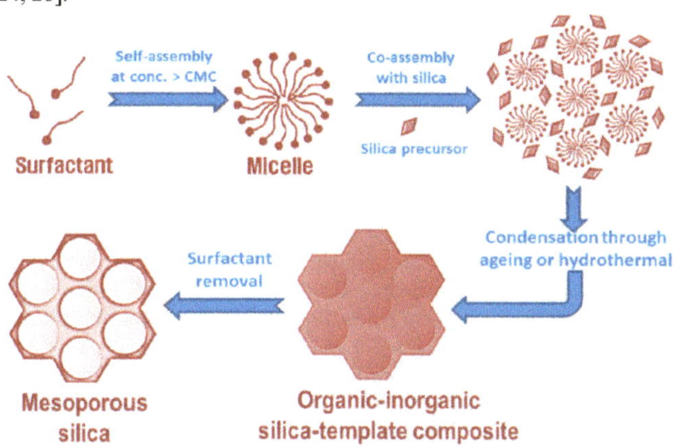

Figure 5. General synthesis route for mesoporous using surfactant-templated. Reproduced with permission from ref. [30]

One of the most widely used techniques due to versatility and efficiency is the Sol-Gel method, which allows obtaining mesoporous materials of oxides, mesoporous carbons, or silicas. This method requires preparing a solution of inorganic precursors with a solvent. Catalysts can modify the hydrolysis and condensation reaction kinetics depending on the reaction between the precursors and the solvent. After condensation and formation of Mx-OMx, a gel starts to form during this step. The template molecule must be added, the gel ages to achieve densification, and the template must be removed by temperature or solvent extraction to obtain a solid porous structure [16, 29].

Impregnation in hard templates is when the precursor material (organometallic or polymeric) is infiltrated containing the well-defined preform. The hard template's material and shape depend upon the size and shape of the desired pore in the final solid porous structure. The infiltration occurs via one of three methods: evaporation, pressure variation, or capillarity. Following

Materials Research Foundations 173 (2025) 160-202
https://doi.org/10.21741/9781644903452-7

infiltration, the precursor material must be solidified through polymerization or hydrolysis, with heat application (sintering or calcination) often necessary to create the desired mesoporous material. The final step is removing the template, which can be achieved through an acidic or basic chemical solution, depending on the chemical nature of the template or the temperature used to calcine the hard template. In the case of carbon- or polymer-based templates, this step can present a challenge to complete removal, which results in elevated costs [28, 29] (Fig. 6).

An alternative approach to synthesizing mesoporous materials is phase separation, which involves the preparation of a mixture of the precursor (inorganic, polymeric, or organic) with a solvent, followed by adding a porogenic agent to induce phase separation. The solvent significantly impacts the phases obtained and the shape of the final mesoporous structure. The separation of the two phases (one phase comprising a higher concentration of the precursors and the other phase containing the solvent in a higher concentration) is achieved by exposing the mixture to a stimulus, such as temperature changes, or by evaporating the solvent from the mixture. Subsequently, the precursor-rich phase must undergo polymerization or rapid cooling to solidify it. Subsequently, the removal of residual porogen or solvent is required. In some cases, this can be achieved through simple washing, while in other instances, dissolution or calcination may be necessary, depending on the composition of the initial mixture [19].

Evaporation-induced self-assembly (EISA) enables the synthesis of mesoporous materials based on silica, metal oxides, or carbon. This technique prepares a solution with the material's precursors, a seed or structuring agent, and a volatile solvent, such as water or hexane. Subsequently, the interior of the mold is filled uniformly, and the evaporation of the volatile solvent commences. The precursors and seed agent undergo a chemical reaction, thereby initiating the formation of the mesoporous material with an ordered structure, which ultimately results in a porous structure. It is essential to regulate solvent evaporation to prevent defects in the solid mesoporous skeleton. Subsequently, the seed agent must be eliminated through high-temperature processes, such as calcination, which is hard templating. The mold or hard template typically comprises silica or polymeric mesoporous materials, exhibiting the desired pore size and arrangement. Subsequently, the precursors are infiltrated into the template by impregnation or casting. This process allows the formation of the mesoporous material, which must then be removed from the template by either chemical dissolution or temperature without affecting the structure of the mesoporous material [24, 26].

The hydrothermal technique is widely utilized as a synthesis method for mesoporous materials, offering the advantages of high purity and well-controlled morphology. A homogeneous solution must be prepared from the precursors and surfactants in the synthesis. This solution is then introduced into an autoclave subjected to elevated temperatures and pressures to promote the solution's crystallization and simultaneously initiate the formation of micelles (molds). These micelles facilitate the interaction between the hydrolyzed and condensed precursors and the mesoporous materials, forming structured and pore-sized materials. Subsequently, the templates must be removed, typically achieved through applying temperature or solvents [23, 31, 32].

Materials Research Forum LLC
https://doi.org/10.21741/9781644903452-7

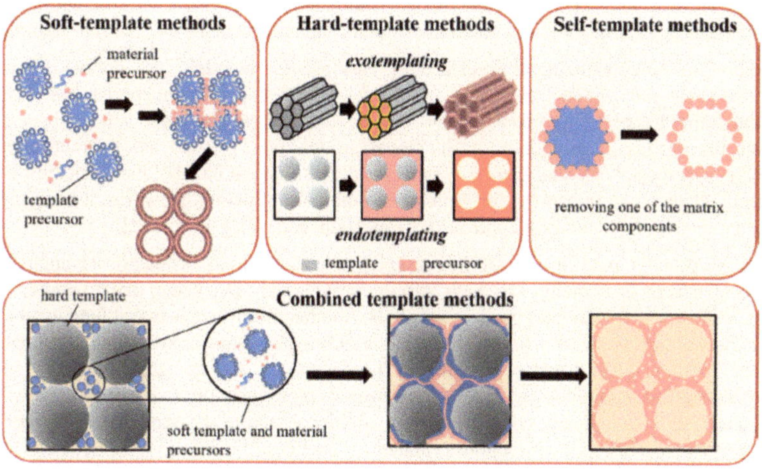

Figure 6. Template synthesis methods. Reproduced with permission from ref.[29]

3.2 Surface Modification

The process of modifying and functionalizing the mesoporous surface by hosting specific molecules or functional groups allows tailoring, customizing, and improving properties such as chemical and thermal stability, adsorption, biocompatibility, or catalysis and thus opening new potential applications and improving performances of mesoporous materials.

Mesoporous materials can be functionalized through various physical and chemical methods. Among the chemical routes is the co-condensation process, which occurs during the synthesis of the material. This method involves the addition of organosilane-type precursors containing the desired functional groups to the solution (precursor and solvent). The resulting material is mesoporous with functional groups integrated into the solid porous skeleton. Another method for surface functionalization via chemical reactions is grafting silanes or polymers. This process entails the interaction of graft molecules with the porous surface, resulting in the formation of covalent bonds and the incorporation of the mesoporous structure [21, 27].

The adsorption property of mesoporous materials is employed to host specific molecules or functional groups within the high pore volume. The molecules interact with the surface via weaker bonds, including van der Waals forces, Coulombic interactions, and charge transfer. Additionally, nanoparticles are incorporated onto the surface of mesoporous materials, leveraging the size of the nanoparticles and the pores. They are deposited and embedded within the solid mesoporous structure.

The enhanced selectivity in the interactions between the mesoporous material and specific molecules is one of the most significant advantages resulting from modifying the mesoporous surface. This selectivity is highly advantageous in applications such as separation by adsorption or catalysis. Furthermore, incorporating specific functional groups into the mesoporous solid structure through modification and functionalization enhances chemical and thermal stability[30].

In addition to functionalizing the surface of mesoporous materials, the option of creating mesoporous materials with a dual-porosity structure comprising two distinct pore types of varying sizes and shapes has been investigated to enhance the surface area. This approach could potentially facilitate the hosting of molecules or functional groups that may increase their efficiency in adsorption, photocatalysis, and detection processes. This method has the disadvantage of being technically challenging to achieve this type of dual porosity structure. However, it has generated significant interest in biomedical applications, detection, and catalysis [17, 32].

As with any process, modification of the mesoporous surface has associated disadvantages. One such disadvantage is reducing the high surface area, which can occur due to molecules or functional groups' partial or total blockage of pores. Another disadvantage is that functional groups may degrade or deactivate over time or if the material is exposed to extreme stimuli (temperature, exposure to light). Furthermore, achieving homogeneous surface area modification and functionalization is a complex process, and ensuring its reproducibility is a significant challenge. Additionally, as it is typically conducted as an additional step to the synthesis, it generates an additional cost, increasing the mesoporous material's final cost[33, 34].

4. Characterization of Mesoporous Materials for Photocatalysis

4.1 Characterization Techniques

Mesoporous materials are structures with pore sizes ranging from 2 to 50 nm, which give them unique properties and wide applications, including water and air remediation. Different characterization techniques are used for these materials to understand and evaluate their properties and to obtain their morphological, structural, electronic, and functional characteristics [35].

4.1.1 Electron Microscopy (SEM, TEM.)

Scanning electron microscopy (SEM) and transmission electron microscopy (TEM) are used to obtain the morphology of materials [36, 37].

The SEM technique uses an electron beam to scan the surface of the materials. When the electrons hit the material's surface, backscattered electrons and x-rays are generated, which are detected by the equipment and used to obtain a high-resolution image of the material's surface under analysis [38].

From this, the following information can be obtained about our mesoporous materials:

- Particle size and morphology, since the SEM shows the particles' size, distribution, and shape in the analyzed sample. The porous structure of the mesoporous materials can also be observed [39].
- Surface topography, pores, and defects on the material's surface can be evaluated. It is observed if the material presents defects in an analyzed area and to observe the uniformity and texture of the material [40].
- Elemental Composition: the SEM equipment can have coupled an Energy Dispersive X-ray Spectroscopy (EDS) equipment, which takes advantage of the X-rays generated by the sample

when the electron beam hits its surface, and the EDS detector analyzes those X-rays, thus being able to detect the chemical elements present in the sample since each component emits X-rays with different energies. The distribution of elements in the material analyzed area can be mapped [41].

Transmission electron microscopy (TEM) is a technique used to obtain the internal structure of nanometric materials to analyze the mesostructure in mesoporous materials, referring to the organization of pores in the material [42]. It provides a two-dimensional image where electrons pass through the sample, generating the internal structure image [43].

To be analyzed by TEM, the sample must be cut into skinny slices to allow the transmission of electrons and to obtain a high-resolution image, so this technique presents a greater difficulty in preparing the sample[43, 44].

Fig. 7 shows images corresponding to mesoporous silica modified with titanium (THMS). In the SEM image (7A), spherical particles with an approximate particle size of 500 nm are observed. In 7B, the mesoporous structure is visible, with pores distributed around the particle, in accordance with the results observed in the TEM image, which provides a more profound image of the structure. In the high-resolution TEM image (7C), the 2 to 3 nm pores are visible, and a highly periodic order is observed. In Figures 7D-7F, the EDS mapping is observed, where the elements present in the THMS (oxygen, silicon, and titanium) are observed.

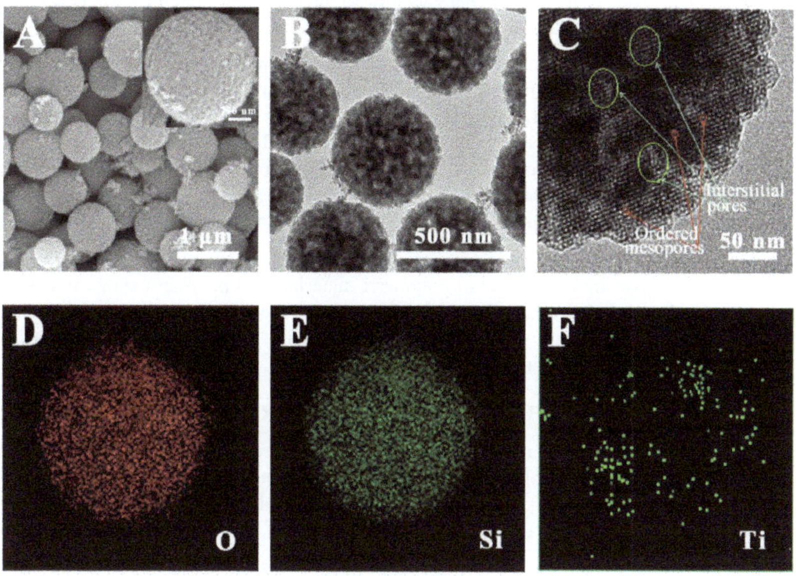

Figure 7. presents scanning electron microscopy (SEM) and transmission electron microscopy (TEM) images of the THMS, accompanied by elemental mapping images (D to F). Reproduced with permission from ref.[45].

Materials Research Foundations 173 (2025) 160-202
https://doi.org/10.21741/9781644903452-7

4.1.2 X-Ray Diffraction (XRD)

The X-Ray Diffraction (XRD) technique can be used to understand the atomic distribution, the percentage of crystallinity, and the pores' size. This technique is fundamental for the characterization of mesoporous materials since we can obtain the material's crystalline phase through the obtained diffraction planes [46]. The procedure consists of making radiation with X-rays fall on the sample so that they diffract when interacting with the sample's atoms when they hit it. The detector detects the angle at which they diffract and the intensity of the scattered rays, thus revealing the crystalline structure of the sample; the obtained diffractogram is compared with the database of different diffractograms being able to obtain the compound to which the diffraction pattern belongs and the crystalline structure it presents; it is also possible to obtain the characteristic distance of the crystalline structure, the pore network and their size, as well as the crystallite size, and the percentage of crystallinity of the analyzed sample [47]. Finally, diffraction patterns are fundamental for characterizing mesoporous materials [48].

4.1.3 Nitrogen Adsorption (BET)

As previously discussed, one of the primary functions of mesoporous materials is the adsorption of contaminants or molecules of interest. Therefore, one of the fundamental characterizations is nitrogen adsorption on the surface, which allows for the calculation of the surface area of the material [49]. This is a crucial characteristic of mesoporous materials, as it determines the material's surface area, which is typically large in these materials [50].

The characterization entails passing an adsorbable gas, in this case N_2, over the mesoporous material (adsorbent) at varying pressures and by differing initial and final pressures, enabling the calculation of the amount remaining in the material's pores and the acquisition of data regarding the gas adsorbed on the material over a specified time [51]. This procedure is conducted at a designated temperature, typically 77K. Once the data have been collected, they are adjusted to the BET adsorption isotherm (so-called by the researchers who developed it, namely Brunauer, Emmet and Teller) at a specific temperature, and the surface area of the mesoporous material is obtained [52].

In light of the growing prevalence of polymeric materials and polymer/nanoparticle composites, characterization techniques have been employed to examine these novel mesoporous materials, as outlined in the preceding sections. These include the use of Fourier Transform Infrared Spectroscopy (FTIR) and Raman Spectroscopy.

4.1.4 Fourier Transform Infrared Spectroscopy (FTIR)

FTIR is a technique that measures the absorption of infrared radiation by molecules in a sample. This radiation generates molecular vibrations that generate infrared (IR) spectrum peaks. The correlation between these peaks and specific functional groups allows for identifying the functional groups present in the sample [51].

In mesoporous materials, FTIR allows the identification of specific functional groups that exist on the surface of the materials, such as hydroxyl (OH) or some added cations [46]. In the case of new mesoporous materials of polymeric fibers [53], it is used to identify the polymer used and whether it forms a bond with any incorporated nanomaterial. Therefore, FTIR can provide information about the surface and composition of mesoporous materials [54].

Materials Research Forum LLC
https://doi.org/10.21741/9781644903452-7

4.1.5 Raman Spectroscopy

Raman spectroscopy is a spectroscopic technique that employs the inelastic scattering of light by molecules to obtain chemical and structural information about a sample.

Raman spectroscopy is a high-resolution photonic technique that provides chemical and structural information about various organic and inorganic materials and chemical compounds. It is based on the inelastic scattering of light generated when a molecule interacts with electromagnetic radiation (typically a laser). As a result of this interaction, some of the radiation's frequencies change, which is characteristic of the material being analyzed [55].

Raman spectroscopy is employed for the following purposes:

- The identification of materials is a key objective of Raman spectroscopy.
- The technique is also employed for the analysis of molecular structures.
- To evaluate chemical bonds.

In mesoporous materials, the technique can be applied to determine the presence of functional groups on pore walls. Evaluating the pores' uniformity and the synthesis's quality is also necessary.

The advantages of this approach are as follows:

- The method is non-destructive, meaning it does not alter the surface of the material under examination.
- The high resolution of this method allows for the detection of even minor alterations in molecular structure.
- The method can be applied directly to the material without any special preparation.

In the case of mesoporous materials, Raman spectroscopy is employed to identify functional groups, evaluate structures, and characterize chemical bonds [55].

4.2 Photocatalytic Properties

4.2.1 Band Gap, UV-VIS (Ultraviolet-Visible) spectroscopy

Mesoporous materials offer a large surface area and an ordered structure, rendering them optimal for catalytic applications. In particular, mesoporous materials exhibit photocatalytic properties due to their porous structure and capacity to interact with light.

The characteristics of these materials render them ideal for a variety of applications. These include their large surface area, which provides a conducive environment for photocatalytic reactions; their porous structure, which allows for the diffusion of reactants and products, thereby enhancing the efficiency of the reactions; which facilitates access to active sites on the surface of the material [56].

Mesoporous materials have been employed in various photocatalytic reactions, including the degradation of organic pollutants in water and air under sunlight, hydrogen production from water by photolysis, and CO_2 reduction.

The potential of mesoporous materials for photocatalysis is a promising avenue for environmental protection and the advancement of clean and renewable energy sources. For these mesoporous materials to facilitate the reaction above mechanisms, they must possess specific characteristics. One such characteristic is that semiconductor materials must present a band gap energy (Eg) or a band gap, which must be adequate to achieve optimum efficiency in photolysis.

The energy gap (or energy difference) between the valence band (VB) and the conduction band (CB) in a semiconductor material is defined as the Eg. A smaller energy gap allows the material to absorb sufficient energy for an electron from the valence band to jump to the conduction band, forming an electron-hole pair. This pair is essential for photocatalysis, as it drives the generation of oxidation and reduction reactions [16]. The Eg of the material affects the efficiency of the photocatalytic process. Lower energies result in higher efficiencies. Additionally, pore size is a significant factor in determining the efficiency of photocatalysis.

Photocatalysis is a process based on the direct or indirect absorption of radiant energy (visible or UV) by a solid (the photocatalyst, typically a broadband semiconductor). In the interfacial region between the excited solid and the solution, the reactions of destruction or removal of the contaminants occur without the catalyst undergoing chemical changes.

The photocatalysis reaction is initiated when the semiconductor absorbs a photon of equal or higher energy than its band gap, transferring an electron from the valence band to the conduction band. This process generates a hole in the valence band and an electron in the conduction band. In this manner, electron-hole pairs are generated, which migrate through the solid and recombine if they reencounter one another. These electron-hole pairs are reducing or oxidizing agents when they encounter surface groups or absorbed molecules on the semiconductor surface [57].

The efficiency of this process is contingent upon several parameters, including the intensity of the light, the nature and concentration of the pollutant, the reaction temperature, and the concentration of the photocatalyst.

4.2.2 Ultraviolet-Visible Spectroscopy (UV-Vis)

Ultraviolet-visible spectroscopy plays a pivotal role in the photocatalysis of mesoporous materials, as the photocatalysis phenomenon occurs at this frequency. Additionally, it enables the identification of the band gap of the mesoporous material to be utilized [56]. Also, UV-Vis warrants the measurement of light absorption by a solution as a function of wavelength [58]. In semiconducting materials, such as titanium dioxide (TiO_2) and zinc oxide (ZnO), a band gap is observed between the valence and conduction bands. The band gap determines the capacity of the material to absorb light and generate electron-hole pairs.

The activation of photocatalysts consists of ultraviolet-visible light that excites electrons in the valence band, thereby creating electron-hole pairs. These carriers engage in redox reactions, such as the oxidation of pollutants or the generation of hydrogen. It is instrumental in the characterization of photocatalytic efficiency and the optimization of materials for specific applications, such as the synthesis of mesoporous materials for photocatalysis. It provides a wealth of information that enables the understanding and enhancement of photocatalysis, facilitating the design of more efficient materials.

4.2.3 Stability and Reusability

The objective is to ascertain the stability and reusability of the material in question. It is crucial to assess the stability and reusability of mesoporous materials in photocatalysis for their practical implementation, as these properties ensure the efficacy and durability of the materials in the degradation of environmental pollutants [59].

The reusability of the mesoporous material in photocatalysis pertains to its capacity to retain its photocatalytic efficacy following multiple cycles of utilization. Experiments are conducted wherein the material is subjected to a radiation source in the presence of a pollutant, which facilitates the photocatalytic reaction to assess this attribute. Subsequently, the material is retrieved and subjected to a new photocatalytic experiment [60]. The following aspects are evaluated:

The number of reaction cycles is defined as the number of times the photocatalytic reaction is performed. Each cycle is referred to as a cycle. Subsequently, the material is washed and recovered after each reaction cycle to remove any residual contamination from the surface [61]. A post-cycle analysis is conducted to assess the photocatalytic activity of the mesoporous material, which is then compared with the activity observed in the initial cycle.

The assessment of the stability of the mesoporous material is contingent upon two factors: the reusability of the material and its structural integrity. The reusability of the material is determined by observing the extent to which its photocatalytic activity is retained following the completion of several cycles of utilization. If the material demonstrates consistent stability or exhibits only a slight decline in stability after multiple cycles, it is deemed stable and reusable. The structural integrity of the mesoporous material is evaluated to ascertain its capacity to maintain its structural configuration. Accordingly, the following criteria are employed to assess the stability of a mesoporous material:

- Structural degradation: The objective is to ascertain whether the material undergoes alterations in its structure (such as pore size, porosity, pore shape, and crystalline structure.) throughout the various cycles of photocatalysis. The loss of activity is also evaluated. The decrease in photocatalytic activity over time is quantified.
- Degradation resistance is assessed by determining whether the material resists chemical or physical degradation induced by light and reactive species generated during photocatalysis [62].

5. Applications in the Degradation of Contaminants

This section shows a patent analysis related to the use of mesoporous materials in pollutant degradation through photocatalysis. The data was primarily sourced from Espacenet, a global database providing access to millions of patent documents worldwide [63]. The purpose was to identify technological trends and key players and assess the state of innovation in photocatalysis. Patents from various regions have been reviewed, focusing on the types of mesoporous materials used, the photocatalysts employed, and specific applications in environmental remediation. The information gathered provides a detailed view of the international patent landscape, offering more profound insights into the opportunities and challenges faced by this ever-evolving technological field.

Figure 10 shows the increase in the number of patent families related to using mesoporous materials to degrade pollutants by photocatalysis over time (x-axis) and the cumulative number of patent families (right axis in blue). Here is an interpretation of the figure: Steady growth from 2005 onwards: Before 2005, the number of patents related to this technology was deficient. From that year onwards, a gradual increase in published patents was observed, i.e., research in photocatalysis with mesoporous materials began to gain interest in the first decade of the 21st century, perhaps due to a greater awareness of the need for sustainable and efficient technologies to combat pollution. Since 2010, the curve shows a much more pronounced growth in the number of patents, indicating a significant increase in the innovation and development of this technology due to advances in the field of nanotechnology, as well as a growing demand for solutions for water treatment and pollutant reduction. From 2016 onwards, the curve shows some fluctuations; this behavior could reflect changes in research trends, funding variations, or environmental regulation adjustments that could have affected patent publication. Still, the overall numbers remain robust compared to previous years.

The blue line shows a continuous cumulative increase, indicating that although the patents published annually may fluctuate, innovation in this field continues to rise. By the end of the period shown (around 2023), the total number of patent families is close to 2,000, highlighting the growing importance of this technology.

Although the graph extends to approximately 2023-2024, the upward trend in the cumulative number suggests that photocatalysis research and the development of mesoporous materials will continue to be an active area of innovation, likely driven by the need for cleaner and more sustainable technologies.

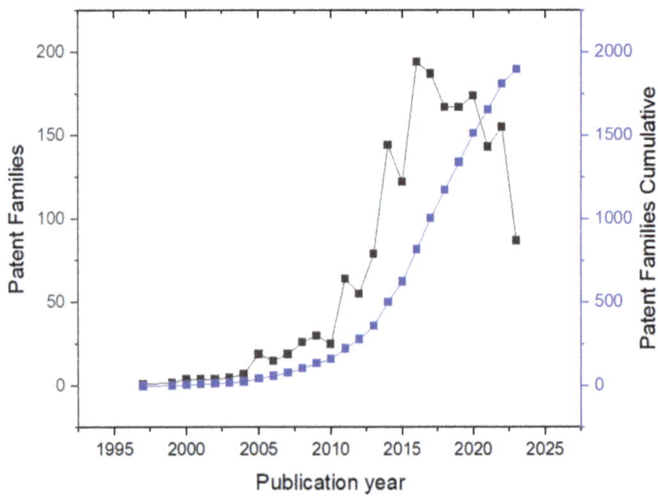

Figure 10. Number of patents by publication year (gray), and accumulative number (blue)

Figure 11 shows the distribution of patents related to contaminant degradation using photocatalysis, classified by the inventors' country. China has a lead in patent documents (150), focusing on developing photocatalysis technologies and mesoporous materials. The country is likely investing considerably in research and development in this area, driven by the need to address environmental pollution and improve water and air quality. The US is in second place, with 120 patents. This demonstrates the continued interest in innovation in sustainable technologies, where mesoporous materials for photocatalysis can be a key answer to environmental and energy challenges, such as wastewater treatment and pollutant reduction. Saudi Arabia (SA), in third place with 78 papers, its investments could be due to advanced technologies to mitigate pollution problems and develop new water treatment forms, which are crucial in arid regions.

Diverse global contribution: Beyond the top three countries, there is a remarkably even distribution of papers in South Korea (KR), France (FR), United Kingdom (GB), Spain (ES), India (IN), and Canada (CA), all with patent counts varying between 20 and 31. This underscores that research and development in photocatalysis is a field of global interest, with active participation from diverse developed and emerging countries in addressing environmental challenges.

Countries such as Brazil, Colombia, and Romania have fewer papers, with six patents each, which could be related to local research initiatives or international collaborations in environmental technologies.

This geographical distribution may also reflect where mesoporous materials and photocatalysis technologies are being developed, driven by advances in nanotechnology and the growing global awareness of sustainability and the need to reduce the environmental footprint.

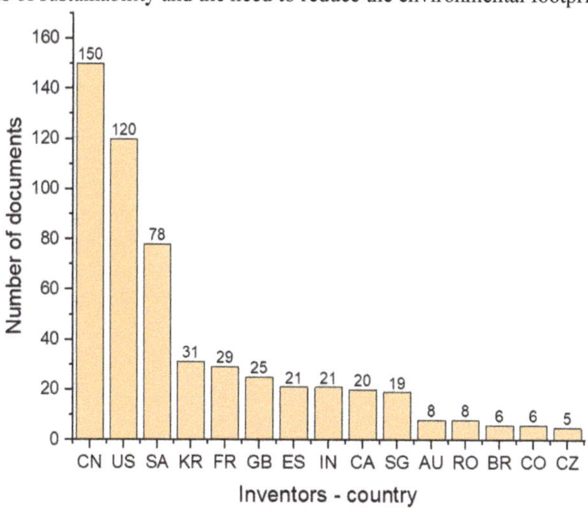

Figure 11. Number of documents by country

Table 1 reflects a broad overview of patents using mesoporous materials to degrade pollutants through photocatalysis, covering various research areas and technological applications. The following interpretation related to this data is presented:

- Broad focus on catalysis and ion exchange (B01J): With 2620 patents, the most represented area is related to catalytic processes and ion exchange. This suggests that research on mesoporous materials for photocatalysis is deeply linked to developing innovative catalysts and optimizing their efficiency.

- Importance of wastewater treatment (C02F): With 2105 patents, water and waste treatment is another critical area, indicating that mesoporous materials are being widely applied in water purification through photocatalysis, a potentially effective solution for the removal of organic and inorganic pollutants.
- Separation Technologies (B01D): With 541 patents in this field, mesoporous materials are also being explored to improve solid-liquid separation techniques. This can be crucial in separating pollutants from aqueous media after photocatalytic degradation.
- Reduction of greenhouse gas emissions (Y02): The 513 patents in this category underscore the relevance of mesoporous materials and their application in green technologies. This focus, possibly on CO2 capture and climate change mitigation through photocatalytic processes, inspires hope for a more sustainable future.
- Nanotechnology (B82Y): With 107 patents, nanotechnology is gaining ground in the manufacturing and development of mesoporous materials. This trend indicates that advances in nanoscale manipulation are crucial to improving the efficiency of these materials in photocatalytic applications, paving the way for exciting future developments.
- Composition of industrial products and coatings (C08K and C09D): The presence of 96 and 85 patents in these fields, respectively, shows that mesoporous materials are also being investigated to improve the formulation of industrial products, including photocatalytic coatings that can break down pollutants.
- Other specialized applications: Areas related to textiles, cement, batteries, and chemical analysis (D06M, C04B, H01M, G01N) are also represented, although to a lesser extent. Mesoporous materials have a broader potential, from manufacturing advanced textiles to creating products for energy storage or more precise chemical analysis.

The table shows that mesoporous materials in photocatalysis have wide-ranging applications. These include water treatment, emissions reduction, green technologies, and improving industrial products and nanotechnology. The versatility of these materials is truly intriguing.

Materials Research Foundations 173 (2025) 160-202
https://doi.org/10.21741/9781644903452-7

Table 1. Principal International Patent Classifications (IPC) to mesoporous materials for photocatalytic degradation of pollutants

Series	Number of Patents	Description
B01J	2620	Covers various aspects of ion exchange and catalyst-related processes.
C02F	2105	Encompasses different methods and processes for treating waste and sewage water.
B01D	541	Deals with separation techniques for solids or liquids.
Y02	513	Includes patents for technologies aimed at reducing greenhouse gas emissions.
C01B and C01P	336	Relates to silicon compounds and their preparation.
B82Y	107	Involves nanotechnology-related patents.
C08K	96	Focuses on industrial product compositions and applications.
C09D	85	Pertains to paint and coating compositions and their uses.
A62D	38	Relates to products for fire and smoke protection.
D01D and D01F	37	Related to textile fiber manufacturing processes.
C04B	32	Involves cement and glass manufacturing technologies.
A61L	26	Covers products for wound treatment and disinfection.
D06M	25	Pertains to textile treatment and manufacturing processes.
C25B	19	Relates to electrolytic processes and devices.
H01M	17	Focuses on electrochemical cells and batteries.
G01N	11	Involves instruments for chemical analysis.

6. Photocatalytic Degradation Mechanisms

Photocatalytic degradation is an oxidation process, which utilizes light energy to induce the degradation of pollutants. The oxidation and hydrolysis of molecules take place by absorption of photons in the visible (vis), ultraviolet (UV) and infrared (IR) electromagnetic regions. Migration of the photo responsive electron from the valence to the conduction band is caused by the activation of the photocatalytic material with light because a photon has equivalent or greater energy than the catalyst band gap energy causing its absorption. This generates pairs of photo-induced electrons (e^-) and holes (h^+), which can react via redox reactions with oxygen, water and hydroxyl groups on the catalyst surface to produce reactive oxygen species (ROS). The pollutants degrade from the result of the interaction between their molecules and the reactive oxygen species [64-68].

Photocatalysts are solid semiconductors with the ability to absorb the light energy from a specific source and induce the photo-carriers to stimulate the photocatalytic reaction [69].

Jabbar et al. [69] describe the photocatalytic reaction as follows: when the semiconductor is illuminated by light energy

$$\text{Photocatalyst} + \text{light energy } (h\nu) \rightarrow e^- + h^+$$

the photo-induced electrons (e^-) migrate from the photocatalyst valence band (VB) to its conduction band (CB), while the holes (h^+) are maintained on its VB. Hydroxyl radicals ($\cdot OH$) can be produced from oxidation reactions between the holes and water molecules

$$h^+ + H_2O \rightarrow \cdot OH + H^+,$$
$$h^+ + OH \rightarrow \cdot OH.$$

Reduction reactions between the photo-excited electrons and the dissolved oxygen on the valence band, generate the hydrogen peroxide (H_2O_2) and the superoxide radicals ($O_2^{\cdot -}$)

$$e^- + O_2 \rightarrow O_2^{\cdot -},$$
$$O_2 + 2H^+ + 2e^- \rightarrow H_2O_2.$$

Subsequently, the created H_2O_2 and/or $O_2^{\cdot -}$ is converted to $\cdot OH$ according to

$$O_2^{\cdot -} + 2H_2O_2HO\cdot + 2HO\cdot + O_2 + H_2O_2 + O_2^{\cdot -} \rightarrow \cdot OH,$$
$$H_2O_2 + e^- \rightarrow \cdot OH,$$
or
$$O_2^{\cdot -} + e^- \rightarrow \cdot OH.$$

After that, the pollutants can be converted to less toxic products by oxidation or reduction established by the free reactive oxygen species (ROS) in the pollutant solution.

$$ROS + pollutant \rightarrow CO_2 + H_2O + intermediates.$$

The photocatalyst efficiency depends on several factors, such as band gap, electron-hole pair separation efficiency, and optical absorption. Notwithstanding, some photocatalysts have wide band gaps between the valence and conduction bands, effective only in the UV region, which demands higher energy. Also, electrons of a semiconductor photocatalyst with a narrow band gap recombine rapidly with photogenerated holes. Another problem relates to the kinetics for recombining photo-generated electrons/holes pairs and the degradation process [64, 67].

6.1 Light-Material-Pollutant Interaction

6.1.1 Reactive oxygen species (ROS) generation

Reactive oxygen species (ROS) contain oxygen atoms that make them chemically active. They are found in the environment, and living forms play an important role in different processes (biological and chemical) that produce specific molecules, ions, and radicals. ROS are important for pollutant degradation because they can scavenge undesired substances from the environment. Due to their strong redox characteristics, ROS can degrade pollutants via photocatalysis [66, 70].

Understanding the principle of photocatalytic molecular oxygen activation and the process of ROS generation is important in improving the efficiency of molecular oxygen activation and increasing the redox capacity of ROS, because they have different properties and application areas [71].

ROS are generated from photochemical reactions of the substances; for example, the photolysis method uses UV light radiation to produce the ions or radicals. Some examples of ROS are hydroxyl radicals ($\cdot OH$, a nonselective oxidant with high oxidation ability. It is frequently considered as the most effective oxidant reactant as it can quickly oxidize most organic compounds and has an important role in advanced oxidation technologies), hydrogen peroxide (H_2O_2, a stable form of ROS with a long lifetime and high oxidation capability), superoxide anion radicals ($O_2^{\cdot -}$, an important intermediate for the formation of other ROS under certain conditions), singlet oxygen ($^1O_2$, an excited state for oxygen molecules), hypochlorite ions (ClO^-), and peroxynitrite ($ONOO^-$), among others [64, 66, 67, 69, 71].

Some examples of chemical reactions for the ROS transformations are: the reduction of O_2 forms $O_2^{\cdot -}$ and the oxidation of H_2O produces H_2O_2. Further reduction of H_2O_2 gives $\cdot OH$, while further oxidation of $O_2^{\cdot -}$ forms $^1O_2$.

Moreover, specific pathway generation for some ROS via photocatalytic oxygen activation is illustrated in Fig. 12 and is described next [71, 72]:

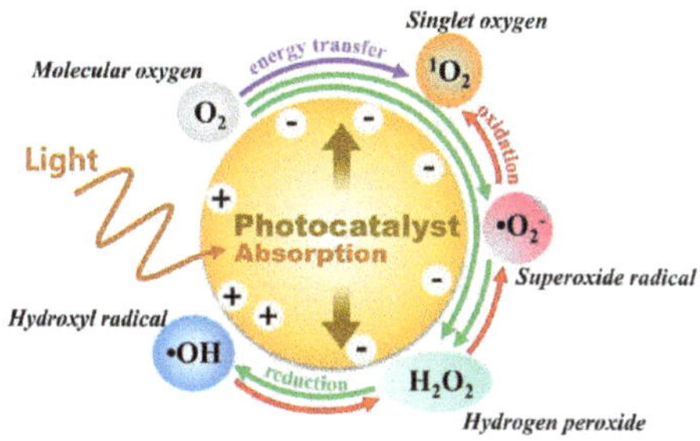

Figure 12. Photocatalytic molecular oxygen pathway [71].

The first step in the electron transfer pathway to generate ROS involves carrier generation in the presence of light, which separates and migrates. The superoxide radical, $O_2^{\cdot -}$, can then be generated by a single electron transfer process which activates the oxygen adsorbed on the surface of the catalyst,

$$e^- + O_2 \rightarrow O_2^{\cdot -},$$

Also, $O_2^{\cdot -}$ can be generated via a redox reaction of H_2O_2 with h^+, then the one-electron process is the main pathway for its production:

$$H_2O_2 + h^+ \rightarrow O_2^{\cdot -} + 2H^+.$$

The photocatalysts that generate $O_2^{\cdot -}$ are TiO_2, CdS, ZnS, and In_2S_3 [72].

Singlet oxygen, $^1O_2$, can be obtained by the energy transfer pathway from molecular oxygen: the ground state sensitizer absorbs energy and becomes a unilinear mention excited state sensitizer, then becomes a lower energy triplet excited state sensitize after undergoing a intersystem crossing process, finally it transfers the energy to the triplet state molecular oxygen ($^3O_2$) to activate it to $^1O_2$.

$^1O_2$ can also be produced via the electron transfer pathway, undergoing successive reduction of O_2 and oxidation of $O_2^{\cdot -}$:

$$O_2 + e^- \rightarrow O_2^{\cdot -} + h^+ + {}^1O_2.$$

The energy transfer process is the main generation path of $^1O_2$.

Hydrogen peroxide, H_2O_2, can be produced by either a consecutive two-step single-electron reduction pathway

$$O_2 + e^- \rightarrow O_2^{\cdot -} + e^- + 2H^+ \rightarrow H_2O_2,$$

or a one-step double-electron reduction pathway

$$O_2 + 2e^- + 2H^+ \rightarrow H_2O_2.$$

Also, H_2O_2 can be synthesized through the coupling of two •OH, in the absence of oxygen

$$\cdot OH + \cdot OH \rightarrow H_2O_2.$$

The hydroxyl radical, $\cdot OH$, can be obtained by the consecutive three-step single-electron transfer pathway

$$O_2 + e^- \rightarrow O_2^{2-} + 2H^+ \rightarrow H_2O_2 + e^- \rightarrow OH^- + \cdot OH.$$

The hydroxyl radical can also be obtained by water oxidation:

$$H_2O + h^+ \rightarrow \cdot OH + H^+.$$

Semiconductors act as catalysts due to the electronic arrangement of the selected metal (identified by an empty conduction band and filled a valence band, divided by a band or energy gap). Advantages of the most common semiconductors used as photocatalysts (ZnO and TiO_2), are that they are inert, photo catalytically stable, easy to obtain and use, moreover, they are less expensive and do not cause risks to human and the environment [65, 68, 69].

In this regard, TiO_2 produces ROS according to the following equations [66]:

$$TiO_2 + h\nu \rightarrow e^- + h^+$$
$$e^- + O_2 \rightarrow O_2^{\cdot -}$$
$$H_2O + h^+ \rightarrow \cdot OH + H^+.$$

More details on generations mechanisms for ROS with TiO_2 are reported in Nosaka and Nosaka [72].

6.1.2 Specific degradation mechanisms

The target water pollutants to degrade via photocatalysis are dyes, pharmaceutically active compounds such as antibiotics, and heavy metals. In addition, photocatalytic oxidation can also be used for degradation or air contaminants, especially volatile organic compounds. Several works focus on specific pollutants, providing degradation mechanisms according to the considered substances in the process. Next, we present some mechanisms available in the literature.

Regarding dyes, the degradation process depends on dyes' adsorption on the photocatalyst's surface and the separation of photogenerated charges.

The mechanism is described in the following equations [65]:

$$\text{Photocatalyst} + h\nu \rightarrow e^- + h^+,$$
$$OH^- + h^+ \rightarrow \cdot OH,$$
$$e^- + O_2 \rightarrow O_2^{\cdot -},$$
$$O_2^{\cdot -} + H^+ \rightarrow HO_2^{\cdot},$$
$$2HO_2^{\cdot} \rightarrow H_2O_2 + O_2,$$
$$H_2O_2 + e^- \rightarrow OH^- + \cdot OH,$$
$$H_2O_2 + h^+ \rightarrow \cdot OH + H^+,$$
$$\text{Dye} + \cdot OH \rightarrow \text{dye decolorization},$$
$$\text{Dye} + h^+ \rightarrow \text{dye oxidation},$$
$$\text{Dye} + e^- \rightarrow \text{dye reduction}.$$

Xie et al. [68] also present the reaction equations which are similar to the above mechanism.

Chacón-García et al. [70] study the use of a photoactive Bi-MOF (SU-101) in the degradation of diclofenac and sulfamethazine. The pathways and mechanisms they determined for these processes are as follows: dechlorination and further hydroxylation of diclofenac results in the formation of 2-[2-(2-chloro-6-hydroxyanilino)phenyl]acetic acid. Then, glycolic acid and fumaric acid are formed. In a second degradation process, some diclofenac is converted into the 2,6-dichloro-N-(2-methylphenyl)aniline by decarboxylation and subsequent C-N excision to give the 4-amino-3,5-dichlorophenol at 4 h. The possible transformation of diclofenac into 1-chloro-8-methyl-9H-carbazole gives a low-intensity fragment at m/z = 215. Regarding the sulfamethazine, the main way occurs by the scission of its S···N bond. The process proceeds through SO_2 extrusion from sulfanilic acid and subsequent hydroxylation to phenol, which was transformed into succinic acid by aromatic ring breakage and decomposed into acetic acid.

Tang et al. [73] report a 2D/2D CeO_2/BiOCl S-cheme photocatalyst constructed by mesoporous CeO_2 and BiOCl nanosheets and tried its effect on four antibiotics: tetracycline, chlortetracycline, doxycycline and oxytetracycline under UV light radiation. They use a heterojunction photocatalyst, which includes two or more semiconductors with different banded structures, boosting photocatalytic performance via charge separation enhancement and light absorption improvement. Consequently, this improves the photocatalytic performance of antibiotic degradation.

More specifically, they provide the tetracycline (TC) photodegradation mechanism over the CeO_2/BiOCl composite: CeO_2 and BiOCl surfaces undergo photoexcitation upon exposure to UV-light irradiation. This generated e^- and h^+. Then the photoinduced e^- on BiOCl CB would complex with the photoinduced h^+ on CeO_2 VB, allowing and efficient pair separation and imparting an important redox capacity to the composite. The e^- and h^+ react with oxygen and H_2O is used to generate ROS, which are the oxidative species that are present during pollutant degradation. The equations for this mechanism are as follows:

$$BiOCl + h\nu \rightarrow e^- + h^+$$
$$CeO_2 + h\nu \rightarrow e^- + h^+$$
$$e^-_{BiOCl} + h^+_{CeO_2} \rightarrow h\nu$$
$$e^-_{CeO_2} + O_2 \rightarrow O_2^{\cdot -}$$
$$h^+_{BiOCl} + H_2O \rightarrow \cdot OH + H^+$$
$$h^+_{BiOCl} + TC \rightarrow \text{degradation products}$$
$$O_2^{\cdot -} + TC \rightarrow \text{degradation products}$$
$$\cdot OH + TC \rightarrow \text{degradation products}$$

Concerning volatile organic compounds (VOCs), Ahmadi et al. [74] present a review focused on photocatalysts for these pollutants. They present a general mechanism of degradation: photoexcited e^-s transfer from the VB to the CB, leaving positive holes (h^+) in the photocatalyst. These e^-/h^+ pairs play an important role in redox reactions during the decomposition of VOCs. They may also react with the surface-adsorbed VOC molecules as they migrate to the surface of the semiconductor. This can produce reactive oxygen species that initiate redox reactions. Finally, pollutant degradation is achieved through the redox reactions over the CB and VB of photocatalysts.

Xu et al. [75] present a study on nitric oxide (NO) removal using TiO_2/NH_2-MIL-125(Ti). The mechanism they propose for the catalytic degradation is as follows: photogenerated electrons are collected at the conduction band of TiO_2, and photogenerated holes are collected at the valence

  https://doi.org/10.21741/9781644903452-7

band of NH2-MIL-125(Ti). The photogenerated electrons collected at the conduction band of TiO_2 are unlikely to produce $O_2^{\cdot-}$, because the conduction band position of TiO_2 is -0.24 eV than the redox potential -0.33 eV ($O_2/O_2^{\cdot-}$) more positive. At the same time, it is not able to generate $\cdot OH$ because the valence band position is significantly lower than $+2.38$ eV ($OH^-/\cdot OH$). The photocatalytic oxidation of NO is achieved using TiO_2/NH_2-MIL-125(Ti).

Other studies on photocatalytic degradation can be consulted in the following references: [76, 77].

6.2 Efficiency and Kinetics

6.2.1 Factors affecting efficiency

Several factors affect the efficiency of pollutant photocatalytic degradation: pollutant concentration, species present in the effluent or environment, the type and quantity of ROS present in the system, photocatalyst characteristics such as the capability of charge carrier separation, surface area, band-gap energy, and electron-hole pair recombination [65, 66, 71, 78]. With respect to the pollutant itself, the degradation process is affected by the nature of the pollutant and the initial concentration of the pollutant. Many authors report that a high initial concentration of the pollutant causes a decrease in the photocatalytic activity due to the saturation of the available sites on the surface of the photocatalyst, and a low concentration increases the pollutant photodegradation [78].

Other factors are related to the presence of ions, compounds, salts, and scavengers in the matrix of the water. Scavengers can decrease the rate of degradation as a result of trapping the $O_2^{\cdot-}$ and $\cdot OH$ radicals [78].

Regarding specific studies, Zhu et al. [79] synthesized mesoporous C@ZnO hollow nanospheres with no treatment, a pickling sample, and a sample via pickling and annealing to study the photocatalytic degradation of Rhodamine B. The degradation efficiency was determined as $\eta = (C_0 - C_t)/C_0 \times 100$ %, where C_0 is the initial concentration of Rhodamine, and C_t is the concentration after t minutes of light exposure. The degradation efficiency of the second and third samples reached 97.16 % and 96.06 % in 30 min, respectively. They conclude that pickling or annealing improves photocatalytic performance by suppressing charge recombination due to the reduced surface or internal crystal defects that are recombination centers of photogenerated carriers.

The review by Ahmadi and Kim [74] present detailed information on environmental factors affecting efficiency regarding photocatalytic degradation of VOC's.

6.3 Kinetic models

Photocatalytic kinetics aims to describe and predict the reaction rate and mechanism alterations that occur during photocatalysis. It considers the following factors: light conditions, catalyst properties, and reactant concentrations, among others. The research on photocatalytic kinetics is important for the development of efficient photocatalysts for the degradation of pollutants [68].

Regarding the kinetics for ROS, the decrease of $O_2^{\cdot-}$ in the equation

$$O_2^{\cdot-} + \cdot O_2H + H_2O \rightarrow H_2O_2 + O_2 + OH^-,$$

Materials Research Foundations 173 (2025) 160-202
https://doi.org/10.21741/9781644903452-7

the decay kinetics is second order. Also, for nitrogen- and sulfur-doped TiO_2 powders, kinetics is second order. Firs order kinetics was reported for TiO_2 films and fractal-like kinetics for TiO_2 powders in aqueous suspensions [72].

For the photocatalytic degradation of tetracycline antibiotics [73], the kinetics were studied by the pseudo-first-order kinetic model. The degradation constant of TC for their $0.4CeO_2/BiOCl$ composite was 0.01524 min^{-1}.

Zhu et al. [79] used the pseudo first-order kinetic equation $\ln(C_0/C_t) = Kt$, where K is the kinetic rate constant and t is the irradiation time. The rate constant for each sample were 0.0978, 0.117, and 0.106 min^{-1}, respectively.

Kinetic models can then be defined as mathematical formulations that describe the time-dependent relationships between the operating conditions of a system and the degradation rate of pollutants in the presence of photocatalysts [80]. In this sense, the proper application of kinetic models for the analysis and interpretation of experimental data can contribute to the optimization of the photocatalytic degradation of contaminants and the design and operation of the photoreactors used in these processes [81, 82].

6.3.1 The Langmuir-Hinshelwood model

Several studies have reported that the kinetics of photocatalytic degradation of pollutants using mesoporous materials can be described through the Langmuir-Hinshelwood (L-H) model or sub-models derived from the L-H model. The degradation rate of pollutants in photocatalytic processes can be calculated with the L-H model for any of the following conditions [83]:

I) A reaction takes place while two species are adsorbed.

II) A reaction takes place between a non-bound radical and an adsorbed organic molecule.

III) A reaction takes place between an adsorbed radical and a free organic molecule arriving at the catalyst surface.

IV) A reaction takes place between two free species in the fluid phase.

The L-H model is based on the following mechanism: (I) adsorption of molecules of a pollutant on the surface of a catalyst, (2) dissociation of the adsorbed molecules, (3) reactions of dissociated molecules to generate products, and (4) desorption of the products from the catalyst [80]. In this case, the photocatalytic degradation of pollutants is expressed according to the Eq. 1 [83]:

$$r = -\frac{dC}{dt} = \frac{k_r K_{ads} C}{1 + K_{ads} C} \quad (1)$$

where r is the degradation rate, k_r is the degradation constant, C is the concentration of the pollutant, t is the time, and K_{ads} is the adsorption equilibrium constant. The application of Eq. 1 depends on the following assumptions: the reaction system is in dynamic equilibrium, the reaction is surface mediated, and the competition for the active surface sites by intermediates and other reactive species is not limiting [84].

Eq. 1 can be solved by using a discrete change for C, from an initial concentration (C_0) to a concentration at a certain time (C_t), which gives:

$$\ln\left(\frac{C_t}{C_0}\right) + K_{ads}(C_t - C_0) = -k_r K_{ads} t \quad (2)$$

For those cases where removal of the pollutant in the dark period for a fixed catalyst loading is low, adsorption phenomena is neglected. This assumption has been reported in studies where the concentration of pollutant is very low [80, 85]. In these cases, the value of K_{ads} is very small and, in turn, $K_{ads}C << 1$. Therefore, Eq. 1 can be written as:

$$-\frac{dC}{dt} = k_r K_{ads} C \tag{3}$$

Eq. 3 can be rearranged and integrated from C_0 to C_t, which gives:

$$\ln\left(\frac{C_t}{C_0}\right) = -k_1 t \tag{4}$$

where k_1 is the apparent first order constant, which is expressed as $k_1 = k_r K_{ads}$. The term *apparent* indicates that all influencing parameters, such as catalyst/degradation agent concentration, irradiation rate, temperature, etc., are constant and indirectly included in k_1 [86, 87]. Considering this, Eq. 4 is commonly referred to as *pseudo-first-order kinetic model.* In this case, the value of k_1 is given by the slope from the linear regression of the term in the left side of Eq. 4 vs. t.

6.3.2 Pseudo-order kinetic models

It has been reported that, in some cases, the photocatalytic degradation of pollutants does not follow a pseudo-first-order kinetic model [88, 89], which, as demonstrated, can be derived from the Langmuir-Hinshelwood model. In these cases, the use of a simplified rate law has been proposed [90]:

$$r = -\frac{dC}{dt} = k_n C^n \tag{5}$$

where k_n is the apparent nth-order constant, and n is the apparent or *pseudo* rate order, typically in the range $0 \leq n \leq 2$. In this case, it can be noticed that, for $n = 1$, rearrangement and integration of Eq. 5 from C_0 to C_t leads to Eq. 4. For $n \neq 1$, on the other hand, integration of Eq. (5) yields to the following general expression:

$$C^{1-n} - C_0^{1-n} = k_n(n-1)t \tag{6}$$

For $n = 2$, Eq. 6 is rewritten as:

$$\frac{1}{C} = \frac{1}{C_0} + k_2 t \tag{7}$$

where k_2 is the apparent second order constant. In this case, Eq. 7 is known as *pseudo-second-order kinetic model.*

For $n = 0$, Eq. 6 becomes:

$$C - C_0 = -k_0 t \tag{8}$$

where k_0 is the apparent zeroth order constant. Eq. 8 is known as *pseudo-zeroth-order kinetic model.*

The values of k_2 and k_0 can be obtained from the linear regression of the term in the left side of Eq. 7 and Eq. 8, respectively, vs. t.

Materials Research Foundations 173 (2025) 160-202
https://doi.org/10.21741/9781644903452-7

6.3.3 Conventional kinetic models in reported studies

Given the importance of kinetic models, several authors have reported important kinetic details in their studies on the photodegradation of diverse pollutants using mesoporous materials. These kinetic details include the reaction order that follows a given photodegradation process and the value of the corresponding apparent constant rate. In this sense, most of the research has focused on the photocatalytic degradation of organic dyes, pharmaceuticals, and pesticides, which are considered three major classes of organic pollutants [91, 92]

Different mesoporous catalysts have been used for the photodegradation of various organics dyes, such as Congo Red (CR) [93], Amaranth [94], Remazol Black XP (RxP) [95], Rhodamine B [96], and Methyl Orange (MO) [97]. In this case, the authors of all the reviewed studies have reported that a pseudo-first-order kinetics model can represent the photocatalytic degradation of the organic dyes. This can also be seen in Table 2, which summarizes the main aspects of some recent studies on this topic. In this case, it has been reported that the value of the apparent constant rate, k_1, depends on some factors, including the concentration of the catalyst [95, 98], and the catalyst's preparation method. According to Erusappan et al. [99], TiO_2 prepared by through a hydrothermal method showed a better performance and higher values of k_1, than TiO_2 prepared using a method based on solution combustion during the degradation of Congo Red (CR) and Reactive Yellow 145 (RY-145).

Different studies have proposed using mesoporous materials for the photodegradation of pharmaceuticals, mainly antibiotics [100-102], although the degradation of calcium channel blockers has also been explored [103]. The main aspects of these studies are presented in Table 3. In this case, it has been reported that the concentration of some reactants used to prepare the mesoporous catalyst influences the degradation rate. For instance, Zeng et al. [101] studied the photodegradation of different antibiotics with B-N/TiO_2. According to the authors, the catalysts prepared with 300 mg of boric acid exhibited higher values of the apparent first-order constant and better photocatalytic activity than those prepared with 50, 100, 200 and 500 mg of boric acid. It is worth mentioning that, even though most of the authors have reported that the photodegradation reaction follows a pseudo-first-order kinetic model, some studies have indicated that a different kinetic model offers a better representation of the process. In this sense, Abbas et al. [89] have reported that the photodegradation of cefazolin using a nanoheterojunction of TiO_2 nanoparticles over ZSM5 is better described by a pseudo-second-order kinetic model.

The main pesticides that have been considered in photodegradation studies where mesoporous catalysts were used include trifluralin, 2,4 dichlorophenoxyacetic acid (2,4-D), glyphosate and chlorpyrifos [104, 105]. In this case, Table 4 summarizes the main aspects of some recent studies on the photocatalytic degradation pf pesticides. In this sense, most of the studies on this topic have reported that the photodegradation of pesticides follow a pseudo-first-order kinetic model. However, Ejeta and Imae [106] have reported that a pseudo-zeroth-order kinetic model better represents the photodegradation of 2,4-D with oxygen-doped graphitic carbon nitride (Og-CN). According to the authors, the apparent kinetic constant, k_0, exhibited a higher value for the experiments performed with UV irradiation compared to those performed with visible light. This indicates that the photocatalytic process does not depend on the concentration of the pollutant but rather on the intensity of irradiation.

Materials Research Foundations 173 (2025) 160-202
https://doi.org/10.21741/9781644903452-7

Table 2. Main aspects of recent studies on photocatalytic degradation of dyes

Catalyst	Pollutant	Kinetic model	Findings	Reference
Fe_2O_3/SiO_2 monoliths. Maximum pore diameter: 2–14 nm.	Remazol black XP (RxP).	Pseudo-first-order kinetics.	A catalyst concentration 0.10 g/L resulted in the highest catalytic efficiency (97.8%). The corresponding value of *k* is 0.0421 min^{-1}.	[95]
Fe_2O_3/g-C_3N_4 monoliths. Pore diameter: < 50 nm.	Rhodamine B (RhB)	Pseudo-first-order kinetics.	A value of 0.0452 min^{-1} was estimated for *k* at a catalytic efficiency of 94.78%, pH = 7, and a catalyst concentration of 0.1 g/L.	[98]
TiO_2/PVDF membranes. Average pore diameter: 2.4–5.1 nm.	Congo Red (CR). Reactive Yellow 145 (RY-145).	Pseudo-first-order kinetics.	TiO_2 prepared through a modified hydrothermal method exhibited a better performance than that prepared using solution combustion synthesis. The highest values of *k* were 0.0156 min^{-1} for CR and 0.0329 min^{-1} for RY-145.	[99]
Au_9 nanoclusters deposited on S-modified TiO_2 (Au_9 NCs/SMTiO_2). Average pore diameter: 4 nm.	Methyl Orange (MO)	Pseudo-first-order kinetics.	Experiments with Au_9 NCs/SMTiO_2 exhibited a higher value of *k* (0.149 min^{-1}) than those with SMTiO_2 and only under UV irradiation (0.0702 and 0.004 min^{-1}, respectively).	[97]

Materials Research Forum LLC
https://doi.org/10.21741/9781644903452-7

Table 3. The main aspects of recent studies on photocatalytic degradation of pharmaceuticals

Catalyst	Pollutant	Kinetic model	Findings	Reference
Carbon quantum dots loaded with g-C_3N_4 (g-C_3N_4/CQDs). Average pore diameter: 12 nm.	Ofloxacin.	Pseudo-first-order kinetics.	The values of *k* gradually improved, from 0.0128 to 0.0350 min^{-1}, following the increase of CQDs, from 0 to 0.5 wt%.	[100]
B-N codoped TiO_2 (B-N/TiO_2). Average pore diameter: 8.727 nm.	Flumequine (FLU). Norfloxacin (NOR). Ofloxacin (OFL).	Pseudo-first-order kinetics.	B-N/TiO_2 prepared with 300 mg of boric acid exhibited the highest photocatalytic activity, compared with B-N/TiO_2 prepared with 50, 100, 200 and 500 mg of boric acid. The values of *k* for FLU, NOR and OFL were 0.0223, 0.0258 and 0.0238 min^{-1}, respectively. These values were higher than those estimated for the photolysis process without catalyst.	[101]
A nanohetero-junction of TiO_2 nanoparticles over ZSM5. Average pore diameter: 33.9 nm.	Cefazolin compound.	Pseudo second order kinetics.	The mesoporosity added many effective sites on the surface of the nanoheterojunction photocatalyst boosted the high decomposition efficiency of the Cefazolin compound. The estimated value of *k* is 0.014 g gm^{-1} min^{-1}.	[89]

Materials Research Foundations 173 (2025) 160-202
https://doi.org/10.21741/9781644903452-7

Table 4. Main aspects of recent studies on photocatalytic degradation of pesticides

Catalyst	Pollutant	Kinetic model	Findings	Reference
Magnetic hexagonal mesoporous silica containing N-doped TiO_2 (FM/TiN). Average pore diameter: 12.137 nm.	Trifluralin. 2,4 dichloro phenoxy acetic acid (2,4-D). Glyphosate.	Pseudo first order kinetics.	FM/TiN prepared with a ratio of urea/titanium tetra isopropoxide (TTIP) of 60 exhibited the best photocatalytic performance, compared with FM/TiN prepared with lower ratios of urea/TTIP. The highest values of *k* were obtained for experiments with UV irradiation, being: 0.1019, 0.1298 and 0.0265 min^{-1} for the trifluralin, 2,4-D and glyphosate, respectively. These values are 12 times higher than those obtained for experiments with visible light.	[104]
Graphene oxide (GO)-Fe_3O_4/TiO_2. Average pore diameter: 4.08 nm.	Chlorpyrifos (CP).	Pseudo first order kinetics.	The estimated value of *k* is 0.0525 min^{-1}, which is higher than that reported in other studies for the photodegradation of CP using different catalysts.	[105]
Oxygen-doped graphitic carbon nitride (Og-CN). Average pore diameter: 26.68 nm.	2,4 dichloro phenoxy acetic acid (2,4-D).	Pseudo zeroth order kinetics.	Og-CN exhibited a value of *k* higher than gCN in the experiments with both UV and visible light. In this case, the highest value of *k* was obtained for the experiments with UV irradiation (0.00975 ppm min^{-1}).	[106]

6.3.4 Non-conventional kinetic models in reported studies

According to some authors, there are certain cases where none of the models (i.e. Langmuir-Hinshelwood model and pseudo-order models) are appropriate to represent the photodegradation of pollutants with mesoporous; for example, those systems where the contribution of adsorption mechanisms is significant, and the concentration of pollutant in the adsorbed phase is substantial.

In this sense, Deng et al. [107] have proposed the use of a modified form of the L-H model to describe the adsorption-photocatalysis synergy for the degradation of Methyl-Orange (MO) using mesoporous TiO_2-BiOBr microspheres. In this case, the proposed model is based on the following assumptions: (1) the rate of degradation is proportional to the concentration on the surface of the catalyst, (2) adsorption equilibrium is achieved throughout the experiment, and (3) the Freundlich adsorption isotherm describes equilibrium. The model is given by Eq. 9:

Materials Research Foundations 173 (2025) 160-202
https://doi.org/10.21741/9781644903452-7

$$r = -\frac{dC}{dt} = \frac{k_f D C^n}{1 + K n D C^{n-1}} \tag{9}$$

where D is the concentration of photocatalyst, k_f is the overall rate constant ($k_f = K$). K and n are known as the fitted Freundlich coefficients. According to the authors, the overall rate constant calculated through the numerical solution of Eq. 9 for the experiments with TiO_2-BiOBr was 3 times that calculated for the experiments with pure BiOBr.

Faycal et al. [108] have also proposed modifying the L-H model to obtain formulation that describes the degradation rate as a function of the specific surface area of the catalyst. In this case, the proposed model has the following form:

$$r_A = \frac{V}{A_{\mathrm{BET}}\, m_c} k_{\mathrm{app}}\, c \tag{10}$$

where r_A is the reaction rate based on the surface area of the catalyst, V is the whole volume of the suspension, A_{BET} is the BET surface area, c is the molar concentration of the probe molecule, m_c is the mass of the catalyst in the reacting suspension, and k_{app} is the apparent rate constant, which is expressed as $k_{\mathrm{app}} = k_V K$. In this case, k_V and K are, respectively, the maximum volumetric reaction rate, and a parameter reflecting the adsorption of the respective probe molecule at the catalyst surface.

The proposed model described the photocatalytic degradation of the herbicide imazapyr and phenol with TiO_2 nanoparticles synthesized through different calcination temperatures. In this case, the apparent degradation rate constant for a specific area was calculated using the proposed model for experiments with the different TiO_2 nanoparticles obtained. According to the authors, k_{app} exhibited a higher value for the catalyst prepared at a calcination temperature of 800 °C than those that were prepared at lower temperatures. Based on these results, the authors concluded that the proposed model is useful to compare different photocatalysts with different surface areas.

Recapitulation

This publication presents the use of mesoporous materials as catalysts in photocatalysis processes, an emerging tool in environmental remediation. These materials, due to their high surface area and adsorption capacity, allow the efficient degradation of pollutants in water and air, offering more sustainable solutions for waste treatment. Semiconductors such as titanium dioxide (TiO_2) and zinc oxide (ZnO) are of great interest due to their ability to generate reactive species under light irradiation.

Photocatalysis, a process based on the activation of semiconductor materials by ultraviolet or visible light, is crucial for generating electron-hole pairs that promote redox reactions on the catalyst surface. This process is key to decomposing organic and inorganic compounds without leaving toxic residues. However, the efficiency of the process is heavily influenced by various factors, such as the intensity and wavelength of the light, the medium's pH, the pollutant's initial concentration, and the amount of photocatalyst used. Optimizing these factors is paramount to maximizing the effectiveness of the process.

In addition, different synthesis methods for mesoporous materials are described, including the sol-gel method, phase separation, and evaporation-induced self-assembly. These methods allow the design of materials with controlled pore sizes and structures, which improves their performance in photocatalysis applications. The ability to modify the surface of these materials through chemical functionalization also opens the door to new applications, both in catalysis and other fields, such as energy and biomedicine.

On the other hand, an analysis of the state of patents is included, highlighting the growing interest in mesoporous materials for photocatalysis at an international level. Significant advances in water and air purification technologies are observed, reflecting these materials' impact in the search for more sustainable solutions.

In conclusion, combining the unique properties of mesoporous materials with advances in photocatalytic technology presents a promising approach to addressing current environmental challenges. These materials have the potential to provide efficient and sustainable solutions for the degradation of pollutants, and research in this field continues to evolve to meet these pressing needs.

References

[1] L. Lin, W. Jiang, L. Chen, P. Xu, H. Wang, Treatment of Produced Water with Photocatalysis: Recent Advances, Affecting Factors and Future Research Prospects, Catalysts, 10 (2020) 924. https://doi.org/10.3390/catal10080924

[2] S.I. Sinar Mashuri, M.L. Ibrahim, M.F. Kasim, M.S. Mastuli, U. Rashid, A.H. Abdullah, A. Islam, N. Asikin Mijan, Y.H. Tan, N. Mansir, N.H. Mohd Kaus, T.-Y. Yun Hin, Photocatalysis for Organic Wastewater Treatment: From the Basis to Current Challenges for Society, Catalysts, 10 (2020) 1260. https://doi.org/10.3390/catal10111260

[3] X. Yang, D. Wang, Photocatalysis: From Fundamental Principles to Materials and Applications, ACS Applied Energy Materials, 1 (2018) 6657-6693. https://doi.org/10.1021/acsaem.8b01345

[4] L. Buzzetti, G.E.M. Crisenza, P. Melchiorre, Mechanistic Studies in Photocatalysis, Angewandte Chemie International Edition, 58 (2019) 3730-3747. https://doi.org/10.1002/anie.201809984

[5] A. Balapure, J.R. Dutta, R. Ganesan, Recent advances in semiconductor heterojunction: a detailed review of fundamentals of the photocatalysis, charge transfer mechanism, and materials, RSC Applied Interfaces, (2023). https://doi.org/10.1039/D3LF00126A

[6] F. Zhang, X. Wang, H. Liu, C. Liu, Y. Wan, Y. Long, Z. Cai, Recent Advances and Applications of Semiconductor Photocatalytic Technology, Applied Sciences, 9 (2019) 2489. https://doi.org/10.3390/app9122489

[7] L. Wang, J. Zhao, H. Liu, J. Huang, Design, modification and application of semiconductor photocatalysts, Journal of the Taiwan Institute of Chemical Engineers, 93 (2018) 590-602. https://doi.org/10.1016/j.jtice.2018.09.004

[8] J. Schneider, M. Matsuoka, M. Takeuchi, J. Zhang, Y. Horiuchi, M. Anpo, D.W. Bahnemann, Understanding TiO2 Photocatalysis: Mechanisms and Materials, Chemical Reviews, 114 (2014) 9919-9986. https://doi.org/10.1021/cr5001892

[9] A. Meng, L. Zhang, B. Cheng, J. Yu, Dual Cocatalysts in TiO2 Photocatalysis, Advanced Materials, 31 (2019) 1807660. https://doi.org/10.1002/adma.201807660

[10] Q. Guo, C. Zhou, Z. Ma, X. Yang, Fundamentals of TiO2 Photocatalysis: Concepts, Mechanisms, and Challenges, Advanced Materials, 31 (2019) 1901997. https://doi.org/10.1002/adma.201901997

[11] X. Chen, Z. Wu, D. Liu, Z. Gao, Preparation of ZnO Photocatalyst for the Efficient and Rapid Photocatalytic Degradation of Azo Dyes, Nanoscale Research Letters, 12 (2017) 143. https://doi.org/10.1186/s11671-017-1904-4

[12] K. Qi, B. Cheng, J. Yu, W. Ho, Review on the improvement of the photocatalytic and antibacterial activities of ZnO, Journal of Alloys and Compounds, 727 (2017) 792-820. https://doi.org/10.1016/j.jallcom.2017.08.142

[13] C.A. Jaramillo-Páez, J.A. Navío, M.C. Hidalgo, M. Macías, ZnO and Pt-ZnO photocatalysts: Characterization and photocatalytic activity assessing by means of three substrates, Catalysis Today, 313 (2018) 12-19. https://doi.org/10.1016/j.cattod.2017.12.009

[14] R. Georgekutty, M.K. Seery, S.C. Pillai, A Highly Efficient Ag-ZnO Photocatalyst: Synthesis, Properties, and Mechanism, The Journal of Physical Chemistry C, 112 (2008) 13563-13570. https://doi.org/10.1021/jp802729a

[15] N. Kolobov, M.G. Goesten, J. Gascon, Metal-organic frameworks: molecules or semiconductors in photocatalysis?, Angewandte Chemie International Edition, 60 (2021) 26038-26052. https://doi.org/10.1002/anie.202106342

[16] B. Niu, X. Wang, K. Wu, X. He, R. Zhang, Mesoporous Titanium Dioxide: Synthesis and Applications in Photocatalysis, Energy and Biology, Materials, 11 (2018) 1910. https://doi.org/10.3390/ma11101910

[17] E. Rastegari, Y.-J. Hsiao, W.-Y. Lai, Y.-H. Lai, T.-C. Yang, S.-J. Chen, P.-I. Huang, S.-H. Chiou, C.-Y. Mou, Y. Chien, An update on mesoporous silica nanoparticle applications in nanomedicine, Pharmaceutics, 13 (2021) 1067. https://doi.org/10.3390/pharmaceutics13071067

[18] W. Li, J. Liu, D. Zhao, Mesoporous materials for energy conversion and storage devices, Nature Reviews Materials, 1 (2016) 16023. https://doi.org/10.1038/natrevmats.2016.23

[19] L. Zhao, H. Qin, R.a. Wu, H. Zou, Recent advances of mesoporous materials in sample preparation, Journal of Chromatography A, 1228 (2012) 193-204. https://doi.org/10.1016/j.chroma.2011.09.051

[20] S.L. Suib, A Review of Recent Developments of Mesoporous Materials, The Chemical Record, 17 (2017) 1169-1183. https://doi.org/10.1002/tcr.201700025

[21] J.A.S. Costa, R.A. de Jesus, D.O. Santos, J.B. Neris, R.T. Figueiredo, C.M. Paranhos, Synthesis, functionalization, and environmental application of silica-based mesoporous materials of the M41S and SBA-n families: A review, Journal of Environmental Chemical Engineering, 9 (2021) 105259. https://doi.org/10.1016/j.jece.2021.105259

[22] R. Lin, Y. Ding, A Review on the Synthesis and Applications of Mesostructured Transition Metal Phosphates, Materials, 6 (2013) 217-243. https://doi.org/10.3390/ma6010217

[23] K. Bai, J. Hao, Y. Yang, A. Qian, The effect of hydrothermal temperature on the properties of SBA-15 materials, Heliyon, 6 (2020). https://doi.org/10.1016/j.heliyon.2020.e04436

[24] P. Innocenzi, Mesoporous ordered films via self-assembly: trends and perspectives, Chemical Science, 13 (2022) 13264-13279. https://doi.org/10.1039/D2SC04828K

[25] B.M. Al-Shehri, A.E.R.S. Khder, S.S. Ashour, M.S. Hamdy, A review: the utilization of mesoporous materials in wastewater treatment, Materials Research Express, 6 (2019) 122002. https://doi.org/10.1088/2053-1591/ab52af

[26] K.-W. Kim, T.Y. Yun, S.-H. You, X. Tang, J. Lee, Y. Seo, Y.-T. Kim, S.H. Kim, H.C. Moon, J.K. Kim, Extremely fast electrochromic supercapacitors based on mesoporous WO3 prepared by an evaporation-induced self-assembly, NPG Asia Materials, 12 (2020) 84. https://doi.org/10.1038/s41427-020-00257-w

[27] Y. Li, J. Shi, Hollow-Structured Mesoporous Materials: Chemical Synthesis, Functionalization and Applications, Advanced Materials, 26 (2014) 3176-3205. https://doi.org/10.1002/adma.201305319

[28] N. Pal, A. Bhaumik, Soft templating strategies for the synthesis of mesoporous materials: Inorganic, organic-inorganic hybrid and purely organic solids, Advances in Colloid and Interface Science, 189-190 (2013) 21-41. https://doi.org/10.1016/j.cis.2012.12.002

[29] I.Y. Kaplin, E.S. Lokteva, E.V. Golubina, V.V. Lunin, Template Synthesis of Porous Ceria-Based Catalysts for Environmental Application, Molecules, 25 (2020) 4242. https://doi.org/10.3390/molecules25184242

[30] N. Pal, J.-H. Lee, E.-B. Cho, Recent Trends in Morphology-Controlled Synthesis and Application of Mesoporous Silica Nanoparticles, Nanomaterials, 10 (2020) 2122. https://doi.org/10.3390/nano10112122

[31] D.S. Kim, S.-Y. Kwak, The hydrothermal synthesis of mesoporous TiO2 with high crystallinity, thermal stability, large surface area, and enhanced photocatalytic activity, Applied Catalysis A: General, 323 (2007) 110-118. https://doi.org/10.1016/j.apcata.2007.02.010

[32] Y. Wang, Y.-J. Hu, X. Hao, P. Peng, J.-Y. Shi, F. Peng, R.-C. Sun, Hydrothermal synthesis and applications of advanced carbonaceous materials from biomass: a review, Advanced Composites and Hybrid Materials, 3 (2020) 267-284. https://doi.org/10.1007/s42114-020-00158-0

[33] A. Simaioforidou, V. Kostas, M.A. Karakassides, M. Louloudi, Surface chemical modification of macroporous and mesoporous carbon materials: Effect on their textural and catalytic properties, Microporous and Mesoporous Materials, 279 (2019) 334-344. https://doi.org/10.1016/j.micromeso.2019.01.005

[34] J. Wang, Q. Ma, Y. Wang, Z. Li, Z. Li, Q. Yuan, New insights into the structure-performance relationships of mesoporous materials in analytical science, Chemical Society Reviews, 47 (2018) 8766-8803. https://doi.org/10.1039/C8CS00658J

[35] Y. Li, M. Dong, W. Song, X. Liang, Y. Chen, Y. Liu, Preparation and Characterization of Paraffin/Mesoporous Silica Shape-Stabilized Phase Change Materials for Building Thermal Insulation, Materials, 14 (2021) 1775. https://doi.org/10.3390/ma14071775

[36] C. Li, Q. Li, Y.V. Kaneti, D. Hou, Y. Yamauchi, Y. Mai, Self-assembly of block copolymers towards mesoporous materials for energy storage and conversion systems, Chemical Society Reviews, 49 (2020) 4681-4736. https://doi.org/10.1039/D0CS00021C

[37] M.H. Amin, Relationship Between the Pore Structure of Mesoporous Silica Supports and the Activity of Nickel Nanocatalysts in the CO2 Reforming of Methane, Catalysts, 10 (2020) 51. https://doi.org/10.3390/catal10010051

[38] Y. Awoke, Y. Chebude, I. Díaz, Controlling Particle Morphology and Pore Size in the Synthesis of Ordered Mesoporous Materials, Molecules, 25 (2020) 4909. https://doi.org/10.3390/molecules25214909

[39] P.E. Imoisili, K.O. Ukoba, T.-C. Jen, Synthesis and characterization of amorphous mesoporous silica from palm kernel shell ash, Boletín de la Sociedad Española de Cerámica y Vidrio, 59 (2020) 159-164. https://doi.org/10.1016/j.bsecv.2019.09.006

[40] S. Sonal, P. Prakash, B.K. Mishra, G.C. Nayak, Synthesis, characterization and sorption studies of a zirconium(iv) impregnated highly functionalized mesoporous activated carbons, RSC Advances, 10 (2020) 13783-13798. https://doi.org/10.1039/C9RA10103A

[41] Y. Yan, G. Chen, P. She, G. Zhong, W. Yan, B.Y. Guan, Y. Yamauchi, Mesoporous Nanoarchitectures for Electrochemical Energy Conversion and Storage, Advanced Materials, 32 (2020) 2004654. https://doi.org/10.1002/adma.202004654

[42] B.K. Singh, Y. Kim, S. Kwon, K. Na, Synthesis of Mesoporous Zeolites and Their Opportunities in Heterogeneous Catalysis, Catalysts, 11 (2021) 1541. https://doi.org/10.3390/catal11121541

[43] B. Singh, J. Na, M. Konarova, T. Wakihara, Y. Yamauchi, C. Salomon, M.B. Gawande, Functional Mesoporous Silica Nanomaterials for Catalysis and Environmental Applications, Bulletin of the Chemical Society of Japan, 93 (2020) 1459-1496. https://doi.org/10.1246/bcsj.20200136

[44] A. Glotov, A. Vutolkina, A. Pimerzin, V. Nedolivko, G. Zasypalov, V. Stytsenko, E. Karakhanov, V. Vinokurov, Ruthenium Catalysts Templated on Mesoporous MCM-41 Type Silica and Natural Clay Nanotubes for Hydrogenation of Benzene to Cyclohexane, Catalysts, 10 (2020) 537. https://doi.org/10.3390/catal10050537

[45] Y. Song, Z. Zhang, L. Yan, L. Zhang, S. Liu, S. Xie, L. Xu, J. Du, Electrodeposition of Ti-Doped Hierarchically Mesoporous Silica Microspheres/Tungsten Oxide Nanocrystallines Hybrid Films and Their Electrochromic Performance, Nanomaterials, 9 (2019) 1795. https://doi.org/10.3390/nano9121795

[46] M. Trejda, A. Jądrzak, A. Nurwita, D. Kryszak, An efficient synthesis of acidic mesoporous materials, Catalysis Today, 354 (2020) 61-66. https://doi.org/10.1016/j.cattod.2019.04.030

[47] M. Dell'Edera, F. Petronella, A. Truppi, L.F. Liotta, N. Gallì, T. Sibillano, C. Giannini, R. Brescia, F. Milano, M. Striccoli, A. Agostiano, M.L. Curri, R. Comparelli, Low Temperature Synthesis of Photocatalytic Mesoporous TiO2 Nanomaterials, Catalysts, 10 (2020) 893. https://doi.org/10.3390/catal10080893

[48] A.L.A. Marinho, F.S. Toniolo, F.B. Noronha, F. Epron, D. Duprez, N. Bion, Highly active and stable Ni dispersed on mesoporous CeO2-Al2O3 catalysts for production of syngas by

dry reforming of methane, Applied Catalysis B: Environmental, 281 (2021) 119459. https://doi.org/10.1016/j.apcatb.2020.119459

[49] M.G. Mohamed, E.C. Atayde, B.M. Matsagar, J. Na, Y. Yamauchi, K.C.W. Wu, S.-W. Kuo, Construction Hierarchically Mesoporous/Microporous Materials Based on Block Copolymer and Covalent Organic Framework, Journal of the Taiwan Institute of Chemical Engineers, 112 (2020) 180-192. https://doi.org/10.1016/j.jtice.2020.06.013

[50] F. Sahel, F. Sebih, S. Bellahouel, A. Bengueddach, R. Hamacha, Synthesis and characterization of highly ordered mesoporous nanomaterials Al-MCM-41 and Al-SBA-15 from bentonite as efficient catalysts for the production of biodiesel MELA and EELA, Research on Chemical Intermediates, 46 (2020) 133-148. https://doi.org/10.1007/s11164-019-03939-5

[51] M. Esperanza Adrover, M. Pedernera, M. Bonne, B. Lebeau, V. Bucalá, L. Gallo, Synthesis and characterization of mesoporous SBA-15 and SBA-16 as carriers to improve albendazole dissolution rate, Saudi Pharmaceutical Journal, 28 (2020) 15-24. https://doi.org/10.1016/j.jsps.2019.11.002

[52] S. Abdulridha, J. Jiang, S. Xu, Z. Zhou, H. Liang, B. Mao, Y. Zhou, A.A. Garforth, Y. Jiao, X. Fan, Cellulose nanocrystals (CNCs) as hard templates for preparing mesoporous zeolite Y assemblies with high catalytic activity, Green Chemistry, 22 (2020) 5115-5122. https://doi.org/10.1039/D0GC01070G

[53] M.T. Colomer, Special Issue "Design, Synthesis and Applications of Macroporous, Mesoporous, and Microporous Materials", International Journal of Molecular Sciences, 25 (2024) 7127. https://doi.org/10.3390/ijms25137127

[54] H. Sanaeishoar, M. Sabbaghan, F. Mohave, Synthesis and characterization of micro-mesoporous MCM-41 using various ionic liquids as co-templates, Microporous and Mesoporous Materials, 217 (2015) 219-224. https://doi.org/10.1016/j.micromeso.2015.06.027

[55] V. Hulea, E. Dumitriu, F. Fajula, Mild Oxidation of Organosulfur Compounds with H2O2 over Metal-Containing Microporous and Mesoporous Catalysts, Catalysts, 11 (2021) 867. https://doi.org/10.3390/catal11070867

[56] A.S. Thill, W.T. Figueiredo, F.O. Lobato, M.O. Vaz, W.P. Fernandes, V.E. Carvalho, E.A. Soares, F. Poletto, S.R. Teixeira, F. Bernardi, New horizons in photocatalysis: the importance of mesopores for cerium oxide, Journal of Materials Chemistry A, 8 (2020) 24752-24762. https://doi.org/10.1039/D0TA08655J

[57] S.Z. Islam, S. Nagpure, D.Y. Kim, S.E. Rankin, Synthesis and Catalytic Applications of Non-Metal Doped Mesoporous Titania, Inorganics, 5 (2017) 15. https://doi.org/10.3390/inorganics5010015

[58] T. Wang, Y. Li, W.-T. Wu, Y.-l. Zhang, L.-g. Wu, H.-l. Chen, Effect of chiral-arrangement on the solar adsorption of black TiO2-SiO2 mesoporous materials for photodegradation and photolysis, Applied Surface Science, 537 (2021) 148025. https://doi.org/10.1016/j.apsusc.2020.148025

[59] H. Hou, G. Shao, W. Yang, W.-Y. Wong, One-dimensional mesoporous inorganic nanostructures and their applications in energy, sensor, catalysis and adsorption, Progress in Materials Science, 113 (2020) 100671. https://doi.org/10.1016/j.pmatsci.2020.100671

[60] H. Wu, Y. Xiao, Y. Guo, S. Miao, Q. Chen, Z. Chen, Functionalization of SBA-15 mesoporous materials with 2-acetylthiophene for adsorption of Cr(III) ions, Microporous and Mesoporous Materials, 292 (2020) 109754. https://doi.org/10.1016/j.micromeso.2019.109754

[61] L. Bouna, A. Ait El Fakir, A. Benlhachemi, K. Draoui, M. Ezahri, B. Bakiz, S. Villain, F. Guinneton, N. Elalem, Synthesis and characterization of mesoporous geopolymer based on Moroccan kaolinite rich clay, Applied Clay Science, 196 (2020) 105764. https://doi.org/10.1016/j.clay.2020.105764

[62] H. Zhang, G. Yuan, H. Guo, H. Li, L. Xiong, X. Chen, Preparation and characterization of mesoporous materials from low-grade palygorskite clay and its applied in composite phase change material, Journal of Energy Storage, 40 (2021) 102791. https://doi.org/10.1016/j.est.2021.102791

[63] EPO, Espacenet - Patent Search, in, European Patent Office, 2024.

[64] D.C.d.S. Alves, B.S. de Farias, C. Breslin, L.A.d.A. Pinto, T.R.S.A. Cadaval, Chapter 18 - Carbon nanotube-based materials for environmental remediation processes, in: D. Giannakoudakis, L. Meili, I. Anastopoulos (Eds.) Advanced Materials for Sustainable Environmental Remediation, Elsevier, 2022, pp. 475-513. https://doi.org/10.1016/B978-0-323-90485-8.00017-5

[65] M.F. Lanjwani, M. Tuzen, M.Y. Khuhawar, T.A. Saleh, Trends in photocatalytic degradation of organic dye pollutants using nanoparticles: A review, Inorganic Chemistry Communications, 159 (2024) 111613. https://doi.org/10.1016/j.inoche.2023.111613

[66] W. Yu, L. Zhao, Chemiluminescence detection of reactive oxygen species generation and potential environmental applications, TrAC Trends in Analytical Chemistry, 136 (2021) 116197. https://doi.org/10.1016/j.trac.2021.116197

[67] A.S. Adday, S.M. Al-Jubouri, Photocatalytic oxidative removal of the organic pollutant from wastewater using recyclable Ag2O@CRA heterojunction photocatalyst, Case Studies in Chemical and Environmental Engineering, 10 (2024) 100852. https://doi.org/10.1016/j.cscee.2024.100852

[68] W. Xie, G. Liu, Y. Liu, Y. Bai, Y. Liao, T. Li, C. Wang, S. Chang, J. Hu, Multidimensional TiO2 photocatalysts for the degradation of organic dyes in wastewater treatment, Journal of Porous Materials, 31 (2024) 1655-1681. https://doi.org/10.1007/s10934-024-01619-3

[69] Z.H. Jabbar, B.H. Graimed, S.H. Ammar, D.A. Sabit, A.A. Najim, A.Y. Radeef, A.G. Taher, The latest progress in the design and application of semiconductor photocatalysis systems for degradation of environmental pollutants in wastewater: Mechanism insight and theoretical calculations, Materials Science in Semiconductor Processing, 173 (2024) 108153. https://doi.org/10.1016/j.mssp.2024.108153

[70] A.J. Chacón-García, S. Rojas, E.S. Grape, F. Salles, T. Willhammar, A.K. Inge, Y. Pérez, P. Horcajada, SU-101 for the removal of pharmaceutical active compounds by the combination

of adsorption/photocatalytic processes, Scientific Reports, 14 (2024) 7882. https://doi.org/10.1038/s41598-024-58014-w

[71] Y. Sun, S.-Q. Guo, L. Fan, J. Cai, W. Han, F. Zhang, Molecular oxygen activation in photocatalysis: Generation, detection and application, Surfaces and Interfaces, 46 (2024) 104033. https://doi.org/10.1016/j.surfin.2024.104033

[72] Y. Nosaka, A.Y. Nosaka, Generation and Detection of Reactive Oxygen Species in Photocatalysis, Chemical Reviews, 117 (2017) 11302-11336. https://doi.org/10.1021/acs.chemrev.7b00161

[73] H. Tang, H. E, C. Yao, X. Wang, J. Zhou, W. Song, Z. Zhang, Boosted antibiotic elimination over 2D/2D mesoporous CeO2/BiOCl S-scheme photocatalyst, Separation and Purification Technology, 354 (2025) 128977. https://doi.org/10.1016/j.seppur.2024.128977

[74] Y. Ahmadi, K.-H. Kim, Modification strategies for visible-light photocatalysts and their performance-enhancing effects on photocatalytic degradation of volatile organic compounds, Renewable and Sustainable Energy Reviews, 189 (2024) 113948. https://doi.org/10.1016/j.rser.2023.113948

[75] M. Xu, P. Zhu, Q. Cai, M. Bu, C. Zhang, H. Wu, Y. He, M. Fu, S. Li, X. Liu, In-situ fabrication of TiO2/NH2−MIL-125(Ti) via MOF-driven strategy to promote efficient interfacial effects for enhancing photocatalytic NO removal activity, Chinese Chemical Letters, 35 (2024) 109524. https://doi.org/10.1016/j.cclet.2024.109524

[76] R. Ma, Y. Su, C. Li, X. Lv, W. Li, J. Yang, W. Zhang, H. Wang, Novel MOF-derived dual Z-scheme g-C3N4/Bi2O2CO3/β-Bi2O3 heterojunctions with enhanced photodegradation of tetracycline hydrochloride, Separation and Purification Technology, 354 (2025) 128580. https://doi.org/10.1016/j.seppur.2024.128580

[77] Z. Jiang, T. Wang, J. Wang, T. Yu, C. Kong, Z. Yang, H. Zhu, Oxygen vacancy-rich BiVO4 modified with mesoporous MIL-88A(Fe) Z-scheme heterojunction for enhanced photocatalytic formaldehyde degradation, Separation and Purification Technology, 353 (2025) 128581. https://doi.org/10.1016/j.seppur.2024.128581

[78] J.O. Eniola, M.O. Ansari, M.A. Barakat, R. Kumar, Chapter 6 - Aerogels in photocatalysis, in: A.A.P. Khan, M.O. Ansari, A. Khan, A.M. Asiri (Eds.) Advances in Aerogel Composites for Environmental Remediation, Elsevier, 2021, pp. 87-108. https://doi.org/10.1016/B978-0-12-820732-1.00006-0

[79] K. Zhu, C. Jin, Z. Jian, Y. Wei, R. Nan, C. Zhang, L. Hu, Significantly enhanced photocatalytic performance of mesoporous C@ZnO hollow nanospheres via suppressing charge recombination, Chemical Physics Letters, 716 (2019) 102-105. https://doi.org/10.1016/j.cplett.2018.12.013

[80] M. Ateia, M.G. Alalm, D. Awfa, M.S. Johnson, C. Yoshimura, Modeling the degradation and disinfection of water pollutants by photocatalysts and composites: A critical review, Science of The Total Environment, 698 (2020) 134197. https://doi.org/10.1016/j.scitotenv.2019.134197

[81] A. Yusuf, H. Oladipo, L. Yildiz Ozer, C. Garlisi, V. Loddo, M.R.M. Abu-Zahra, G. Palmisano, Modelling of a recirculating photocatalytic microreactor implementing

mesoporous N-TiO2 modified with graphene, Chemical Engineering Journal, 391 (2020) 123574. https://doi.org/10.1016/j.cej.2019.123574

[82] A. Yusuf, G. Palmisano, Three-dimensional CFD modelling of a photocatalytic parallel-channel microreactor, Chemical Engineering Science, 229 (2021) 116051. https://doi.org/10.1016/j.ces.2020.116051

[83] C.S. Turchi, D.F. Ollis, Photocatalytic degradation of organic water contaminants: Mechanisms involving hydroxyl radical attack, Journal of Catalysis, 122 (1990) 178-192. https://doi.org/10.1016/0021-9517(90)90269-P

[84] M.N. Chong, B. Jin, H.Y. Zhu, C.W.K. Chow, C. Saint, Application of H-titanate nanofibers for degradation of Congo Red in an annular slurry photoreactor, Chemical Engineering Journal, 150 (2009) 49-54. https://doi.org/10.1016/j.cej.2008.12.002

[85] P.K. Jaseela, K.O. Shamsheera, A. Joseph, Mesoporous Titania-Silica nanocomposite as an effective material for the degradation of Bisphenol A under visible light, Journal of Saudi Chemical Society, 24 (2020) 651-662. https://doi.org/10.1016/j.jscs.2020.05.004

[86] P.M. Rendel, G. Rytwo, Degradation kinetics of caffeine in water by UV/H2O2 and UV/TiO2, Desalination and Water Treatment, 173 (2020) 231-242. https://doi.org/10.5004/dwt.2020.24693

[87] G. Rytwo, A.L. Zelkind, Evaluation of Kinetic Pseudo-Order in the Photocatalytic Degradation of Ofloxacin, Catalysts, 12 (2022) 24. https://doi.org/10.3390/catal12010024

[88] P. Pourdayhimi, P.W. Koh, H. Nur, S.L. Lee, Highly Crystalline Zinc Oxide/Mesoporous Hollow Silica Composites Synthesized at Low Temperature for the Photocatalytic Degradation of Sodium Dodecylbenzenesulfonate, Australian Journal of Chemistry, 72 (2019) 252-259. https://doi.org/10.1071/CH18175

[89] K.K. Abbas, K.M. Shabeeb, A.A.A. Aljanabi, A.M.H.A. Al-Ghaban, Photocatalytic degradation of Cefazolin over spherical nanoparticles of TiO2/ZSM-5 mesoporous nanoheterojunction under simulated solar light, Environmental Technology & Innovation, 20 (2020) 101070. https://doi.org/10.1016/j.eti.2020.101070

[90] D.F. Ollis, Kinetics of photocatalyzed reactions: five lessons learned, Frontiers in chemistry, 6 (2018) 378. https://doi.org/10.3389/fchem.2018.00378

[91] K. Bisaria, S. Sinha, R. Singh, H.M.N. Iqbal, Recent advances in structural modifications of photo-catalysts for organic pollutants degradation - A comprehensive review, Chemosphere, 284 (2021) 131263. https://doi.org/10.1016/j.chemosphere.2021.131263

[92] A. Intisar, A. Ramzan, S. Hafeez, N. Hussain, M. Irfan, N. Shakeel, K.A. Gill, A. Iqbal, M. Janczarek, T. Jesionowski, Adsorptive and photocatalytic degradation potential of porous polymeric materials for removal of pesticides, pharmaceuticals, and dyes-based emerging contaminants from water, Chemosphere, 336 (2023) 139203. https://doi.org/10.1016/j.chemosphere.2023.139203

[93] T.H. Sia, S. Dai, B. Jin, M. Biggs, M.N. Chong, Hybridising nitrogen doped titania with kaolinite: A feasible catalyst for a semi-continuous photo-degradation reactor system, Chemical Engineering Journal, 279 (2015) 939-947. https://doi.org/10.1016/j.cej.2015.05.101

[94] A.P. Naik, A.V. Salkar, M.S. Majik, P.P. Morajkar, Enhanced photocatalytic degradation of Amaranth dye on mesoporous anatase TiO2: evidence of C-N, N=N bond cleavage and identification of new intermediates, Photochemical & Photobiological Sciences, 16 (2017) 1126-1138. https://doi.org/10.1039/c7pp00090a

[95] J. Singh, M. Sharma, S. Basu, Heavy metal ions adsorption and photodegradation of remazol black XP by iron oxide/silica monoliths: Kinetic and equilibrium modelling, Advanced Powder Technology, 29 (2018) 2268-2279. https://doi.org/10.1016/j.apt.2018.06.011

[96] M. Sharma, J. Singh, S. Basu, Efficient metal ion adsorption and photodegradation of Rhodamine-B by hierarchical porous Fe-Ni@SiO2 monolith, Microchemical Journal, 145 (2019) 708-717. https://doi.org/10.1016/j.microc.2018.11.042

[97] A. Motamedisade, A. Heydari, D.J. Osborn, A.S. Alotabi, G.G. Andersson, Au9 clusters deposited as co-catalysts on S-modified mesoporous TiO2 for photocatalytic degradation of methyl orange, Applied Surface Science, 655 (2024) 159475. https://doi.org/10.1016/j.apsusc.2024.159475

[98] J. Singh, S. Basu, Synthesis of mesoporous magnetic Fe2O3/g-C3N4 monoliths for Rhodamine B removal, Microporous and Mesoporous Materials, 303 (2020) 110299. https://doi.org/10.1016/j.micromeso.2020.110299

[99] E. Erusappan, S. Thiripuranthagan, R. Radhakrishnan, M. Durai, S. Kumaravel, T. Vembuli, N.J. Kaleekkal, Fabrication of mesoporous TiO2/PVDF photocatalytic membranes for efficient photocatalytic degradation of synthetic dyes, Journal of Environmental Chemical Engineering, 9 (2021) 105776. https://doi.org/10.1016/j.jece.2021.105776

[100] Y. Wang, F. Wang, Y. Feng, Z. Xie, Q. Zhang, X. Jin, H. Liu, Y. Liu, W. Lv, G. Liu, Facile synthesis of carbon quantum dots loaded with mesoporous gC 3 N 4 for synergistic absorption and visible light photodegradation of fluoroquinolone antibiotics, Dalton Transactions, 47 (2018) 1284-1293. https://doi.org/10.1039/C7DT04360K

[101] X. Zeng, X. Sun, Y. Yu, H. Wang, Y. Wang, Photocatalytic degradation of flumequine with B/N codoped TiO2 catalyst: Kinetics, main active species, intermediates and pathways, Chemical Engineering Journal, 378 (2019) 122226. https://doi.org/10.1016/j.cej.2019.122226

[102] A.A. Fauzi, A.A. Jalil, C.N.C. Hitam, F.F.A. Aziz, N. Chanlek, Superior sulfate radicals-induced visible-light-driven photodegradation of pharmaceuticals by appropriate Ce loading on fibrous silica ceria, Journal of Environmental Chemical Engineering, 8 (2020) 104484. https://doi.org/10.1016/j.jece.2020.104484

[103] S. Rabhi, L. Mahtout, M. Bououdina, L. Khezami, H. Belkacemi, A. Kerrami, E. Sakher, Tuning the photocatalytic activity of TiO2 by Ag loading: Experimental and modelling studies for the degradation of amlodipine besylate drug, Ceramics International, 47 (2021) 21509-21521. https://doi.org/10.1016/j.ceramint.2021.04.162

[104] N. Hosseini, M.R. Toosi, Combined adsorption process and photocatalytic degradation of some commercial herbicides over N-doped TiO2 particles supported on recyclable magnetic hexagonal mesoporous silica, Separation Science and Technology, 54 (2019) 1697-1709. https://doi.org/10.1080/01496395.2018.1539105

[105] M. Zangiabadi, T. Shamspur, A. Saljooqi, A. Mostafavi, Evaluating the efficiency of the GO-Fe3O4/TiO2 mesoporous photocatalyst for degradation of chlorpyrifos pesticide under visible light irradiation, Applied Organometallic Chemistry, 33 (2019) e4813. https://doi.org/10.1002/aoc.4813

[106] S.Y. Ejeta, T. Imae, Photodegradation of pollutant pesticide by oxidized graphitic carbon nitride catalysts, Journal of Photochemistry and Photobiology A: Chemistry, 404 (2021) 112955. https://doi.org/10.1016/j.jphotochem.2020.112955

[107] W. Deng, F. Pan, B. Batchelor, B. Jung, P. Zhang, A. Abdel-Wahab, H. Zhou, Y. Li, Mesoporous TiO2-BiOBr microspheres with tailorable adsorption capacities for photodegradation of organic water pollutants: probing adsorption-photocatalysis synergy by combining experiments and kinetic modeling, Environmental Science: Water Research & Technology, 5 (2019) 769-781. https://doi.org/10.1039/C8EW00922H

[108] M. Faycal Atitar, A.A. Ismail, R. Dillert, D.W. Bahnemann, Photodegradation of Herbicide Imazapyr and Phenol over Mesoporous Bicrystalline Phases TiO2: A Kinetic Study, Catalysts, 9 (2019) 640. https://doi.org/10.3390/catal9080640

Materials Research Forum LLC
https://doi.org/10.21741/9781644903452-8

Chapter 8

Mesoporous Materials for Solid-Phase Extraction

Madhur Babu Singh[1], Nutan Sharma[2], Sapna Raghav[3], Rajesh Kumar[4], Pallavi Jain[1*]

[1]Department of Chemistry, SRM Institute of Science and Technology, Delhi-NCR Campus, Modinagar, India

[2]Department of Chemistry, Vivekananda Global University, Jaipur, India

[3]Department of Chemistry, Mewar University, Chittorgarh, Rajasthan, India

[4]Department of Chemistry, R.D.S College, B.R.A Bihar University, Muzaffarpur, India

*palli24@gmail.com

Abstract

Mesoporous materials (MM) have gained more attention for their application in solid-phase extraction (SPE). These materials, characterized by their large and uniform pore volumes, offer exceptional stability, selectivity, and catalytic activity, making them ideal sorbents for SPE. Their wide range of applications spans environmental analysis, biomedical and pharmaceutical fields, and the food and beverage industry. The selection of mesoporous materials as sorbents in SPE involves considerations of selectivity, catalytic activity, reproducibility, and extraction efficiency. Despite their benefits, challenges such as high synthesis costs and the difficulty in shaping pores for specific applications persist. Nonetheless, the versatile nature of mesoporous materials continues to drive innovation in various applicative fields, including electrochemistry, semiconductors, and optical properties.

Keywords

Mesoporous Materials, Solid-Phase Extraction, Sorbents, Selectivity, Catalytic Activity

Contents

Materials Research Foundations 173 (2025) 203-221
https://doi.org/10.21741/9781644903452-8

1. Introduction

Microporous zeolites are renowned as highly effective and extensively utilized porous materials in scientific and industrial domains. Their distinctive attribute lies in the consistent, molecularly-sized pores they possess, which endow them with exceptional stability, selectivity, and catalytic activity owing to their crystalline structure. Additionally, their versatility extends to the incorporation of diverse hetero elements within their framework, enhancing their utility in a variety of processes such as catalysis, sorption, and membrane separations. These unique characteristics underscore zeolites' pivotal role in numerous scientific and industrial applications [1]. In 1990, Yanagisawa et al. achieved a landmark in materials science with their finding of the initial mesoporous silica. Their innovative approach involved the synthesis of an alkyltrimethylammonium–kanemite complex through the chemical reaction between alkyltrimethylammonium chloride solutions and single-layered polysilicate kanemite. Their experimental findings elucidated a remarkable transformation: the condensation of SiO_2 layers within these complexes led to the emergence of intricate three-dimensional SiO_2 networks. These materials exhibit regularly arranged cylindrical pores with adjustable diameters varying from 2-30 nm. As a result, they boast large surface areas from 700-1500 m^2/g. Additionally, their silica composition ensures high chemical and thermal stability, coupled with facile functionalization capabilities. These attributes make them highly appropriate for many utilized in many places such as adsorption, catalysis, chemical separations, and biotechnological devices [2]. This seminal research not only unveiled a new class of porous materials but also laid the foundation for extensive exploration and application across diverse scientific disciplines and industrial sectors [3].

In accordance with the classification by the IUPAC, MM are defined as substances possessing pores with diameters typically varying between 0.2-2.0 nm. In a ground breaking development reported in 1990, Yanagisawa et al. had effectively produced mesoporous silica (MS) with a remarkably even distribution of pore sizes. This achievement was made possible by transforming the layered polysilicate kanemite into a novel material known as FSM-16 (Folded Sheet Materials), marking a remarkable advancement in materials science and nanotechnology [3][4]. These findings represent a significant stride forward in materials science, offering unprecedented potential for applications in various technological domains requiring advanced porous materials with tailored structural properties [5][6][7]. MM are defined by IUPAC have size of pore between

Materials Research Forum LLC
https://doi.org/10.21741/9781644903452-8

2-50 nm [8][9].Mesopore size calculations typically utilize the Kelvin equation to derive their dimensions.

$$\frac{1}{r_1}+\frac{1}{r_2}=\frac{-RT}{\sigma^{1g}v}\ln\left(\frac{P}{P^0}\right)$$

Based on the major radii of curvature of the liquid meniscus within a pore, r_1 and r_2, the relative pressure

P/P^0 at which condensation occurs the Kelvin equation may be used to determine this. Information like the molar volume (v) surface tension of the liquid condensate (σ'^g) are included in it. Applying the equation locally to each element of the liquid surface is the standard procedure. One must make two assumptions in order to use this approach to find the pore width or radius: I) the meniscus's curvature is linearly proportional to the pore width, and (ii) a slit-shaped or cylindrical pore model is assumed. In cylindrical pores, the meniscus is hemispherical ($r_1 = r_2$), while in slit-shaped pores, the meniscus is hemicylindrical r_1 = width of the slit, $r_2 = \infty$ [10].

Table 1: *Methods of Synthesizing Mesoporous Materials*

Method	Precursor Substances	Interactions	Example	References
Soft Templating	Metal salts, triethoxysilane, organosilanes	Chemical reactions, hydrogen bonding, electrostatic interactions	Cationic surfactant (e.g., cetyltrimethylammonium bromide) in alkaline conditions 37	[11]
Sol-Gel Process	Metal/metalloid precursors with reactive ligands	Particle size and shape control through gelation	Silica-based mesoporous materials synthesized by gelation followed by solvent removal	[12][13]
Exo-template (Hard Templating)	Porous solid as template	Interaction with solid template	Nanocasting using hard templates to create ordered mesoporous structures 13	[14]
Microwave Assisted	Deionized water, HCl, 12-bis(triethoxysilyl)ethane	High-temperature reactions facilitated by microwaves	Silica precursor heated in a microwave oven at 100°C to produce mesoporous materials 14	[15]
Endo-template (Soft Templating)	Surfactants as structure-directing agents	Formation of ordered mesoporous materials through surfactant interaction	Surfactants used to guide the formation of mesostructured nanomaterials without hard templates 13	[14]

Mesoporous materials may be synthesized using a number of different processes; one of them is the sol-gel approach (**Table 1**). One of the most well-known wet chemical techniques in ceramic

Materials Research Foundations 173 (2025) 203-221 https://doi.org/10.21741/9781644903452-8

engineering and materials science is the sol-gel method, which is also known by name of chemical solution deposition. The first step is to make a sol, which is a colloidal solution of precursor metals or metalloids encased in reactive ligands. Gelation then transforms the sol into a continuous liquid phase (gel), forming an inorganic network. The transition from sol to gel, achieved by removing solvent, dictates particle size and shape control in the resulting dispersible oxide materials [12][13][2]. Another commonly used technique is the template assisted technique (**Figure 1**). Its preference over other technique is due to its cheap cost. There are two main types of templating procedures used in the synthesis of MM: exo-template, which involves hard matter, and endo-template, which involves soft matter. To help create ordered MM, a surfactant is used in endo-templating as a structure-directing agent. The approach is called the "soft matter templating method" since in this hard template was not used. By contrast, the exo-template technique (also called "nanocasting" or "hard-matter templating") use a porous solid rather than a surfactant as the model[14]. Microwave assisted technique is another method that can be used in the synthesis of MM. This was known to be in existence since 1992. In this technique, An agent that directs the structure is mixed with DI water and swirled at 400°C for 4-6 hours. The next step is to add a consistently mixed solution of distilled water, HCl, and 1,2-bis(triethoxysilyl)ethane, a precursor to silica containing ethoxy groups, while stirring constantly. After that, the liquid is poured into Teflon containers and microwaved at 100°C until it is cooked through. [15][16][17].

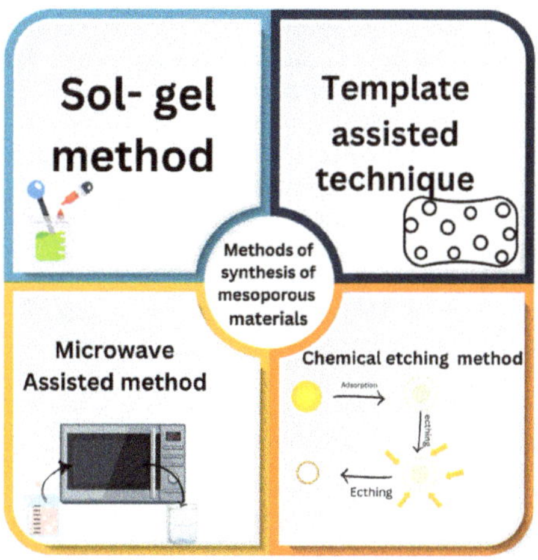

Figure 1 *Methods of Synthesizing Mesoporous Materials*

2. Fundamentals of SPE

SPE is a vital technique in analytical chemistry, facilitating the selective removal, concentration, and purification of target analytes from sample matrices prior to analysis. Employing a solid phase housed in cartridges, disks, or membranes, SPE separates analytes based on their physical and chemical properties. Sorbent materials, like silica-based phases and polymer-based phases, are chosen for their compatibility with diverse analytes and matrices. The SPE process involves conditioning the sorbent with a specific solvent, loading the sample to allow analyte retention through interactions like hydrophobicity or ion exchange, washing to remove interfering substances, and elution to release concentrated analytes for subsequent analysis methods. This method enhances sensitivity by concentrating analytes, minimizes matrix interference, and improves analytical reproducibility. SPE finds wide application in environmental monitoring, pharmaceutical analysis, food safety, and forensic sciences due to its versatility and effectiveness in handling complex samples, underscoring its crucial role in modern analytical workflows.

2.1 Overview of SPE

SPE is a sophisticated technique widely employed in chemistry for the precise isolation and purification of specific substances, known as analytes, from complex mixtures. This method relies on the interaction between a solid material, referred to as the sorbent, and a liquid solvent that flows through it. During SPE, analytes are selectively extracted from the mixture and immobilized on the sorbent surface. This step ensures the concentration and purification of analytes from interfering compounds present in the original sample matrix. Following extraction, the next crucial step in SPE involves eluting or releasing the trapped analytes from the sorbent using a solvent that has a strong tendency for the analytes. This selective elution process allows for the separation and recovery of analytes in a purified form, ready for further analysis. By leveraging the differential affinities of substances for the sorbent and eluting solvents, SPE enables scientists to achieve high selectivity and sensitivity in detecting and quantifying target analytes. SPE used the immobilization of a "liquid" phase onto a solid substrate. This process facilitates the partitioning of analytes between the liquid phase, which represents the sample matrix, and the modified solid surface. The selection of a specific solid phase is determined by its ability to selectively retain analytes possessing particular functional groups [18]. Chladek and Marano conducted a comparative study between a SPE procedure and the conventional EPA method 625, which involves liquid-liquid extraction. Their findings revealed that the average recovery rate of compound C8 using SPE was notably superior, exhibiting an increase of 20% compared to that achieved using the standard method. This underscores the enhanced efficiency and effectiveness of SPE over traditional liquid-liquid extraction techniques, particularly in isolating and recovering specific compounds from complex matrices. Such comparative analyses contribute valuable insights into optimizing analytical methodologies for improved accuracy and reliability in environmental and analytical chemistry applications [19].

Overall, SPE plays a pivotal role in analytical chemistry, offering scientists a powerful tool for sample preparation by effectively concentrating analytes of interest while minimizing interference from other components present in complex mixtures. SPE plays a crucial role in analytical chemistry by facilitating the isolation and purification of target analytes from complex matrices. These analytes are often embedded within matrices containing numerous interfering species. Direct analysis of such samples without an extraction step can lead to significant background interference, complicating the accurate quantification of the target analyte. SPE operates on the principle of selectivity, allowing for the differential treatment of analytes and interfering species. During the loading step, the sample is put onto the SPE column, where the analyte of interest selectively interacts and becomes retained, while interfering species may pass through or are minimally retained. Subsequently, through a rinse step, any loosely bound interfering species can be removed from the column, further enhancing the purity of the retained analyte. Moreover, the

Materials Research Forum LLC
https://doi.org/10.21741/9781644903452-8

elution step in SPE enables the selective release of the analyte from the solid phase matrix using an appropriate solvent. This controlled elution process ensures that interfering species, if retained, remain on the column or are eluted separately, thereby increasing the specificity and sensitivity of subsequent analytical measurements. In essence, SPE serves as a pivotal pre-concentration and purification technique in analytical workflows, effectively minimizing background noise and enhancing the accuracy and reliability of quantifying trace levels of analytes in complex samples [18].

2.2 SPE Phases and Mechanisms

The effectiveness of SPE hinges primarily on the choice of sorbent material employed within the extraction column. While hydrophobic and H-bonding energies are relatively modest in comparison to electrostatic interactions, the latter play a pivotal role in enhancing selectivity during the extraction process. Utilizing higher energy interactions can lead to more precise and selective extraction outcomes. These sorbents, whether silica-based or organic resin-based, are selected based on their specific physical characteristics and chemical properties, tailored to the targeted analytes and sample matrices. The inherent composition of the base material and the incorporation of functional groups significantly influence the performance and versatility of these sorbents in SPE applications. Fundamentally, sorbents used in SPE are characterized by their three-dimensional polymeric structure, engineered under controlled conditions to achieve a highly porous yet rigid material with an extensive surface area. This structural design not only facilitates efficient analyte adsorption but also ensures robust performance across a wide range of analytical scenarios, contributing to the reliability and effectiveness of SPE as a purification and pre-concentration technique in modern analytical chemistry practices [20]. Silica gel is highly effective in adsorption due to its non-swelling, mechanically robust nature and ability to withstand heat treatment. Its heterogeneous surface features various silanol groups. While unmodified silica can serve as an SPE sorbent, its utility is broadened by modifying its surface with diverse functional groups. These modifications include non-polar (e.g., C18), polar (e.g., NH2), ionic, or mixed-mode (e.g., C8/cation exchange) functionalities. SPE sorbents are categorized based on their application in washing and elution solvents: reversed-phase, normal-phase, and ion-exchange [21]. Choosing the right SPE extraction sorbent hinges on grasping how it interacts with the analyte. The solute's and the sorbent's hydrophobic, polar, and ionogenic properties provide the groundwork for this comprehension. Interactions between dipoles ("polar interactions"), H-bonds, van der Waals forces ("non-polar interactions"), and cations and anions ("ionic interactions") provide the bulk of SPE's retention mechanisms. These interactions dictate how effectively the sorbent can selectively retain the analyte of interest during the extraction process. Different sorbents offer different set of properties. Once these interactions are disrupted, the analyte can be eluted from the reversed-phase SPE disk or tube using a non-polar solvent. Ion exchange is another mechanism. Ion exchange SPE employs specific functional groups bonded to silica surfaces to selectively isolate compounds based on their charge in solution. Anionic compounds interact with aliphatic quaternary amine groups, while cationic compounds interact with aliphatic sulfonic acid groups. This electrostatic interaction ensures robust retention of target compounds, enhancing the purification and concentration capabilities of the technique in analytical workflows [22]. In SPE, pH plays a critical role. Silica-based packings typically tolerate pH levels between 2 and 7.5, as extremes can hydrolyze or detach the bonded phase from the silica surface, or even dissolve the silica itself. However, since SPE involves brief contact between solutions and the sorbent, these effects are mitigated during the procedure [23][24].

3. Mesoporous Materials as SPE Sorbents

Mesoporous materials have revolutionized SPE as sorbents by virtue of their well-defined pore structures (2-50 nm), high surface areas, and versatile surface chemistry. These materials provide efficient analyte capture from complex matrices such as biological fluids and environmental samples due to their abundant active sites and uniform pore size distribution, ensuring consistent retention and elution profiles for enhanced reproducibility. Their tuneable surface functionalities through organic or inorganic modifications enable selective interactions with target analytes based on polarity, charge, or specific functional groups, thereby enhancing selectivity and sensitivity in analytical methods. Furthermore, MM exhibit excellent stability against chemical and thermal degradation, ensuring robust performance across a wide pH range and in the presence of various solvents, making them adaptable to diverse analytes and extraction conditions. Recent advancements in synthesis techniques have expanded the range of available MM, including silica-based MCM-41 and SBA-15, MOFs, and carbonaceous materials, each offering unique benefits such as increased adsorption capacities, tuneable pore sizes, or improved compatibility with specific analytes and solvents. Mesoporous materials represent a pivotal advancement in SPE technology, providing enhanced efficiency, versatility, and reliability for the extraction and purification of analytes in analytical and environmental monitoring applications.

3.1 Advantages of Mesoporous Materials in SPE

In contrast to more traditional materials, ordered MS has unique properties such as macroscopic tunneling effects, quantum size effects, volume effects, and surface effects. Being able to control their physical and chemical characteristics as well as their regular pore structure and homogeneous pore size makes them ideal carriers for macromolecules. Their capacity to selectively adsorb bioactive and therapeutic compounds is further demonstrated by their functionalization with organic groups. This unique combination of properties positions ordered MS as a versatile material with significant potential in various scientific and industrial applications, particularly in fields requiring precise control over molecular interactions and transport phenomena at nanoscale dimensions [25]. Ordered MS materials, ranging in particle sizes from micrometres to nanometres, have found extensive applications across diverse disciplines including adsorption [26], separation [27], energy storage [28], bioimaging [29], biopharmaceuticals [30], catalysis [31] and semiconductor technologies [32]. These materials are highly regarded as exceptional adsorbents for SPE owing to their precisely ordered mesoporous structure, substantial surface area (up to 1600 m^2/g), and facile capability for functionalization. Their unique combination of structural characteristics and chemical properties renders them versatile tools in scientific research and industrial applications, facilitating advancements in fields requiring efficient molecular adsorption and separation processes [25]. A work was done by the Farjadian et al. in which they used hexagonal MS in SPE of drugs. A highly sensitive SPE method using HMS nanoparticles detected low concentrations of 5-fluorouracil(5-FU). HMS-NH2, synthesized and characterized, showed a hexagonal structure in SEM and dispersed uniformly (10 nm diameter) in atomic force microscopy. HPLC confirmed effective 5-FU extraction from aqueous and plasma using varying HMS concentrations (5, 10, and 20 mg/ml). Recoveries ranged from 63.17% to 93.38% in aqueous and 59.49% to 81.69% in plasma, proving aminopropyl-functionalized HMS as a promising tool for drug bioanalysis [33]. Citrus flavanones were quantitatively analyzed in a separate work by Cao et al. using miniaturized SPE based on mesoporous molecular sieves [34].

Materials Research Foundations 173 (2025) 203-221
https://doi.org/10.21741/9781644903452-8

3.2 Comparison with Traditional Sorbents

Porous materials are highly regarded as promising adsorbents primarily due to their inherent capability to provide expansive interfaces, which are crucial for effective adsorption processes. This is largely attributed to their significantly elevated specific surface areas, underscoring their potential to efficiently capture and interact with adsorbates across various applications. Therefore, the paramount criterion for an optimal adsorbent remains the maximization of interface size, a quality that porous materials exemplify through their structurally diverse and intricately interconnected network of voids and channels. Traditionally, silica gels and activated carbons have been widely utilized as adsorbents but are primarily microporous, limiting their effectiveness for adsorbing large molecules. However, there is ongoing research and application focus on developing advanced adsorbents capable of enhanced performance, including magnetic separation, molecular sieving, and molecular recognition. Materials with regular and sizable mesopores are increasingly crucial in this context. OMMs with silica and carbon frameworks in particular have attracted a lot of attention from scientists all over the world because of their large pore volumes, stable interconnected structures, and active surfaces that can be easily modified or functionalized. They also have a high specific surface area. Consequently, research into OMMs for adsorption purposes has seen a rapid increase [35]. Stathi et al. conducted research to determine how well the hybrid MCM-41 material absorbed metals in. When compared to the reference material, MCM-41, the hybrid MCM-41 AEDTC material shows much improved metal uptake at pH 7. The absorption of lead (Pb) virtually doubled, while that of zinc (Zn) and copper (Cu) improved by around 300 percent. These results highlight the hybrid material's potential as an efficient sorbent for Pb, Zn, and Cu in environmental applications, particularly in natural waters [36]. in another study done by Yokoi et al. in which Iron (III) ions was bound to amino-functionalized MCM-41to be used as an adsorbent of toxic oxyanions. Upon comparing the data, it was found that the adsorption capacities for arsenate are recorded at 1.56 mmol g^{-1} and 1.31 mmol g^{-1} for Fe/NN-MCM-41 and re-Fe/NN-MCM-41, respectively. This suggests a restoration of 84% of the initial adsorption capacity. Similarly, for chromate, selenate, and molybdate, the degree of restoration in adsorption capacities is noted as 83%, 86%, and 85%, respectively. These findings underscore the effective recuperative capability of the re-Fe/NN-MCM-41 material across various anionic contaminants, positioning it favourably for potential applications in remediation and environmental management strategies [37].

4. Synthesis and Functionalization of Mesoporous SPE Sorbents

Mesoporous materials, characterized by their pore sizes varying from 2-50 nm, have gained significant focus for their application in various fields, particularly in SPE sorbents. SPE is largely used method for the isolation, purification, and preconcentration of analytes in complex matrices, offering high selectivity, efficiency, and reproducibility. The development of mesoporous SPE sorbents provides enhanced surface area, uniform pore distribution, and tunable pore structures, making them ideal candidates for various separation applications. This section will delve into the synthesis and functionalization techniques used to create and optimize mesoporous SPE sorbents for different analytical and industrial purposes. The synthesis of mesoporous SPE sorbents involves various techniques that allow for proper control over the structure, morphology, and pore size of the materials.

4.1 Template-Based Synthesis

The templating approach entails the direct preparation of porous materials by the employed of block copolymers or surfactants as structure-directing agents. In this procedure, precursor chemicals, such as metal salts for metal oxide nanomaterials or triethoxysilane and organosilanes for SiO_2-based nanomaterials, are added to create mesostructured nanomaterials. The production of nanoparticles is influenced by the interactions between the SDAs and molecular precursors, which might include chemical reactions, H-bonding, and electrostatic interactions. The shape of MM is primarily shaped by interactions occurring between organic and inorganic components at their interface. For instance, in alkaline environments, a cationic surfactant like cetyltrimethylammonium bromide can interact with a negatively charged molecular precursor. In contrast, under highly acidic conditions, interactions between organic and inorganic substances rely on weaker hydrogen bonds.This process enables the polymerization, cross-linking, and assembly of inorganic species with the aid of surfactants [11]. A notable drawback of the soft templating process is the limited precision in controlling the condensation reaction, which hinges on sol-gel chemistry and can lead to variability in the final product due to sensitivity to laboratory conditions. Due to the limitations of soft-templating processes in achieving certain mesostructured materials, researchers have turned to using ordered mesoporous silicas as templates, a method known as nano casting [38]. Originally developed with porous Al_2O_3 membranes, nano casting enables the creation of novel mesostructured materials by putting precursors into pore channels through techniques like electrodeposition or CVD, followed by template etching. Mesoporous silicas offer highly organized nanostructures and uniform pore sizes, making them ideal hard templates for synthesizing mesostructured materials such as metal, metal oxide, and metal sulphide nanostructures [39]. The removal of the template constitutes a critical and nuanced phase in the fabrication of MM through nanocasting. Preserving the structural integrity of MOX's mesoporous framework necessitates avoiding harsh conditions during template removal, such as employing chemical treatments involving HF or highly concentrated NaOH solutions (with concentrations exceeding 2 M). These stringent measures are essential to ensure that the desired mesoporous architecture remains intact throughout the processing steps [11].

4.2 Sol-Gel Process

The sol-gel method is recognized for its flexibility and capability to generate materials with exceptional purity and uniformity. One key advantage is its ability to occur at relatively low temperatures, which is particularly beneficial when incorporating organic molecules or thermally sensitive components into the mesoporous structure. Additionally, the sol gel technique allows the integration of various functional groups within the silica matrix, enhancing the sorbent's selectivity and affinity for specific analytes. A crucial step in the sol-gel method involves the selection of surfactants that shape the mesoporous structure [40]. Common surfactants such as cetyltrimethylammonium bromide (CTAB) or block copolymers like Pluronic P123 are often employed. These surfactants self-assemble into micelles or other organized configurations within the solution, acting as structural templates for the developing silica network. The size and morphology of the mesopores can be adjusted by manipulating the surfactant type and concentration, alongside modifying sol-gel parameters like pH and temperature [41]. Following the making of the silica network, it is essential to eliminate the surfactant template to unveil the mesoporous structure. This is typically achieved through calcination, where high temperatures are applied to burn away the organic surfactants, or by solvent extraction, where the surfactant is

Materials Research Forum LLC
https://doi.org/10.21741/9781644903452-8

dissolved and subsequently washed out. The MS obtained from this process is characterized by a high surface area, uniform pore size distribution and a well-defined pore structure, making it an excellent sorbent for SPE. The mesoporous nature of silica sorbents provides multiple benefits for SPE applications [42]. The high surface area facilitates greater analyte adsorption, improving extraction efficiency. Additionally, the uniform pore size distribution ensures reliable and reproducible extraction performance, while tunable pore sizes enable the selective extraction of analytes based on size and shape. Further, functionalizing the MS with different chemical groups enhances selectivity for specific analytes, improving performance in SPE processes. Overall, the sol-gel technique is a highly effective approach for synthesizing mesoporous SPE sorbents. Its ability to offer precise hold over the composition and structure of the final material, combined with its versatility, makes it an optimal technique for producing high-performance sorbents across diverse applications. By strategically choosing surfactant templates and optimizing sol-gel conditions MS materials with tailored properties can be produced, yielding significant improvements in extraction efficiency, selectivity, and consistency.

4.3 Surface Functionalization Techniques

Two main approaches have historically been used to modify MM by means of covalent linkages in the midst of chemical groups and silica structure: direct integration via co-condensation and grafting after synthesis. Periodic mesoporous organo-silicates are a mesoporous material that have lately attracted increasing attention. These materials are notable for embedding functional groups directly within the framework walls, which offers promising opportunities for customizing their properties and expanding their applications across various scientific and industrial domains. The grafting technique is commonly employed to modify mesoporous silica, facilitating the introduction of diverse functional groups such as amino, thiol, and alkyl groups [43]. In a work done by the Zhang et al. Sinapinaldehyde has been employed to modify SBA-15 MS to target Pb^{2+} ions. The molecule's hydroxyl and ether groups allow it to form complexes with Pb^{2+} ions at electron-rich sites. This functionalization enhances the material's selectivity for Pb^{2+} over other heavy metals due to SA's preference for complexing with hard Lewis acid-type Pb(II) cations rather than soft acid-type metals [44]. Another study was by Walcarius et al. in which they synthesized different types of mesoporous silicates functionalized with aminopropyl or mercaptopropyl groups. They studied how Cu ions interact with aminopropyl-functionalized silica and how Hg ions interact with mercaptopropyl-functionalized silica to assess guest molecule accessibility [45]. Lie et al. selected TMCES(Tris-(methoxy)carboxylethylsilane) and TMAPS(Tris-(methoxy)aminopropylsilane) for functionalization due to their carboxyl (-COOH) and amino (-NH2) groups, which interact effectively via electrostatics, hydrogen bonding, and hydrophilicity with charged amino acid residues in proteins. The organic functionalization (-CH2-CH2-COOH) created a favourable microenvironment around the enzyme, increasing its affinity and significantly enhancing specific activity [46]. In the co-condensation approach, organosilicates bearing both organic functional groups and condensable moieties are polymerized alongside silica precursors like tetramethyl orthosilicate and tetraethyl orthosilicate. This process permits for the direct fabrication of MS integrated with organic functionalities in a unified synthesis step, ensuring even distribution of these functional groups throughout the silica matrix [43]. A study done by Nooney et al. in which they used 3-mercaptopropyltrimethoxysilane (MPTS) to functionalize MS as it was found suitable for removal of mercury ions [47]. The material exhibited strong selectivity for Hg and other heavy metals. MPTS proved better due to its sulfur component, which readily forms covalent bonds with soft metals in the d-block. The functionalized material showed a high distribution coefficient and sorption capacity, attributed to MCM-41's large surface area and MPTS's ability to form a densely packed monolayer. The uniform pore structure facilitated unhindered diffusion of mercury throughout the material without blockages [48].

5. Applications of Mesoporous SPE Sorbents

Different applications of Mesoporous SPE were shown in **Figure 2.**

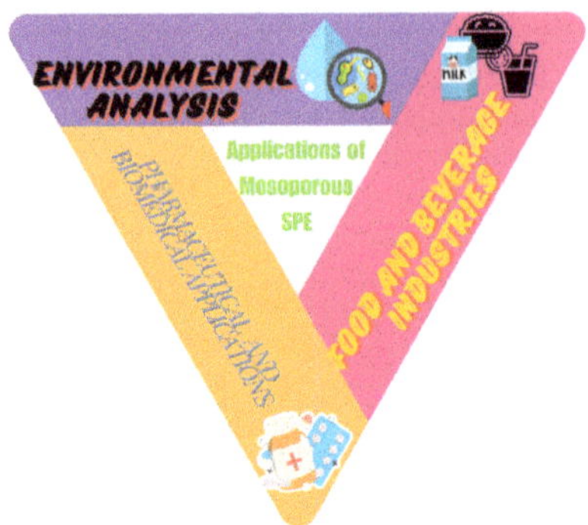

Figure 2 *Applications of Mesoporous SPE*

5.1 Environmental Analysis

In the past few years, there has been a notable rise in the uses of mesoporous SPE materials as sorbents. The study of pharmaceuticals in water samples, Dahane et al. showed the use of MCM-41 MS as a SPE sorbent in conjunction with micro-liquid chromatography-quadrupole-mass spectrometry. They collected five water samples from various sources in Almería, Spain, including river, dam and fountain waters. Specifically, two river samples were taken: one from the Nacimiento River, located in a sparsely populated area, and another from the Andarax River, which flows through a densely populated region and receives discharge from the Nacimiento River. The findings indicated that the MCM-41-SPE method is cost-effective, user-friendly, and consistently efficient in extraction processes. Additionally, the developed micro-LC method reduces analysis time and organic solvent usage, aligning with the principles of Green Analytical Methods [49]. In a study done by Castell et al. a Ti-doped UVM-7 MS was made and employed for SPE of various organophosphorus pesticides from environmental water samples. Organophosphorus pesticides are largely utilized in agriculture for their insecticidal properties. However, residues can contaminate drinking and surface waters post-application, posing health risks due to their carcinogenic and mutagenic effects [50]. Kharbouche et al. investigated closely the production of MSU-1 MS and its use as a sorbent in SPE. From samples of ground and river water, their study aimed to focus on thirteen pesticides of different polarity. Their work focused on evaluating the efficiency of MSU-1 MS in isolating and pre-concentrating these pesticides for subsequent analysis. According to the results the adaptability in synthesizing MM in the lab enables tailored designs to address specific challenges in sample treatment, like concentrating polar pesticides [51].

5.2 Pharmaceutical and Biomedical Applications

In the pharmaceutical industry, MM have shown great promise as solid-phase extraction (SPE) sorbents. In a notable study led by Li et al., dendrimer-functionalized MS (KIT-6) was employed as a dual-mode anion exchange sorbent to concentrate and purify specific acidic pharmaceuticals from human urine samples. The initial evaluation of this application came from this research. The micrometer scale SEM pictures showed that the particle sizes there are appropriate for use in SPE systems. TEM scans revealed that the dendrimer-functionalized KIT-6 (DF-KIT-6) has a three-dimensional pore architecture, expected to be accessible. Eliminating matrix interferences, the high selectivity of the optimal cleaning step and the rather more adsorption efficency helped to efficiently recover acidic nonsteroidal anti-inflammatory medications (NSAIDs) from the complex urine matrix. Good recoveries and repeatability were obtained from the improved approach used for the examination of human urine. [52] Additionally, MM have found applications in drug delivery systems. In a separate investigation conducted by Li et al., hollow MS nanoparticles (HMSNs) with adjustable architectures were utilized for precise drug administration. Experimental findings demonstrated that the release of doxorubicin (Dox) from HMSNs adhered to the Fickian diffusion model. Confocal fluorescence microscopy and MTT assays validated the efficacy of HMSNs as proficient drug conveyors. This strategy has the potential to markedly enhance the development and real-world deployment of HMSNs in drug delivery methodologies.[30].

5.3 Food and Beverage Industry

In the study done by Cao et al. they employed molecular sieve as the sorbent material in chromatography for the analysis of flavonoids found in fruit peels. This study developed an eco-friendly MSPD–UPLC method using SBA-15 molecular sieve to analyse flavonoids in orange peels. It offers precise results with low sample and sorbent use, minimal solvent consumption, and no clean-up steps, all at low cost [53]. In a study done by Cao et al., mesoporous molecular sieves were synthesized for the Matrix solid-phase dispersion in a reduced scale isolation of bioactive flavonoids from toothpaste, plant materials, and saliva. KIT-6 demonstrated excellent extraction capabilities, achieving precise quantification with high recovery rates and low limits of detection (LOD). Compared to traditional silica-based sorbents such as C18, KIT-6's cubic mesostructures and large surface area provided superior interaction with analytes, making it a promising technique in detection of the organic flavonoids in complex samples using a commercial silica-based material, namely UVM-7 and MCM-41 silica, Cejuela et al. used MS materials having double mode pore systems to precisely isolate phospholipids (PLs). Phosphatidylcholine was the test chemical. The morphological characteristics of these materials, including TEM, surface area measurements, and pore size assessments, were thoroughly examined. The interaction mechanisms between the target analytes and MSMs were also investigated. We improved several experimental parameters influencing the SPE process: loading and elution solvents, saturation volume, stocking competence, and recyclability. PLs from human milk fat samples were effectively extracted using the advised technique [54]. In another work by Du et al., a miniaturized SPE technique was developed to concurrently purify and concentrate macrolide antibiotics (MACs) including azithromycin, lincomycin, and roxithromycin from honey and skim milk. Rapid automated extraction, enrichment, and analyte purification made possible by the mini-SPE approach greatly reduced sample and solvent use, therefore lowering contamination and costs [55].

6. Performance Evaluation and Optimization

In the realm of scientific research focused on MM for SPE, rigorous performance evaluation and optimization are crucial for enhancing their efficacy in analytical chemistry. Evaluation encompasses assessing adsorption capacity, selectivity, reproducibility, and stability under varying conditions, utilizing techniques such as BET surface area analysis, SEM, TEM, and FTIR spectroscopy. Optimization strategies involve refining synthesis methods to control pore size, surface area, and surface chemistry, thereby improving adsorption properties and specificity for target analytes. Computational modeling aids in predicting and optimizing material performance, guiding the development of tailored MM for SPE that push the boundaries of analytical capabilities and expand their applications in diverse scientific and industrial contexts.

6.1 Key Performance Indicators

The significance of the sorbent material in SPE cannot be overstated, as it plays a pivotal role. Therefore, the ongoing pursuit to develop novel sorbent materials that exhibit exceptional selectivity, stability, and extraction efficiency is a matter of considerable interest among analysts and researchers alike. A study was done by Sreenu et al. in which a newly synthesized mesoporous hybrid material was created through the surface modification of SBA-15 by grafting 2-mercaptobenzaldehyde (2MB) onto its framework for its utilization in the SPE of the Cd^{2+} from water and food samples. The findings indicated that out of the three acids tested, the use of 5 mL of 1 M nitric acid solution resulted in the recovery of over 95% of cadmium ions from the functionalized SBA-15. In contrast, the efficacy of other acids tested was insufficient for complete elution of these ions. Moreover, the maximum adsorption capacity (N_{fmax}) of an adsorbent is crucial as it shows the quantity of adsorbent needed to fully recover a fixed quantity of metal ions from a solution. To determine the N_{fmax} for Cd(II), column experiments were conducted using 2MB-SBA-15. The functionalization of SBA-15 resulted in a notable enhancement in binding capacity for Cd(II), achieving an N_{fmax} value of 0.94 ± 0.01. This underscores the superior Cd(II) adsorption capabilities of various hybrid silica-based materials(10.1039/c5ay03031e). In another investigation conducted by Liu et al., Bis-pyrazolyl-modified mesoporous SBA-15 was employed to extract Cr^{3+} and detect Cr^{4+} in simulated jewelry samples. The made adsorbent material demonstrates strong chemical and mechanical durability, as well as a significant adsorption capability of 72.5 mg/g and exceptional preference for Cr^{3+}. As a result, DPP@SBA-15 utilized as an SPE adsorbent for capturing Cr^{3+} and quantifying Cr^{4+}. The method exhibited excellent linearity across a concentration of 1–20 μg/mL (R2 = 0.9999) and achieved a low detection threshold of 0.028 μg/mL. Moreover, it effectively detected Cr(VI) released from artificial jewelry, achieving substantial recovery rates between 97.6% and 103.0%.Given the inherent complexity of natural samples, which often contain diverse matrices and background interference from various species such as metal ions, the impact of these ions on the adsorption capabilities of DPP@SBA-15 was thoroughly assessed. This underscores the potential versatility of the proposed technique across a range of natural samples, as the adsorption properties of DPP@SBA-15 remain unaffected even in the presence of relatively high concentrations of alkali and alkaline earth metals. Additionally, the enhanced preference of this sorbent material for Cr^{3+} ions is credited to the incorporation of bis-pyrazolyl functionality.[56].

Materials Research Foundations 173 (2025) 203-221 https://doi.org/10.21741/9781644903452-8

7. Challenges and Future Perspectives

Due to their outstanding attributes, they find utility across a broad spectrum of applications. however, numerous obstacles persist. Synthesizing MM typically involves time-consuming processes with complex reactions and steps, primarily confined to laboratory settings, creating a significant disparity between lab-scale synthesis and industrial production. Despite the favourable properties exhibited by some MM, such as their efficacy as catalyst in chemistry and their utility in gas separation as adsorbents, their regeneracy remains a challenge. Moreover, the high cost associated with their synthesis further complicates widespread adoption. Simplifying these synthesis processes and reducing costs are essential to enable large-scale production effectively. In addition to that, determining a robust theory or method is essential for synthesizing MM with ideal structural characteristics. Designing and producing these materials effectively, particularly mesoporous polymers with spacious pores, excellent mobility, and minimal volume ratios, holds significant scientific and technological importance. These advancements are pivotal for addressing critical challenges and advancing applications across diverse fields of study [57]. The heightened effectiveness of USY zeolite is credited to the presence of more potent Bronsted acid sites within the zeolitic framework.[58]. This approach enhances the versatility and functionality of the resulting materials, opening new avenues for their application in various fields. In a research conducted by Liu et al., they introduced an effective and adaptable approach to produce micro-meso-macroporous polymers functionalized with a various of functional groups. The method involves a direct alkylation process of small functional molecules with 1,4-bis(chloromethyl)benzene, eliminating the necessity for organic templates. These polymers, known as MOPs, demonstrate expansive BET surface areas and plentiful micro-meso-macropores, featuring adjustable and versatile active centers. These attributes enhance their outstanding catalytic capabilities and ability to be reused in converting inexpensive biomass into environmentally friendly biofuels and fine chemicals. These qualities highlight their significant potential for advancing applications in the realms of eco-friendly and sustainable chemistry.[59].

Conclusion

In recent years the utilization of MM as the sorbents have emerged exponentially. It is due its large sized pores which enables them towards the stability, selectivity and the catalytic activities. One such application of the MM is seen in SPE as the sorbents. In the SPE, the sorbents play a very vital role as they interact with the solid phase when the liquid solvent flows through it. So due the large and uniform pore volume of the MM, it makes them the ideal candidate to be used as the sorbents. The various studies mentioned above the wide range of application shown by the MM when used as the SPE sorbents. The wide variety of application includes the one in the environmental analysis. Biomedical and pharmaceutical applications, food and beverage industries etc. Also, there are some criteria which are to be kept in mind while selecting the materials the sorbents materials such as their selectivity, catalytic activity, reproducibility, extraction efficiency etc. However, some challenges are also faced by the MM when used as the sorbents. For example, its synthesis cost is high when compared to the other zeolites. Also, in the medicine transportation it is required to have a particularly shaped pore which is very difficult in the preparation of MM. In conclusion it can be said that the MM embraces towards in the various applicative fields such as in electrochemistry, semiconductors, optical properties etc.

References

[1]V. Meynen, P. Cool, E.F. Vansant, Verified syntheses of mesoporous materials, Microporous Mesoporous Mater. 125 (2009) 170–223. https://doi.org/https://doi.org/10.1016/j.micromeso.2009.03.046

[2]S. Kumar, M.M. Malik, R. Purohit, Synthesis Methods of Mesoporous Silica Materials, Mater. Today Proc. 4 (2017) 350–357. https://doi.org/https://doi.org/10.1016/j.matpr.2017.01.032

[3]T. Yanagisawa, T. Shimizu, K. Kuroda, C. Kato, The Preparation of Alkyltrimethylammonium–Kanemite Complexes and Their Conversion to Microporous Materials, Bull. Chem. Soc. Jpn. 63 (1990) 988–992. https://doi.org/10.1246/bcsj.63.988

[4]A. Vinu, T. Mori, K. Ariga, New families of mesoporous materials, Sci. Technol. Adv. Mater. 7 (2006) 753–771. https://doi.org/10.1016/j.stam.2006.10.007

[5]C.T. Kresge, M.E. Leonowicz, W.J. Roth, J.C. Vartuli, J.S. Beck, Ordered mesoporous molecular sieves synthesized by a liquid-crystal template mechanism, Nature. 359 (1992) 710–712. https://doi.org/10.1038/359710a0

[6]J.C. Vartuli, K.D. Schmitt, C.T. Kresge, W.J. Roth, M.E. Leonowicz, S.B. McCullen, S.D. Hellring, J.S. Beck, J.L. Schlenker, Effect of surfactant/silica molar ratios on the formation of mesoporous molecular sieves: inorganic mimicry of surfactant liquid-crystal phases and mechanistic implications, Chem. Mater. 6 (1994) 2317–2326.

[7]M. Dubois, T. Gulik-Krzywicki, B. Cabane, Growth of silica polymers in a lamellar mesophase, Langmuir. 9 (1993) 673–680.

[8]M. Moritz, M. Geszke-Moritz, Mesoporous materials as multifunctional tools in biosciences: Principles and applications, Mater. Sci. Eng. C. 49 (2015) 114–151. https://doi.org/https://doi.org/10.1016/j.msec.2014.12.079

[9]K.S.W. Sing, No Title, Pure Appl. Chem. 57 (1985) 603–619. https://doi.org/doi:10.1351/pac198557040603

[10] K.S.W. Sing, Reporting physisorption data for gas/solid systems with special reference to the determination of surface area and porosity (Recommendations 1984), 57 (1985) 603–619. https://doi.org/doi:10.1351/pac198557040603

[11] S. Savic, K. Vojisavljevic, M.P. uč a-Nešić, K. Zivojevic, M. Mladenovic, N. Knezevic, Hard template synthesis of nanomaterials based on mesoporous silica, Metall. Mater. Eng. 24 (2018).

[12] H.K. Schmidt, E. Geiter, M. Mennig, H. Krug, C. Becker, R.-P. Winkler, The Sol-Gel Process for Nano-Technologies: New Nanocomposites with Interesting Optical and Mechanical Properties, J. Sol-Gel Sci. Technol. 13 (1998) 397–404. https://doi.org/10.1023/A:1008660909108

[13] Y. Qi, D. Wu, J. Wei, K. Ding, H. Wang, Y. Zhang, X. Qian, Y. Guan, Selective extraction of low molecular weight proteins by mesoporous silica particles with modified internal and external surfaces, Anal. Bioanal. Chem. 398 (2010) 1715–1722.

[14] F. Hoffmann, M. Cornelius, J. Morell, M. Fröba, Silica-Based Mesoporous Organic–Inorganic Hybrid Materials, Angew. Chemie Int. Ed. 45 (2006) 3216–3251. https://doi.org/https://doi.org/10.1002/anie.200503075

[15] B.L. Newalkar, S. Komarneni, H. Katsuki, Rapid synthesis of mesoporous SBA-15 molecular sieve by a microwave–hydrothermal process, Chem. Commun. (2000) 2389–2390. https://doi.org/10.1039/B007441L

[16] B.E. Grabicka, M. Jaroniec, Microwave-assisted synthesis of periodic mesoporous organosilicas with ethane and disulfide groups, Microporous Mesoporous Mater. 119 (2009) 144–149. https://doi.org/https://doi.org/10.1016/j.micromeso.2008.10.006

[17] T. Sánchez, P. Salagre, Y. Cesteros, Ultrasounds and microwave-assisted synthesis of mesoporous hectorites, Microporous Mesoporous Mater. 171 (2013) 24–34. https://doi.org/https://doi.org/10.1016/j.micromeso.2013.01.001

[18] M.E. Raisglid, M.F. Burke, Fundamentals of solid phase extraction and its application to environmental analyses, in: A.B.T.-S. in S.S. and C. Dąbrowski (Ed.), Adsorpt. Its Appl. Ind. Environ. Prot., Elsevier, 1999: pp. 37–75. https://doi.org/https://doi.org/10.1016/S0167-2991(99)80356-5

[19] E. Chladek, R.S. Marano, Use of Bonded Phase Silica Sorbents for the Sampling of Priority Pollutants in Wastewaters*, J. Chromatogr. Sci. 22 (1984) 313–320. https://doi.org/10.1093/chromsci/22.8.313

[20] Solid Phase Extraction Technique - Trends, Opportunities and Applications, Polish J. Environ. Stud. 15 (2006) 677–690.

[21] J. Yu, C. Wu, J. Xing, Development of new solid-phase microextraction fibers by sol–gel technology for the determination of organophosphorus pesticide multiresidues in food, J. Chromatogr. A. 1036 (2004) 101–111. https://doi.org/https://doi.org/10.1016/j.chroma.2004.02.081

[22] S. Ötles, C. Kartal, Solid-Phase Extraction (SPE): Principles and Applications in Food Samples, Acta Sci. Pol. Technol. Aliment. 15 (2016) 5–15. https://doi.org/10.17306/J.AFS.2016..1

[23] G. Font, J. Mañes, J.C. Moltó, Y. Picó, Solid-phase extraction in multi-residue pesticide analysis of water, J. Chromatogr. A. 642 (1993) 135–161. https://doi.org/https://doi.org/10.1016/0021-9673(93)80083-K

[24] H. Sabik, R. Jeannot, B. Rondeau, Multiresidue methods using solid-phase extraction techniques for monitoring priority pesticides, including triazines and degradation products, in ground and surface waters, J. Chromatogr. A. 885 (2000) 217–236. https://doi.org/https://doi.org/10.1016/S0021-9673(99)01084-5

[25] D. Wang, X. Chen, J. Feng, M. Sun, Recent advances of ordered mesoporous silica materials for solid-phase extraction, J. Chromatogr. A. 1675 (2022) 463157. https://doi.org/https://doi.org/10.1016/j.chroma.2022.463157

[26] S. Chatterjee, H. Gohil, I. Raval, S. Chatterjee, A.R. Paital, An Anthracene Excimer Fluorescence Probe on Mesoporous Silica for Dual Functions of Detection and Adsorption of

Mercury (II) and Copper (II) with Biological In Vivo Applications, Small. 15 (2019) 1804749. https://doi.org/https://doi.org/10.1002/smll.201804749

[27] Y. Wang, B. Wang, Q. Wang, J. Di, S. Miao, J. Yu, Amino-Functionalized Porous Nanofibrous Membranes for Simultaneous Removal of Oil and Heavy-Metal Ions from Wastewater, ACS Appl. Mater. Interfaces. 11 (2019) 1672–1679. https://doi.org/10.1021/acsami.8b18066

[28] A. Shkatulov, R. Joosten, H. Fischer, H. Huinink, Core–Shell Encapsulation of Salt Hydrates into Mesoporous Silica Shells for Thermochemical Energy Storage, ACS Appl. Energy Mater. 3 (2020) 6860–6869. https://doi.org/10.1021/acsaem.0c00971

[29] J. V Jokerst, M. Thangaraj, P.J. Kempen, R. Sinclair, S.S. Gambhir, Photoacoustic Imaging of Mesenchymal Stem Cells in Living Mice via Silica-Coated Gold Nanorods, ACS Nano. 6 (2012) 5920–5930. https://doi.org/10.1021/nn302042y

[30] Y. Li, N. Li, W. Pan, Z. Yu, L. Yang, B. Tang, Hollow Mesoporous Silica Nanoparticles with Tunable Structures for Controlled Drug Delivery, ACS Appl. Mater. Interfaces. 9 (2017) 2123–2129. https://doi.org/10.1021/acsami.6b13876

[31] X. Zhang, L. Jing, L. Wei, F. Zhang, H. Yang, Semipermeable Organic–Inorganic Hybrid Microreactors for Highly Efficient and Size-Selective Asymmetric Catalysis, ACS Catal. 7 (2017) 6711–6718. https://doi.org/10.1021/acscatal.7b01659

[32] B. Shen, L. Huang, J. Shen, L. Meng, E.J. Kluender, C. Wolverton, B. Tian, C.A. Mirkin, Synthesis of Metal-Capped Semiconductor Nanowires from Heterodimer Nanoparticle Catalysts, J. Am. Chem. Soc. 142 (2020) 18324–18329. https://doi.org/10.1021/jacs.0c09222

[33] F. Farjadian, S. Azadi, S. Mohammadi-Samani, H. Ashrafi, A. Azadi, A novel approach to the application of hexagonal mesoporous silica in solid-phase extraction of drugs, Heliyon. 4 (2018) e00930. https://doi.org/https://doi.org/10.1016/j.heliyon.2018.e00930

[34] W. Cao, L.-H. Ye, J. Cao, J.-J. Xu, L.-Q. Peng, Q.-Y. Zhu, Q.-Y. Zhang, S.-S. Hu, Quantitative analysis of flavanones from citrus fruits by using mesoporous molecular sieve-based miniaturized solid phase extraction coupled to ultrahigh-performance liquid chromatography and quadrupole time-of-flight mass spectrometry, J. Chromatogr. A. 1406 (2015) 68–77. https://doi.org/https://doi.org/10.1016/j.chroma.2015.06.035

[35] Z. Wu, D. Zhao, Ordered mesoporous materials as adsorbents, Chem. Commun. 47 (2011) 3332–3338. https://doi.org/10.1039/C0CC04909C

[36] P. Stathi, K. Dimos, M.A. Karakassides, Y. Deligiannakis, Mechanism of heavy metal uptake by a hybrid MCM-41 material: Surface complexation and EPR spectroscopic study, J. Colloid Interface Sci. 343 (2010) 374–380. https://doi.org/https://doi.org/10.1016/j.jcis.2009.11.029

[37] T. Yokoi, T. Tatsumi, H. Yoshitake, Fe3+ coordinated to amino-functionalized MCM-41: an adsorbent for the toxic oxyanions with high capacity, resistibility to inhibiting anions, and reusability after a simple treatment, J. Colloid Interface Sci. 274 (2004) 451–457. https://doi.org/https://doi.org/10.1016/j.jcis.2004.02.037

[38] C.R. Martin, Nanomaterials: A Membrane-Based Synthetic Approach, Science (80-.). 266 (1994) 1961–1966. https://doi.org/10.1126/science.266.5193.1961

[39] C.M. Zelenski, P.K. Dorhout, Template Synthesis of Near-Monodisperse1 Microscale Nanofibers and Nanotubules of MoS2, J. Am. Chem. Soc. 120 (1998) 734–742. https://doi.org/10.1021/ja972170q

[40] Y. Han, L. Zhang, W. Yang, Synthesis of Mesoporous Silica Using the Sol–Gel Approach: Adjusting Architecture and Composition for Novel Applications, Nanomaterials. 14 (2024). https://doi.org/10.3390/nano14110903

[41] M.D.G. García, M.M. Galera, M.J.M. Bueno, Silica Based Mesoporous Materials as Sorbents for Solid-Phase Extraction of Organic Pollutants in Water, in: M.M.M. Elnashar, S. Karakuş (Eds.), Water Purif., IntechOpen, Rijeka, 2023. https://doi.org/10.5772/intechopen.112010

[42] Y. Chang, C. Wang, T. Liang, C. Zhao, X. Luo, T. Guo, J. Gong, H. Wu, Sol–gel synthesis of mesoporous spherical zirconia, RSC Adv. 5 (2015) 104629–104634. https://doi.org/10.1039/C5RA23782C

[43] A. Vinu, K.Z. Hossain, K. Ariga, Recent advances in functionalization of mesoporous silica, J. Nanosci. Nanotechnol. 5 (2005) 347–371.

[44] C.Y.R.C. Lijun Zhang Xun Hu, A. Yu, Preparation of sinapinaldehyde modified mesoporous silica materials and their application in selective extraction of trace Pb(II), Int. J. Environ. Anal. Chem. 93 (2013) 1274–1285. https://doi.org/10.1080/03067319.2012.736974

[45] A. Walcarius, M. Etienne, B. Lebeau, Rate of Access to the Binding Sites in Organically Modified Silicates. 2. Ordered Mesoporous Silicas Grafted with Amine or Thiol Groups, Chem. Mater. 15 (2003) 2161–2173. https://doi.org/10.1021/cm021310e

[46] C. Lei, Y. Shin, J. Liu, E.J. Ackerman, Entrapping Enzyme in a Functionalized Nanoporous Support, J. Am. Chem. Soc. 124 (2002) 11242–11243. https://doi.org/10.1021/ja026855o

[47] X. Feng, G.E. Fryxell, L.-Q. Wang, A.Y. Kim, J. Liu, K.M. Kemner, Functionalized Monolayers on Ordered Mesoporous Supports, Science (80-.). 276 (1997) 923–926. https://doi.org/10.1126/science.276.5314.923

[48] R.I. Nooney, M. Kalyanaraman, G. Kennedy, E.J. Maginn, Heavy Metal Remediation Using Functionalized Mesoporous Silicas with Controlled Macrostructure, Langmuir. 17 (2001) 528–533. https://doi.org/10.1021/la000720j

[49] S. Dahane, M. Martínez Galera, M.E. Marchionni, M.M. Socías Viciana, A. Derdour, M.D. Gil García, Mesoporous silica based MCM-41 as solid-phase extraction sorbent combined with micro-liquid chromatography–quadrupole-mass spectrometry for the analysis of pharmaceuticals in waters, Talanta. 152 (2016) 378–391. https://doi.org/https://doi.org/10.1016/j.talanta.2016.02.013

[50] E. Pellicer-Castell, C. Belenguer-Sapiña, P. Amorós, J. El Haskouri, J.M. Herrero-Martínez, A. Mauri-Aucejo, Study of silica-structured materials as sorbents for organophosphorus pesticides determination in environmental water samples, Talanta. 189 (2018) 560–567. https://doi.org/https://doi.org/10.1016/j.talanta.2018.07.044

[51] L. Kharbouche, M.D. Gil García, A. Lozano, H. Hamaizi, M.M. Galera, Solid phase extraction of pesticides from environmental waters using an MSU-1 mesoporous material and

determination by UPLC-MS/MS, Talanta. 199 (2019) 612–619. https://doi.org/https://doi.org/10.1016/j.talanta.2019.02.092

[52] Y. Li, J. Yang, C. Huang, L. Wang, J. Wang, J. Chen, Dendrimer-functionalized mesoporous silica as a reversed-phase/anion-exchange mixed-mode sorbent for solid phase extraction of acid drugs in human urine, J. Chromatogr. A. 1392 (2015) 28–36. https://doi.org/https://doi.org/10.1016/j.chroma.2015.03.003

[53] W. Cao, S.-S. Hu, L.-H. Ye, J. Cao, X.-Q. Pang, J.-J. Xu, Trace matrix solid phase dispersion using a molecular sieve as the sorbent for the determination of flavonoids in fruit peels by ultra-performance liquid chromatography, Food Chem. 190 (2016) 474–480. https://doi.org/https://doi.org/10.1016/j.foodchem.2015.05.133

[54] H. Martínez Pérez-Cejuela, I. Ten-Doménech, J. El Haskouri, P. Amorós, E.F. Simó-Alfonso, J.M. Herrero-Martínez, Solid-phase extraction of phospholipids using mesoporous silica nanoparticles: application to human milk samples, Anal. Bioanal. Chem. 410 (2018) 4847–4854. https://doi.org/10.1007/s00216-018-1121-8

[55] L.-J. Du, L. Yi, L.-H. Ye, Y.-B. Chen, J. Cao, L.-Q. Peng, Y.-T. Shi, Q.-Y. Wang, Y.-H. Hu, Miniaturized solid-phase extraction of macrolide antibiotics in honey and bovine milk using mesoporous MCM-41 silica as sorbent, J. Chromatogr. A. 1537 (2018) 10–20. https://doi.org/https://doi.org/10.1016/j.chroma.2018.01.005

[56] S. Liu, H.-Z. Cui, Y.-L. Li, A.-L. Yang, J.-F. Zhang, R. Zhong, Q. Zhou, M. Lin, X.-F. Hou, Bis-pyrazolyl functionalized mesoporous SBA-15 for the extraction of Cr(III) and detection of Cr(VI) in artificial jewelry samples, Microchem. J. 131 (2017) 130–136. https://doi.org/https://doi.org/10.1016/j.microc.2016.12.003

[57] Z. Li, L. Liu, Z. Wang, P. Gao, G.K. Li, Synthesis and Application of Mesoporous Materials: Process Status, Technical Problems, and Development Prospects: A Mini-Review, Energy & Fuels. 37 (2023) 3413–3427. https://doi.org/10.1021/acs.energyfuels.2c03882

[58] A. Corma, M.S. Grande, V. Gonzalez-Alfaro, A. V Orchilles, Cracking Activity and Hydrothermal Stability of MCM-41 and Its Comparison with Amorphous Silica-Alumina and a USY Zeolite, J. Catal. 159 (1996) 375–382. https://doi.org/https://doi.org/10.1006/jcat.1996.0100

[59] Z. Yang, M. Xu, Y. Liu, F. He, F. Gao, Y. Su, H. Wei, Y. Zhang, Nitrogen-doped, carbon-rich, highly photoluminescent carbon dots from ammonium citrate, Nanoscale. 6 (2014) 1890–1895.

Materials Research Foundations 173 (2025) 222-264 https://doi.org/10.21741/9781644903452-9

Chapter 9

Cutting-Edge Mesoporous Materials for Biotechnological Applications in Separation Science

Nethravathi[1], Sneha Nayak[2], Arun M. Isloor[1]*

[1] Membrane and Separation Technology Laboratory, Department of Chemistry, National Institute of Technology Karnataka, Surathkal, Mangalore, Karnataka-575025, India

[2]Department of Biotechnology Engineering, Nitte (Deemed to Be University), NMAM Institute of Technology, Nitte, Karnataka, India

*isloor@yahoo.com

Abstract

Mesoporous materials (MMs) have emerged as biocompatible and versatile material for biotechnological applications, particularly for protein separation and purification. This review provides a one stop overview of MMs, their types and their synthesis approaches. In addition, the existing challenges and limitations faced for large-scale production of MMs, such as the complexity of synthesis processes, the need for sophisticated characterization techniques, and the difficulties in achieving reproducible MMs with structural integrity at a large scale is emphasized. Additionally, the review stresses the future perspectives of using artificial intelligence (AI) and machine learning (ML) techniques in the design and optimization of MMs. These AI driven technologies can act as crucial tools for enhancing the performance of MMs by predicting their structural and mechanical properties based on characterization data using predictive modeling, potentially reducing the reliance on expensive, high-end equipment's. Finally, the future prospects of MMs in biotechnological protein separation, purification and enzyme immobilization applications. Furthermore, the use of MMs as effective drug delivery carriers and innovative reusable adsorbents for adsorption of heavy metals and organic contaminants (dyes and antibiotics) are also emphasized. The ongoing advancements in AI and ML, coupled with innovative material design strategies by optimization strategies, are likely to expand the scalability and practical utility of MMs, making them indispensable tools in modern biotechnology.

Keywords

Mesoporous, Separation, Biotechnology, Proteins, Purification

Contents

Materials Research Foundations 173 (2025) 222-264 https://doi.org/10.21741/9781644903452-9

1. Introduction

Incredible advances in life science and biotechnology industries have enhanced the requirement for smart and effective bio-separation advances. Rampant development of biotechnology products obtained from strategies like recombinant DNA technology (r-DNA) and cell culturing definitely needs rapid and cost-effective purification and recovery approaches which removes maximum impurities without losing the sample bioactivity [1]. Undoubtedly, separation science plays a crucial role in biotechnological processes, since it majorly involved in improving the purity aspects of the products thus making it market ready [2]. Use of minimum separation techniques during product formulation is necessary to avoid product losses and improve yield [3]. For this reason, use of simple techniques which are scalable and could be easily mimicked from laboratory-scale production to industrial-scale manufacturing is gaining a lot of attention [4]. Commonly used separation techniques in biotechnological applications include electrophoresis [5], chromatography [6], centrifugation [7], precipitation [8] and membrane filtration [2] to name a few. Use of multiple techniques in combinations might be necessary during downstream processing of biotechnological products to make it market ready. Therefore optimization in steps is essential to minimize losses as well as human errors thus maximizing profits [9].

Materials Research Foundations 173 (2025) 222-264 https://doi.org/10.21741/9781644903452-9

In the recent year's use of novel materials have been explored by various researchers for separation applications since conventional zeolites and microporous materials are not capable to meet the practical demand of effective mass transfer of large molecules [10]. Novel materials of mesoporous nature have been explored for wide range of applications due to their unaverred potential, intermediate pore size (2-50nm), high surface area, and availability of mesoporous channels [11]. Some of the major areas include membrane based [12] separation [13] / diffusion [14], adsorption [15], catalysis [16], biomedical sector for kits development [17], energy related applications [18] and biotechnology [19].

Various synthetic techniques including sol-gel process [20], template-assisted synthesis [21], soft-templating methods and hard-templating methods [22], hydrothermal and solvothermal synthesis [23], electrochemical deposition [24], spray drying techniques [25] and microwave assisted synthesis [26] have been experimented for mesoporous material production. Among the mesoporous materials, silica and its variants [27], carbon based materials [28] and Crystalline TiO_2 [29] have been widely explored for biotechnological applications which include enzyme immobilization [30], protein storage and delivery [31], DNA/RNA separation [32] , cell separation sorting and analysis [33] to name a few. However, recent works have also explored various agricultural biomasses for cutting edge mesoporous material synthesis which could be explored for separation of biological moieties [34].

The current review thus intends to bridge the gap and provide one point information on sustainable mesoporous materials (MM) that could be explored for separation applications of biotechnological products. Additionally, various synthesis techniques with its pros and cons and recent advances in synthesis methods have been elucidated for reader's benefit. Major emphasis on challenges and limitations of these novel MM and future perspectives are stressed upon for benefit of researchers who are willing to explore these materials for cutting edge biotechnological applications. There exists a research gap in exploiting the various functionalization strategies that can enhance the applicability of these mesoporous materials in biotechnology product separation which needs to be explored more in future studies. Further, the scalability of the lab synthesized mesoporous materials can be increased by focusing on more facile synthetic routes. Also, there is a huge gap between material chemistry and the practical biotechnological application of it which needs to be bridged. The tunability of the porosity and pore size distribution of the mesoporous materials required for desired biotechnological separation need to be deepened.

2. Synthesis of mesoporous materials

2.1 Overview of synthesis methods

The highlighted properties of the mesoporous materials such as higher surface area, sharp distribution of pore sizes, large pore volume, appreciable thermal and chemical stability, and functionalities can be tuned by choosing an appropriate synthesis route. Also, the particle shape and size can be designed desirably by adopting certain specific synthetic strategies. The facile and versatile discovery of mesoporous materials was first discovered by Kuroda and the group by developing a folded sheet mechanism of synthesis [35]. Customized architecture of the pore and morphologies such as spheres, fibers, sheets, and rods have been focused on while tailoring the mesoporous materials.

Materials Research Foundations 173 (2025) 222-264
https://doi.org/10.21741/9781644903452-9

There are many synthetic strategies employed to obtain mesoporous materials having the desired properties. Templating methods (hard and soft), the Stober method, spray pyrolysis, and sol-gel methods are a few to navigate through. In the Stober method, tetraethylorthosilicate is hydrolyzed at an alcohol-water interface generating spherical colloids. Over the past few years, mesoporous silica has been synthesized with this method having slight modification, wherein the researcher's usecetyltrimethylammoniabromide (CTAB). Spray pyrolysis was first reported in the 1980s when a precursor was atomized forming aerosols. Each droplet in the aerosol undergoes different physical and chemical changes and behaves like a microreactor here. The steps followed here are (i) solvent evaporation (ii) solute precipitation (iii) thermolysis (iv) porous architect formation (v) sintering. Chemical etching and microwave techniques are employed in a few cases as well. Chemical etching can be used to synthesize heterogeneous hollow nanostructures for inorganic nanocrystals like Au, Fe_3O_4, and Fe_2O_3. The microwave technique can sometimes reduce pore size than expected via rapid heating. The remaining methods of mesoporous material design are explained in detail in the upcoming part. The different synthetic techniques have their advantages and disadvantages. However, the templating method is more widely employed to achieve precise controlled pore and particle size.

Focus on understanding the mechanism operating in the individual methods and the correlation of the different synthetic techniques with the properties of the final mesoporous product is necessary in this regard. Further, the scalability and the facile synthesis to get predefined morphologies in the mesoporous materials are to be explored more deeply. At the nanoscale level, the pore morphologies can be tuned better by using developing new copolymers and surfactants [36].

Polymer membranes, as mesoporous materials, are characterized by their pore sizes ranging from 2 to 100 nanometers, bridging the gap between microporous and macroporous materials. These membranes are engineered to possess a high surface area and controlled pore distribution, which makes them particularly effective for selective separation processes. The mesoporous structure facilitates efficient transport of molecules and ions, enhancing their performance in applications such as gas separation, water purification, and catalytic reactions. The versatility and tunability of polymer membranes allow for their adaptation to various industrial and environmental needs, making them a vital component in advanced material science. The membrane process can easily be integrated with other separation techniques. Upscaling to meet industrial requirements is also easy through this method [37]. Membranes can be fabricated in various forms [38], namely flat sheet [39], tubular, and hollow fiber forms. Out of these the two most important ones are flat sheet and hollow fibers. In flat sheet type, a sheet form of polymeric membrane is fabricated via non-solvent induced [40] or thermal induced or vapour-phase induced phase separation method. These sheets later can be rolled into the modules during separation. The driving force for the separation here can be concentration gradient, pressure gradient, or chemical potential gradient. Depending on the size of the particle to be separated and the pressure gradient there are four types of membranes [41]. Microfiltration (MF) [42] blocks particles of sizes 10 to 0.1 micron at the pressure range of 0.2 to 5 bar. Ultrafiltration (UF) [43] blocks particles of sizes up to 0.01 micron at the pressure range of 1 to 10. Nanofiltration [44] (NF) blocks particles of sizes up to 0.001 microns at the pressure range of 5 to 10 bar. Reverse osmosis (RO) blocks particles of size 0.0001 micron at the pressure range of 10 to 150 bar. The transport through the membrane can be of three types. Active transport is the one where the flow takes place against the conjugative driving force. Passive transport is where the flow takes place in the direction of the driving force. Facilitated or carrier-mediated transport includes the transport of permeate-carrier complex across the membrane [45].

Membranes can be classified in various ways. Based on the structure and morphology, there are symmetric, asymmetric [46], and composite membranes. Based on the charge there are anion exchange, cation exchange, and bipolar membranes. Filtration through membranes can be done either via dead-end mode or via cross-flow mode. In dead-end filtration, the permeate passes through the membrane leaving behind the contaminants. This mode requires regular cleansing to avoid fouling. In cross-flow mode, the permeate passes along the membrane. The constant tangential flow helps to avoid the accumulation of particles on the membrane surface which would otherwise eventually lead to fouling [47].

HF membranes give a better flux [48] compared to flat sheets due to their high surface area. But they are very delicate to handle. HF membranes will usually be fabricated via the dry-wet spinning technique. Initially, the prepared dope solution will be allowed to run through the annular part of the spinneret [49]. Deionized water runs as bore liquid via the internal tube of the spinneret. A sufficient distance between the non-solvent bath and the spinneret will be maintained to have the dry-wet process [50]. At the tip of the spinneret, bore liquid and the dope solution meet which is further guided to run through the coagulation bath. A roller drum mounted at the last part the machine collects the HF membranes.

Membranes are fabricated from natural as well as synthetic polymers. Polyamides, Polysulfones [51], Teflon, PVDF, Polypropylene, Polyphenyl sulfones, etc. are a few synthetic polymers. Silk, Polylactic acid, polyhydroxyalkanoates, Cellulose-based polymers like starch, alginate polyisoprene, products of bacterial fermentation such as serine, collagen, Chitin, etc. are a few bio-polymers available for the fabrication of membranes. The hollow fiber form of the membrane has drawn attention worldwide lately. Hollow fibers provide excellent mass transfer because of their high surface-to-volume ratio. Phase inversion is the most commonly employed process for its fabrication. The phase inversion can be achieved by varying the temperature or by inducing diffusion. The main tricky process in the fabrication of hollow fibers is the spinning step which generally follows four steps. The very first step involves the preparation of a dope solution. In this step desired blending, structural modifications, and addition of nanoparticles that enhance the membrane properties need to be done. This is followed by extrusion where the fibers are formed. The next step involves the coagulation of the spun fibers. Finally, the coagulated fibers need to be cured by immersing them in a non-solvent bath. The spinning parameters are very crucial to decide the nature and applicability of the fibers. Type and amount of the base polymers used, additives incorporated into the dope solution, solvents, the flow rate of the dope solution and the bore fluid, spinning velocity, temperature of the coagulation bath, etc. decide the performance of the hollow fiber membrane. The distinctive physicochemical properties like exceptional mechanical strength and favorable surface area make hollow fibers a better candidate for diverse applications such as ultrafiltration, dialysis, gas-separation, cell culture, tissue engineering, and membrane bioreactors.

2.2 Sol-gel process

So-gel chemistry, also called as chemical solution deposition technique involves the synthesis of network structure called gels using liquid precursors which will consist metals or metalloids surrounded by many reactive ligands. These precursors when comes in contact with water forms sol which gradually forms gel with the loss of liquid. Calcination finally yields required network of mesoporous material here. Hydrolysis followed by metal-alkoxide condensation are the main types of reactions involved in sol-gel method. Synthesis of mesoporous materials via this method various structure directing templates such as surfactants, small organic molecules, and triblock copolymers can also be used [52].

Lavrova and coworkers synthesized silica-polyethylene glycol based composites (biocatalyst) using sol-gel method [53]. Liu et al. synthesized mesoporous metal oxides under strongly acidic condition on which metal nanoparticles such as Pt, Au, and Pd were grown. The mesoporous Metal oxides studied in this study were TiO_2, ZrO_2, SiO_2, and Al_2O_3 [54]. Adam et. al synthesized mesoporous silica (average diameter = 50.9 nm and BET surface area of 245 m^2g^{-1}, pore size distribution = 5.6 to 9.6 nm) with spherical morphology from rice husk [55]. Tessema et al. synthesized modified silica from Teff straw ash with the same sol-gel method which had a BET surface area of 909.25 m^2/g, pore volume of 2.670 nm, and pore size of 0.5422cm^3/g [56]. Zhang et al. used ionic liquids as structure directing templates for synthesizing mesoporous silica by sol-gel method. The pore diameter in the material was 3 – 12 nm, BET surface area of 800 m^2/g, and pore volume of 0.4-1.3 cm^3/g [57]. Li et al. synthesized ordered mesoporous carbons for desalination via sol-gel method which had a BET surface area of 1491 m^2g^{-1} with pore diameter of 3.7 nm [58]. Vazquez et al. used TEOS and CTAB for synthesizing mesoporous silica for drug delivery application. These mesoporous silica nanoparticles had specific BET surface area of 1480 m^2/g with pore size in the range 2.5 2.8 nm. They concluded that the surfactant concentration plays vital role in deciding the particle morphology varying to agglomerates from the nanospheres. Also, with sol-dilution they obtained rod-like morphologies [59]. Yoo et al. used ionic liquids for synthesizing mesoporous TiO_2. [Bmim][PF_6] (Bmim – 1-butyl-3-methylimidazolium hexafluorophosphate) was employed as the templating materials due to the poor dispersibility of TiO_2 in water. They found that the cationic part of the ionic liquid use played important role in deciding the pore volume and pore size of the material [60].

2.3 Template-assisted synthesis of mesoporous materials

This synthesis route helps in obtaining predefined morphology, controlled pore dimension in the mesoporous materials. The first step here is the synthesis of template followed by the fabrication of materials for which methods like elimination, addition, isomerization, or substitution can be used. In the next step, the used template is removed by calcination or dissolution. There are two types of templating, one is soft templating and the other is hard templating.

2.3.1 Soft-templating method

Soft-templating technique can produce versatile architectures with controlled surface morphology and the templates after the process can be removal easily. Templates used here such as soft matters like surfactants, block copolymers, and flexible organic molecules generally experience intermolecular interactions. Whereas the interaction between the precursor and the template will usually be electrostatic, Van der Waals, or hydrogen bonding. The method can widely be used for synthesizing inorganic metal-oxides, silica, zeolite, and mesoporous carbons. For synthesizing mesoporous carbons, the fundamental requirements are (i) presence of pore former and carbon yielding species. (ii) self-assembling capacity of precursor molecules (iii) temperature withstanding capacity of the pore former (iv) the carbon precursor used must have the strong cross-linking capacity so that the mesoporous framework stays intact even after the removal of the template [61]. Finally, after the reaction, the soft template can be removed by chemical treatment at low temperature or even pyrolysis. Generally, the soft templating involves the following steps (i) cooperative assembly – This is the initial step wherein the soft matter and the small precursor

Materials Research Forum LLC
https://doi.org/10.21741/9781644903452-9

groups come together by the interaction mentioned previously. Liquid crystals engulfed by the precursor will precipitate out into the solution here. (ii) Liquid-crystal templating – This step demands certain parameters like high surfactant concentration and high solution viscosity. (iii) Evaporation-induced self-assembly (EISA) – This is mainly used for metal oxide synthesis starting with a dilute solution containing precursors, volatile solvent and the soft matter. Further, micelle is formed upon the evaporation of the solvent.

***Table 1:** Reported soft-templated mesoporous carbons*

Precursor	Template	Surface area (m^2g^{-1})	Reference
Dopamine	PS-B-PEO	229 - 450	[62]
RF	Pluronic F127	494 - 563	[63]
Dopamine	Pluronic F127	414 - 687	[64]
PDA	PS-b-PEO	343 - 363	[61]
TEOS	CTAC	598-681	[65]
TEOS	CTAB	620-888	[66]
DAP	Pluronic F127	1797	[67]
Resol	Pluronic F127	413-607	[68]

2.3.2 Hard-templating method

Hard-templating is also called nanocasting method where the pores or the surfaces of the template will be covered with the precursor molecules. After the completion of the reaction the template is removed. There are different types of templates used, such as macroscopic structures, for example, powders, films, or fibers, biological organic macromolecules, tissues, anodized aluminum oxide (AAO). Sometimes, the precursor molecules may undergo chemical or physical conversion in situ yielding templates. (i) Silica based templates: During the synthesis of mesoporous carbons, generally the carbon precursor molecules get infiltrated in the silica pores followed by carbonization at temperatures >800 °C forming a porous framework. Further, the after the removal of the template, the silica occupied space becomes the pores.MCM-48, HMS, SBA-15, and MSU-H are a few silica templates that have been used in the past research. Mokaya et al. used SBA-15 template to synthesize a mesoporous carbon with temperature-controlled pore size in the range 2 to 4.3 nm [69]. (ii) Polymer templates: These are more popular among the hard templates due to their capacity to guide controlled synthesis of porous materials and are also easily removable by simple techniques like etching or calcination. Polystyrene (PS) and poly (methyl methacrylate) (PMMA) are commonly used in the literature. Further, if PS is used, it can easily be removed by decomposing over 400 °C or etched using N, N-dimethylformamide. Further, metal-organic framework (MOFs) are also the significant class of materials which can be used as hard template due to their attractive architecture and microporosity. Among the various MOFs, Fe^{3+}, Cr^{3+}, Ti^{4+}, Zr^{4+} centered ones are popular for forming more stable structures with carboxylate-based ligands. Also, MOFs with triazolate, imidazoles, and pyrozolates [70]. (iii) Metal-based hard templates:

Materials Research Forum LLC
https://doi.org/10.21741/9781644903452-9

MgO, Fe_2O_3, MnO_2 are used in a few literatures. ZnO is low cost, commercially feasible metal template because of its easy recovery. ZnO templates can be obtained from pyrolyzing zinc containing organic or inorganic compounds [71].

Table 2: *Reported Hard-templated mesoporous carbons*

Precursor	Template	Surface area (m^2g^{-1})	Reference
Furfuryl alcohol	MCM-48	1278 - 2034	[72]
RF	Pluronic F127	1822 - 1928	[62]
Isobutene	Silica (SBA-15)	173.6 - 971	[73]
RF	Silica spheres	1667 - 2170	[74]
Ferrocene	MSS	1708 - 2396	[75]
Aniline	SiO_2 spheres	53 - 2239	[76]
Phenolic resin	SiO_2 spheres	22.7 - 1520	[77]

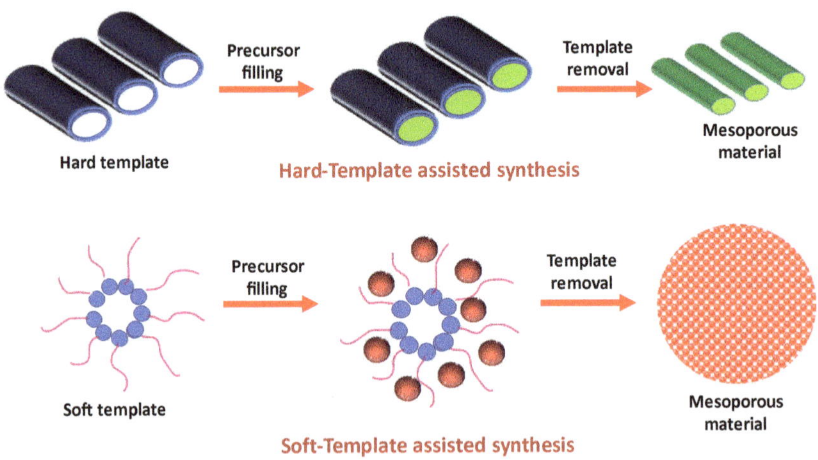

Fig 1*: Template assisted mesoporous material synthesis*

2.3.3 Template removal techniques

Broad categorization of template removal consists of physical and chemical methods. Supercritical fluid, calcination, microwave, ozone treatment, ultrasonic and plasma techniques are a few

Materials Research Foundations 173 (2025) 222-264 https://doi.org/10.21741/9781644903452-9

physical techniques. Solvent extraction, ionic liquid treatment, and oxidation are the chemical methods.

Calcination in the presence of catalyst is another interesting technique where the surfactant molecules will be removed by oxidative elimination. The temperature requirement here will be expectedly less due to the presence of catalyst. Goworek et al. evaluated the removal of surfactant from MCM-41 by calcinating upto 250 °C. The biproducts of calcination were hexadecane and hexadecane-N,N-dimethylamine [78]. The advantages of supercritical extraction are that we can control the viscosity, diffusivity, density and mass transfer in a better way. CO_2 attracted more attention in past research due to its low toxicity, non-flammability, and cost effectiveness. Huang et al. tried extraction of surfactant templates from MCM-48, MCM-41, SBA-3, and SBA-1 using methanol modified supercritical carbon dioxide [79]. Lai et al. studied the removal of surfactant templates from SBA-15 via microwave assisted extraction (MAE) and were able to achieve the removal within 6 minutes [80]. Under ultrasonic treatment a powerful irradiation with 20 kHz to 10MHz frequencies will be done to remove the templates. Energy released after the acoustic cavitation will be a high local-temperatures of 5000 K with a pressure of 1000 bar [81]. Under ozone treatment, sourcing mercury lamps of low-pressure with two peaks, one at 254 nm of energy 472 kJ/mol and another at 185 nm with energy of 647 kJ/mol are used. Since most of the organic bond energies like, C-H (413 kJ/mol), C-O (352 kJ/mol), C-C (348 kJ/mol) lie below the mercury wavelength energy, the treatment can easily remove the template breaking the organic bonds[82]. Dielectric barrier discharge and glow discharge (GD) are the two popular types of nonthermal or cold plasma technique which can easily remove the organic templates like zeolite and other mesoporous materials like silica [83].

Chemical methods of template removal consist of solvent extraction using solvents like water, dichloromethane, acetonitrile, acetone, ethanol, and methanol where water usually has low extraction capacity. Zhuang et al. used sulfuric acid to remove PEO-PPO-PEO templates from a co-assembly of TEOS, phenolic resins and F127 (triblock copolymer) [84]. Chemical oxidation is often done with H_2O_2 treatment, which can decompose the surfactants by in situ generated $OH^{\cdot}$ Radical. Biomass delignification was the inspiration for removal of templates using ILs. The advantages of this method are less framework shrinkage and requires low temperatures (120 °C) compared to others. However, IL removal approach is quite costly.

2.4 Comparison of synthetic techniques

Sol-gel method

- This technique helps synthesizing high purity materials.
- The technique is cost effective.
- A few silicates synthesized via sol-gel method are soluble in body fluids hence probably hinder the targeted biotechnological applications.

Template-assisted synthesis

- This is a greener approach compared the sol-gel method. The current hotspot in here is exploring nontoxic, inexpensive templates with good stability.
- Hard template-assisted technique

- Previously prepared templates such AAO, mesoporous carbon, templates, or polymer microspheres.
- In the next step, the nanomaterials get deposited or chemically grows into the nanopores of the templates which finally involves template removal by techniques mentioned earlier.
- Specifications and the dimensions of the product can be precisely controlled.
- Templates will have high stability and reproducibility
- Template removal stage has to be carried out carefully, as it may cause damage to the mesoporous materials synthesized.

Soft template-assisted technique

- Here, an organic-inorganic phase is formed between the high polymers, biopolymers, surfactants, and the target production.
- Soft templates are in situ prepared during mesoporous material synthesis.
- Mesoporous materials with specific structures can be produced.
- Soft templates can be removed easily compared to hard templates.

2.5 Recent advances in the synthetic methods of mesoporous materials

Although different methods are available for the synthesis of mesoporous materials, optimization of the specific techniques in order to achieve desirable morphology and dimension is very essential. CTAB can produce highly ordered structure under basic condition whereas, under acidic conditions it yields cubic shaped mesopores. Several efforts have been made recently optimizing the type of the surfactant used, and silica precursor, counter ions, solvent, temperature, pH of the system. Wang and coworkers explored intermetallic mesoporous compound synthesis with controlled structure and compositions [85]. Karafiludis et al. presented engineering of the mesoporosity of transition metal phosphates without using any templates. The phosphate precursor used was M-struvite i.e., $NH_4MPO_4.6H_2O$, where M may be Ni^{2+}, Co^{2+}, $N_xCo_{1-x}^{2+}$. The pore volumes and surface areas were 0.32 $cm^{-3}g^{-1}$ 240 m^2g^{-1} for Mg and 0.13 $cm^{-3}g^{-1}$ and 90 m^2g^{-1} for Ni respectively [86]. Cocuzza and his coworkers synthesized a spongy textured silica. They studied the influence of three different types of mesoporous silica compounds as support for lactate dehydrogenase enzyme. Using the traditional TEOS and SBA they synthesized the silica which was post synthetically modified with glyoxyl and amine groups [87]. Dendritic mesoporous silica is an emerging class of porous materials which is explored for enzyme immobilization. There are more advanced techniques like evaporation induced self-assembly (EISA), micelle-based synthesis, 3D printing, biomimetic synthesis.

Garcia et al. designed mesostructured cobalt-silicon based catalyst which was used for H_2S production. The synthesis method involved simple mixing of precursor molecules and drying at room temperature forming a thin membrane of the material followed by calcination [88]. Witecka et al. studied the assembly of a polylon via electrostatic complexation of block polymers and a cationic surfactant cetylpyridium chloride. The resulting micelles were used as structure directing agents yielding mesoporous silica by sol-gel strategy [89]. Biomimetic synthesis of hierarchical mesostructures follow dynamic assemblies where in situ growth occurs on the template. Mesostructures resembling lotus leaves, shells, or cupped buds were grown on mesh templates

[90]. Recent light-based 3D-printing techniques such two-photon polymerization are able to yield high resolution complex mesoporous structures. However, despite of the advancements in this area, mesoporous materials with precise pore diameter in the range 2 to 50 nm cannot be printed. Hybrid precursors like tetramethyl orthosilicate (TMOS), tetraethyl orthosilicate (TEOS), trimethoxymethylsilane (MTMS), etc. are used to control on the formation of desired shapes. Hybrid printing has the advantages of both fast radical polymerization and slow hydrolysis-condensation sol-gel chemistry [91]. Zhao et al. worked on a simple and fast way of synthesizing mesoporous silica combining the digital light processing (DLP) and sol-gel chemistry [92].

3. Surface modification of mesoporous materials

Surface modifications onto a mesoporous material can enhance its properties extending its applications in adsorption, catalysis, sensing and biomedical application. These modifications only vary the chemical properties keeping intact the mesoporous nature. Attaching surface functional groups like -NH_2, -COOH, -SH etc., grafting larger molecules like polymers, impregnation which is the deposition of nanoparticles or metal ions on the mesoporous material surface, silylation, calcination, plasma treatment which can release free radicals on the surface are a few explore surface modifications. Schwartz, et al. monitored the in situ changes in the living cell undecylenic acid treated porous silica [93].

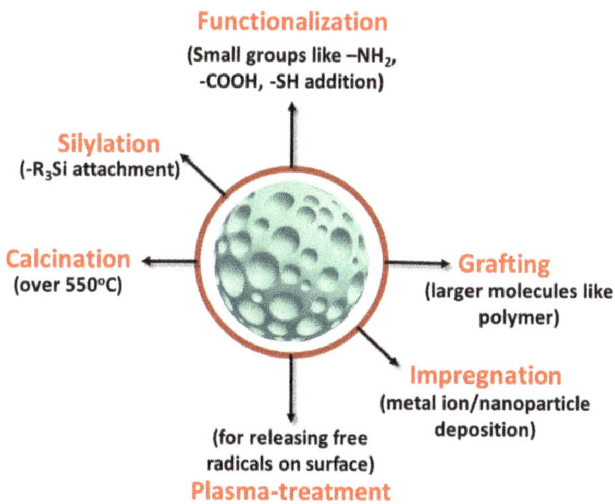

Fig 2: Surface modifications of mesoporous materials

Lakshmikumar et al. studied the low temperature stabilization of mesoporous silica by photo-assisted reaction with acetylene. HF/H_2O_2 electrolyte and a current density of 80 mA/cm^2 was used for the synthesis of porous silica which were then treated with acetylene optically illuminating it for quartz-halogen lamp [94]. In one study, the silicon surfaces were modified for the DNA attachment and hybridization. The surface was modified by amine group using free end of ACP

Materials Research Foundations 173 (2025) 222-264 https://doi.org/10.21741/9781644903452-9

(amine 1-amino-3-cyclopentene) and one more molecule with protecting group attached (B-1-BOC-amino-3-cyclopentene (BACP)) [93].

Porrang and co-workers synthesized mesoporous silica from rice husk and modified its surface by plasma discharge in modes, direct and hybrid. The dielectric barrier-discharge plasma induced uniform free radical polymerization over the surface. They tested the modified apoptosis study and found that the surface modification by direct mode induced high level apoptosis [95]. Moreno et al. synthesized MCM-41 and silica-gel and treated them with hexamethylenedisiloxane (HMDSO), oxygen, ethanol vapors and acrylonitrile under plasma of frequency 20 W for 30 minutes. The plasma modification with HMDSO and AN changed the color, shape, and size of MCM-41 and silica [96]. Xu et al. studied the activation of Ni/MCM-41 by ethanol and cold plasma coupling. The support destruction by the plasma irradiation was inhibited by the addition of ethanol [97].

Wai et al. trapped peroxophosphotungstate into mesoporous SBA-15 with the help of silylation using different agents having C_3, C_5, and C_8 arms out of which C_8 acted as a promising material for the epoxide synthesis [98]. Li et al. functionalized mesoporous silica synthesized by the co-condensation of vinyltriethoxysilane (VTES) and tetraethoxysilane. The results showed high surface area for the silylated silica, improved humidity resistance, and high absorbance for p-xylene [99].

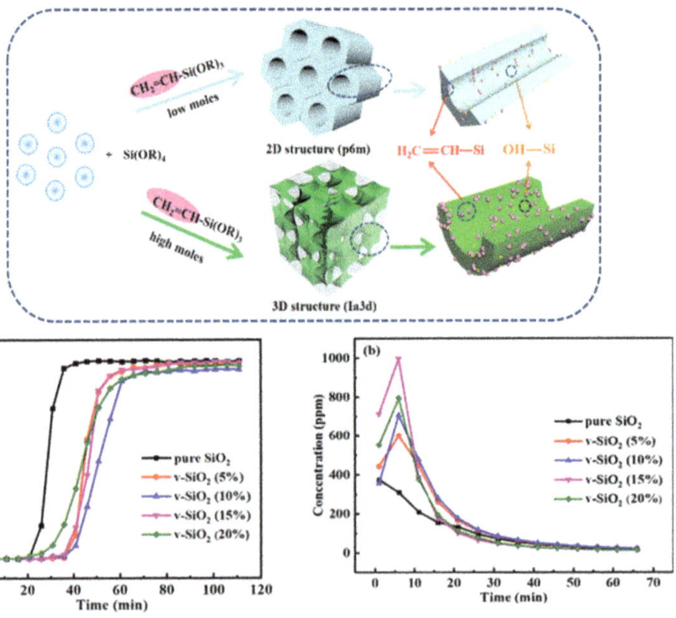

Fig 3: *Silyl functionalized mesoporous silica used for p-xylene [99]. Reproduced from journal of physical chemical C, ACS publications, 2020*

Materials Research Forum LLC
https://doi.org/10.21741/9781644903452-9

4. Characterization techniques

Once the mesoporous material has been synthesized, then comes the crucial stage of characterization of the synthesized materials. The pore morphology, structure, particle size, and surface information are a few important parameters which need to analyzed. Techniques like powder X-ray diffraction (PXRD), scanning electron microscopy (SEM), transmission electron microscopy (TEM), energy dispersive X-ray (EDX), nitrogen adsorption-desorption, Fourier transform infrared spectroscopy, nuclear magnetic resonance (NMR), thermogravimetric analysis (TGA), differential scanning calorimetry (DSC), and X-Ray photoelectron spectroscopy (XPS).

Table 3: *Mesoporous materials with their reported pore sizes and structures [100].*

Material	Structure	Pore size reported
CMK-1	3-D regular mesopores, disordered carbon framework	3.01
DMS	Disordered, narrow and bimodal	2.56
FSM	Highly ordered, hexagonal, wall made of double SiO_4 layer	2.2
HMS	Hexagonal, uniform pore size, disordered, cross-linked framework	2.8
KIT	3D wormlike channels, fully disordered, uniform pore size.	3.52
MAMCM	highly ordered, hybrid material	3.17
MCM-48	3D cubic, highly ordered	3.49
Meso-fly ash	hexagonal	2.74
MSF	2D, highly ordered, uniform pore size	1.6
MTS	Mesopores rea formed by the space between the monodispersed particles	3.42

SEM and TEM images are the direct approaches to obtain the mesoporous nature of the materials synthesis. The structural order of these materials can be determined by diffraction techniques and TEM. Morphology and particle size can be revealed by SEM. The surface area and porosity and can be measured by nitrogen adsorption studies. EDX gives qualitative and quantitative information about the chemical composition on the surface.

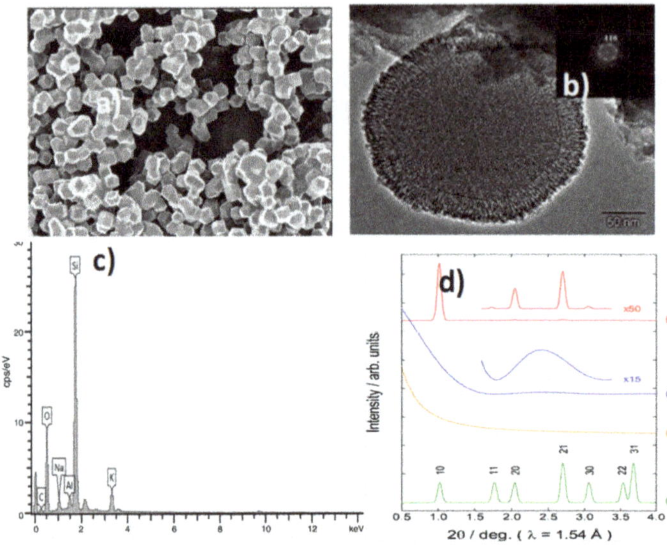

Fig 4: *a) SEM image of cubic mesoporous silica b) High resolution TEM image of mesoporous AMS-1 c) EDX of silica obtained from Palm kernel shell d) XRD pattern of mesoporous silica [101], [102], [103], [104].*

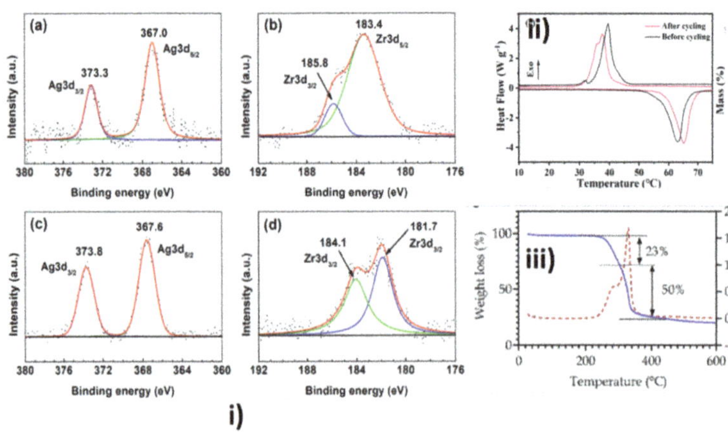

Fig 5: *i) XPS spectra of a) Ag@SBA-15 b) Zr@SBA-15 c) Ag-3d of AgZr@SBA-15 d) Zr-3d of AgZr@SBA-15 ii) DSC of mesoporous silica iii) TGA of mesoporous silica [102], [105], [106].*

Materials Research Forum LLC
https://doi.org/10.21741/9781644903452-9

5. Applications in Separation Science

MM are perfect carriers for analyte molecules, therefore widely explored in analytical science. The exceptional mesoporous structure offers superior properties including larger surface area, outstanding pore connectivity and tunable pore size. The structural and functional properties of these materials have been used to improvise existing analytical methods. For example, the enormous specific surface area of MM can provide copious active sites on MM and increase the possibility of analyte contact to produce sturdier signals, thus leading to enhancement in detection sensitivity.

Likewise, materials of mesoporous nature are employed in membrane technology for wide range of biotechnological applications. In addition to their ability to functionalize with specific molecules for tailored interactions, they enhance the separation efficiency in various biotechnological processes. An overview of various mesoporous materials and its variants used in protein separation/ purification or enzyme immobilization is shown in table 3

Table 4: An overview of various mesoporous materials and its variants used in separation science

Type of Mesoporous Material	Mechanism	Application	Reference
FDU-12 membrane	Adsorption	Dairy Proteins separation : α-Lactalbumin and β-lactoglobulin	[101]
Magnetic mesoporous core/shell heteronanostructure (Fe_3O_4@$NiSiO_3$)	Magnetic separation	His-tagged proteins separation from a cell lysate of E. *coli*	[102]
Hexagonal mesoporous silica (Met-SBA-15).	Extraction	Hydrophobic membrane protein separation	[103]
Polymeric ionic liquid	Adsorption	Ovalbumin separation	[104]
SBA-15	Covalent bonding	Enzyme immobilization	[105]
Functionalized SBA 15	Adsorption	Lipase immobilization	[106]
Amine and aldehyde functionalized mesoporous silica on magnetic nanoparticles	Physical adsorption and covalent attachment	Enhanced lipase immobilization for biodiesel production	[107]
Mesoporous materials of silica MCM-41 and SBA-15 and carbon CMK-3	Adsorption	Enrichment of marine biofunctional peptides	[108]
Functionalized SBA-15	Adsorption	Selective separation of proteins	[109]

An highlight on various areas of biotechnology where MM have been explored is highlighted below

Materials Research Foundations 173 (2025) 222-264 https://doi.org/10.21741/9781644903452-9

5.1 Protein separation

Over the years separation of proteins with similar charges and sizes has been a major challenge. For this reason, development of sustainable strategies to overcome these challenges has been a major focus. For this purpose, novel cholinium-based amino acid [Cho][Pro] ionic liquids have been used as trace additives which can be physically loaded on a mesoporous TiO_2 for effective separation of lysozyme and cytochrome c (similar proteins). The mechanism of selective adsorption can be explained by the partial dissociation of ions under different pH conditions and modifications in hydration properties. As the pH is incrementally increased from 5.0 to 9.8, the degree of hydration of ionic liquids also increases, thus weakening the interactive strength of the lysozymes with the mesoporous TiO_2 substrate, thus leading to their effective recovery. These findings were effectively utilized for separation of protein mixture in a high-performance liquid chromatography set up [110]. Ovalbumin (glycoprotein) from eggs shell have been successfully isolated using novel mesoporous hydrophilic polymeric ionic liquids. Adsorption followed langmuir isotherm with excellent capacity of 2468.0 mg/g. Circular dichroism spectral studies also revealed that the adsorption followed by elution process did not cause any conformational variations in the adsorbed Ovalbumin, indicating the biocompatible nature of mesoporous polymeric ionic liquids [104]. Panchami et al. Synthesized P(MBAm-co-DMAPDS-co-VA) zwitterionic nanoparticles via distillation-precipitation-polymerization (DPP) technique and incorporated them into ultrafilteration polyphenylsulfone (PPSU) matrix. These membranes exhibited 92.1%, 60.4%, 80.3% rejection for bovine serum albumin (BSA), proteins pepsin and egg almumin respectively [111]. In another work the same group synthesized P(MBAm-co-MAA) by DPP technique to embed in the PPSU matrix that gave rejections of 86.9% for egg albumin, 68.4% for pepsin, and 94.8% for BSA [112].

Chondroitin sulfate (CS) covalently conjugated to dendritic mesoporous silica nanoparticles with amino terminal were synthesized for selective adsorption of low density lipoprotein (LDL) which was further tested in LDL simulated pateint serum samples suffering from hypercholesterolemia. Selective adsorption was observed in simulated samples indicating the potential of CS conjugated dendritic mesoporous silica nanoparticles for real time applications in healthcare [113]. Protein glycosylation of glycopeptide, an essential step in biomarkers discovery requires a lot of enrichment steps before its use by high end anaytical instruments (MS detection). This could be overcome by the use of MM with higher specific surface areas, abundant pores, for their use as innovative enrichment and separation approach [114]. The general mechanism of selective separation through adsoption by smart mesoporous materials is depicted in fig 6.

Materials Research Foundations 173 (2025) 222-264 https://doi.org/10.21741/9781644903452-9

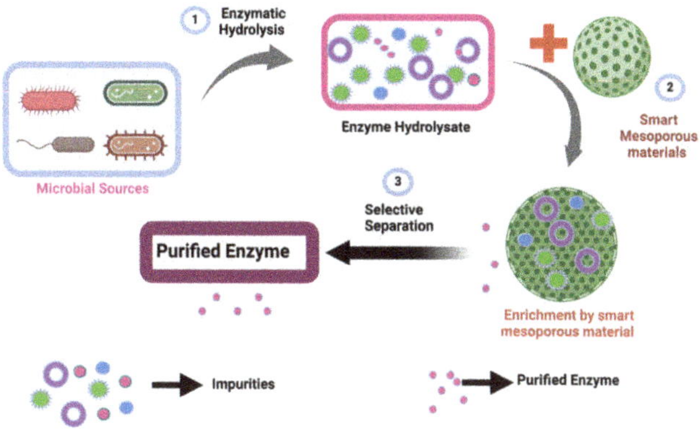

Fig 6: *The general mechanism of selective protein separation by smart mesoporous materials.*

5.2 Applications in protein purification

Use of MM for protein purification applications has been explored by researchers in the recent past. For instance, isolation of marine biofunctional peptides from marine worm *Urechis unicinctus* was tried with enzyme hydrolysis followed by purification using three different mesoporous materials MCM-41, SBA-15 and CMK-3. Variation in elemental composition and pore size lead to changes in surface area and hydrophobicity. Therefore, maximum adsorption capacities was seen for CMK-3 (626.3 mg/g), followed by MCM-41 (459.5 mg/g) and SBA-15 (431.3 mg/g). This study thus demonstrated the skyrocketing potential of mesoporous materials for the enrichment of marine biofunctional low molecular weight peptides [108]. Proteins from blood (a complex biological fluid) have been purified using innovative oderely porous materials (OPMs). These materials possess properties such as (exceptional surface area, good adsorption capacity and superior heamocompatibility) for blood purification.

Likewise, hemoperfusion, a physical separation strategy for impurities removal from blood using adsorbing agents, offers an efficient method for blood purification. OPMs have shown potential to replace the traditional hemoperfusion agents which are having poor adsorption capabilities and not hemocompatibility. No doubt, OPMs show groundbreaking potential for further research in the field of biomedical engineering [115].

5.3 Enhancements in yield and purity.

Another way of enhancing the catalytic capability of enzymes could be its effective immobilization in suitable MMs and its subsequent use in production of products of industrial importance. For instance, L-ribose isomerase has been successfully immobilized in MMs (MCM41 and SBA15) for significantly improving its biocatalytic efficiency by multiple folds (1.2 fold). Increased catalytic efficiency after suitable immobilization strategy could be accounted to the suitable enzyme orientation exposing the active sites as well as the improved substrate binding accessibility. The immobilization of L-ribose isomerase on mesoporous silica support could thus positively improve the biocatalytic potential, immobilized enzyme storage stability and its subsequent reusability. This approach achieved over 4 reuse cycles, retaining > 50% of catalytic activity, and efficiently converting L-ribulose to L-ribose (a rare sugar) within 2 hrs with a yield of 22%. In particular, the immobilization yield of L-ribose isomerase on MCM41 was 60 % and SBA15 was 45 % under the optimal conditions. The immobilized biocatalyst was further characterized by various sophisticated analytical techniques (HR-TEM, EDS, FTIR, TGA and BET), thus signifying the importance of characterization for analyzing the biocatalyst surface in effective product formation applications [116].

As an example, solvothermal synthesis of nickel ferrite magnetic nanoparticles ($NiFe_2O_4$ MNPs) have offered unique advantages in both immobilization and purification of enzymes. The binding sites on the surface of the magnetic nanoparticles (MNPs) which have nickels enable the direct rapid attachment of three His-tagged enzymes. Moreover, the enzymatic activities of all immobilized samples, which were directly prepared from cell lysates, were found to be comparable to the activity of immobilized purified enzyme using the conventional method. When comparing both the conventional and current method using MNPs, utilizing $NiFe_2O_4$ MNPs for direct enzyme immobilization from cell lysates significantly reduces the number of processing steps, time, and unnecessary chemical usage [117]. Magnetic nanoparticles (as core) capped with, and mesoporous silica (shell) functionalized with amines have been explored for magnetic separation of recombinant human growth hormone produced by *Pichia pastoris* with exceptional adsorption capacity of 335.21 μg/mg.

The pictorial representation of magnetic separation of proteins by such magnetic nanoparticles capped with amine functionalized silica mesoporous materials is shown in fig 7.

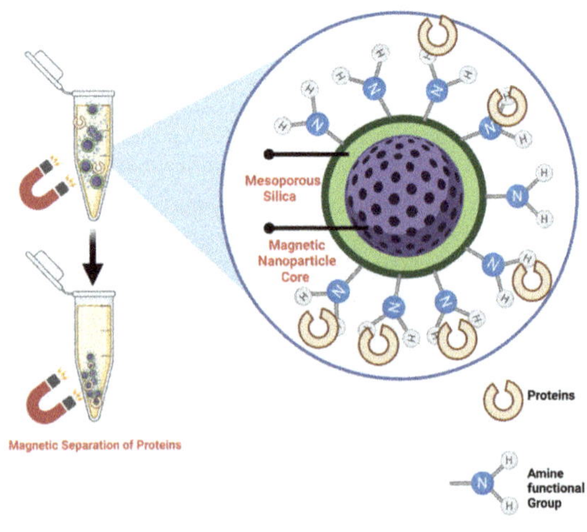

Fig 7: *Illustration of Protein Magnetic Separation Using Magnetic Nanoparticles Coated with Amine-Functionalized Silica Mesoporous Materials*

Materials Research Foundations 173 (2025) 222-264 https://doi.org/10.21741/9781644903452-9

Recently, renewable silica sources (rice husk) have been utilized for mesoporous silica nanoparticle synthesis using the soft templating methods. The synthesized mesoporous nanocarriers were further characterized with sophisticated techniques like adsorption –desorption isotherms, X-ray diffraction (XRD), atomic force microscopy (AFM), molecular docking and thermos gravimetric analysis (TGA) to understand the surface properties of the synthesized nanobiocatlyst for degradation of lignocellulosic agricultural wastes (rice husk). Xylanase immobilization on mesoporous silica nanoparticle improved the stability (better enzyme performance than free enzyme) and recyclability (9 cycles) confirming the potential of the sustainable nanobiocatalyst for lignocellulosic biomasses saccharification [118] .

5.4 Enzyme immobilization and separation

Enzyme immobilization on mesoporous materials and its subsequent use in various biotechnological applications have enhanced the performance as effective biocatalysts. For instance, lipase B from *Candida antarctica* has been successfully immobilized on surface of mesoporous SBA-15 and subsequently used for esterification of levulinic acid which has application in fields of flavor and fragrance. Highest conversion efficiency of 94% was achieved within 2 h at temperature of 50 °C indicating the synthesized biocatalyst had exceptional operational stability and reusability up to 4 cycles [106]. Similarly, mesoporous material coated on magnetic nanoparticle could enhance the enzyme loading efficiency from 67.8 to 82.4% [107]. Among the enzymes, lipases have been effectively explored for its immobilization potential. Overview of advantages of enzyme immobilization on mesoporous materials for effective product formation through enhanced properties of enzymes is shown in fig 8. Likewise, an summary of microbial lipase sources and their immobilization on mesoporous materials for specific applications have been presented in table 4 4.

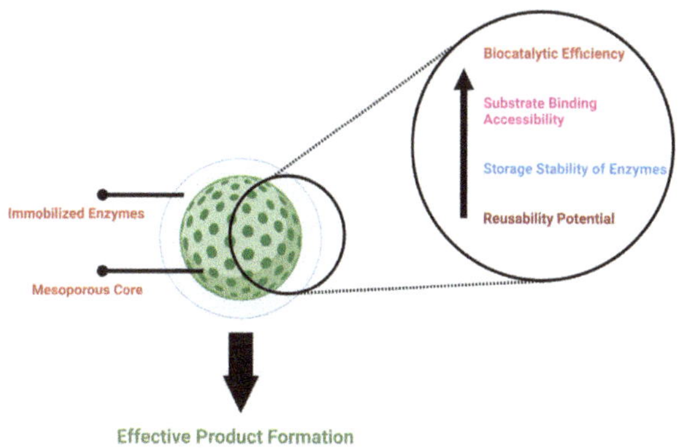

Fig 8 : *Overview of enzyme immobilization for effective product formation*

Materials Research Foundations 173 (2025) 222-264 https://doi.org/10.21741/9781644903452-9

***Table 5:** Summary of microbial lipase sources and their immobilization on mesoporous materials for specific applicatins*

5.5 Benefits in enzymatic reactions and separation processes

Enzyme Source	Type of Mesoporous Material	Improved Property	Applications	Reference
Candida rugosa	Hydrophobic hierarchical porous hollow silica microsphere (HPHSM-C3)	Higher activity, thermal and pH stability	Enzymatic enrichment of ω-3 PUFAs in food industries	[119]
Candida antarctica	Meso-MIL-88A	Thermal and pH stability and better dynamic performance	Synthesis of phosphatidyl-DHA required for brain growth	[120]
Thermomyces lanuginosus	Uncalcined SBA-15 (SBAUC) and calcined SBA-15 (SBAC)	Thermally and operationally stable	Immobilized enzyme for food, chemical, and pharmaceutical sectors	[121]
Candida rugosa	SBA-15	Storage stability and reusability	Synthesis of food flavour ester isoamyl acetate	[122]
Rhizomucor miehei	Silica SBA-15	Thermostable, Improved enzymatic activity and higher conversion rates	Diacylglycerols production	[123]
Pichia pastoris	Pore-expanded SBA-15	Thermostable, High conversions and Storage stability	Secondary alcohol production	[124]

In addition to the use of innovative mesoporous materials for applications related to nucleic acid and protein separation and purification, They have been extensively explored as scaffolds for cell separation, encapsulating material for effective immobilization of enzymes and sorting of cells based on size / charge by pore size modifications. These materials are not only used for above applications they can also be used as medium where enzymatic reactions could be carried out thereby assisting separation processes. For instance, methylated well-ordered hexagonal mesoporous silica (Met-SBA-15) have been used as nanoreactors for selective adsorption of hydrophobic membrane proteins from bacteria (*E.coli*) which are less in concentration. These nanoreactors not only adsorb the membrane proteins but help in providing a scaffold for selective in situ digestion of these proteins by trypsin, for it to be effectively analyzed using mass spectroscopic techniques[103]. Additionally, mesoporous silica nanoparticles (< 200nm) have been utilized as potential scaffolds for proten encapsulation and delivery. Dimensionional analysis of these particles, carried out using the nitrogen adsorption–desorption isotherms revealed exceptional encapsulation efficiency and controllable release profiles for a variety of proteins, enabling potential applications in fields of drug delivery and protein therapy. It was seen that the variations in the surface chemistry of the MSNs, coupled with other properties (isoelectric point, molecular weight, ionic strength of the buffer) impacted the rate and amount of protein uptake and release [125]. Silica and titania nanoparticles deposited on mesoporous thin films have been effectively use as scaffolds for cell growth [19].

Additionally, use of bioderived materials *(Acacia mangium* tannin extract) combined with triblock copolymer Pluronic have been utilized for hierarchical porous TiO_2 Synthesis. It was observed that tannic acid sulfonic groups and phenolic hydroxyl groups were involved in controlling the hydrolysis rate of titanium sulfate. By adjusting the tannins and copymer ratios, the pore structure of hierarchical porous TiO_2 could be adjusted. The synthesized hierarchical porous TiO_2 were utilized for selective adsorption of lyzozyme (adsorption capacity of 42 mg/g). Based on the potential of these materials, they have been further utilized for adsorption of biomacromolecues from leather industry wastewater [126].

5.6 Benefits as effective drug carrier vehicles

MMs have shown promising prospects as efficient drug carrier molecules for drug delivery applications [127]. The advantages of using MMs include increased drug dissolution rates, increased bioavailability, improved stability, sustained and controlled release, minimal side effects and biocompatible nature [128]. No doubt, the performance of MMs can be further improved by material modifications and optimizing the synthesis/ operating conditions Among the MMs, mesoporous silica nanoparticles (MSNs), mesoporous carbon nanoparticles (MCNs), metal organic frameworks (MOFs) and mesoporous hydroxyapatite (MHAPs) have been extensively used for drug delivery applications. Functionalization of MMs with amine and phenyl groups has shown to enhance the drug loading efficiency due to the formation of hydrogen and Π- Π-interactions between drug and MMs[127]. Moreover reduction in surface area of MMs as evident from BET results after drug encapsulation is due to the filling of pores by drug molecules [129]. An overview of MMs used for drug delivery applications is described in table 1.

***Table 6:** An overview of MMs used for drug delivery applications*

Mesoporous Material	Functionalizing agent	Characterization	Drug	Mechanism of drug adsorption /drug delivery	Reference
SBA-15	Phenyl and amine	TEM , IR, Nitrogen adsorption desorption, TGA, NMR, BET	Naproxen sodium salt	Hydrogen bonds and π-π interactions	[127]
SBA-15	Amine	XRD, BET, SEM, EDX, TEM, FTIR, and TGA	Curcumin	Electrostatic interactions	[129]
SBA-15	Amidoxime		Dox anticancer drug	pH-responsive drug delivery	[130]
PES membrane modified by SBA-15	Glutamine	SEM,water contact angle and Tensile strength	Azithromycin	Membrane top layer: Sustained release MMs porous sublayer : Drug reservoir	[131]
Magnetic SBA-15 nanocomposites	Superparamagnetic iron oxide nanoparticles	XRD, TGA	Doxorubicin	Targeted drug delivery	[132]
SBA-15	12-tungstophosphoric acid (TPA)	TGA, FT-IR spectroscopy, XRD, N_2 adsorption measurements, HRTEM, and SEM	Alendronate sodium	Controlled release	[133]
Polyimide-COF/amino-functionalized MCM-41 nanohybrid	Polyimide-COF and amine	XRD, FTIR, and TGA	Erythromycin	Prolonged and sustained drug release	[134]
MCM-48 nanoparticles	Polyoxometalates	Elemental analysis, FT-IR, TGA, HRTEM and N_2 adsorption measurements	Doxorubicin Hydrochloride	Controlled release	[128]
MCM-41	cobalt ferrites-oleic acid	SEM and HR-TEM	Doxorubicin	pH sensitivity and controlled drug release	[135]
HAP Incorporated with Fe_3O_4@MCM-41 Core–Shell	Hydroxyapatite Incorporated with Fe_3O_4	XRD, FTIR, zeta potential, DLS, VSM, SEM, TEM	Teriparatide	pH-responsive release	[136]

(SBA-15: Santa Barbara Amorphous No. 15, PES: Asymmetric polyethersulfone, MCM-48: Mobil Composition of Matter No. 48, MCM-41: Mobil Composition of Matter No. 41, Polyimde COF: Polyimide-Covalent Organic Framework, HAP: Hydroxyapetite, TEM: Transmission Electron Microscopy, IR: Infrared Spectroscopy, TGA: Thermogravimetric Analysis, NMR: Nuclear Magnetic Resonance, BET: Brunauer–Emmett–Teller, XRD: X-ray Diffraction, SEM: Scanning Electron Microscope, EDX: Energy Dispersive X-ray Spectroscopy, FTIR: Fourier

Materials Research Foundations 173 (2025) 222-264
https://doi.org/10.21741/9781644903452-9

Transform Infrared Spectroscopy, VSM: Vibrating Sample Magnetometer, HRTEM: High-Resolution Transmission Electron Microscopy, DLS: **Dynamic Light Scattering**)

5.7 Benefits in heavy metal adsorption

Functionalized mesoporous silica materials modified using chemical modification techniques have been explored for heavy metal adsorption applications due to their promising metal binding capabilities [137]. For instance, functionalized hexagonal mesoporous materials like MCM-48 [138], HMS [139], SBA-15 [140] and MCM-41[141] have been explored for its heavy metal adsorption capabilities. MCM-41 prepared from agricultural residue (rice husk) have been utilized for adsorption of Co(II) and Cu(II) ions in aqueous solutions due to its exceptional specific surface area of 548.92 m^2/g and total pore volume of 126.12 cm^3/g respectively [141]. Identically, SDS functionalized MCM-41 have been used for selective adsorption of Cu^{2+}, Cd^{2+}, and Zn^{2+} from synthetic aqueous solutions. It was evident that the adsorption capability of MCM-41 improved at neutral pH than at acidic pH conditions and the material could be regenerated (up to 6 cycles) with favourable adsorption capacity of 70% . This confirms that the solution pH has significant contribution on the adsorption capability of mesoporous materials [142]. Amine functionalized mesoporous silica (MCM-48) have shown excellent adsorption capability of 90% for Cu(II), 98% for Hg(II) and 97% for Ce(III) respectively after 180 min. It was evident from characterization results (SEM, XRD, FTIR, NMR) that amine functionalization did not modify the properties of the mesoporous material [143]. Likewise, negatively charged mesoporous hydroxyapatite synthesized using microwave assisted method showed excellent adsorption capabilities (1438.85 mg/g) for Pb^{+2} in aqueous systems [144]. Similarly, Guanidine functionalized mesoporous SBA-15-Gd have been coupled with Fe_2O_3 nanoabsorbent to minimize the agglomeration effect in Fe_2O_3 at nanoscale. This approach of combining mesoporous materials with nano-adsorbents significantly improved the available surface area for interaction and subsequent removal of heavy metals ((Pb (II) and As (V)). The adsorption process was explained with Freundlich model, and the kinetics of reaction was explained by pseudo-second-order model. Moreover, the magnetic adsorbents (Fe_3O_4@SBA-15-Gd) could be easily recovered using a magnet and subsequently used for 3 subsequent cycles without a momentous decline in adsorption efficiency which proves its regeneration capabilities [145]. A brief overview of diverse mesoporous materials used for adsorption of heavy metals from aqueous solution is presented in Table 2.

Table 7: *A brief overview of diverse mesoporous materials used for adsorption of heavy metals from aqueous solution*

Mesoporous Material	Targeted Metal	Adsorption Capacity (mg/g)	Reference
HAP	Pb(II)	1438.85	[144]
MCM-48	Cd(II)	82.7	[146]
	Pb(II)	119.24	
MCM-48	Cr(VI)	17.9	[147]
MCM-48-NH_2	Pb (II)	75.2	[148]
MCM-41 silica	Cd(II)	21.1	
	Cu(II)	39.3	
	Ni(II)	33.67	[149]
Fe/Mg-MCM-41	As(V)	71.53	
Melamine-functionalized silica-SBA-15	Cr(VI)	50	[150]
Biogenic cyclam-functionalized homologous SBA-15	Ni(II)	266.28	[151]
Guanidine-functionalized SBA-15/Fe_3O_4	Cu(II)	370	
	Pb(II)	333	[145]
$MnFe_2O_4$@SBA-15-$(CH_2)_3$-adenine	As(V)	21.99	
	Cd (II)	29.62	[152]
	Pb(II)	31.01	

(HMS : Hexagonal Mesoporous Silica, HAP: Hydroxyapetite, MCM-41: Mobil Composition of Matter No. 41, MCM-48: Mobil Composition of Matter No. 48, SBA-15: Santa Barbara Amorphous No. 15)

5.8 Benefits in separation of organic contaminants

Mesoporous materials have been explored as effective adsorbing agents for separation of organic moieties (dyes and antibiotics) in polluted water samples. MMs basically serve as microreactors, by increasing the available sites for reaction between the mesoporous materials and the water contaminants, thus anticipating tremendous potential in environmental remediation applications. Various MMs and functionalized MMs, which include SBA-16. MCM-41, CMK-5, OMC, mesoporous Fe_3O_4, SBA-15 NH_2 and many more have been successfully utilized by various research groups. The reason for functionalizing the MMs with active groups like amino or carboxylate is basically to increase the affinity between the contaminants and the mesoporous materials. It is a known fact that amino functionalization of MMs leads to protonation of amino groups (NH_3^+) which helps in effective electrostatic interaction with negatively charged contaminants. For instance, pentahexane functionalized SBA-3 has shown promising potential to adsorb acid dyes with adsorption capacity of 479.55 mg/g. Likewise, amino functionalized MCM-41 have shown adsorption capability of 300mg/g for direct blue dye. However, $COOH^-$ functionalization of MMs have helped in successful removal of cationic dyes under basic medium due to deprotonation of $COOH^-$ to COO^- and its electrostatic interaction with cationic dyes (Positively charged ones). Additionally, the effect of parameters which has got significant impact on the adsorption capabilities of MMs (initial concentration of the dye, pH, temperature and time) has been focused for improving the adsorption capabilities of MMs. Consequently, use of optimization techniques like response surface methodology has further helped in enhancing the parameters for effective removal of dyes. For instance, mesoporous SBA-16 was able to remove 98.6% of methyl orange with absorption capacity of 37.73mg/g under optimal conditions of pH: 4.07, Temp: 50 °C, Time: 35 minutes, initial concentration of dye: 10mg/L with 0.04g of adsorbent quantity as seen by response surface methodology approach. For this reason, mesoporous SBA-16 can be used as an ideal candidate for removing coloured pollutants from wastewater coming from textile industries [153].

Presence of organic contaminants and their derivatives in water bodies due to overuse of antibiotics, its widespread dispersion capabilities, harmful bioaccumulation potential and mineralization resistance has posed a significant threat to mankind [154]. This has forced researchers to focus on the development and use of novel materials as adsorbing agents for removal of these organics. In fact, amine functionalized mesoporous SBA-15 have shown successful removal of ampicillin antibiotic with a adsorption capacity of 333mg/g. Electrostatic interaction between the protonated amino group and negatively charged carboxylic group of ampicillin is the reason for enhanced removal capabilities [155]. Over the years, mesoporous carbon materials have outperformed for its capabilities in organics removal by adsorption phenomenon [28]. Mesoporous carbon materials are preferred due to their ideal properties like hydrophobicity, flexible functional group regulation capabilities, and magnetic nature of particles which improves the regeneration capability for successive cycle use [156]. For all these reasons, mesoporous carbon materials have been used for water purification applications. Biological/chemical activated mesoporous carbon has been used as carriers for enzyme laccase immobilization and its subsequent removal of antibiotic (oxytetracycline). It was evident that HCl treated carrier showed maximum enzyme loading and activity. This may be due to the electrostatic interaction between carrier and enzyme at acidic pH conditions. In comparison with free enzyme for oxytetracycline removal the immobilized enzyme with mesoporous carbon (modified enzyme catalyst) showed highest capacity. The efficiency of removal improved from 0.88 L/mgh (free enzyme) to 201.6 L/mgh (immobilized enzyme on mesoporous carbon) indicating enhanced removal capabilities with the help of carrier molecule [157].

Innovative reusable magnetite mesoporous carbons a green adsorbent have been used for three antibiotics removal (metronidazole, sulfamethoxazole and ciprofloxacin). The adsorption capacity for metronidazole, sulfamethoxazole and ciprofloxacin was 76 mg/g, 70 mg/g and 44 mg/g respectively. Moreover the adsorbent could be regenerated successfully for 3 consecutive cycles with 95% adsorption capabilities thus proving the reusable potential of the novel adsorbent [158]. Likewise, ordered mesoporous carbon (OMC) with narrow pore size of 4.8nm could effectively remove/exclude sulfamethoxazole antibiotic in the presence of humic acid (an organic interference molecule). Additionally, it was clear that smaller the pore size of OMC, better was the size exclusion capability which could be further enhanced by increasing the molecular weight of humic acid [159]. Following this, D-glucose based mesoporous carbon synthesized using pyrolysis @ 500°C, 3h with N_2 gas flow rate of 10°C/min was subsequently used for tetracycline removal from water. The adsorption followed pseudo second order kinetics with exceptional adsorption capacity of 286.63 mg/g at pH 5 and temperature of 30°C. However at higher pH, reduced adsorption capacity was seen. This could be ascribed to zwitter ion formation that poses significant diffusion barrier thus reducing the adsorption potential [160]. Similarly, mesoporous carbon doped with cobalt and N element was used for tetracycline removal. Various interactions between the antibiotic and the adsorbent at acidic pH which includes electrostatic attraction, π- π interactions, lewis acid base interactions and metal complexation was the reason for achieving enhanced adsorption capability of 344.83 mg/g [161].

6. Challenges & Limitations

Mesoporous materials (MMs) have gained significant attention in biotechnological separation applications due to their exceptional surface area, tunable pore architecture, crystalline structure, diverse chemical composition, and versatile potential [162]. However, despite these advantages, several challenges and limitations hinder their widespread use. The synthesis of MMs, particularly achieving precise control over size and structure, often demands sophisticated methods. This complexity, coupled with variability in performance and consistency, can impede their industrial-scale application [163]. The synthesis process typically involves multiple steps, the use of expensive templates, and extended processing times, making it costly and susceptible to sampling errors [164].

The structural stability of some MMs at high temperatures (low thermal stability) and their fragility under mechanical stress further limit their practical applications [165]. Consequently, batch-to-batch variability and high production costs pose significant challenges for scaling up these materials [166]. However, the use of waste biomass, such as agricultural residues, as raw materials for MM synthesis, could help address the issue of high production costs [126]. Environmental concerns are also significant, particularly regarding the use of toxic precursors, solvents, and surfactants in MM synthesis. These raise questions about safe disposal and the long-term environmental impact of MMs [141]. The sustainability of the raw materials and the hazardous nature of any by-products formed during synthesis further exacerbate these concerns. Additionally, the challenges of selective catalysis and functionalization of MMs often require extensive research to optimize synthesis conditions and improve characterization, including crystalline structure and pore architecture [167].

The effectiveness of MMs over time can diminish due to factors like pore blocking, contamination, and structural collapse, which reduce their lifespan [162]. If MMs are fouled by chemical or biological agents, their regeneration potential for multiple cycles may be compromised [168]. Furthermore, studies on the long-term degradation of MMs are limited, and their environmental impact over extended periods is not fully understood. For these reasons, there is a pressing need for ongoing research and innovation to overcome these limitations and optimize the performance of MMs for various applications.

7. Future Perspectives

Indeed, there has been a tremendous demand for novel separation mediums /separation methods with high selection capability, simple in operation, cost effective and works at mild operating conditions for the extraction of peptides, proteins, enzymes and antibodies of low abundance in complex samples [110]. Research with likely focus on tailoring the surface properties of MMs (pore size, surface chemistry, and functional groups) to achieve higher specificity and selectivity for targetted proteins, remains a significant area of exploration. This could further lead to one step separation, purification, and detection processes thus making the entire process error free with minimal steps. No doubt, developing MMs with enhanced biocompatibility and stability is crucial for their application in biotechnological fields. Moreover creating materials that can maintain their structural integrity and functionality over prolonged periods is crucial. Similarly, integration of mesoporous materials with other nanomaterials, such as metal-organic frameworks (MOFs/ magnetic nanoparticles or quantum dots, could further enhance their multifunctionality potential for combined capabilities, such as simultaneous protein separation and detection [169]. Nanoreactors of methylated well-ordered hexagonal mesoporous silica (Met-SBA-15) are expected to be utilized for studies related to microbiome and proteomics, which includes microbe–host interactions and microbial membrane protein biomarker identification for vaccine development [103]. Nevertheless, smart materials provide a facile expandable surface for functional integration. For instance, high specificity which sigificantly improves the separation selectivity. For all these reasons, smart and novel bio-separation materials are ready to accelerate for advances in bio-separation, enrichment and purification applications.
Therefore the following perspectives must be applied for future studies which focus on

- Development of scalable, cost-effective, and environmentally friendly synthesis methods which definitely taps green chemistry approaches.
- Approaches for utilizing agricultural wastes (underutilized biomass) for scalable manufacturing of carbonaceous and silica MMs.
- Recent advances in the leveraging artificial intelligence (AI) and machine learning for MM design and discovery for acceleration in the discovery of heterogenous catalyst thus optimizing their performance in protein applications.
- Development of predictive AI models that help in crafting materials for specific proteins or biological applications.
- Design and operational strategies for continuous flow pilot-scale production of highly porous MMs with high production rate for effective scale up.
- Design and development of characterization methods for MMs which are rapid, cost effective and user friendly in nature.
- Life cycle assessment of MMs to study the significant impact of these materials on the environment.
- Studies which focus on improving the antifouling and antimicrobial properties of MMs

Conclusions

This review positively addresses the use of MMs for biotechnological protein separation and as innovative carrier molecules for drug delivery and reusable adsorbents for removal of heavy metals and organics from aqueous systems. Various types of MMs (silica, carbon based, and biomass derived) are discussed for specific applications with examples. Moreover, challenges and limitations of MMs for effective scale up along with future prospects of these materials has been discussed. Following this, recent advances in leveraging artificial intelligence (AI) and machine learning for MM design in the discovery of heterogenous catalyst has been stressed, for optimizing their performance in protein applications. Particularly, future research which focuses on the development of predictive AI models that utilizes the characterization results of MMs and interprets its structural and mechanical properties, thus reducing the need for sophisticated expensive high end equipment could be a boon for its successful implementation and scale up in the fields. Research on optimizing the conditions for MM synthesis with exceptional properties (tunable pores, high adsorption capability, sustainability, reusability) is already in progress by various research groups. Major emphasis on challenges and limitations of these novel MM and future perspectives are stressed upon for the benefit of researchers who are willing to explore these materials for cutting edge biotechnological applications. In summary, the future prospects of mesoporous materials for biotechnological protein applications and its use as drug carriers and innovative adsorbent for heavy metals and organics (Dyes and antibiotics) is bright, with ongoing advancements likely to expand their capabilities and applications across diverse fields.

References

[1]I. Modasiya, P. Mori, and S. Goswami, "Microbial Production of r-DNA Products," *Microbial Products for Health and Nutrition*, pp. 131–157, 2024. https://doi.org/10.1007/978-981-97-4235-6_6

[2]M. B. Falasconi, A. Bertino, L. Mazzeo, F. Caputo, and V. Piemonte, "Membranes for the downstream treating of biotechnology processes," *Current Trends and Future Developments on (Bio-) Membranes: Advances on Membrane Engineering*, pp. 519–547, Jan. 2024. https://doi.org/10.1016/B978-0-323-90258-8.00028-6

[3]F. N. U. Akshit *et al.*, "Technological challenges in production of goat milk products and strategies to overcome them: a review," *International Journal of Food Science & Technology*, vol. 59, no. 1, pp. 6–16, Jan. 2024. https://doi.org/10.1111/IJFS.16782

[4]A. Schmidt, A. Hengelbrock, and J. Strube, "Continuous biomanufacturing in upstream and downstream processing," *Physical Sciences Reviews*, Dec. 2023. https://doi.org/10.1515/PSR-2022-0106/XML

[5]I. I. Amarakoon, C. L. Hamilton, S. A. Mitchell, P. F. Tennant, and M. E. Roye, "Biotechnology: principles and applications," *Pharmacognosy: Fundamentals, Applications, and Strategies, Second Edition*, pp. 627–645, Jan. 2023. https://doi.org/10.1016/B978-0-443-18657-8.00017-7

[6]K. Chakraborty, S. Dhara, A. Saijan, N. Alex, and K. P. Shilpa, "Chromatographic methods to isolate marine natural products from seaweeds," 2024

[7]K. Shanu, S. Choudhary, S. Kumari, K. Anu, and S. Devi, "Downstream Processing for Bio-product Recovery and Purification," *Recent Advances in Bioprocess Engineering and Bioreactor Design*, pp. 139–169, 2024. https://doi.org/10.1007/978-981-97-1451-3_7

[8]S. Kendir and M. Franzreb, "Synergies of pH-induced calcium phosphate precipitation and magnetic separation for energy-efficient harvesting of freshwater microalgae," *Bioresource Technology*, vol. 391, p. 129964, Jan. 2024. https://doi.org/10.1016/J.BIORTECH.2023.129964

[9]M. A. Alam, T. Yuan, X. Wenlong, B. Zhang, Y. Lv, and J. Xu, "Process optimization for the production of high-concentration ethanol with Scenedesmus raciborskii biomass," *Bioresource Technology*, vol. 294, p. 122219, Dec. 2019. https://doi.org/10.1016/J.BIORTECH.2019.122219

[10] E. G. Fawaz, D. A. Salam, S. S. Rigolet, and T. J. Daou, "Hierarchical Zeolites as Catalysts for Biodiesel Production from Waste Frying Oils to Overcome Mass Transfer Limitations," *Molecules 2021, Vol. 26, Page 4879*, vol. 26, no. 16, p. 4879, Aug. 2021. https://doi.org/10.3390/MOLECULES26164879

[11] N. Mohassel Yazdi and M. R. Naimi-Jamal, "One-pot synthesis of quinazolinone heterocyclic compounds using functionalized SBA-15 with natural material ellagic acid as a novel nanocatalyst," *Scientific Reports 2024 14:1*, vol. 14, no. 1, pp. 1–15, May 2024. https://doi.org/10.1038/s41598-024-61803-y

[12] P. Satishkumar, A. M. Isloor, L. N. Rao, and R. Farnood, "Fabrication of 2D Vanadium MXene Polyphenylsulfone Ultrafiltration Membrane for Enhancing the Water Flux and for Effective Separation of Humic Acid and Dyes from Wastewater," *ACS Omega*, vol. 9, no. 24, pp. 25766–25778, Jun. 2024. https://doi.org/10.1021/acsomega.3c10078

[13] M. Kumar, A. M. Isloor, S. R. Todeti, A. F. Ismail, and R. Farnood, "Hydrophilic nano-aluminum oxide containing polyphenylsulfone hollow fiber membranes for the extraction of arsenic (As-V) from drinking water," *Journal of Water Process Engineering*, vol. 44, p. 102357, Dec. 2021. https://doi.org/10.1016/J.JWPE.2021.102357

[14] M. Cui *et al.*, "Corrosion protection of Aluminium Alloy 2024 through an epoxy coating embedded with smart microcapsules: The responses of smart microcapsules to corrosive entities," *Corrosion Communications*, vol. 1, pp. 1–9, Mar. 2021. https://doi.org/10.1016/J.CORCOM.2021.06.001

[15] F. G. H. S. Pinto, V. P. da S. Caldeira, J. Villarroel-Rocha, K. Sapag, S. B. C. Pergher, and A. G. D. Santos, "Al/SBA-15 Mesoporous Material: A Study of pH Influence over Aluminum Insertion into the Framework," *Nanomaterials 2024, Vol. 14, Page 208*, vol. 14, no. 2, p. 208, Jan. 2024. https://doi.org/10.3390/NANO14020208

[16] Y. Wu, Y. Zhang, J. Zhou, and D. Gu, "Recent progress on functional mesoporous materials as catalysts in organic synthesis," *Emergent Materials*, vol. 3, no. 3, pp. 247–266, Jun. 2020. https://doi.org/10.1007/S42247-020-00086-1/METRICS

[17] I. A. Spiridon, I. D. Căruntu, I. Spiridon, and R. Brăescu, "Insight into Potential Biomedical Application of Mesoporous Materials," *Pharmaceutics 2022, Vol. 14, Page 2382*, vol. 14, no. 11, p. 2382, Nov. 2022. https://doi.org/10.3390/PHARMACEUTICS14112382

[18] L. Zu *et al.*, "Mesoporous Materials for Electrochemical Energy Storage and Conversion," *Advanced Energy Materials*, vol. 10, no. 38, p. 2002152, Oct. 2020. https://doi.org/10.1002/AENM.202002152

[19] B. Sartori, H. Amenitsch, and B. Marmiroli, "Functionalized Mesoporous Thin Films for Biotechnology," *Micromachines 2021, Vol. 12, Page 740*, vol. 12, no. 7, p. 740, Jun. 2021. https://doi.org/10.3390/MI12070740

[20] R. Lyu *et al.*, "Efficient adsorption of methylene blue by mesoporous silica prepared using sol-gel method employing hydroxyethyl cellulose as a template," *Colloids and Surfaces A: Physicochemical and Engineering Aspects*, vol. 606, p. 125425, Dec. 2020. https://doi.org/10.1016/J.COLSURFA.2020.125425

[21] K. W. Kim, B. Park, J. Kim, C. Jo, and J. K. Kim, "Recent progress in block copo lymer s oft-template-assisted synthesis of versatile mesoporous materials for energy storage systems," *Journal of Materials Chemistry A*, vol. 11, no. 14, pp. 7358–7386, Apr. 2023. https://doi.org/10.1039/D2TA09353G

[22] T. Haynes, O. Bougnouch, V. Dubois, and S. Hermans, "Preparation of mesoporous silica nanocapsules with a high specific surface area by hard and soft dual templating approach: Application to biomass valorization catalysis," *Microporous and Mesoporous Materials*, vol. 306, p. 110400, Oct. 2020. https://doi.org/10.1016/J.MICROMESO.2020.110400

[23] L. Ndlwana *et al.*, "Sustainable Hydrothermal and Solvothermal Synthesis of Advanced Carbon Materials in Multidimensional Applications: A Review," *Materials 2021, Vol. 14, Page 5094*, vol. 14, no. 17, p. 5094, Sep. 2021. https://doi.org/10.3390/MA14175094

[24] S. Akbar, J. M. Elliott, A. M. Squires, and A. Anwar, "Optimum conditions for electrochemical deposition of 3-D mesoporous platinum framework," *Journal of Nanoparticle Research*, vol. 22, no. 6, pp. 1–9, Jun. 2020. https://doi.org/10.1007/S11051-020-04912-9/FIGURES/6

[25] S. Belbekhouche *et al.*, "Fabrication of large pore mesoporous silica microspheres by salt-assisted spray-drying method for enhanced antibacterial activity and pancreatic cancer treatment," *International Journal of Pharmaceutics*, vol. 590, p. 119930, Nov. 2020. https://doi.org/10.1016/J.IJPHARM.2020.119930

[26] B. Díaz de Greñu *et al.*, "Fast Microwave-Assisted Synthesis, Calcination and Functionalization of a Silica Mesoporous Nanomaterial: UVM-7," *ChemSusChem*, vol. 16, no. 12, p. e202300123, Jun. 2023. https://doi.org/10.1002/CSSC.202300123

[27] Z. Li, L. Liu, Z. Wang, P. Gao, and G. K. Li, "Synthesis and Application of Mesoporous Materials: Process Status, Technical Problems, and Development Prospects: A Mini-Review," *Energy and Fuels*, vol. 37, no. 5, pp. 3413–3427, Mar. 2023. https://doi.org/10.1021/ACS.ENERGYFUELS.2C03882/ASSET/IMAGES/MEDIUM/EF2C03882_0009.GIF

[28] S. Mehdipour-Ataei and E. Aram, "Mesoporous Carbon-Based Materials: A Review of Synthesis, Modification, and Applications," vol. 13, no. 2, 2022. https://doi.org/10.3390/catal13010002

[29] J. Li, R. Li, W. Wang, K. Lan, and D. Zhao, "Ordered Mesoporous Crystalline Frameworks Toward Promising Energy Applications," *Advanced Materials*, vol. 36, no. 16, p. 2311460, Apr. 2024. https://doi.org/10.1002/ADMA.202311460

[30] F. Jia *et al.*, "Immobilization of Enzymes on Cyclodextrin-Anchored Dehiscent Mesoporous TiO2 for Efficient Photoenzymatic Hydroxylation," *ACS Applied Materials and Interfaces*, vol. 15, no. 6, pp. 7928–7938, Feb. 2023. https://doi.org/10.1021/ACSAMI.2C17971/SUPPL_FILE/AM2C17971_SI_001.PDF

[31] A. Rocío Hernández *et al.*, "Disordered mesoporous silica particles: an emerging platform to deliver proteins to the lungs," *Drug Delivery*, vol. 31, no. 1, Dec. 2024. https://doi.org/10.1080/10717544.2024.2381340

[32] L. Yu *et al.*, "Nanoporous Crystalline Materials for the Recognition and Applications of Nucleic Acids," *Advanced Materials*, p. 2305171, 2023. https://doi.org/10.1002/ADMA.202305171

[33] Q. Lei *et al.*, "Sol–Gel-Based Advanced Porous Silica Materials for Biomedical Applications," *Advanced Functional Materials*, vol. 30, no. 41, p. 1909539, Oct. 2020. https://doi.org/10.1002/ADFM.201909539

[34] A. G. Gebretatios, A. R. Kadiri Kanakka Pillantakath, T. Witoon, J. W. Lim, F. Banat, and C. K. Cheng, "Rice husk waste into various template-engineered mesoporous silica materials for different applications: A comprehensive review on recent developments," *Chemosphere*, vol. 310, p. 136843, Jan. 2023. https://doi.org/10.1016/J.CHEMOSPHERE.2022.136843

[35] A. E. Garcia-Bennett, "Synthesis, Toxicology and Potential of Ordered Mesoporous Materials in Nanomedicine," *Nanomedicine*, vol. 6, no. 5, pp. 867–877, Jul. 2011. https://doi.org/10.2217/nnm.11.82

[36] A. Saleem, Y. Zhang, M. Usman, M. Haris, and P. Li, "Tailored architectures of mesoporous carbon nanostructures: From synthesis to applications," *Nano Today*, vol. 46, p. 101607, Oct. 2022. https://doi.org/10.1016/j.nantod.2022.101607

[37] M. Chandrashekhar Nayak, A. M. Isloor, Inamuddin, B. Lakshmi, H. M. Marwani, and I. Khan, "Polyphenylsulfone/multiwalled carbon nanotubes mixed ultrafiltration membranes: Fabrication, characterization and removal of heavy metals Pb2+, Hg2+, and Cd2+ from aqueous solutions," *Arabian Journal of Chemistry*, vol. 13, no. 3, pp. 4661–4672, Mar. 2020. https://doi.org/10.1016/J.ARABJC.2019.10.007

[38] V. R. Pereira, A. M. Isloor, U. K. Bhat, A. F. Ismail, A. Obaid, and H. K. Fun, "Preparation and performance studies of polysulfone-sulfated nano-titania (S-TiO 2) nanofiltration membranes for dye removal," *RSC Advances*, vol. 5, no. 66, pp. 53874–53885, Jun. 2015. https://doi.org/10.1039/C5RA07994B

[39] S. I. Gnani Peer Mohamed, A. M. Isloor, and R. Farnood, "Catalyst- and Stabilizer-Free Rational Synthesis of Ionic Polymer Nanoparticles in One Step for Oil/Water Separation Membranes," *ACS Appl. Mater. Interfaces*, vol. 14, no. 40, pp. 45800–45809, Oct. 2022. https://doi.org/10.1021/acsami.2c11814

[40] G. P. Syed Ibrahim, A. M. Isloor, M. Bavarian, and S. Nejati, "Integration of Zwitterionic Polymer Nanoparticles in Interfacial Polymerization for Ion Separation," *ACS Applied Polymer Materials*, vol. 2, no. 4, pp. 1508–1517, Apr. 2020. https://doi.org/10.1021/ACSAPM.9B01192/SUPPL_FILE/AP9B01192_SI_001.PDF

[41] R. S. Hebbar, A. M. Isloor, and A. F. Ismail, "Preparation and evaluation of heavy metal rejection properties of polyetherimide/porous activated bentonite clay nanocomposite membrane," *RSC Advances*, vol. 4, no. 88, pp. 47240–47248, Sep. 2014. https://doi.org/10.1039/C4RA09018G

[42] M. Padaki, A. M. Isloor, K. K. Nagaraja, H. S. Nagaraja, and M. Pattabi, "Conversion of microfiltration membrane into nanofiltration membrane by vapour phase deposition of aluminium for desalination application," *Desalination*, vol. 274, no. 1–3, pp. 177–181, Jul. 2011. https://doi.org/10.1016/J.DESAL.2011.02.007

[43] I. Moideen K, A. M. Isloor, A. F. Ismail, A. Obaid, and H. K. Fun, "Fabrication and characterization of new PSF/PPSU UF blend membrane for heavy metal rejection," *New pub: Balaban*, vol. 57, no. 42, pp. 19810–19819, Sep. 2015. https://doi.org/10.1080/19443994.2015.1106985

[44] M. Padaki, A. M. Isloor, P. Wanichapichart, and A. F. Ismail, "Preparation and characterization of sulfonated polysulfone and N-phthloyl chitosan blend composite cation-exchange membrane for desalination," *Desalination*, vol. 298, pp. 42–48, Jul. 2012. https://doi.org/10.1016/J.DESAL.2012.04.025

[45] R. S. Hebbar, A. M. Isloor, B. Prabhu, Inamuddin, A. M. Asiri, and A. F. Ismail, "Removal of metal ions and humic acids through polyetherimide membrane with grafted bentonite clay," *Scientific Reports 2018 8:1*, vol. 8, no. 1, pp. 1–16, Mar. 2018. https://doi.org/10.1038/s41598-018-22837-1

[46] R. S. Hebbar, A. M. Isloor, Inamuddin, Mohd. S. Abdullah, A. F. Ismail, and A. M. Asiri, "Fabrication of polyetherimide nanocomposite membrane with amine functionalised halloysite nanotubes for effective removal of cationic dye effluents," *Journal of the Taiwan Institute of Chemical Engineers*, vol. 93, pp. 42–53, Dec. 2018. https://doi.org/10.1016/j.jtice.2018.07.032

[47] R. S. Hebbar, A. M. Isloor, and A. F. Ismail, "Preparation of antifouling polyetherimide/hydrolysed PIAM blend nanofiltration membranes for salt rejection applications," *RSC Advances*, vol. 4, no. 99, pp. 55773–55780, Oct. 2014. https://doi.org/10.1039/C4RA05791K

[48] G. P. Syed Ibrahim, A. M. Isloor, A. F. Ismail, and R. Farnood, "One-step synthesis of zwitterionic graphene oxide nanohybrid: Application to polysulfone tight ultrafiltration hollow fiber membrane," *Scientific Reports 2020 10:1*, vol. 10, no. 1, pp. 1–13, Apr. 2020. https://doi.org/10.1038/s41598-020-63356-2

[49] G. P. S. Ibrahim, A. M. Isloor, Inamuddin, A. M. Asiri, and R. Farnood, "Tuning the surface properties of Fe3O4 by zwitterionic sulfobetaine: application to antifouling and dye removal membrane," *Int. J. Environ. Sci. Technol.*, vol. 17, no. 9, pp. 4047–4060, Sep. 2020. https://doi.org/10.1007/s13762-020-02730-z

[50] R. S. Hebbar, A. M. Isloor, A. K. Zulhairun, M. Sohaimi Abdullah, and A. F. Ismail, "Efficient treatment of hazardous reactive dye effluents through antifouling polyetherimide hollow fiber membrane embedded with functionalized halloysite nanotubes," *Journal of the Taiwan Institute of Chemical Engineers*, vol. 72, pp. 244–252, 2017. https://doi.org/10.1016/j.jtice.2017.01.022

[51] M. Kumar *et al.*, "Use of cellulose acetate/polyphenylsulfone derivatives to fabricate ultrafiltration hollow fiber membranes for the removal of arsenic from drinking water," *International Journal of Biological Macromolecules*, vol. 129, pp. 715–727, May 2019. https://doi.org/10.1016/J.IJBIOMAC.2019.02.017

[52] S. Kumar, M. M. Malik, and R. Purohit, "Synthesis Methods of Mesoporous Silica Materials," *Materials Today: Proceedings*, vol. 4, no. 2, Part A, pp. 350–357, Jan. 2017. https://doi.org/10.1016/j.matpr.2017.01.032

[53] D. G. Lavrova *et al.*, "Biocompatible Silica-Polyethylene Glycol-Based Composites for Immobilization of Microbial Cells by Sol-Gel Synthesis," *Polymers*, vol. 15, no. 2, Art. no. 2, Jan. 2023. https://doi.org/10.3390/polym15020458

[54] J. Liu *et al.*, "A general synthesis of mesoporous metal oxides with well-dispersed metal nanoparticles via a versatile sol–gel process," *Journal of Materials Chemistry A*, vol. 1, no. 12, pp. 4038–4047, 2013. https://doi.org/10.1039/C3TA00570D

[55] F. Adam, T.-S. Chew, and J. Andas, "A simple template-free sol–gel synthesis of spherical nanosilica from agricultural biomass," *J Sol-Gel Sci Technol*, vol. 59, no. 3, pp. 580–583, Sep. 2011. https://doi.org/10.1007/s10971-011-2531-7

[56] B. Tessema, G. Gonfa, S. M. Hailegiorgis, and S. V. Prabhu, "Synthesis and Characterization of Modified Silica Gel from Teff Straw Ash Using Sol-gel Method," *Next Materials*, vol. 3, p. 100146, Apr. 2024. https://doi.org/10.1016/j.nxmate.2024.100146

[57] J. Zhang, Y. Ma, F. Shi, L. Liu, and Y. Deng, "Room temperature ionic liquids as templates in the synthesis of mesoporous silica via a sol–gel method," *Microporous and Mesoporous Materials*, vol. 119, no. 1, pp. 97–103, Mar. 2009. https://doi.org/10.1016/j.micromeso.2008.10.003

[58] "Ordered mesoporous carbons synthesized by a modified sol–gel process for electrosorptive removal of sodium chloride - ScienceDirect." Accessed: Aug. 06, 2024. [Online]. Available: https://www.sciencedirect.com/science/article/pii/S000862230800609X

[59] "Synthesis of mesoporous silica nanoparticles by sol–gel as nanocontainer for future drug delivery applications - ScienceDirect." Accessed: Aug. 07, 2024. [Online]. Available: https://www.sciencedirect.com/science/article/pii/S0366317517300225

[60] K. S. Yoo, T. G. Lee, and J. Kim, "Preparation and characterization of mesoporous TiO2 particles by modified sol–gel method using ionic liquids," *Microporous and Mesoporous Materials*, vol. 84, no. 1, pp. 211–217, Sep. 2005. https://doi.org/10.1016/j.micromeso.2005.05.029

[61] M. R. Benzigar *et al.*, "Recent advances in functionalized micro and mesoporous carbon materials: synthesis and applications," *Chemical Society Reviews*, vol. 47, no. 8, pp. 2680–2721, 2018. https://doi.org/10.1039/C7CS00787F

[62] "Quantitative Control of Pore Size of Mesoporous Carbon Nanospheres through the Self-Assembly of Diblock Copolymer Micelles in Solution - Tian - 2016 - Small - Wiley Online Library." Accessed: Aug. 08, 2024. [Online]. Available: https://onlinelibrary.wiley.com/doi/abs/10.1002/smll.201600364

[63] D. Liu *et al.*, "Surfactant Assembly within Pickering Emulsion Droplets for Fabrication of Interior-Structured Mesoporous Carbon Microspheres," *Angewandte Chemie*, vol. 130, no. 34, pp. 11065–11070, 2018. https://doi.org/10.1002/ange.201805022

[64] L. Peng *et al.*, "Programmable synthesis of radially gradient-structured mesoporous carbon nanospheres with tunable core-shell architectures," *Chem*, vol. 7, no. 4, pp. 1020–1032, Apr. 2021. https://doi.org/10.1016/j.chempr.2021.01.001

[65] "Biphase Stratification Approach to Three-Dimensional Dendritic Biodegradable Mesoporous Silica Nanospheres | Nano Letters." Accessed: Aug. 09, 2024. [Online]. Available: https://pubs.acs.org/doi/full/10.1021/nl404316v

[66] "Synthesis of Ordered Mesoporous Silica with Tunable Morphologies and Pore Sizes via a Nonpolar Solvent-Assisted Stöber Method | Chemistry of Materials." Accessed: Aug. 09, 2024. [Online]. Available: https://pubs.acs.org/doi/full/10.1021/acs.chemmater.6b00499

[67] "Space-Confined Polymerization: Controlled Fabrication of Nitrogen-Doped Polymer and Carbon Microspheres with Refined Hierarchical Architectures - Wang - 2019 - Advanced Materials - Wiley Online Library." Accessed: Aug. 09, 2024. [Online]. Available: https://onlinelibrary.wiley.com/doi/abs/10.1002/adma.201807876

[68] "Ionic Liquid-Assisted Synthesis of Mesoporous Carbons with Surface-Enriched Nitrogen for the Hydrogen Evolution Reaction | ACS Applied Materials & Interfaces." Accessed: Aug. 09, 2024. [Online]. Available: https://pubs.acs.org/doi/full/10.1021/acsami.7b14919

[69] "Generalized and Facile Synthesis Approach to N-Doped Highly Graphitic Mesoporous Carbon Materials | Chemistry of Materials." Accessed: Aug. 12, 2024. [Online]. Available: https://pubs.acs.org/doi/full/10.1021/cm048057y

[70] "Facile preparation of hierarchically porous carbons from metal-organic gels and their application in energy storage | Scientific Reports." Accessed: Aug. 12, 2024. [Online]. Available: https://www.nature.com/articles/srep01935

[71] "Review on porous carbon materials engineered by ZnO templates: Design, synthesis and capacitance performance," *Materials & Design*, vol. 201, p. 109518, Mar. 2021. https://doi.org/10.1016/j.matdes.2021.109518

[72] "Structurally Ordered Mesoporous Carbon Nanoparticles as Transmembrane Delivery Vehicle in Human Cancer Cells | Nano Letters." Accessed: Aug. 12, 2024. [Online]. Available: https://pubs.acs.org/doi/abs/10.1021/nl801976m

[73] "Facile synthesis of hierarchically porous carbonaceous materials derived from olefin/aldehyde precursors using silica as templates - RSC Advances (RSC Publishing) DOI:10.1039/C8RA01270A." Accessed: Aug. 12, 2024. [Online]. Available: https://pubs.rsc.org/en/content/articlehtml/2018/ra/c8ra01270a

[74] D. Bhattacharjya, M.-S. Kim, T.-S. Bae, and J.-S. Yu, “High performance supercapacitor prepared from hollow mesoporous carbon capsules with hierarchical nanoarchitecture,” *Journal of Power Sources*, vol. 244, pp. 799–805, Dec. 2013. https://doi.org/10.1016/j.jpowsour.2013.01.112

[75] “Mesoporous carbon nanospheres with an excellent electrocapacitive performance - Journal of Materials Chemistry (RSC Publishing) DOI:10.1039/C0JM03322G.” Accessed: Aug. 12, 2024. [Online]. Available: https://pubs.rsc.org/en/content/articlehtml/2011/jm/c0jm03322g

[76] Z. Lei, Z. Chen, and X. S. Zhao, “Growth of Polyaniline on Hollow Carbon Spheres for Enhancing Electrocapacitance,” *J. Phys. Chem. C*, vol. 114, no. 46, pp. 19867–19874, Nov. 2010. https://doi.org/10.1021/jp1084026

[77] K. Zhang, Q. Zhao, Z. Tao, and J. Chen, “Composite of sulfur impregnated in porous hollow carbon spheres as the cathode of Li-S batteries with high performance,” *Nano Res.*, vol. 6, no. 1, pp. 38–46, Jan. 2013. https://doi.org/10.1007/s12274-012-0279-1

[78] J. Goworek, A. Kierys, and R. Kusak, “Isothermal template removal from MCM-41 in hydrogen flow,” *Microporous and Mesoporous Materials*, vol. 98, no. 1, pp. 242–248, Jan. 2007. https://doi.org/10.1016/j.micromeso.2006.09.011

[79] L. Huang, S. Kawi, C. Poh, K. Hidajat, and S. C. Ng, “Extraction of cationic surfactant templates from mesoporous materials by CH3OH-modified CO2 supercritical fluid,” *Talanta*, vol. 66, no. 4, pp. 943–951, May 2005. https://doi.org/10.1016/j.talanta.2004.12.057

[80] “Rapid removal of organic template from SBA-15 with microwave assisted extraction - ScienceDirect.” Accessed: Aug. 13, 2024. [Online]. Available: https://www.sciencedirect.com/science/article/pii/S0167577X09003553

[81] H. Xu, B. W. Zeiger, and K. S. Suslick, “Sonochemical synthesis of nanomaterials,” *Chemical Society Reviews*, vol. 42, no. 7, pp. 2555–2567, 2013. https://doi.org/10.1039/C2CS35282F

[82] A. Hozumi *et al.*, “Photocalcination of Mesoporous Silica Films Using Vacuum Ultraviolet Light,” *Advanced Materials*, vol. 12, no. 13, pp. 985–987, Jun. 2000. https://doi.org/10.1002/1521-4095(200006)12:13<985::AID-ADMA985>3.0.CO;2-#

[83] “IJMS | Free Full-Text | Influence of ‘Glow Discharge Plasma’ as an External Stimulus on the Self-Assembly, Morphology and Binding Affinity of Gold Nanoparticle-Streptavidin Conjugates.” Accessed: Aug. 13, 2024. [Online]. Available: https://www.mdpi.com/1422-0067/13/6/6534

[84] X. Zhuang, X. Qian, J. Lv, and Y. Wan, “An alternative method to remove PEO–PPO–PEO template in organic–inorganic mesoporous nanocomposites by sulfuric acid extraction,” *Applied Surface Science*, vol. 256, no. 17, pp. 5343–5348, Jun. 2010. https://doi.org/10.1016/j.apsusc.2009.12.074

[85] H. Lv, Y. Wang, L. Sun, Y. Yamauchi, and B. Liu, “A general protocol for precise syntheses of ordered mesoporous intermetallic nanoparticles,” *Nat Protoc*, vol. 18, no. 10, pp. 3126–3154, Oct. 2023. https://doi.org/10.1038/s41596-023-00872-1

[86] S. Karafiludis *et al.*, "Template-free synthesis of mesoporous and amorphous transition metal phosphate materials," *Nanoscale*, vol. 15, no. 8, pp. 3952–3966, 2023. https://doi.org/10.1039/D2NR05630E

[87] C. Cocuzza *et al.*, "Synthesis and characterization of mesoporous silicas with dendritic and spongy-like structures: Potential supports for human lactate dehydrogenase-based microreactors aimed at anticancer inhibitor screening," *Microporous and Mesoporous Materials*, vol. 376, p. 113182, Aug. 2024. https://doi.org/10.1016/j.micromeso.2024.113182

[88] E. Aguilar-García *et al.*, "Hydrogen Production by Hydrogen Sulfide Decomposition Using Cobalt Catalysts Doped in SBA-15 Synthetized by EISA Method," *Catal Lett*, vol. 154, no. 7, pp. 3724–3737, Jul. 2024. https://doi.org/10.1007/s10562-023-04515-8

[89] A. Witecka, J. Schmitt, M. Courtien, C. Gérardin, and G. Rydzek, "Hybrid mesoporous silica materials templated with surfactant polyion complex (SPIC) micelles for pH-triggered drug release," *Microporous and Mesoporous Materials*, vol. 365, p. 112913, Feb. 2024. https://doi.org/10.1016/j.micromeso.2023.112913

[90] G. Strukov, G. Strukova, M. Leonard, and M. M. Kuklja, "Biomimetic In Situ Self-Assembly of Metal Nanoparticles into Hierarchical 3D Mesostructures: Synthesis, Analysis, and Prospects," *Crystal Growth & Design*, vol. 24, no. 14, pp. 5865–5882, Jul. 2024. https://doi.org/10.1021/acs.cgd.3c01539

[91] J. Gluns *et al.*, "3D Printing of Ordered Mesoporous Silica Using Light-Induced Sol-Gel Chemistry," *Advanced Functional Materials*, vol. n/a, no. n/a, p. 2405511. https://doi.org/10.1002/adfm.202405511

[92] L. Zhao *et al.*, "Printing of In Situ Functionalized Mesoporous Silica with Digital Light Processing for Combinatorial Sensing," *Small*, vol. 20, no. 28, p. 2311121, 2024. https://doi.org/10.1002/smll.202311121

[93] J. Salonen and V.-P. Lehto, "Fabrication and chemical surface modification of mesoporous silicon for biomedical applications," *Chemical Engineering Journal*, vol. 137, no. 1, pp. 162–172, Mar. 2008. https://doi.org/10.1016/j.cej.2007.09.001

[94] S. T. Lakshmikumar and P. K. Singh, "Stabilization of porous silicon surface by low temperature photoassisted reaction with acetylene," *Current Applied Physics*, vol. 3, no. 2, pp. 185–189, Apr. 2003. https://doi.org/10.1016/S1567-1739(02)00198-0

[95] S. Porrang, N. Rahemi, S. Davaran, M. Mahdavi, B. Hassanzadeh, and A. M. Gholipour, "Direct surface modification of mesoporous silica nanoparticles by DBD plasma as a green approach to prepare dual-responsive drug delivery system," *Journal of the Taiwan Institute of Chemical Engineers*, vol. 123, pp. 47–58, Jun. 2021. https://doi.org/10.1016/j.jtice.2021.05.024

[96] J. A. JUAREZ-MORENO *et al.*, "Effect of inductively coupled plasma surface treatment on silica gel and mesoporous MCM-41 particles," *Plasma Sci. Technol.*, vol. 20, no. 6, p. 065506, Apr. 2018. https://doi.org/10.1088/2058-6272/aaabb5

[97] J. Xu *et al.*, "Cold plasma and ethanol coupling-activated Ni/MCM-41 catalysts with highly stability and activity for dry reforming of methane," *Microporous and Mesoporous Materials*, vol. 364, p. 112847, Jan. 2024. https://doi.org/10.1016/j.micromeso.2023.112847

[98] P. T. Wai, P. Jiang, Y. Shen, P. Zhang, and Q. Gu, "Entrapment of peroxophosphotungstate in SBA-15 by silylation and its catalytic efficiency in the epoxidation of soybean oil," *Applied Catalysis A: General*, vol. 596, p. 117537, Apr. 2020. https://doi.org/10.1016/j.apcata.2020.117537

[99] X. Li, J. Yuan, J. Du, H. Sui, and L. He, "Functionalized Ordered Mesoporous Silica by Vinyltriethoxysilane for the Removal of Volatile Organic Compounds through Adsorption/Desorption Process," *Ind. Eng. Chem. Res.*, vol. 59, no. 8, pp. 3511–3520, Feb. 2020. https://doi.org/10.1021/acs.iecr.9b06062

[100] C. G. Sonwane and S. K. Bhatia, "Characterization of Pore Size Distributions of Mesoporous Materials from Adsorption Isotherms," *J. Phys. Chem. B*, vol. 104, no. 39, pp. 9099–9110, Oct. 2000. https://doi.org/10.1021/jp000907j

[101] M. N. Sarvi, T. Budianto Bee, C. K. Gooi, B. W. Woonton, M. L. Gee, and A. J. O'Connor, "Development of functionalized mesoporous silica for adsorption and separation of dairy proteins," *Chemical Engineering Journal*, vol. 235, pp. 244–251, Jan. 2014. https://doi.org/10.1016/J.CEJ.2013.09.036

[102] Z. Liu, M. Li, X. Yang, M. Yin, J. Ren, and X. Qu, "The use of multifunctional magnetic mesoporous core/shell heteronanostructures in a biomolecule separation system," *Biomaterials*, vol. 32, no. 21, pp. 4683–4690, Jul. 2011. https://doi.org/10.1016/J.BIOMATERIALS.2011.03.038

[103] J. Zhao *et al.*, "Mesoporous Silica as Sorbents and Enzymatic Nanoreactors for Microbial Membrane Proteomics," *ACS Applied Materials and Interfaces*, vol. 13, no. 10, pp. 11571–11578, Mar. 2021. https://doi.org/10.1021/ACSAMI.0C19528/SUPPL_FILE/AM0C19528_SI_009.XLSX

[104] S. Qi, C. Zhang, Z. Hu, and X. Chen, "Biocompatible mesoporous polymeric ionic liquid with high adsorption capacity for selective isolation of ovalbumin," *Journal of Materials Science*, vol. 56, no. 34, pp. 19283–19295, Dec. 2021. https://doi.org/10.1007/S10853-021-06529-2/FIGURES/7

[105] S. S. Ding, J. P. Zhu, Y. Wang, Y. Yu, and Z. Zhao, "Recent progress in magnetic nanoparticles and mesoporous materials for enzyme immobilization: an update," *Brazilian Journal of Biology*, vol. 82, p. e244496, Jun. 2021. https://doi.org/10.1590/1519-6984.244496

[106] H. M. Salvi and G. D. Yadav, "Surface functionalization of SBA-15 for immobilization of lipase and its application in synthesis of alkyl levulinates: Optimization and kinetics," *Biocatalysis and Agricultural Biotechnology*, vol. 18, p. 101038, Mar. 2019. https://doi.org/10.1016/J.BCAB.2019.101038

[107] F. Esmi, T. Nematian, Z. Salehi, A. A. Khodadadi, and A. K. Dalai, "Amine and aldehyde functionalized mesoporous silica on magnetic nanoparticles for enhanced lipase immobilization, biodiesel production, and facile separation," *Fuel*, vol. 291, p. 120126, May 2021. https://doi.org/10.1016/J.FUEL.2021.120126

[108] X. Li *et al.*, "The efficient enrichment of marine peptides from the protein hydrolysate of the marine worm Urechis unicinctus by using mesoporous materials MCM-41, SBA-15 and CMK-3," *Analytical Methods*, vol. 13, no. 21, pp. 2405–2414, Jun. 2021. https://doi.org/10.1039/D1AY00616A

[109] X. Liu, X. Shi, H. Wang, and H. Zhang, "Atom transfer radical polymerization of diverse functional SBA-15 for selective separation of proteins," *Microporous and Mesoporous Materials*, vol. 200, pp. 165–173, Dec. 2014. https://doi.org/10.1016/J.MICROMESO.2014.08.027

[110] Y. Dong, A. Laaksonen, M. Gong, R. An, and X. Ji, "Selective Separation of Highly Similar Proteins on Ionic Liquid-Loaded Mesoporous TiO2," *Langmuir*, vol. 38, no. 10, pp. 3202–3211, Mar. 2022. https://doi.org/10.1021/ACS.LANGMUIR.1C03277/ASSET/IMAGES/LARGE/LA1C03277_0007.JPEG

[111] H. R. Panchami, A. M. Isloor, and A. F. Ismail, "Improved hydrophilic and antifouling performance of nanocomposite ultrafiltration zwitterionic polyphenylsulfone membrane for protein rejection applications," *Journal of Nanostructure in Chemistry 2021*, pp. 1–22, Aug. 2021. https://doi.org/10.1007/S40097-021-00416-7

[112] P. Heggabailu Rajashekhara, A. M. Isloor, and A. Fauzi Ismail, "One-step synthesis and characterization of hydrophilic polymer microspheresimmobilized with polyphenylsulfone ultraltration membranes for protein rejection application," 2022. https://doi.org/10.21203/rs.3.rs-1331802/v1

[113] J. F. Cao, W. Xu, Y. Y. Zhang, Y. Shu, and J. H. Wang, "Chondroitin sulfate-functionalized 3D hierarchical flower-type mesoporous silica with a superior capacity for selective isolation of low density lipoprotein," *Analytica Chimica Acta*, vol. 1104, pp. 78–86, Apr. 2020. https://doi.org/10.1016/J.ACA.2019.12.075

[114] Y. Luo, B. Wang, L. Yi, C. F. Ding, C. Deng, and Y. Yan, "Mesoporous materials for glycopeptide separation," *TrAC Trends in Analytical Chemistry*, vol. 167, p. 117234, Oct. 2023. https://doi.org/10.1016/J.TRAC.2023.117234

[115] M. Zhang, X. Liu, W. Zhou, X. Zheng, S. Wang, and L. Zhou, "Ordered porous materials for blood purification," *Separation and Purification Technology*, vol. 327, p. 124844, Dec. 2023. https://doi.org/10.1016/J.SEPPUR.2023.124844

[116] A. Singh and S. K. Yadav, "Immobilization of L-ribose isomerase on the surface of activated mesoporous MCM41 and SBA15 for the synthesis of L-ribose," *Journal of Biotechnology*, vol. 362, pp. 45–53, Jan. 2023. https://doi.org/10.1016/J.JBIOTEC.2022.12.010

[117] E. C. H. T. Lau *et al.*, "Direct purification and immobilization of his-tagged enzymes using unmodified nickel ferrite NiFe2O4 magnetic nanoparticles," *Scientific Reports 2023 13:1*, vol. 13, no. 1, pp. 1–10, Dec. 2023. https://doi.org/10.1038/s41598-023-48795-x

[118] S. Ariaeenejad *et al.*, "An efficient nano-biocatalyst for lignocellulosic biomass hydrolysis: Xylanase immobilization on organically modified biogenic mesoporous silica nanoparticles," *International Journal of Biological Macromolecules*, vol. 164, pp. 3462–3473, Dec. 2020. https://doi.org/10.1016/J.IJBIOMAC.2020.08.211

[119] Z. Dong *et al.*, "Fabrication of immobilized lipases from Candida rugosa on hierarchical mesoporous silica for enzymatic enrichment of ω-3 polyunsaturated fatty acids by selective hydrolysis," *Food Chemistry: X*, vol. 22, p. 101434, Jun. 2024. https://doi.org/10.1016/J.FOCHX.2024.101434

[120] Y. Li, Z. Liu, L. Dai, D. Liu, and W. Du, "Synthesis and application of mesoporous MIL-88A for enhanced lipase immobilization and high value oil conversion," *Chemical Engineering Journal*, vol. 496, p. 153802, Sep. 2024. https://doi.org/10.1016/J.CEJ.2024.153802

[121] P. J. M. Lima, N. S. Rios, E. Vilarrasa-García, J. A. Cecilia, E. Rodríguez-Castellón, and L. R. B. Gonçalves, "Preparation of a heterogeneous biocatalyst through Thermomyces lanuginosus lipase immobilization on pore-expanded SBA-15," *International Journal of Biological Macromolecules*, vol. 274, p. 133359, Aug. 2024. https://doi.org/10.1016/J.IJBIOMAC.2024.133359

[122] Y. Zhang *et al.*, "Immobilized lipase based on SBA-15 adsorption and gel embedding for catalytic synthesis of isoamyl acetate," *Food Bioscience*, vol. 60, p. 104427, Aug. 2024. https://doi.org/10.1016/J.FBIO.2024.104427

[123] N. Zhong, W. Chen, L. Liu, and H. Chen, "Immobilization of Rhizomucor miehei lipase onto the organic functionalized SBA-15: Their enzymatic properties and glycerolysis efficiencies for diacylglycerols production," *Food Chemistry*, vol. 271, pp. 739–746, Jan. 2019. https://doi.org/10.1016/J.FOODCHEM.2018.07.185

[124] N. S. Rios *et al.*, "Strategies of covalent immobilization of a recombinant Candida antarctica lipase B on pore-expanded SBA-15 and its application in the kinetic resolution of (R,S)-Phenylethyl acetate," *Journal of Molecular Catalysis B: Enzymatic*, vol. 133, pp. 246–258, Nov. 2016. https://doi.org/10.1016/J.MOLCATB.2016.08.009

[125] J. Tu *et al.*, "Mesoporous Silica Nanoparticles with Large Pores for the Encapsulation and Release of Proteins," *ACS Applied Materials and Interfaces*, vol. 8, no. 47, pp. 32211–32219, Nov. 2016. https://doi.org/10.1021/ACSAMI.6B11324/SUPPL_FILE/AM6B11324_SI_002.MPG

[126] T. Qiang, Y. Song, R. Zhu, and W. Yuan, "Biomass material derived hierarchical porous TiO2: Adjustable pore size for protein adsorption," *Journal of Alloys and Compounds*, vol. 829, p. 154512, Jul. 2020. https://doi.org/10.1016/J.JALLCOM.2020.154512

[127] Ľ. Zauška *et al.*, "Adsorption and Release Properties of Drug Delivery System Naproxen-SBA-15: Effect of Surface Polarity, Sodium/Acid Drug Form and pH," *Journal of Functional Biomaterials*, vol. 13, no. 4, p. 275, Dec. 2022. https://doi.org/10.3390/JFB13040275/S1

[128] D. Dasgupta, M. Das, S. Thakore, A. Patel, S. Kumar, and S. Seshadri, "Development of a controlled sustainable anticancer drug delivery nanosystem comprising doxorubicin and functionalized MCM-48," *Journal of Drug Delivery Science and Technology*, vol. 72, p. 103419, Jun. 2022. https://doi.org/10.1016/J.JDDST.2022.103419

[129] N. A. Atiyah, T. M. Albayati, and M. A. Atiya, "Interaction behavior of curcumin encapsulated onto functionalized SBA-15 as an efficient carrier and release in drug delivery," *Journal of Molecular Structure*, vol. 1260, p. 132879, Jul. 2022. https://doi.org/10.1016/J.MOLSTRUC.2022.132879

[130] S. Ulagesan *et al.*, "Mesoporous silica (SBA-15) with enriched amidoxime functionalities for pH-controlled anticancer drug delivery," *Inorganic Chemistry Communications*, vol. 146, p. 110132, Dec. 2022. https://doi.org/10.1016/J.INOCHE.2022.110132

[131] M. Samari, S. Kashanian, S. Zinadini, and H. Derakhshankhah, "Designing of a new transdermal antibiotic delivery polymeric membrane modified by functionalized SBA-15 mesoporous filler," *Scientific Reports 2024 14:1*, vol. 14, no. 1, pp. 1–14, May 2024. https://doi.org/10.1038/s41598-024-60727-x

[132] N. V. Mdlovu, K. S. Lin, M. T. Weng, Y. S. Lin, and S. Y. Liu, "Preparation and in-vitro/in-vivo evaluation of doxorubicin-loaded magnetic SBA-15 nanocomposites from rice husk for enhancing therapeutic efficacy," *Colloids and Surfaces B: Biointerfaces*, vol. 220, p. 112923, Dec. 2022. https://doi.org/10.1016/J.COLSURFB.2022.112923

[133] R. Li, / Rscpharma, A. Patel, and S. Mehta, "Functionalized SBA-15: engineering, detailed study on release and kinetics of alendronate as well as its anti-tumour properties towards osteosarcoma," *RSC Pharmaceutics*, 2024. https://doi.org/10.1039/D4PM00078A

[134] R. Foulady-Dehaghi, S. Sohrabnezhad, and M. Hadavi, "Drug delivery with solvent-free synthesized polyimide-COF/amino-functionalized MCM-41 nanohybrid," *Journal of Drug Delivery Science and Technology*, vol. 81, p. 104283, Mar. 2023. https://doi.org/10.1016/J.JDDST.2023.104283

[135] S. Septian Dwitya, K. S. Lin, M. T. Weng, N. Vukile Mdlovu, M. T. Yang, and C. M. Wu, "Synthesis and characterization of oleic acid-stabilized cobalt ferrite @MCM-41/nanocomposites for pH-responsive drug delivery," *Journal of Industrial and Engineering Chemistry*, Aug. 2024. https://doi.org/10.1016/J.JIEC.2024.08.036

[136] H. R. Hosseini, M. Abdouss, and M. Golshekan, "Hydroxyapatite Incorporated with Fe3O4@MCM-41 Core-Shell: A Promising Nanocomposite for Teriparatide Delivery in Bone Tissue Regeneration," *ACS Omega*, vol. 8, no. 44, pp. 41363–41373, Nov. 2023. https://doi.org/10.1021/ACSOMEGA.3C04931/ASSET/IMAGES/LARGE/AO3C04931_0007.JPEG

[137] E. Da'na, "Adsorption of heavy metals on functionalized-mesoporous silica: A review," *Microporous and Mesoporous Materials*, vol. 247, pp. 145–157, Jul. 2017. https://doi.org/10.1016/J.MICROMESO.2017.03.050

[138] M. Anbia, K. Kargosha, and S. Khoshbooei, "Heavy metal ions removal from aqueous media by modified magnetic mesoporous silica MCM-48," *Chemical Engineering Research and Design*, vol. 93, pp. 779–788, Jan. 2015. https://doi.org/10.1016/J.CHERD.2014.07.018

[139] H. Javadian, B. B. Koutenaei, E. Shekarian, F. Z. Sorkhrodi, R. Khatti, and M. Toosi, "Application of functionalized nano HMS type mesoporous silica with N-(2-aminoethyl)-3-aminopropyl methyldimethoxysilane as a suitable adsorbent for removal of Pb (II) from aqueous media and industrial wastewater," *Journal of Saudi Chemical Society*, vol. 21, pp. S219–S230, Jan. 2017. https://doi.org/10.1016/J.JSCS.2014.01.007

[140] R. Ezzatkhah, Z. Rahimi-Ahar, A. Babapoor, M. Nemati Tasandeh, S. Yaghoubi, and F. Valizadeh Harzand, "Urea-Functionalized mesoporous silica SBA-15 for heavy metal Adsorption: Synthesis, Characterization, and optimization," *Materials Science and Engineering: B*, vol. 297, p. 116799, Nov. 2023. https://doi.org/10.1016/J.MSEB.2023.116799

[141] J. V. Ros-Lis, S. Vetter, and P. Smith, "A comparative life cycle assessment of the synthesis of mesoporous silica materials on a small and a large scale," *Green Chemistry*, 2024. https://doi.org/10.1039/D4GC02347A

[142] W. Kaewprachum, S. Wongsakulphasatch, W. Kiatkittipong, A. Striolo, C. K. Cheng, and S. Assabumrungrat, "SDS modified mesoporous silica MCM-41 for the adsorption of Cu2+, Cd2+, Zn2+ from aqueous systems," *Journal of Environmental Chemical Engineering*, vol. 8, no. 1, p. 102920, Feb. 2020. https://doi.org/10.1016/J.JECE.2019.102920

[143] M. Pandey, S. Shabuddhin, N. Tsunoji, S. Das, and M. Bandyopadhyay, "Extraction of heavy metals from wastewater using amine-modified mesoporous silica," *Environmental Science and Pollution Research*, vol. 30, no. 53, pp. 113409–113423, Nov. 2023. https://doi.org/10.1007/S11356-023-30092-9/METRICS

[144] J. Jiang, Y. Long, X. Hu, J. Hu, M. Zhu, and S. Zhou, "A facile microwave-assisted synthesis of mesoporous hydroxyapatite as an efficient adsorbent for Pb2+ adsorption," *Journal of Solid State Chemistry*, vol. 289, p. 121491, Sep. 2020. https://doi.org/10.1016/J.JSSC.2020.121491

[145] F. Hassanzadeh-Afruzi, F. Esmailzadeh, S. Asgharnasl, F. Ganjali, R. Taheri-Ledari, and A. Maleki, "Efficient removal of Pb(II)/Cu(II) from aqueous samples by a guanidine-functionalized SBA-15/Fe3O4," *Separation and Purification Technology*, vol. 291, p. 120956, Jun. 2022. https://doi.org/10.1016/J.SEPPUR.2022.120956

[146] H. Vatandoust, H. Younesi, Z. Mehraban, A. Heidari, and H. Khakpour, "Comparative Adsorption of Cd (II) and Pb (II) by MCM-48 and Amine-Grafted MCM-48 in Single and Binary Component Systems," *Water Conservation Science and Engineering 2021 6:2*, vol. 6, no. 2, pp. 67–78, May 2021. https://doi.org/10.1007/S41101-021-00103-2

[147] H. Li, S. Wang, R. Li, Y. Zhang, and H. Wang, "Preparation of Sulfhydryl Functionalized MCM-48 and Its Adsorption Performance for Cr(VI) in Water," *Water (Switzerland)*, vol. 15, no. 3, p. 524, Feb. 2023. https://doi.org/10.3390/W15030524/S1

[148] A. A. Malhis, S. H. Arar, M. K. Fayyad, and H. A. Hodali, "Amino- and thiol-modified microporous silicalite-1 and mesoporous MCM-48 materials as potential effective adsorbents for Pb(II) in polluted aquatic systems," *Adsorption Science and Technology*, vol. 36, no. 1–2, pp. 270–286, Feb. 2018. https://doi.org/10.1177/0263617416689270/ASSET/IMAGES/LARGE/10.1177_0263617416689270-FIG10.JPEG

[149] S. Li *et al.*, "Nanoconfined polyethyleneimine in mesoporous MCM-41 silica for heavy metal ions removal," *Separation and Purification Technology*, vol. 353, p. 128421, Jan. 2025. https://doi.org/10.1016/J.SEPPUR.2024.128421

[150] Y. Song *et al.*, "Uptake of arsenic(V) using iron and magnesium functionalized highly ordered mesoporous MCM-41 (Fe/Mg-MCM-41) as an effective adsorbent," *Science of The Total Environment*, vol. 833, p. 154858, Aug. 2022. https://doi.org/10.1016/J.SCITOTENV.2022.154858

[151] W. Yang, S. Shirazian, R. Soltani, and M. H. Zare, "Bio-originated mesosilicate SBA-15: synthesis, characterization, and application for heavy metal removal," *npj Clean Water 2024 7:1*, vol. 7, no. 1, pp. 1–15, Jun. 2024. https://doi.org/10.1038/s41545-024-00340-7

[152] F. Badvi Loulic, R. Haji Seyed Mohammad Shirazi, M. Miralinaghi, H. Ahmad Panahi, and E. Moniri, "Highly efficient removal of toxic As(V), Cd (II), and Pb(II) ions from water samples using MnFe2O4@SBA-15-(CH2)3-adenine as a recyclable bio-nanoadsorbent," *Microporous and Mesoporous Materials*, vol. 356, p. 112567, Jun. 2023. https://doi.org/10.1016/J.MICROMESO.2023.112567

[153] M. H. Fekri, S. Soleymani, M. Razavi Mehr, and F. Saki, "Removal of methyl orange dye from aqueous solutions by SBA-16 nano mesopore and optimization of effective parameters using response surface method," *Iranian Journal of Health and Environment*, vol. 16, no. 2, pp. 339–356, 2023

[154] R. Ma, Y. Xue, Q. Ma, Y. Chen, S. Yuan, and J. Fan, "Recent Advances in Carbon-Based Materials for Adsorptive and Photocatalytic Antibiotic Removal," *Nanomaterials (Basel, Switzerland)*, vol. 12, no. 22, Nov. 2022. https://doi.org/10.3390/NANO12224045

[155] V. Nairi, L. Medda, M. Monduzzi, and A. Salis, "Adsorption and release of ampicillin antibiotic from ordered mesoporous silica," *Journal of colloid and interface science*, vol. 497, pp. 217–225, Jul. 2017. https://doi.org/10.1016/J.JCIS.2017.03.021

[156] E. Zhang *et al.*, "On the origin of mesopore collapse in functionalized porous carbons," *Carbon*, vol. 149, pp. 743–749, Aug. 2019. https://doi.org/10.1016/J.CARBON.2019.04.116

[157] J. Blair-González, E. Contreras-Villacura, A. Carvajal Guevara, and C. Palma Toloza, "Oxytetracycline removal by biological/chemical activated mesoporous carbon," *Microporous and Mesoporous Materials*, vol. 327, p. 111384, Nov. 2021. https://doi.org/10.1016/J.MICROMESO.2021.111384

[158] T. Ariyanto, N. Y. Pradana, M. H. N. Saif, B. A. Prasetyo, I. Prasetyo, and M. Munoz, "Reusable adsorbent of magnetite in mesoporous carbon for antibiotic removal," *Environmental Science and Pollution Research*, vol. 31, no. 24, pp. 35824–35834, May 2024. https://doi.org/10.1007/S11356-024-33658-3/METRICS

[159] X. Hu *et al.*, "Insights on Size-Exclusion Effect of Ordered Mesoporous Carbon for Selective Antibiotics Adsorption Under the Interference of Natural Organic Matter," *SSRN Electronic Journal*, Nov. 2022. https://doi.org/10.2139/SSRN.4281035

[160] S. Ao *et al.*, "Sheet-like graphitized glucose-based mesoporous carbon for aqueous adsorption of tetracycline antibiotic," *DRM*, vol. 141, p. 110718, Jan. 2024. https://doi.org/10.1016/J.DIAMOND.2023.110718

[161] X. Chen *et al.*, "Insights into high-efficient removal of tetracycline by a codoped mesoporous carbon adsorbent," *Chinese Journal of Chemical Engineering*, vol. 44, no. 4, p. 148, Apr. 2022. https://doi.org/10.1016/J.CJCHE.2021.07.023

[162] P. V. Mane, R. M. Rego, P. L. Yap, D. Losic, and M. D. Kurkuri, "Unveiling cutting-edge advances in high surface area porous materials for the efficient removal of toxic metal ions from water," *Progress in Materials Science*, vol. 146, p. 101314, Dec. 2024. https://doi.org/10.1016/J.PMATSCI.2024.101314

[163] G. E. Gomez *et al.*, "Advances in Nanomaterials and Composites Based on Mesoporous Materials as Antimicrobial Agents: Relevant Applications in Human Health," *Antibiotics*

2024, Vol. 13, Page 173, vol. 13, no. 2, p. 173, Feb. 2024. https://doi.org/10.3390/ANTIBIOTICS13020173

[164] J. ming Gao, S. Ma, B. Wang, Z. Ma, Y. Guo, and F. Cheng, "Template-free facile preparation of mesoporous silica from fly ash for shaped composite phase change materials," *Journal of Cleaner Production*, vol. 384, p. 135583, Jan. 2023. https://doi.org/10.1016/J.JCLEPRO.2022.135583

[165] A. Grisolia *et al.*, "Hybrid Polymer-Silica Nanostructured Materials for Environmental Remediation," *Molecules 2023, Vol. 28, Page 5105*, vol. 28, no. 13, p. 5105, Jun. 2023. https://doi.org/10.3390/MOLECULES28135105

[166] R. B. Jundale, J. R. Sonawane, A. V. Palghadmal, H. K. Jaiswal, H. S. Deore, and A. A. Kulkarni, "Scaling-up continuous production of mesoporous silica particles at kg scale: design & operational strategies," *Reaction Chemistry & Engineering*, vol. 9, no. 7, pp. 1914–1923, Jun. 2024. https://doi.org/10.1039/D3RE00707C

[167] A. Maurya, S. Singh, and N. P. Pathak, "The Importance of Mesoporous Materials (Silica, Alumina, and Zeolite) as Solid Supports for Metal Complex Catalysts in Organic Transformations," *Journal of Inorganic and Organometallic Polymers and Materials 2024*, pp. 1–21, Jul. 2024. https://doi.org/10.1007/S10904-024-03249-3

[168] S. Banerjee, A. Das, R. Das, and C. Bhattacharjee, "Mesoporous Mg–Al–Ti composite oxide incorporated mixed matrix ultrafiltration membrane with superior multi-heavy metal removal capacity, anti-fouling & anti-microbial property," *Journal of Membrane Science*, vol. 703, p. 122836, Jun. 2024. https://doi.org/10.1016/J.MEMSCI.2024.122836

[169] Z. Tavakoli, B. Rasekh, F. Yazdian, A. Maghsoudi, M. Soleimani, and J. Mohammadnejad, "One-step separation of the recombinant protein by using the amine-functionalized magnetic mesoporous silica nanoparticles; an efficient and facile approach," *International Journal of Biological Macromolecules*, vol. 135, pp. 600–608, Aug. 2019. https://doi.org/10.1016/J.IJBIOMAC.2019.05.114

 Materials Research Forum LLC
https://doi.org/10.21741/9781644903452-10

Chapter 10

Advanced Mesoporous Materials Based Nanocomposites as High-performance Catalysts towards Electrochemical Sensing and Environmental Application

Balaji Parasuraman[1], Nivetha Venkatesan[1], Shanmugam Paramasivam[2] and Pazhanivel Thangavelu [1*]

[1]Smart Materials Laboratory, Department of Physics, Periyar University, Salem-636011 Tamil Nadu, India

[2]Department of Chemistry, Faculty of Science and Technology, Thammasat University, Pathum, Thani 12120, Thailand

*pazhanit@gmail.com

Abstract

Mesoporous materials have gained significant attention due to their unique properties such as high surface area, tunable pore size, and excellent chemical stability. These materials, when incorporated into nanocomposites, offer remarkable potential for high-performance catalytic applications in fields like electrochemical sensing and environmental applications. This review focuses on the development and utilization of advanced mesoporous material-based nanocomposites as catalysts in these applications, this chapter also covers their role in biosensors for detecting biological molecules and in environmental systems for water and air purification, highlighting their potential for scalable and sustainable solutions to global environmental and health-related issues.

Keywords

Nanotechnology, Mesoporous Materials, Electrochemical Sensing, Adsorption, Environmental Pollution

Contents

Materials Research Foundations 173 (2025) 265-301 https://doi.org/10.21741/9781644903452-10

1. Introduction

For the past two decades, mesoporous carbon materials have drawn interest from the fields of physics, chemistry, engineering and materials sciences due to their distinctive future, physiochemical characteristics, superior chemical and mechanistic stability, and strong electrical conductivity along with the affordable price [1]. Mesoporous carbon structures were originally created at the interstices between carbon particles, such in carbon aerogels, or by widening micropores by oxidation during the activation process, as in the case of activated carbons [2]. The yield of carbon was significantly reduced by the first procedure, and it was challenging to regulate the shape of the second one [3]. Consequently, novel techniques for creating mesoporous carbons have been devised. Accordingly, Ryoo et al. published the first investigation on the production of mesoporous carbon materials in 1999 [4]. They created a mesoporous structure with high specific surface areas and adjustable pore diameters between 2 and 50 nm by using mesoporous silica as a hard template and sucrose as a carbon source [5], for the preparation and potential application of mesoporous carbon materials, this invention has attracted much attention from researchers [6]. Dai and Zhao's groups have made rapid progress in the synthesis of mesoporous carbon materials by using copolymer molecular arrays and carbon precursors [7]. In Fig. 1, displayed the several methods that can be utilized for sensing and environmental technology. The mesoporous nanomaterials have garnered significant interest in various applications, notably in electrochemical sensing[8] and environmental degradation applications, because of some unique properties like, high surface area, tunable pore size, and good electrical conductivity, which make them ideal candidates for enhancing the performance of sensors and improving the degradation efficiency of hazardous pollutants [9].

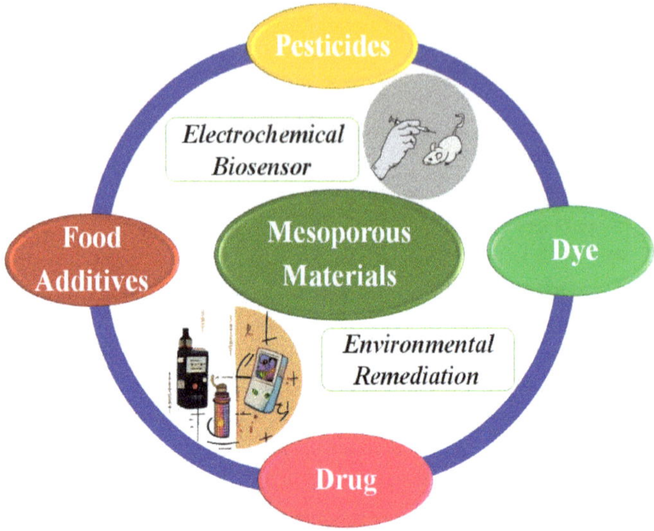

Figure. 1 illustrates the mesoporous materials employed in several of applications.

In particular, the versatility of mesoporous materials allows for the incorporation of some functional agents, which can improve the selective absorption and chemical reaction, in sensing applications [10], these materials can be engineered to detect specific analytes such as food additives, drug molecule, small amount of dye molecule and pesticides detection due to their enhanced absorption capacity and surface activity, whether utilized in gas sensor, biosensor, humidity sensor, optical sensor and electrical devices, mesoporous materials provide the rapid response, good selectivity, repeatability, and superior stability [11]. For example, in their work on MTF-based electrodes, María L. Scala-Benuzzi and colleagues adopt an integrated perspective, drawing insights from material science, physical chemistry, and analytical chemistry to explore their sensing applications. The paper begins with an overview of the processes that produce mesoporous thin films, highlighting how these electrodes are fabricated [12]. It then offers a detailed comparison of the two most commonly used synthesis methods: Electro-Assisted Self-Assembly (EASA) [13] and Evaporation Induced Self-Assembly (EISA) [14]. These techniques enable the creation of uniform thin films with strong adhesion to substrates, leading to modified electrodes that exhibit both mechanical and chemical durability for electrochemical sensing applications [15]. In this context, photocatalytic technology refers to a process that uses light to accelerate a chemical reaction in the presence of photocatalysts, this technology has been used in various applications like, environmental remediation [16], water purification [17], energy production [18] and carbon dioxide reduction [19]. In particular, photocatalytic technology uses light energy and photocatalytic materials to break pollutants in wastewater, it can degrade numerous types of contaminates including, various dyes [20], antibiotics [21], pesticides [22], organic acids [23], crude oil [24]and microbes [24]. This photocatalytic technology has a several

Materials Research Foundations 173 (2025) 265-301 https://doi.org/10.21741/9781644903452-10

advantaged due to working at normal temperatures, being inexpensive, having no secondary waste, being easily assessed, and being readily available [25]. For example, Shivatharsiny Rasalingam et al employed a straightforward, rapid approach using mesoporous siliceous and titanium-loaded materials [26]. The effective adsorption of dye onto a catalyst enhances its photocatalytic degradation, with a higher dye uptake correlating to increased degradation efficiency [27]. Therefore, in addition to studying Si-MCM-48 and Si-MCM-41 materials, a comprehensive investigation into the photocatalytic degradation performance of titanium-based adsorbents like Ti-MCM-48 and Ti-MCM-41 was conducted [28]. Titania-loaded MCM catalysts were synthesized with a relatively high titania content, achieving a Si/Ti ratio of 10. The study found that while surface area, pore volume, and surface charges play a significant role in the adsorption of rhodamine B, the coordination of Ti^{4+} species is crucial for its photocatalytic degradation [29]. Additionally, a detailed examination of the degradation process of rhodamine B dye is presented, utilizing electrospray ionization-mass spectrometry (ESI-MS) analysis [30]. This research offers new insights into how the coordination of Ti^{4+} species influences the photocatalytic breakdown of rhodamine B on periodic mesoporous materials [31]. In this present chapter, we emphasize the ongoing research and innovations in the field, showcasing the versatility and potential of mesoporous materials in various scientific and industrial applications. This comprehensive overview aims to provide readers with a solid understanding of the significance and functionality of mesoporous materials in contemporary research and technology. In this chapter, we will emphasize the importance of various mesoporous materials and their applications in fields such as electrochemical sensing and environmental remediation. The unique benefits of these materials make them reliable and effective for the fabrication of biosensors and the detection of analytes like food additives, drug molecules, and dyes. Additionally, we will discuss their role in the degradation and removal of hazardous organic pollutants, providing a detailed overview of different types of synthetic mesoporous materials.

2. Synthesis methods of mesoporous materials

Mesoporous materials are advanced materials characterized by pores with diameters between 2 and 50 nanometers, offering high surface areas and tunable pore structures. Their unique properties, including large pore volumes and ordered pore arrangements, make them suitable for various applications in catalysis, adsorption, drug delivery, and energy storage. Common types include mesoporous silica, carbon, metal oxides, and polymers, each with distinct advantages for specific industries. These materials are synthesized using techniques like soft-templating, hard-templating, and sol-gel methods, allowing control over pore size and structure, making them essential in both research and industrial sectors.

2.1 Soft-Templating Method

A soft-templating method is a versatile approach used in the preparations of porous materials, including mesoporous structures, which has gained greater attention due to its simplicity, ecofriendly, and adaptability [32]. In this method, a soft template, typically an organic molecule like a surfactant or triblock copolymer, directs the formation of the desired material's structure [33]. The process involves the self-assembly of these organic molecules, which interact with the precursor materials, leading to the formation of a well-ordered meso-structure. Finally, the calcination process is used to remove the template, and thus obtain a mesoporous structure are revealed in Fig.2 . Compared with hard templating, which involves solid sacrificial molds, soft-templating is less complex and allows for better control over pore size, shape, and distribution. Recently, Xinyu et al. reported the synthesis of highly ordered mesoporous carbon through a solvent-free soft template method, facilitated by the self-assembly promoted by $Mg(NO_3)_2$ [34]. Similarly, Ewelina et al. reported the fabrication of mesoporous alumina enriched with lanthanum nanoparticles using a soft template method. The resulting mesoporous materials were utilized for a diclofenac delivery system [35]. The soft-template method offers the flexibility to create various nanostructured materials including silica, carbon, and metal oxides, by modifying the chemical properties of the template and the interaction with the inorganic precursors. This method is particularly advantageous for producing materials with high surface areas, tunable pore diameters, and enhanced performance in applications such as catalysis, energy storage, and environmental remediation.

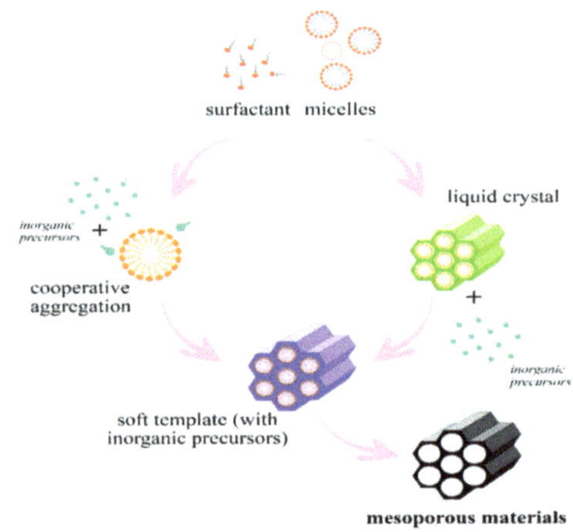

Figure 2, Synthesis Procedure for Mesoporous Materials Using the Soft-Templating Method.

2.2 Hard-Templating Method (Nanocasting)

The hard-templating method, also a well-known method for the fabrication of nano-casting, is a robust technique for synthesizing porous and nanostructured materials with precise control over morphology. In this approach produces a rigid template like silica, zeolites, or other solid frameworks, is helps to construct the desired shape and structure. The template is first impregnated with a precursor material, which fills its pores or voids are depicted in Fig.3. The following chemical processes including polymerization, crystallization, or deposition help to convert the precursor into the final desired product [36]. Once this is achieved, the hard template is removed, typically through chemical etching or dissolution, leaving behind a porous replica of the template. Zhang et al investigated the synthesis of crystalline mesoporous TiO_2 for an efficient photocatalyst for hydrogen evolution reaction. The mesoporous TiO_2 was prepared by a hard template method and obtained a high surface area line of 207 $m^2\ g^{-1}$ [37]. Unlike soft-templating methods, hard-templating offers superior control over the final material's architecture, allowing for the replication of complex nanostructures with high fidelity. This method is widely used in the production of materials such as mesoporous carbons, metal oxides, and metal nanoparticles, which exhibit high surface areas, uniform pore sizes, and tailored functionalities. Based on the above precision, the hard-templating method is highly valued in applications requiring structural integrity and specific surface properties, such as catalysis, drug delivery, and energy storage devices [38]. However, the multi-step process and the need to remove the rigid template can increase production costs and complexity.

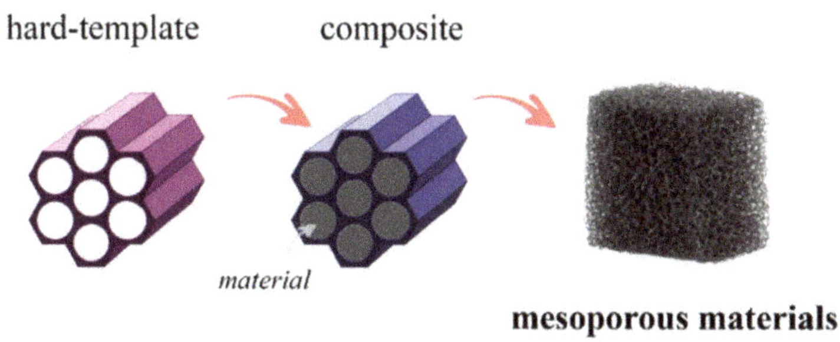

Figure 3. Synthesis Procedure for mesoporous materials via hard-templating (nanocasting) method.

2.3 Sol-Gel Method

The sol-gel method is a commonly employed nanocasting technique for creating various nanostructured mesoporous materials, especially metal oxides, with precise control over their composition and morphology [39]. This process involves a solution (sol) of metal precursors undergoing hydrolysis and polymerization reactions to form a gel-like network, which is subsequently dried and calcined to yield a solid material are shown in Fig.4. The sol-gel method is particularly useful for creating porous materials because the gel network can be manipulated at the molecular level to achieve specific pore sizes and structures. A.K. Dikshit investigated the preparations of mesoporous phase materials via the Sol-gel process using Indian cementitious raw material. This study presents a novel approach to the synthesis of zeolites through a green sol-gel route, with a particular emphasis on significantly reducing the use of organic templates and potentially eliminating solvents [40]. When combined with nanocasting, a hard template including silica or carbon framework-is filled with the sol, then allows to formation of a solid material that mimics the well-structured shape and pore size template [41]. After the formation gelation process, the hard template is removed by thermal process, leaving behind a porous nanomaterial with a highly ordered structure. The sol-gel process offers significant advantages, such as low-temperature synthesis, ease of controlling the chemical composition, and the ability to form complex nanostructures. This method is particularly versatile, enabling the synthesis of materials for a wide range of applications, including catalysis, catalyst support, adsorbent, sensors, and energy storage [40,41]. To alter the chemical composition and reaction conditions to achieve the fine tube the materials properties including porosity, surface area, and functionality. Despite these advantages, challenges remain, particularly in controlling shrinkage and cracking during the drying process, which can affect the material's structural integrity. However, when combined with

Materials Research Foundations 173 (2025) 265-301
https://doi.org/10.21741/9781644903452-10

nanocasting, the sol-gel method remains a powerful tool for fabricating highly ordered porous materials with excellent reproducibility and tailored performance characteristics [42].

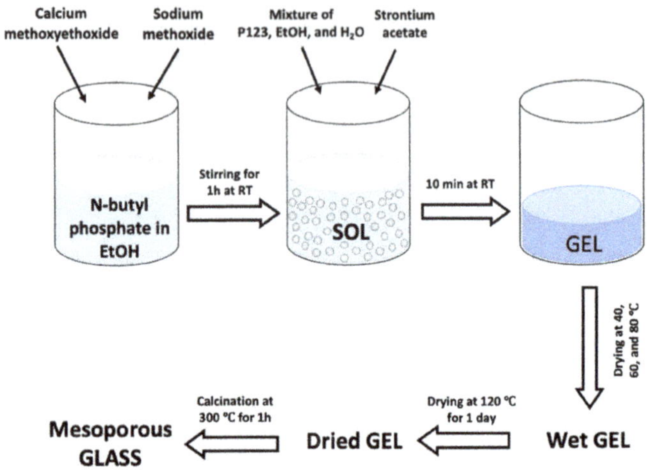

Figure 4. Synthesis Procedure for mesoporous materials through sol-gel method.

2.4 Evaporation-Induced Self-Assembly (EISA)

Evaporation-Induced Self-Assembly (EISA) is a highly effective method for synthesizing mesoporous materials, thus helping to control the pore structure and surface area. From this technique, a solution containing surfactants or block copolymers (as organic templates) and inorganic precursors is cast into a film or solution are depicted in Fig.5 [43]. The concentration of surfactant was increased by solvent evaporation, thus helping to form the self-assembly of micelles into ordered mesophases. These micelles template the inorganic precursors, which condense around them to form a structured material. Finally, the organic template was removed by calcination or solvent extraction method, and then highly ordered mesoporous structured materials were formed. A similar study was reported by Alexandre et al., who prepared nanostructured alumina-based mixed metal oxides with tailored porosity using evaporation-induced self-assembly [44]. EISA is favored for its simplicity, scalability, and flexibility in tuning pore size and geometry by adjusting parameters such as surfactant type, evaporation rate, and precursor concentration. This technique has been widely used to synthesize mesoporous materials such as silica, metal oxides, and hybrid organic-inorganic frameworks. These materials, characterized by their high surface areas and uniform pore structures, are valuable in applications like catalysis, drug delivery, and energy storage [45,46]. Despite the method's advantages, careful control of evaporation conditions is crucial to avoid defects or non-uniformity in the final structure, making it a technique that requires precise optimization.

Materials Research Foundations 173 (2025) 265-301 https://doi.org/10.21741/9781644903452-10

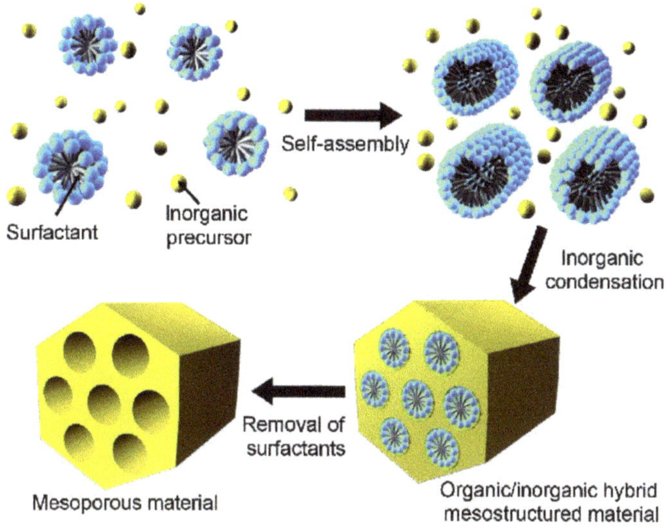

Figure 5. Synthesis Procedure for mesoporous materials through Evaporation-Induced Self-Assembly (EISA).

2.5 Microemulsion Method

The microemulsion method is a versatile approach for synthesizing mesoporous materials, offering control of pore structure, particle size, and morphology [47]. In this process, a microemulsion- a thermodynamically stable mixture of water, oil, and surfactants-forms nanometer-sized water droplets dispersed within the oil phase are depicted in Fig. 6. These droplets/emulsions act as nanoreactors, constraining the inorganic precursors and controlling the nucleation and growth of mesoporous structures [48]. In this case, the surfactant molecules play a dual role by stabilizing the microemulsion and playing a template for the mesoporous network formation. After the precursor material solidifies within the droplets, the surfactants are typically removed via calcination or solvent extraction, leaving behind a porous material. The microemulsion method helps to control pore size and distribution by adjusting the water-to-oil ratio, surfactant concentration, and reaction conditions. A similar study was reported by Muthusami et al., who prepared mesoporous silica nanoparticles with a lychee-like morphology and dual pore arrangement and investigated their application in biomimetic activity through functionalization with a copper (II) complex [49]. The microemulsion method is particularly effective for synthesizing mesoporous materials such as silica, metal oxides, and polymers. The microemulsion method-derived material has a high surface area, uniform pore size and enhanced catalytic or adsorption properties, making them ideal for applications in catalysis, drug delivery, and separation technologies [49–51]. Another important advantage of this technique is its ability to produce mesoporous materials with tunable pore characteristics. However, the need for large amounts of surfactants and the challenge of fully removing them can add complexity to the process.

Materials Research Foundations 173 (2025) 265-301
https://doi.org/10.21741/9781644903452-10

Nonetheless, the microemulsion method remains a highly efficient route for generating mesoporous materials with well-defined structures.

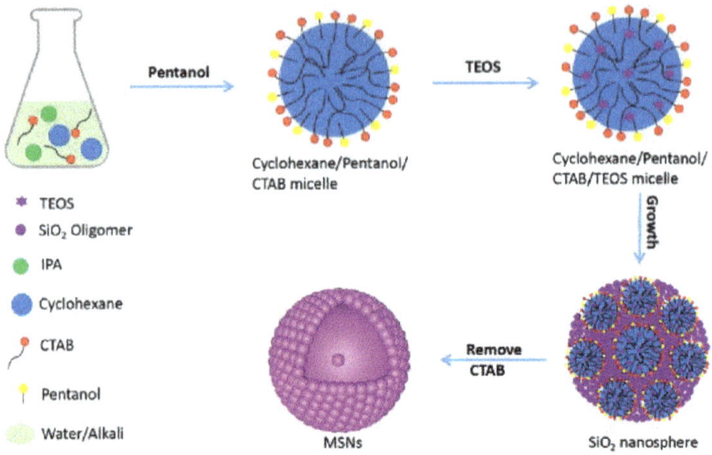

Figure 6. Synthesis Procedure for mesoporous materials via microemulsion method

2.6 Hydrothermal Synthesis

The hydrothermal process is another important method for the fabrication of mesoporous materials and thus produces high crystalline structures with well-controlled porosity [52]. This technique involves the use of an aqueous solution containing inorganic precursors, which is heated in a sealed autoclave under high pressure and temperature conditions. The elevated temperature and pressure promote the dissolution and reprecipitation of the precursors, facilitating the formation of a mesoporous structure. This process allows for the growth of materials with unique morphologies including nanorods, nanowires, nanostars, and nanospheres, without the addition of polymers, surfactants or other additional templates is revealed in Fig.7s. The production of well-defined pore structures and controlled crystal size is one of the major advantages [53,54]. To alter the reaction parameters including temperature, pressure, precursor concentration, and reaction time, can control the pore size and distribution in the mesoporous material. Hydrothermal synthesis is particularly useful for the production of mesoporous materials like metal oxides, zeolites, and carbon-based materials, which are highly valued in applications such as catalysis, energy storage, and environmental remediation [50,55]. Yu et al., reported the synthesis of mesoporous silica spheres using a hydrothermal process and obtained the silica spheres were utilized for the effect of the cooling process [56]. The hydrothermal process is also relatively eco-friendly, as it operates under milder conditions compared to other methods, and the use of water as a solvent reduces the need for hazardous chemicals. However, the requirement for specialized autoclave equipment and the need to control high-pressure conditions can add complexity to the procedure. Nonetheless, the hydrothermal process remains a key technique for synthesizing mesoporous materials with superior crystallinity, tunable pore sizes, and diverse structural morphologies.

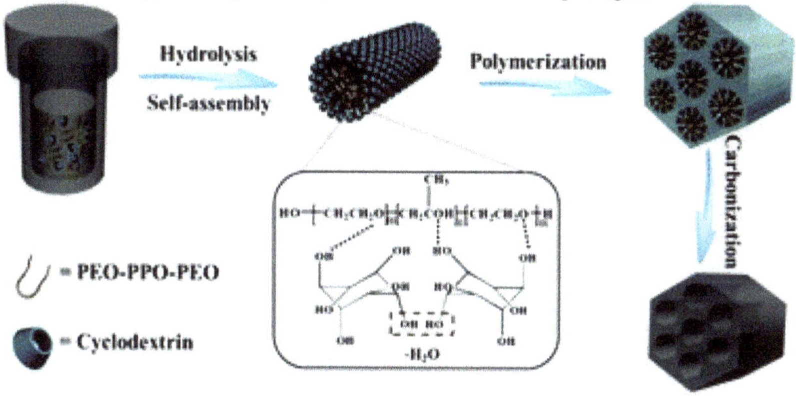

Figure 7. Synthesis Procedure for mesoporous materials via hydrothermal method.

3. Applications of mesoporous materials

A fascinating and quickly developing area of study is the use of mesoporous materials in photocatalytic degradation and electrochemical sensing. This field has a lot of promise since it combines state-of-the-art analytical tools with environmental sustainability concepts. In particular, mesoporous materials are essential for addressing environmental issues because of their capacity to effectively break down hazardous compounds like pesticides, drugs, and organic dyes when exposed to light. This makes them useful for water treatment and pollution management. Mesoporous materials aid in the decomposition of harmful pollutants, improving ecosystems and surrounding conditions. Additionally, in the realm of electrochemical sensing, mesoporous materials enhance the sensitivity and selectivity of sensors used for detecting a wide range of substances, including heavy metals, gases, and biological markers. Their high surface area allows for increased adsorption of analytes, which improves detection limits and response times [57]. This capability is particularly important for applications in environmental monitoring, health diagnostics, and safety assessments due to the tunable pore sizes and high surface areas, facilitating the optimization of both photocatalytic and sensing processes. For instance, the porous nature enables better access for reactants in photocatalytic reactions, while also promoting rapid electron transfer in electrochemical sensors. Eventfully, the use of mesoporous materials in these applications exemplifies the convergence of scientific innovation with practical solutions for environmental and analytical challenges, paving the way for advancements that support sustainable development and improved public health [58].

3.1 Photocatalytic technology

Photocatalytic technology utilizes light-activated catalysts to facilitate chemical processes that degrade pollutants or generate energy. Mesoporous materials are particularly well-suited for this application due to their unique structure, high surface area, and effective light absorption properties. In this context, photocatalytic technology can be classified based on its use with three types of model pollutants as

- Dye Degradation
- Drug Degradation
- Pesticides Degradation

3.1.1 Dye Degradation

The increasing use of synthetic dyes in various industries has led to significant environmental pollution, particularly in water bodies. Traditional wastewater treatment methods often fail to efficiently remove these contaminants, necessitating the development of advanced technologies. Photocatalytic dye degradation using mesoporous materials has emerged as a promising solution due to their unique structural properties and effectiveness in breaking down pollutants under light irradiation. The mesoporous materials area synthesized by their approximate pore size ranging from 2 to 50 nm, thus providing a high surface area that facilitates the absorption of dye molecules commonly the mesoporous materials include silica-based frame works, metal-organic frameworks, carbon based mesoporous structures [59]. The mesoporous materials have a high surface area, light absorption, charge carrier separation and twisted pore size, thus properties can be modified to absorb a broader spectrum of light, including visible light, thereby increasing the potential for enhanced photocatalytic activity. In additionally, the several kinds of mesoporous materials used for dye degradation applications, like silica, carbon and MOF-based nanomaterials to possess specific catalytic properties and have shown promise in dye degradation applications due to their large surface area and tunable porosity, these mesoporous materials can efficiency degrade the wide range of dye molecules, including azo, anthraquinone and methylene blue dyes making them suitable for treating complex wastewater are revealed in Fig. 8.

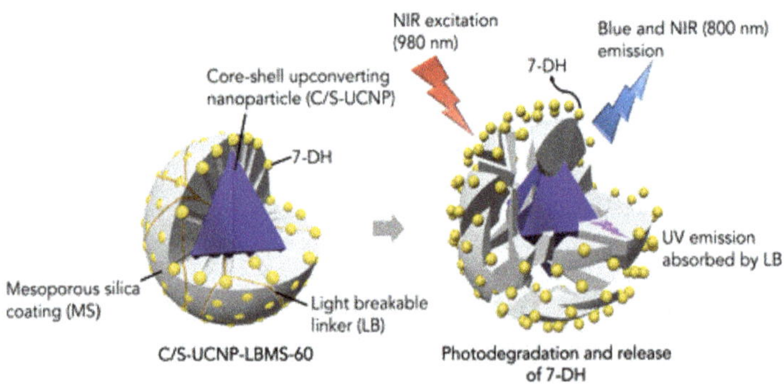

Figure 8. Shows the photocatalytic dye degradation for using mesoporous materials.

For example, Binita Nanda et al was synthesized MnO_2-modified MCM 41 by in situ synthesized method. This material was adopted for dye degradation of different dye molecules Rd 6G, MB, MG, and Rh B, the meso- MnO_2-MCM-41 nanocomposites demonstrated complete photocatalytic

Materials Research Foundations 173 (2025) 265-301
https://doi.org/10.21741/9781644903452-10

degradation and adsorption of a 100-ppm dye solution in 60-90 minutes, respectively [60]. **Table 1** shows the previous studies have reported the use of various basic mesoporous nanomaterials in the degradation of dyes. Additionally, Mian Zahid Hussain et al. reported the synthesis of ZnO nanoparticles embedded in porous carbon using a simple carbonization method. The resulting ZnO/C composite was characterized using various physicochemical techniques, including XRD, SEM, TEM, and XPS analysis, confirming the carbonization of MOF-5 in water. This ZnO/C composite was then utilized for MB dye adsorption and photocatalytic degradation, demonstrating excellent degradation efficiency under visible light irradiation [61].

***Table 1** shows the comparison table of the mesoporous materials used for dye degradation.*

S. No	Name of the Photocatalyst	Module Pollutant	Light Source	Degradation Time	Degradation Efficiency (%)	Reference
1	V_2O_5@g-C_3N_4	Rh B	UV light	60 min	93.4	[62]
2	WO_3, SiO_2-WO_3	MB	72 W, under visible light	300 min	87.5	[63]
3	$CoFe_2O_4$-gC_3N_4	MB, MO and CR	150 W Xenon lamp	30 min	91.2	[64]
4	MnO_2 modified MCM-41	Rh-6G, MB, MG and Rh B	visible light	60 min	100	[65]
5	Pt/WO_3-GO	MB	visible light	70 min	94	[66]
6	WO_3-GO	MB	visible light	70 min	90	[66]
7	TiO_2/ZSM-5	MO	Solar irradiation	180 min	95.16	[67]
8	g-C_3N_4/MIL-125(Ti)	Rh B	visible light		70	[68]

3.1.2 Drug Degradation

The use of pharmaceutical compounds loses their potency and effectiveness over time in water bodies, the wastewater treatment facilities aim to degrade the pharmaceutical compounds before the water is released back into the environment. This is crucial for protecting water bodies from contamination [69]. In this regard, the mesoporous materials used to degrade in water can lead to the development of more environmentally friendly pharmaceutical drugs that break down naturally mesoporous materials are very desirable drug repositories because of their physiochemical characteristics which include large pore volume, pore diameter, adjustable pore size, high surface area and ability to modify the surface utilizing a verity of chemical functional [70]. For instance, Mishal W. Ibrahim et al. (2021) reported on a ZnO photocatalyst supported by mesoporous silica material, synthesized through a mechanochemical approach [71].

Materials Research Forum LLC
https://doi.org/10.21741/9781644903452-10

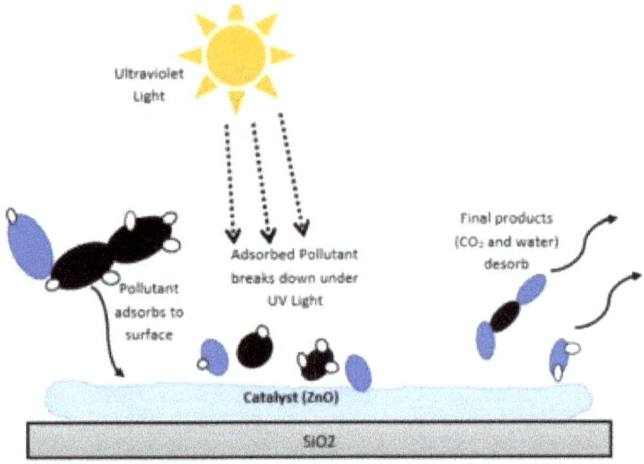

Figure 9. Shows the photocatalytic drug degradation for the using mesoporous materials.

This MS/ZnO photocatalyst demonstrated an impressive degradation efficiency of 83.6% for ibuprofen (IBP) under UV-Visible light irradiation over 4 hours are displayed in Fig. 9. This study highlights the potential of mesoporous materials in enhancing the efficiency of photocatalytic drug degradation. T**able 2** shows the previous studies have reported the use of various basic mesoporous nanomaterials in the degradation of drug molecules.

***Table 2** shows the comparison table of the mesoporous materials used for drug degradation.*

S. No	Name of the Photocatalyst	Module Pollutant	Light Source	Degradation Time	Degradation Efficiency (%)	Reference
1	Mesoporous TiO_2	Ibuprofen	UV light	60 min	95	[72]
2	Mesoporous TiO_2	Paracetamol	UV light	50 min	90	[73]
3	Mesoporous SiO_2	Ciprofloxacin	UV light	70 min	85	[74]
4	SBA/TiO_2	Tetracycline	UV light	60 min	89	[72,75]
5	Mesoporous HfO_2	Erythromycin	UV - Visible light	45 min	92	[72]

3.1.3 Pesticides Degradation

Pesticide degradation is a process by which a chemical reaction called hydrolysis, photolysis and biodegradation converts a pesticide into simpler components like water, carbon-di-oxide, an ammonia, the chemical substance called pesticides are frequently employed in agricultural goods to keep pests under control crops from spoiling [76]. However, there will always be more pesticide residue in foods when pesticides are applied widely. Therefore, it is essential to find efficient ways to reduce pesticide residue on fresh products to avoid detrimental effects on human health. In recent times, many researchers studied on the pesticide degradation by using mesoporous materials under photocatalytic technology [77]. For example, Weiyao Yan et al synthesized mesoporous cooper sulfidic nanocapsule by electrocoating method, the PPS@HmCuS-CTs NPs assessed by Curcumin and physcion pesticides degradation under visible light irradiation. The obtained results demonstrate that PPS@HmCuS-CTs NPs efficiency was more than 1.47 times those of pure materials are revealed in Fig. 10. Therefore, the mesoporous-based nanomaterials have tremendous features in the pesticide degradation system. In **table 3** shows the previous studies have reported the use of various basic mesoporous nanomaterials in the degradation of pesticides. [78]

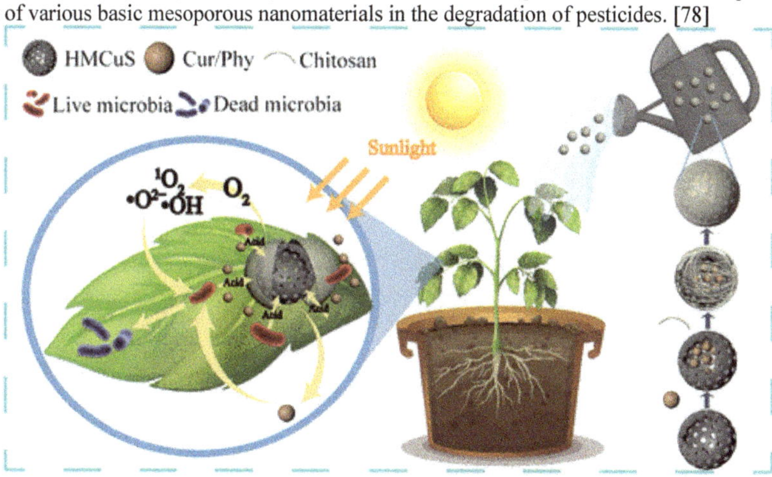

Figure 10. Shows the degradation of the photocatalytic pesticide for using mesoporous materials.

Materials Research Foundations 173 (2025) 265-301 https://doi.org/10.21741/9781644903452-10

Table 3 *shows the comparison table of the mesoporous materials used for pesticide degradation.*

S. No	Name of the Photocatalyst	Module Pollutant	Light Source	Degradation Time	Degradation Efficiency (%)	Ref.
1	SBA-15-supported TiO_2	Isoproturon	solar light	1 hr	100	[79]
2	TiO_2-decorated SBA-15	Amicarbazone	UV	2 hr	100	[80]
3	Co_3O_4/MCM-41 nanocomposite	Omethoate	Visible light	2 hr	100	[81]
4	TiO_2-decorated SBA-15	Dimethoate	Simulated solar light	7 hr	100	[82]
5	Fe_3O_4 and WS_2 nanoparticles immobilized /inKIT-6	Chlorpyrifos	Visible light	0.9 hr	92.1	[83]
6	Co_3O_4/MCM-41 nanocomposite	Acephate	visible light	1 hr	100	[84]
7	Zinc (II)phthalocyanines immobilizedin Al-MCM-41	Fenamiphos pentachloropheno l	Polychromatic light	3h	70- 80	[85]
8	MCM-41 immobilized halogenated porphyrins	Fenamiphos	polychromatic light	5h	60	[86]
9	Phosphotungstic acid supported on MCM-41	Imidacloprid	monochromati c light (365nm)	5h	58	[87]

3.2 Electrochemical sensing technology

Electrochemical sensor technology utilizing mesoporous materials has garnered significant attention in recent times. In mesoporous materials have a significant pore size and high surface area and tunable pore size for the absorption of some analytes, enhancing their sensitivity [88]. The mesoporous structure facilitates the faster diffusion rates of analyte to the enhance the sensitivity interface due to improve the response and recovery times and also mesoporous materials can be easily modified in the various functional groups to enhance the selectivity for specific analyte, In additionally the mesoporous materials can immobilize biomolecule for detecting biomolecules (or) biological analyte, they can be used to detect pollutant such as heavy metals, Pesticides, dye, food additives and drug molecules in water and Food samples [89]. Cheerfully, the mesoporous materials are revolutions performance through their unique structural characteristics, continued research and development will likely lead to innovative applications across various fields from health care to environmental monitoring. In this context, the electrochemical sensor can be classified based on its use with five model pollutants such as

Materials Research Foundations 173 (2025) 265-301
https://doi.org/10.21741/9781644903452-10

- Dye detection
- Drug detection
- Food additives detection
- Heavy metals detection
- Pesticides detection.

For example, Maria L. Scala-Benuzzi, et al reviewed on the entitled on ordered mesoporous electrodes for sensing applications, mesoporous thin film (MTF) has great attention in the sensing sector because of the signal modulation, molecular sieving, and steric excursion also. It is central to enhancing the reproducible rate for sensitive, selective, reproducible and stable. Furthermore, it is necessary to understand the complex mass and charge transport process that takes place through the film (mesopore, pore surface, and interface). In this group, reviewed and summarized the current application in electrochemical sensing and electrochemical sensing applications, MTF-based mesoporous materials have a significant role in electrochemical sensing applications [90].

3.2.1 Dye detection

Electrochemical sensors have proven to be highly effective in detecting dyes across various environments due to their ability to measure redox reactions involving dye molecules [91]. This effectiveness is significantly enhanced by the use of mesoporous materials. These materials have unique structural properties, such as high surface area and tunable pore size, which greatly improve the sensitivity and selectivity of electrochemical sensors. The high surface area and tunable pore size of mesoporous materials facilitate the adsorption and diffusion of dye molecules, allowing for better interaction with the sensor's active sites. For example, mesoporous silica nanoparticles can encapsulate dyes, enabling the specific detection of various analytes, including metal ions and organic pollutants, with low detection limits [92]. Additionally, the integration of enzymes with mesoporous materials further enhances the performance of biosensors, making them more efficient in detecting target substances. Aisha Ganesh et al. synthesized copper nanoparticles (CuNPs) using an electrodeposition method with cyclic voltammetry. The resulting CuO nanoparticles were coated onto a Pt electrode, and the materials were characterized using various techniques. The modified Pt/CuO nanoparticles electrode was tested with Congo Red (CR) dye, showing an impressive recovery and response time in wastewater samples, ranging from 95% to 104%. This modified electrode demonstrated excellent recyclability and reusability is displayed [93] in Fig. 11. This study highlights the potential of mesoporous materials in enhancing the efficiency of photocatalytic dye molecule detection. In **table 4** shows the previous studies have reported the use of various basic mesoporous nanomaterials in the detection of dye molecules.

Materials Research Foundations 173 (2025) 265-301 https://doi.org/10.21741/9781644903452-10

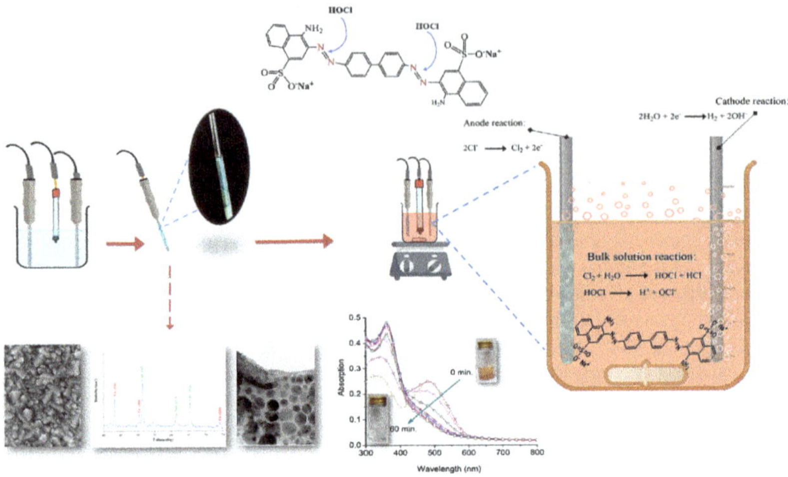

Figure 11. Shows the electrochemical dye detection for using mesoporous materials.

***Table 4** shows the comparison table of the mesoporous materials used for dye detection.*

S. No	Name of the Electrocatalysts	Module Pollutant (dye)	Wide Linear Range	LOD	Reference
1	Ag-doped SBA-16	Hair Dying Cream	20 µM to 8 mM	8 to 20 mM	[94]
2	NH_2-SBA15/CPE	Catechol	1.0 – 140 µmol L^{-1}	0.5 µmol L^{-1}	[95]
3	NH_2-SBA15/CPE	Resorcinol	2.0– 160 µmol L^{-1}	0.8 µmol L^{-1}	[95]
4	NH_2-SBA15/CPE	Hydroquinone	0.8– 160 µmol L−1	0.3 µmol L^{-1}	[95]
5	AuNps/GCE	Resorcinol	0.5 to 21.0 µM	.15 µM	[96]
6	MSN@MOF	Cholesterol	0–10 µg/mL	0.89 nM	[97]
7	ZIF-8/g-C_3N_4/Co	D&C Red 33 and Patent Blue V	0.08–10 µM	0.034 µM	[98]
8	PdNPs@SBA-15-PrEn/ GCE	Methanol	20–1000 µM	12 µM	[99]
9	NH_2-SBA15/CPE	1-naphthol and 2-naphthol	1.0×10^{-8}–1.0×10^{-6} M	1.0×10^{-9} M ($S/N = 3$)	[100]
10	MB-SO_3H-MSM	Riboflavin	10.0 nM to 50 µM	5.0 nM	[101]

3.2.2 Drug detection

Electrochemical sensors have garnered considerable interest recently due to their capability to detect drugs within complex biological matrices. The incorporation of mesoporous materials into these sensors significantly enhances their sensitivity and selectivity, making them effective for monitoring various pharmaceutical substances. The unique mesoporous structure promotes efficient electron transfer and provides sample active sites for drug molecules to interact, resulting in improved detection limits and faster response times [102]. These sensors are particularly beneficial in clinical environments for monitoring drug concentrations in biological fluids, which ensures accurate therapeutic dosing and timely identification of potential overdoses. Overall, the integration of mesoporous materials into electrochemical sensors represents a promising strategy to enhance drug detection capabilities, offering improved sensitivity, specificity, and anti-fouling properties essential for analyzing complex biological samples [103]. For example, Hanaa El-Desoky et al was mesoporous silica (SBA-15) prepared by simple separation and catalysis process a hard template method revealed in Fig. 12. The synthesized SBA-15 was analyzed using XRD, SEM, EDX, TEM, and nitrogen sorption isotherm (BET) analysis. The silica SBA-15/CPE sample was evaluated using electrochemical impedance, cyclic voltammetry, and stripping voltammetry techniques [104]. This fabricated sensor exhibited also superior electrocatalytic activity towards dapoxetine hydrochloride drug, the proposed sensor limit of detection (LOD) value of 2.45x 10^{-10} M and a wide linear range (9 x 10^{-10} -1x10^{-8} M). The results were compared with the previously reported electrochemical sensor for the determination of a variety of mesoporous materials in various drug molecules (Table 5).

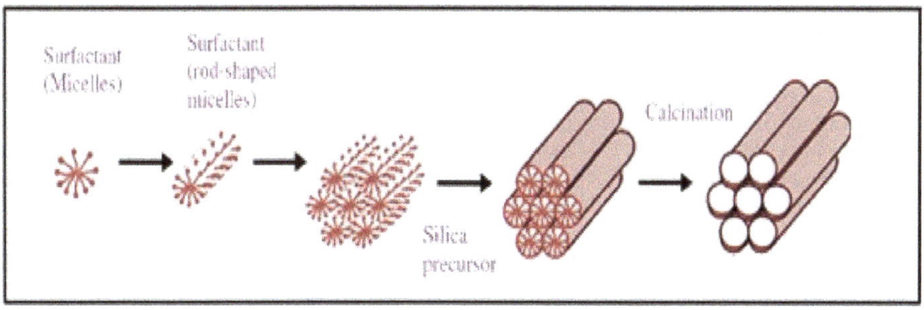

Figure.12. shows the electrochemical sensing platform of drug detection for using mesoporous materials.

Table 5. *shows the comparison table of the mesoporous materials used for drug detection.*

S. No	Name of the electrocatalysts	Module Pollutant (drug)	Linear Range	LOD	Reference
1	g-C3N4/N-CNO /gold	Oxycodone	0.05–150 μM	16 nM	[105]
2	Nafion-SWNT	Dopamine	0.02 μM to 6.0 μM 5.00 nM	0.02 μM to 6.0 μM 5.00 nM	[106]
3	MWCNTs	Paracetamol	0.01 μM to 20 μM 10 nM	0.01 μM to 20 μM 10 nM	[107]
4	SWCNTs	Acetaminophen	5 nM to 80 μM	4.3 nM	[108]
5	MWCNTs	Methimazole	0.074 μM to 63.5 μM	0.056 μM	[109]
6	ERG/Ni_2O_3-NiO	acetaminophen	0.04 μM to 100 μM	0.02 μM	[110]
7	MWCNT/GO	Dopamine	0.2 μM to 400 μM	22 nM	[111]
9	CuNPs/graphene	Glucose	0.5 μM to 4.5 mM	0.5 μM	[112]

3.2.3 Food additives detection

Mesoporous materials are highly efficient in electrochemical sensors for detecting food additives, to their large surface area, adjustable pore sizes, and excellent conductivity. These characteristics significantly enhance the sensitivity and reproducibility of the sensors [113]. For example, mesoporous carbon materials have been utilized to detect additives such as nitrites and sulfites in food products. The mesoporous structure promotes efficient electron transfer and offers numerous active sites for the interaction with food additive molecules, resulting in improved detection performance, such sensors play a vital role in ensuring food safety by monitoring and regulating the levels of various additives in food products are revealed in Fig. 13. The integration of mesoporous materials into electrochemical sensors substantially enhances their capability to detect food additives by improving sensitivity, selectivity, and response times, consequently playing a crucial role in maintaining food safety. For instance, Sancz Hamzeh et al were praseodymium oxide nanoparticles contractual by using a simple sonochemical method. The Pr_6O_{11} nanoparticles have been analyzed by different kinds of physiochemical techniques [113]. Furthermore, this Pr_6O_{11} nanoparticle was adopted to measure the different electrochemical techniques like chronoamperometry cyclic voltammetry, and differential pulse voltammetry has been used to determine the electrochemical behavior of carmoisine.

As a result, the carmoisine a low detection limit of 12nm in a linear range of 0.09-135 μm was achieved by this proposed sensor, consequently, the Pr_6O_{11} sample has good stability significant selectivity, and enhanced reproducibility [114]. Finally, the mesoporous-based nanomaterials are widely used in the food additive detection for the colorant sector. T**able 6** shows the previous studies have reported the use of various basic mesoporous nanomaterials in the degradation of pesticides.

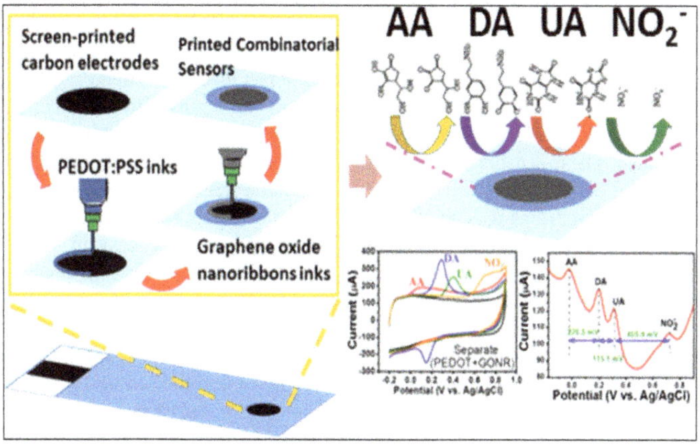

Figure.13. shows the electrochemical sensing platform of food additive detection for using mesoporous materials.

Table 6. *shows the comparison table of the mesoporous materials used for food additive detection.*

S. No	Name of the electrocatalysts	Module Pollutant (Food Additives)	Linear Range	LOD	Reference
1	rGO-AuNPs	Ascorbic acid, uric acid,	20~150, 10~500 μM	9.4, 10.9 μM	[115]
2	MAC-Co-SPE	Sudan I	0.3–400.0 μM	9.0 μM	[116]
3	PB-rGO/TCD/AuNPs	Quercetin	0.005~0.4 μM	1.83 μM	[117]
4	Pd–Au/PEDOT/rGO	Caffeic acid	0.001~55 μM	0.00037 μM	[118]
5	FeO_3 @MIL-100(Fe)	Chlorogenic acid	0.1~460 μM	0.05 μM	[119]
6	PMB@Ni–TPA	Cafeic acid	0.25~15.0 μM	0.2 μM	[120]
7	ZrO_2/CoO3 4/rGO	Gallic acid, cafeic acid, protocatechuic acid	0.00624~0.47768, 0.00248~0.52490, 0.00540~0.42496 μM	0.00156, 0.00062, 0.00135 μM	[121]
8	MMIP-FeO_3 @SiO_4 @NH	Rosmarinic acid	0.1~500 μM	0.085 μM	[122]

Materials Research Foundations 173 (2025) 265-301 https://doi.org/10.21741/9781644903452-10

3.2.4 Heavy metals detection

Mesoporous materials are highly effective in electrochemical sensors for detecting heavy metals due to their extensive surface area, adjustable pore sizes, and superior conductivity, which collectively enhance the sensors sensitivity and selectivity. For instance, mesoporous carbon aerogels doped with bismuth oxide have been utilized to detect lead (Pb^{2+}) and cadmium (Cd^{2+}) at picomolar concentrations [123]. The mesoporous structure facilitates efficient electron transfer and provides numerous active sites for heavy metal ions to interact, resulting in exceptional detection performance. These sensors are crucial for environmental monitoring and ensuring water safety by identifying trace levels of heavy metals in water, mesoporous materials into electrochemical sensors significantly enhances their ability to detect heavy metals by improving sensitivity, selectivity, and response times, thereby playing a vital role in environmental monitoring and safety. For instance, Sufang He and colleagues synthesized a functionalized highly ordered mesoporous material, Al 10 -SBA-15, which demonstrates exceptional efficacy as an adsorbent for removing Arsenic(V) from water. This material serves as an active adsorption site for as (V) across a wide pH range of 2.0 to 8.2, even in solutions with low concentrations of As (V)[124]. The maximum adsorption capacity for As (V) was determined to be 197.7 mg/g, while the order of effectiveness for coexisting anions in the removal of As(V) under similar conditions was found to be $NO^{3\ -}$ & lt; F^- & lt; PO_4 $^{3-}$. Consequently, Al10-SBA-15 is considered a highly effective adsorbent for treating arsenate-contaminated water are displayed in Fig.s 14. In a **table,** shows that previous studies that have reported the use of various basic mesoporous nanomaterials in the degradation of pesticides.

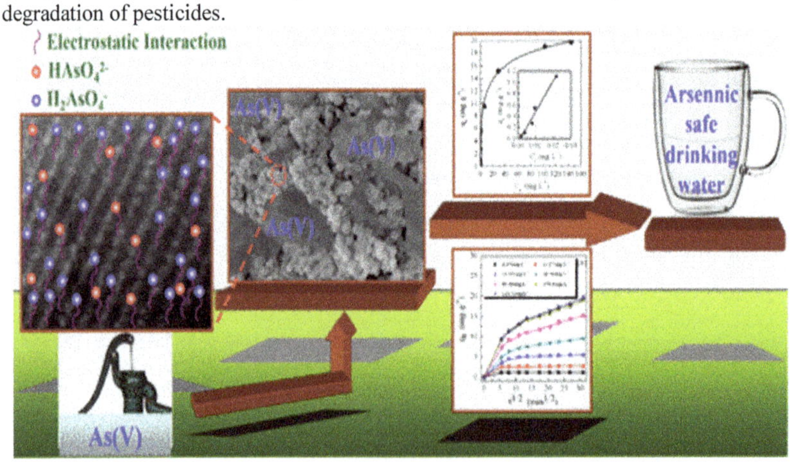

Figure.14. shows the electrochemical sensing platform of heavy metal detection for the using mesoporous materials.

Materials Research Foundations 173 (2025) 265-301
https://doi.org/10.21741/9781644903452-10

Table 7. *shows the comparison table of the mesoporous materials using for dye detection.*

S. No	Name of the electrocatalysts	Model Heavy Metals	Detection Limit	Response Time	Reference
1	Mesoporous Silica	Hg (II), Pb (II)	0.1 nM	5 s	[125]
2	Mesoporous Carbon	Cd (II), Pb (II)	0.05 ppb	3 s	[126]
3	Z-8@GO	Pb^{2+}	1.076 μM	~10 s	[127]
4	Mesoporous Silica	Hg^{2+}, Cu^{2+}, Zn^{2+}	Varies	15 s	[128]
5	MXenes	Various HMIs	Varies	5 s	[129]
6	Graphene Oxide	Pb^{2+}, Cd^{2+}	Varies	8s	[130]
7	ZnO Nanoparticles	Cd^{2+}	0.2 μM	6s	[131]

3.2.5 Pesticides detection

Electrochemical sensors have become a valuable tool for detecting pesticides in complex matrices, to their high sensitivity, selectivity, and rapid response times. The incorporation of mesoporous materials into these sensors has greatly improved their effectiveness in pesticide analysis because of their extensive surface area, adjustable pore sizes, and superior conductivity, which collectively boost the sensor's sensitivity and selectivity[132]. The mesoporous structure enhances electron transfer efficiency and offers abundant active sites for pesticide molecules to interact, resulting in superior detection performance. These sensors are essential for environmental monitoring and food safety, as they detect pesticide residues in water and agricultural products are depicted in Fig.15. The integration of mesoporous materials into electrochemical sensors offers a promising strategy to enhance pesticide detection capabilities, providing improved sensitivity, selectivity, and anti-fouling properties that are crucial for analyzing complex environmental samples. For example, Vinitha Mariyappan et al synthesized Gadolinium Sulfide Rod-Embedded RGO for Electrochemical Sensor Based Sensing of Carbofuran pesticides. Gd_2S_3/RGO composite was prepared by the facile hydrothermal method. Electrochemical studies revealed that the Gd_2S_3/RGO sensor demonstrated high sensitivity and selectivity for CF detection. Furthermore, this sensor exhibited a low detection limit (LOD) and linear ranges of 0.0128 and 0.001−1381 μM for CF under optimized experimental conditions. Additionally, we explored the practical applicability of the sensor for detecting CF in environmental and food samples. The potential of the Gd_2S_3/RGO composite as an electrochemical sensor, opens up opportunities for its application in diverse areas such as pesticide monitoring and environmental protection. Its exceptional sensitivity, selectivity, and ability to detect at low concentrations make it a promising asset in analytical chemistry research.

Materials Research Foundations 173 (2025) 265-301
https://doi.org/10.21741/9781644903452-10

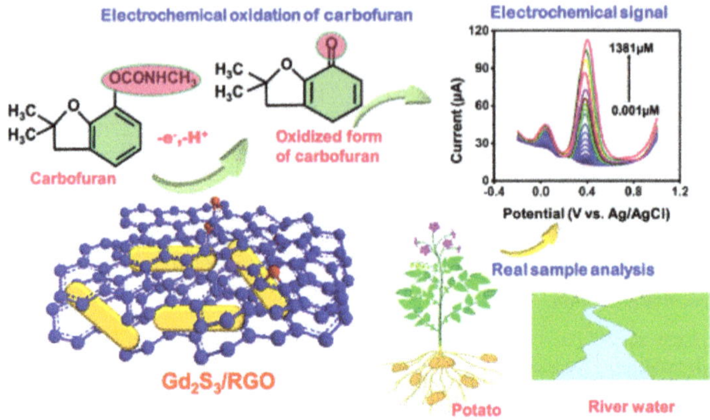

Figure.15. shows the electrochemical sensing platform of heavy metal detection for using mesoporous materials.

Table 8. *shows the comparison table of the mesoporous materials used for dye detection.*

S. No	Name of the electrocatalysts	Module Pollutant (Pesticides)	Linear Range	LOD	Reference
1	CuOx@mC/GCE	Glyphosate	1.0 × 10-15 to 1.0 × 10-4 mol L-1	7.69 × 10-16 mol L-1	[133]
2	Ag@Meso-C/Hematite Ore	Imidacloprid	63 – 870 μM	0.257 μM	[134]
3	Pd/ZnO/APC	carbaryl	0.01–5.0 μM	0.01 μM	[135]
4	$Co_1Zn_{0.5}$-P/CoO/ZnO	carbofuran (CF) and diuron (DU)	2–429 μM and 2–149 μM	2.98 and 2.52 nM	[136]
5	Fe-ZnO NPs	Endrin	0.065 μM	0.019 μM	[137]
6	CuO/ZnO	organophosphorus	-	1.367 nM	[138]
7	$TiSi_4$Pic+Cl–	pentachlorophenol	0.99 and 4.21 μmol L−1,	29 nmol L-1	[139]
8	AChE–CdS–G–CHIT	Carbaryl	2 ng/mL–2 g/mL	0.7 ng/mL	[140]
9	NF/AChE-CS/PtNPs-CGR-NF	Methyl parathion	10−13–10−10 M	5× 10−14 M	[[141]
10	CLDH-AChE/GN-AuNPs	Chlorpyrifos	0.05–150 μg/L	0.05 μg/L	[142]

Materials Research Foundations 173 (2025) 265-301 https://doi.org/10.21741/9781644903452-10

Conclusion

Mesoporous materials-based nanocomposites have emerged as highly effective catalysts for electrochemical sensing and environmental applications due to their unique structural and functional properties. The integration of mesoporous materials with nanoparticles and other active components in nanocomposites enhances the overall performance, providing superior catalytic activity, selectivity, and stability. These nanocomposites offer large surface areas, tunable pore sizes, and enhanced mass transfer, which are critical for efficient catalytic reactions. Their ability to be functionalized further broadens their applicability in detecting biological and chemical species with high sensitivity and in facilitating environmental remediation processes. The advancement of mesoporous materials-based nanocomposites represents a significant step towards developing next-generation catalysts that are not only high-performing but also versatile in a variety of challenging applications.

Feature Outlook

The future development of mesoporous materials-based nanocomposites as high-performance catalysts is expected to focus on several key areas:

- Continued research will aim at optimizing the pore structure and surface properties of mesoporous materials to maximize catalytic efficiency. This includes the development of hierarchical pore systems and the precise control of pore size to improve reactant accessibility and diffusion.
- Functionalization of mesoporous materials with novel active sites, such as metal nanoparticles, enzymes, or other catalytic species, will be a major area of focus. This will enable the creation of highly specialized catalysts tailored for specific electrochemical sensing and environmental applications.
- There will be an increased emphasis on sustainable synthesis methods for mesoporous nanocomposites, including green chemistry approaches that minimize environmental impact. The use of renewable resources and non-toxic reagents in the synthesis process will be prioritized.
- Mesoporous nanocomposites are expected to be integrated with emerging technologies such as microfluidics, lab-on-a-chip devices, and smart sensor systems. This integration will enhance their application in real-time electrochemical sensing and environmental monitoring.
- Efforts will be made to scale up the production of mesoporous materials-based nanocomposites, making them more accessible for commercial applications. This includes developing cost-effective synthesis techniques and improving the reproducibility of the nanocomposite properties.
- The versatility of mesoporous materials-based nanocomposites will be further explored to develop multi-functional catalysts capable of performing a range of tasks, such as simultaneous detection of multiple analytes or combined catalytic degradation of various environmental pollutants.
- The outlook for mesoporous materials-based nanocomposites is highly promising, with the potential to significantly impact electrochemical sensing and environmental technologies.

Continued innovation in this field will likely lead to the development of even more sophisticated and effective catalytic systems.

References

[1] S. Mehdipour-Ataei, E. Aram, Mesoporous Carbon-Based Materials: A Review of Synthesis, Modification, and Applications, Catalysts 13 (2022) 2. https://doi.org/10.3390/catal13010002

[2] N.R. Khalili, M. Campbell, G. Sandi, J. Golaś, Production of micro- and mesoporous activated carbon from paper mill sludge, Carbon N Y 38 (2000) 1905-1915. https://doi.org/10.1016/S0008-6223(00)00043-9

[3] X. Zhang, S.-X. Guo, K.A. Gandionco, A.M. Bond, J. Zhang, Electrocatalytic carbon dioxide reduction: from fundamental principles to catalyst design, Mater Today Adv 7 (2020) 100074. https://doi.org/10.1016/j.mtadv.2020.100074

[4] Y. Zhang, L. Han, L.-L. Hu, Y.-Q. Chang, R.-H. He, M.-L. Chen, Y. Shu, J.-H. Wang, Mesoporous carbon nanoparticles capped with polyacrylic acid as drug carrier for bi-trigger continuous drug release, J Mater Chem B 4 (2016) 5178-5184. https://doi.org/10.1039/C6TB00987E

[5] N. Carlsson, H. Gustafsson, C. Thörn, L. Olsson, K. Holmberg, B. Åkerman, Enzymes immobilized in mesoporous silica: A physical-chemical perspective, Adv Colloid Interface Sci 205 (2014) 339-360. https://doi.org/10.1016/j.cis.2013.08.010

[6] L. Cui, X. Ren, M. Sun, H. Liu, L. Xia, Carbon Dots: Synthesis, Properties and Applications, Nanomaterials 11 (2021) 3419. https://doi.org/10.3390/nano11123419

[7] C. Liang, S. Dai, Synthesis of Mesoporous Carbon Materials via Enhanced Hydrogen-Bonding Interaction, J Am Chem Soc 128 (2006) 5316-5317. https://doi.org/10.1021/ja060242k

[8] P. Xie, Z. Liu, S. Huang, J. Chen, Y. Yan, N. Li, M. Zhang, M. Jin, L. Shui, A sensitive electrochemical sensor based on wrinkled mesoporous carbon nanomaterials for rapid and reliable assay of 17β-estradiol, Electrochim Acta 408 (2022) 139960. https://doi.org/10.1016/j.electacta.2022.139960

[9] W. Wang, B. Ibarlucea, C. Huang, R. Dong, M. Al Aiti, S. Huang, G. Cuniberti, Multi-metallic MOF based composites for environmental applications: synergizing metal centers and interactions, Nanoscale Horiz 9 (2024) 1432-1474. https://doi.org/10.1039/D4NH00140K

[10] M.L. Scala-Benuzzi, S.N. Fernández, G. Giménez, G. Ybarra, G.J.A.A. Soler-Illia, Ordered Mesoporous Electrodes for Sensing Applications, ACS Omega 8 (2023) 24128-24152. https://doi.org/10.1021/acsomega.3c02013

[11] H. Kaur, S.S. Siwal, G. Chauhan, A.K. Saini, A. Kumari, V.K. Thakur, Recent advances in electrochemical-based sensors amplified with carbon-based nanomaterials (CNMs) for sensing pharmaceutical and food pollutants, Chemosphere 304 (2022) 135182. https://doi.org/10.1016/j.chemosphere.2022.135182

[12] M.L. Scala-Benuzzi, G.J.A.A. Soler-Illia, J. Raba, F. Battaglini, R.J. Schneider, S. V. Pereira, G.A. Messina, Immunosensor based on porous gold and reduced graphene platform

Materials Research Forum LLC
https://doi.org/10.21741/9781644903452-10

for the determination of EE2 by electrochemical impedance spectroscopy, Journal of Electroanalytical Chemistry 897 (2021) 115604. https://doi.org/10.1016/j.jelechem.2021.115604

[13] A. Goux, M. Etienne, E. Aubert, C. Lecomte, J. Ghanbaja, A. Walcarius, Oriented Mesoporous Silica Films Obtained by Electro-Assisted Self-Assembly (EASA), Chemistry of Materials 21 (2009) 731-741. https://doi.org/10.1021/cm8029664

[14] T. Kimura, Evaporation-induced Self-assembly Process Controlled for Obtaining Highly Ordered Mesoporous Materials with Demanded Morphologies, The Chemical Record 16 (2016) 445-457. https://doi.org/10.1002/tcr.201500262

[15] F. Rechotnek, H.D.M. Follmann, R. Silva, Mesoporous silica decorated with L-cysteine as active hybrid materials for electrochemical sensing of heavy metals, J Environ Chem Eng 9 (2021) 106492. https://doi.org/10.1016/j.jece.2021.106492

[16] F. Mohamadpour, A.M. Amani, Photocatalytic systems: reactions, mechanism, and applications, RSC Adv 14 (2024) 20609-20645. https://doi.org/10.1039/D4RA03259D

[17] C. Cooper, R. Burch, Mesoporous materials for water treatment processes, Water Res 33 (1999) 3689-3694. https://doi.org/10.1016/S0043-1354(99)00095-0

[18] N. Linares, A.M. Silvestre-Albero, E. Serrano, J. Silvestre-Albero, J. García-Martínez, Mesoporous materials for clean energy technologies, Chem. Soc. Rev. 43 (2014) 7681-7717. https://doi.org/10.1039/C3CS60435G

[19] G. Fadillah, T.A. Saleh, Advances in mesoporous material for adsorption and photoconversion of CO2 in environmental pollution: Clean environment and clean energy, Sustain Chem Pharm 29 (2022) 100812. https://doi.org/10.1016/j.scp.2022.100812

[20] P. Haghighi, S. Alijani, A. Bazyari, L.T. Thompson, Visible light dye degradation over fluorinated mesoporous TiO2 − WO3 − Bi2O3/SiO2 nanocomposite photocatalyst-adsorbent using immersion well reactor, J Photochem Photobiol A Chem 426 (2022) 113790. https://doi.org/10.1016/j.jphotochem.2022.113790

[21] Z. Pei, H. Guo, L. Zhu, C. Li, Z. Fu, J. Xu, Photocatalytic degradation of various antibiotics under visible light irradiation by CdS-doped SiO2@BiOX (X = Br, Cl) prepared by mixed solvothermal method, Materials Science and Engineering: B 287 (2023) 116134. https://doi.org/10.1016/j.mseb.2022.116134

[22] X.-P. Kong, B.-H. Zhang, J. Wang, Multiple Roles of Mesoporous Silica in Safe Pesticide Application by Nanotechnology: A Review, J Agric Food Chem 69 (2021) 6735-6754. https://doi.org/10.1021/acs.jafc.1c01091

[23] B. Cinlar, B.H. Shanks, Characterization of the acidic sites in organic acid functionalized mesoporous silica in an aqueous media, Appl Catal A Gen 396 (2011) 76-84. https://doi.org/10.1016/j.apcata.2011.01.044

[24] Y. Sakata, M. Azhar Uddin, A. Muto, Y. Kanada, K. Koizumi, K. Murata, Catalytic degradation of polyethylene into fuel oil over mesoporous silica (KFS-16) catalyst, J Anal Appl Pyrolysis 43 (1997) 15-25.. https://doi.org/10.1016/S0165-2370(97)00052-1

[25] M.A. Al-Nuaim, A.A. Alwasiti, Z.Y. Shnain, The photocatalytic process in the treatment of polluted water, Chemical Papers 77 (2023) 677-701. https://doi.org/10.1007/s11696-022-02468-7

[26] L. Mahoney, S. Rasalingam, C.-M. Wu, R. Koodali, Nanocasting of Periodic Mesoporous Materials as an Effective Strategy to Prepare Mixed Phases of Titania, Molecules 20 (2015) 21881-21895. https://doi.org/10.3390/molecules201219812

[27] A. Akinyemi, O. Agboola, E. Alagbe, E. Igbokwe, The role of catalyst in the adsorption of dye: Homogeneous catalyst, heterogeneous catalyst, and advanced catalytic activated carbon, critical review, Desalination Water Treat 320 (2024) 100780. https://doi.org/10.1016/j.dwt.2024.100780

[28] N.T.T. Ha, H.L. Ngo, T.B. Pham, N. Hoang Hao, C.T. Bui, T.L. Phung, L.M. Cam, N. Ngoc Ha, Comprehensive Study on the Adsorption and Degradation ofDichlorodiphenyltrichloroethane on Bifunctional Adsorption-Photocatalysis Material TiO 2 /MCM-41 Using Quantum Chemical Methods, ACS Omega (2024).

[29] S. Singh, A.K. Atri, I. Qadir, S. Sharma, U. Manhas, D. Singh, Role of Different Fuels and Sintering Temperatures in the Structural, Optical, Magnetic, and Photocatalytic Properties of Chromium-Containing Nickel Ferrite: Kinetic Study of Photocatalytic Degradation of Rhodamine B Dye, ACS Omega 8 (2023) 6302-6317. https://doi.org/10.1021/acsomega.2c06249

[30] D. Deng, Y. Li, M. Wu, Y. Song, Q. Huang, Y. Duan, Y. Chang, Y. Zhao, C. He, Electrocatalytic Degradation of Rhodamine B on the Sb-Doped SnO 2 /Ti Electrode in Alkaline Medium, ACS Omega 8 (2023) 48480-48490. https://doi.org/10.1021/acsomega.3c08391

[31] S. Rasalingam, R. Peng, R.T. Koodali, An insight into the adsorption and photocatalytic degradation of rhodamine B in periodic mesoporous materials, Appl Catal B 174-175 (2015) 49-59. https://doi.org/10.1016/j.apcatb.2015.02.040

[32] N. Pal, A. Bhaumik, Soft templating strategies for the synthesis of mesoporous materials: Inorganic, organic-inorganic hybrid and purely organic solids, Adv Colloid Interface Sci 189-190 (2013) 21-41. https://doi.org/10.1016/j.cis.2012.12.002

[33] X. Deng, K. Chen, H. Tüysüz, Protocol for the Nanocasting Method: Preparation of Ordered Mesoporous Metal Oxides, Chemistry of Materials 29 (2017) 40-52. https://doi.org/10.1021/acs.chemmater.6b02645

[34] X. Li, H. Yoshikawa, K. Ishihara, K. Miyake, Y. Uchida, N. Nishiyama, Solvent-Free Soft-Template Synthesis of Highly Ordered Mesoporous Carbons via Self-Assembly Promoted by Mg(NO 3) 2, Langmuir 39 (2023) 2036-2042. https://doi.org/10.1021/acs.langmuir.2c03197

[35] A. Lolli, R. Amadori, C. Lucarelli, M.G. Cutrufello, E. Rombi, F. Cavani, S. Albonetti, Hard-template preparation of Au/CeO 2 mesostructured catalysts and their activity for the selective oxidation of 5-hydroxymethylfurfural to 2,5-furandicarboxylic acid, Microporous and Mesoporous Materials 226 (2016) 466-475. https://doi.org/10.1016/j.micromeso.2016.02.014

Materials Research Forum LLC
https://doi.org/10.21741/9781644903452-10

[36] M. Ortega-Franqueza, S. Ivanova, M.I. Domínguez, M.Á. Centeno, Mesoporous Carbon Production by Nanocasting Technique Using Boehmite as a Template, Catalysts 11 (2021) 1132. https://doi.org/10.3390/catal11091132

[37] Z. Zhang, F. Zuo, P. Feng, Hard template synthesis of crystalline mesoporous anatase TiO2 for photocatalytic hydrogen evolution, J Mater Chem 20 (2010) 2206. https://doi.org/10.1039/b921157h

[38] A. Lolli, R. Amadori, C. Lucarelli, M.G. Cutrufello, E. Rombi, F. Cavani, S. Albonetti, Hard-template preparation of Au/CeO 2 mesostructured catalysts and their activity for the selective oxidation of 5-hydroxymethylfurfural to 2,5-furandicarboxylic acid, Microporous and Mesoporous Materials 226 (2016) 466-475. https://doi.org/10.1016/j.micromeso.2016.02.014

[39] D.B. Emrie, Sol-Gel Synthesis of Nanostructured Mesoporous Silica Powder and Thin Films, J Nanomater 2024 (2024) 1-16. https://doi.org/10.1155/2024/6109770

[40] A.K. Dikshit, K. Chugh, S.K. Chaturvedi, B.N. Mohapatra, Synthesis of mesoporous phase materials via Sol-gel process using Indian cementitious raw material, Mater Today Proc 62 (2022) 1132-1138. https://doi.org/10.1016/j.matpr.2022.04.326

[41] N.I. Vazquez, Z. Gonzalez, B. Ferrari, Y. Castro, Synthesis of mesoporous silica nanoparticles by sol-gel as nanocontainer for future drug delivery applications, Boletín de La Sociedad Española de Cerámica y Vidrio 56 (2017) 139-145. https://doi.org/10.1016/j.bsecv.2017.03.002

[42] R. Ryoo, S.H. Joo, S. Jun, Synthesis of Highly Ordered Carbon Molecular Sieves via Template-Mediated Structural Transformation, J Phys Chem B 103 (1999) 7743-7746. https://doi.org/10.1021/jp991673a

[43] J.E.S. van der Hoeven, A. V. Shneidman, N.J. Nicolas, J. Aizenberg, Evaporation-Induced Self-Assembly of Metal Oxide Inverse Opals: From Synthesis to Applications, Acc Chem Res 55 (2022) 1809-1820. https://doi.org/10.1021/acs.accounts.2c00087

[44] A.A.S. Gonçalves, M. Jaroniec, Evaporation-induced self-assembly synthesis of nanostructured alumina-based mixed metal oxides with tailored porosity, J Colloid Interface Sci 537 (2019) 725-735. https://doi.org/10.1016/j.jcis.2018.11.044

[45] J. Fan, M. Guerrero, A. Carretero-Genevrier, M.D. Baró, S. Suriñach, E. Pellicer, J. Sort, Evaporation-induced self-assembly synthesis of Ni-doped mesoporous SnO 2 thin films with tunable room temperature magnetic properties, J Mater Chem C Mater 5 (2017) 5517-5527. https://doi.org/10.1039/C7TC01128H

[46] J.-C. Seo, H. Kim, Y.-L. Lee, S. Nam, H.-S. Roh, K. Lee, S. Bin Park, One-Pot Synthesis of Full-Featured Mesoporous Ni/Al 2 O 3 Catalysts via a Spray Pyrolysis-Assisted Evaporation-Induced Self-Assembly Method for Dry Reforming of Methane, ACS Sustain Chem Eng 9 (2021) 894-904. https://doi.org/10.1021/acssuschemeng.0c07927

[47] M.H. Aboonasr Shiraz, M. Rezaei, F. Meshkani, Microemulsion synthesis method for preparation of mesoporous nanocrystalline γ-Al2O3 powders as catalyst carrier for nickel catalyst in dry reforming reaction, Int J Hydrogen Energy 41 (2016) 6353-6361. https://doi.org/10.1016/j.ijhydene.2016.03.017

[48] M. Bielec, A. Huang, Y. Xia, H. Dai, Microemulsion Synthesis of Mesoporous β-tricalcium Phosphate Powder with a Novel System, Journal of Wuhan University of Technology-Mater. Sci. Ed. 37 (2022) 773-778 https://doi.org/10.1007/s11595-022-2595-y

[49] R. Muthusami, A. Kesavan, V. Ramachandran, V. Vasudevan, K. Irena, R. Rangappan, Synthesis of mesoporous silica nanoparticles with a lychee-like morphology and dual pore arrangement and its application towards biomimetic activity via functionalization with copper(II) complex, Microporous and Mesoporous Materials 294 (2020) 109910. https://doi.org/10.1016/j.micromeso.2019.109910

[50] Y. Wang, Q. Zhao, N. Han, L. Bai, J. Li, J. Liu, E. Che, L. Hu, Q. Zhang, T. Jiang, S. Wang, Mesoporous silica nanoparticles in drug delivery and biomedical applications, Nanomedicine 11 (2015) 313-327. https://doi.org/10.1016/j.nano.2014.09.014

[51] J.L. Vivero-Escoto, I.I. Slowing, B.G. Trewyn, V.S. -Y. Lin, Mesoporous Silica Nanoparticles for Intracellular Controlled Drug Delivery, Small 6 (2010) 1952-1967. https://doi.org/10.1002/smll.200901789

[52] P. Shanmugam, S.M. Smith, S. Boonyuen, A. Luengnaruemitchai, In-situ development of boron doped g-C3N4 supported SBA-15 nanocomposites for photocatalytic degradation of tetracycline, Environ Res 224 (2023) 115496. https://doi.org/10.1016/j.envres.2023.115496

[53] Y.E. Kim, M.Y. Byun, K.-Y. Lee, M.S. Lee, Hydrothermal synthesis of mesoporous TiO2 using β-diketonate stabilizing agents for photocatalytic degradation of methyl violet 2B under visible light, Catal Today 411-412 (2023) 113954. https://doi.org/10.1016/j.cattod.2022.11.015

[54] Q. Li, J. Zhang, Q. Lu, J. Lu, J. Li, C. Dong, Q. Zhu, Hydrothermal synthesis and characterization of ordered mesoporous magnesium silicate-silica for dyes adsorption, Mater Lett 170 (2016) 167-170. https://doi.org/10.1016/j.matlet.2016.02.029

[55] H. Ahmed, S.S. Gomte, E. Prathyusha, P. A, M. Agrawal, A. Alexander, Biomedical applications of mesoporous silica nanoparticles as a drug delivery carrier, J Drug Deliv Sci Technol 76 (2022) 103729. https://doi.org/10.1016/j.jddst.2022.103729

[56] Y.E. Kim, M.Y. Byun, K.-Y. Lee, M.S. Lee, Hydrothermal synthesis of mesoporous TiO2 using β-diketonate stabilizing agents for photocatalytic degradation of methyl violet 2B under visible light, Catal Today 411-412 (2023) 113954. https://doi.org/10.1016/j.cattod.2022.11.015

[57] S.L. Suib, A Review of Recent Developments of Mesoporous Materials, The Chemical Record 17 (2017) 1169-1183. https://doi.org/10.1002/tcr.201700025

[58] M. Moritz, M. Geszke-Moritz, Mesoporous materials as multifunctional tools in biosciences: Principles and applications, Materials Science and Engineering: C 49 (2015) 114-151. https://doi.org/10.1016/j.msec.2014.12.079

[59] D. V. Wellia, Y. Kusumawati, L.J. Diguna, N. Pratiwi, R.A. Putri, M.I. Amal, Mesoporous Materials for Degradation of Textile Dyes, in: 2020: pp. 255-288. https://doi.org/10.1007/978-3-030-16427-0_10

[60] B. Nanda, D. Rath, K.M. Parida, M. Das, Fabrication of Mesoporous MnO 2 -MCM-41 Composite: A Promising Material for Photocatalytic Degradation of Organic Dyes Under Visible Light, Adv Sci Lett 20 (2014) 579-583. https://doi.org/10.1166/asl.2014.5403

[61] M.Z. Hussain, G.S. Pawar, Z. Huang, A.A. Tahir, R.A. Fischer, Y. Zhu, Y. Xia, Porous ZnO/Carbon nanocomposites derived from metal organic frameworks for highly efficient photocatalytic applications: A correlational study, Carbon N Y 146 (2019) 348-363. https://doi.org/10.1016/j.carbon.2019.02.013

[62] S.M. Saleh, A.E.A.E. Albadri, M.A. Ben Aissa, A. Modwi, Fabrication of Mesoporous V2O5@g-C3N4 Nanocomposite as Photocatalyst for Dye Degradation, Crystals (Basel) 12 (2022) 1766. https://doi.org/10.3390/cryst12121766

[63] D.P. DePuccio, P. Botella, B. O'Rourke, C.C. Landry, Degradation of Methylene Blue Using Porous WO 3 , SiO 2 -WO 3 , and Their Au-Loaded Analogs: Adsorption and Photocatalytic Studies, ACS Appl Mater Interfaces 7 (2015) 1987-1996. https://doi.org/10.1021/am507806a

[64] D. Gogoi, P. Makkar, N.N. Ghosh, Solar Light-Irradiated Photocatalytic Degradation of Model Dyes and Industrial Dyes by a Magnetic CoFe 2 O 4 -gC 3 N 4 S-Scheme Heterojunction Photocatalyst, ACS Omega 6 (2021) 4831-4841. https://doi.org/10.1021/acsomega.0c05809

[65] B. Nanda, A.C. Pradhan, K.M. Parida, A comparative study on adsorption and photocatalytic dye degradation under visible light irradiation by mesoporous MnO 2 modified MCM-41 nanocomposite, Microporous and Mesoporous Materials 226 (2016) 229-242. https://doi.org/10.1016/j.micromeso.2015.12.027

[66] A.A. Ismail, M. Faisal, A. Al-Haddad, Mesoporous WO 3 -graphene photocatalyst for photocatalytic degradation of Methylene Blue dye under visible light illumination, Journal of Environmental Sciences 66 (2018) 328-337.. https://doi.org/10.1016/j.jes.2017.05.001

[67] H. Znad, K. Abbas, S. Hena, Md.R. Awual, Synthesis a novel multilamellar mesoporous TiO2/ZSM-5 for photo-catalytic degradation of methyl orange dye in aqueous media, J Environ Chem Eng 6 (2018) 218-227 https://doi.org/10.1016/j.jece.2017.11.077

[68] H. Wang, X. Yuan, Y. Wu, G. Zeng, X. Chen, L. Leng, H. Li, Synthesis and applications of novel graphitic carbon nitride/metal-organic frameworks mesoporous photocatalyst for dyes removal, Appl Catal B 174-175 (2015) 445-454.. https://doi.org/10.1016/j.apcatb.2015.03.037

[69] C.S. Katsiotis, M. Strømme, K. Welch, Processability of mesoporous materials in fused deposition modeling for drug delivery of a model thermolabile drug, Int J Pharm X 5 (2023) 100149. https://doi.org/10.1016/j.ijpx.2022.100149

[70] S.E. Moya, R.R. Hernández, P.C. Angelomé, Degradation of Mesoporous Silica Materials in Biological Milieu: The Gateway for Therapeutic Applications, Adv Nanobiomed Res (2024).. https://doi.org/10.1002/anbr.202400005

[71] M.W. Ibrahim, Y. Khane, Y.T. Mahmood, A. Schulz, H. Kosslick, Mesoporous aluminosilicate materials supported zinc oxide photocatalytic degradation of pharmaceutical pollutants, Desalination Water Treat 320 (2024) 100588. https://doi.org/10.1016/j.dwt.2024.100588

[72] B. Kaushik, G. Rao, D. Vaya, Photocatalytic Degradation of Drugs, in: Handbook of Green and Sustainable Nanotechnology, Springer International Publishing, Cham, 2023: pp. 797-825. https://doi.org/10.1007/978-3-031-16101-8_6

[73] M.A. Hassaan, M.A. El-Nemr, M.R. Elkatory, S. Ragab, V.-C. Niculescu, A. El Nemr, Principles of Photocatalysts and Their Different Applications: A Review, Top Curr Chem 381 (2023) 31. https://doi.org/10.1007/s41061-023-00444-7

[74] H.S. Naik, P.M. Sah, R.W. Raut, Nanotechnology-Based Photocatalytic Degradation of Pharmaceuticals, in: Modern Nanotechnology, Springer International Publishing, Cham, 2023: pp. 465-486. https://doi.org/10.1007/978-3-031-31111-6_19

[75] A. Ali, M. Shoeb, Y. Li, B. Li, M.A. Khan, Enhanced photocatalytic degradation of antibiotic drug and dye pollutants by graphene-ordered mesoporous silica (SBA 15)/TiO2 nanocomposite under visible-light irradiation, J Mol Liq 324 (2021) 114696. https://doi.org/10.1016/j.molliq.2020.114696

[76] J. Feng, J. Yang, Y. Shen, W. Deng, W. Chen, Y. Ma, Z. Chen, S. Dong, Mesoporous silica nanoparticles prepared via a one-pot method for controlled release of abamectin: Properties and applications, Microporous and Mesoporous Materials 311 (2021) 110688. https://doi.org/10.1016/j.micromeso.2020.110688

[77] M. Šťastný, V. Štengl, J. Henych, J. Tolasz, P. Vomáčka, J. Ederer, Mesoporous manganese oxide for the degradation of organophosphates pesticides, J Mater Sci 51 (2016) 2634-2642. https://doi.org/10.1007/s10853-015-9577-9

[78] W. Yan, G. Hu, J. Xiao, Y. Liu, G. Tang, Z. Zhou, Y. Huang, X. Zhang, G. Yan, J. Wang, Y. Cao, Fabrication of Hollow Mesoporous Copper Sulfide Nanocapsules Loaded with Natural Photoactivated Pesticides for Sustainable Plant Disease Management, ACS Sustain Chem Eng 12 (2024) 1207-1220. https://doi.org/10.1021/acssuschemeng.3c04782

[79] M.V. Phanikrishna Sharma, V. Durga Kumari, M. Subrahmanyam, TiO2 supported over SBA-15: An efficient photocatalyst for the pesticide degradation using solar light, Chemosphere 73 (2008) 1562-1569. https://doi.org/10.1016/j.chemosphere.2008.07.081

[80] D.S. Conceição, C.A.L. Graça, D.P. Ferreira, A.M. Ferraria, I.M. Fonseca, A.M. Botelho do Rego, A.C.S.C. Teixeira, L.F. Vieira Ferreira, Photochemical insights of TiO2 decorated mesoporous SBA-15 materials and their influence on the photodegradation of organic contaminants, Microporous and Mesoporous Materials 253 (2017) 203-214. https://doi.org/10.1016/j.micromeso.2017.07.013

[81] A.S. Mohamed, M.R. Abukkhadra, E.A. Abdallah, A.M. El-Sherbeeny, R.K. Mahmoud, The photocatalytic performance of silica fume based Co3O4/MCM-41 green nanocomposite for instantaneous degradation of Omethoate pesticide under visible light, J Photochem Photobiol A Chem 392 (2020) 112434. https://doi.org/10.1016/j.jphotochem.2020.112434. https://doi.org/10.1016/j.jphotochem.2020.112434

[82] G. Li, B. Wang, W.Q. Xu, Y. Han, Q. Sun, Rapid TiO 2 /SBA-15 synthesis from ilmenite and use in photocatalytic degradation of dimethoate under simulated solar light, Dyes and Pigments 155 (2018) 265-275. https://doi.org/10.1016/j.dyepig.2018.03.058

[83] S. Merci, A. Saljooqi, T. Shamspur, A. Mostafavi, Investigation of photocatalytic chlorpyrifos degradation by a new silica mesoporous material immobilized by WS 2 and Fe 3 O 4 nanoparticles: Application of response surface methodology, Appl Organomet Chem 34 (2020). https://doi.org/10.1002/aoc.5343

[84] M.R. AbuKhadra, A.S. Mohamed, A.M. El-Sherbeeny, M.A. Elmeligy, Enhanced photocatalytic degradation of acephate pesticide over MCM-41/Co3O4 nanocomposite synthesized from rice husk silica gel and Peach leaves, J Hazard Mater 389 (2020) 122129. https://doi.org/10.1016/j.jhazmat.2020.122129

[85] M. Silva, M.J.F. Calvete, N.P.F. Gonçalves, H.D. Burrows, M. Sarakha, A. Fernandes, M.F. Ribeiro, M.E. Azenha, M.M. Pereira, Zinc(II) phthalocyanines immobilized in mesoporous silica Al-MCM-41 and their applications in photocatalytic degradation of pesticides, J Hazard Mater 233-234 (2012) 79-88. https://doi.org/10.1016/j.jhazmat.2012.06.059

[86] M. Silva, M.E. Azenha, M.M. Pereira, H.D. Burrows, M. Sarakha, C. Forano, M.F. Ribeiro, A. Fernandes, Immobilization of halogenated porphyrins and their copper complexes in MCM-41: Environmentally friendly photocatalysts for the degradation of pesticides, Appl Catal B 100 (2010) 1-9. https://doi.org/10.1016/j.apcatb.2010.07.033

[87] C. Feng, Y. Li, X. Liu, Photocatalytic Degradation of Imidacloprid by Phosphotungstic Acid Supported on a Mesoporous Sieve MCM-41, Chin J Chem 30 (2012) 127-132. https://doi.org/10.1002/cjoc.201180453

[88] M.L. Scala-Benuzzi, S.N. Fernández, G. Giménez, G. Ybarra, G.J.A.A. Soler-Illia, Ordered Mesoporous Electrodes for Sensing Applications, ACS Omega 8 (2023) 24128-24152. https://doi.org/10.1021/acsomega.3c02013

[89] A. Walcarius, Mesoporous materials and electrochemistry, Chem Soc Rev 42 (2013) 4098. https://doi.org/10.1039/c2cs35322a

[90] M.L. Scala-Benuzzi, S.N. Fernández, G. Giménez, G. Ybarra, G.J.A.A. Soler-Illia, Ordered Mesoporous Electrodes for Sensing Applications, ACS Omega 8 (2023) 24128-24152. https://doi.org/10.1021/acsomega.3c02013

[91] R. Yahya, A. Shah, T. Kokab, N. Ullah, M.K. Hakeem, M. Hayat, A. Haleem, I. Shah, Electrochemical Sensor for Detection and Degradation Studies of Ethyl Violet Dye, ACS Omega 7 (2022) 34154-34165. https://doi.org/10.1021/acsomega.2c03472

[92] M. Irfan, A. Shah, F.J. Iftikhar, M. Hayat, M.N. Ashiq, I. Shah, Electrochemical Sensing Platform Based on Functionalized Multi-Walled Carbon Nanotubes and Metal Oxide for the Detection and Degradation Studies of Orange II Dye, ACS Omega 7 (2022) 32302-32312. https://doi.org/10.1021/acsomega.2c03641

[93] A. Ganash, L. Alajlani, E. Ganash, A. Al-Moubaraki, Efficient electrochemical degradation of congo red dye by Pt/CuNPs electrode with its attractive performance, energy consumption, and mechanism: Experimental and theoretical approaches, Journal of Water Process Engineering 56 (2023) 104497. https://doi.org/10.1016/j.jwpe.2023.104497

[94] S.N. Azizi, S. Ghasemi, A. Samadi-Maybodi, M. Ranjbar-Azad, A new modified electrode based on Ag-doped mesoporous SBA-16 nanoparticles as a non-enzymatic sensor for

hydrogen peroxide, Sens Actuators B Chem 216 (2015) 271-278. https://doi.org/10.1016/j.snb.2015.03.078

[95] X. Zhang, S. Duan, X. Xu, S. Xu, C. Zhou, Electrochemical behavior and simultaneous determination of dihydroxybenzene isomers at a functionalized SBA-15 mesoporous silica modified carbon paste electrode, Electrochim Acta 56 (2011) 1981-1987. https://doi.org/10.1016/j.electacta.2010.11.048

[96] T. Zhong, Q. Guo, X. Zhu, R. Liu, S. Huang, Based on Gold Nanoparticles-l-Tyr-Amino Functionalized Mesoporous Materials-Polyphenol Oxidase Modified Biosensor for the Detection of Resorcinol, Analytical Sciences 37 (2021) 817-823. https://doi.org/10.2116/analsci.20P288

[97] K. Wang, H. Ren, N. Li, X. Tan, F. Dang, Ratiometric fluorescence sensor based on cholesterol oxidase-functionalized mesoporous silica nanoparticle@ZIF-8 core-shell nanocomposites for detection of cholesterol, Talanta 188 (2018) 708-713. https://doi.org/10.1016/j.talanta.2018.06.019

[98] H. Karimi-Maleh, R. Darabi, M. Shabani-Nooshabadi, M. Baghayeri, F. Karimi, J. Rouhi, M. Alizadeh, O. Karaman, Y. Vasseghian, C. Karaman, Determination of D&C Red 33 and Patent Blue V Azo dyes using an impressive electrochemical sensor based on carbon paste electrode modified with ZIF-8/g-C3N4/Co and ionic liquid in mouthwash and toothpaste as real samples, Food and Chemical Toxicology 162 (2022) 112907. https://doi.org/10.1016/j.fct.2022.112907

[99] Z. Karimi, M. Shamsipur, M.A. Tabrizi, S. Rostamnia, A highly sensitive electrochemical sensor for the determination of methanol based on PdNPs@SBA-15-PrEn modified electrode, Anal Biochem 548 (2018) 32-37. https://doi.org/10.1016/j.ab.2018.01.033

[100] X. Zheng, S. Duan, S. Liu, M. Wei, F. Xia, D. Tian, C. Zhou, Sensitive and simultaneous method for the determination of naphthol isomers by an amino-functionalized, SBA-15-modified carbon paste electrode, Analytical Methods 7 (2015) 3063-3071. https://doi.org/10.1039/C5AY00027K

[101] R. Gupta, P.K. Rastogi, U. Srivastava, V. Ganesan, P.K. Sonkar, D.K. Yadav, Methylene blue incorporated mesoporous silica microsphere-based sensing scaffold for the selective voltammetric determination of riboflavin, RSC Adv 6 (2016) 65779-65788. https://doi.org/10.1039/C6RA12336H

[102] T. Nasir, G. Herzog, M. Hébrant, C. Despas, L. Liu, A. Walcarius, Mesoporous Silica Thin Films for Improved Electrochemical Detection of Paraquat, ACS Sens 3 (2018) 484-493. https://doi.org/10.1021/acssensors.7b00920

[103] M. Vallet-Regí, F. Balas, D. Arcos, Mesoporous Materials for Drug Delivery, Angewandte Chemie International Edition 46 (2007) 7548-7558. https://doi.org/10.1002/anie.200604488

[104] H. El-Desoky, M. Abdel-Galeil, A. Khalifa, Mesoporous SiO2 (SBA-15) modified graphite electrode as highly sensitive sensor for ultra trace level determination of Dapoxetine hydrochloride drug in human plasma, Journal of Electroanalytical Chemistry 846 (2019) 113157. https://doi.org/10.1016/j.jelechem.2019.05.039

[105] M. Ghalkhani, E. Sohouli, Z.S. Dehkordi, Electrochemical sensor based on mesoporous g-C3N4/N-CNO/gold nanoparticles for measuring oxycodone, Sci Rep 14 (2024) 17221. https://doi.org/10.1038/s41598-024-68310-0

[106] H.-S. Wang, T.-H. Li, W.-L. Jia, H.-Y. Xu, Highly selective and sensitive determination of dopamine using a Nafion/carbon nanotubes coated poly(3-methylthiophene) modified electrode, Biosens Bioelectron 22 (2006) 664-669. https://doi.org/10.1016/j.bios.2006.02.007

[107] R.T. Kachoosangi, G.G. Wildgoose, R.G. Compton, Sensitive adsorptive stripping voltammetric determination of paracetamol at multiwalled carbon nanotube modified basal plane pyrolytic graphite electrode, Anal Chim Acta 618 (2008) 54-60. https://doi.org/10.1016/j.aca.2008.04.053

[108] B.-R. Adhikari, M. Govindhan, A. Chen, Carbon Nanomaterials Based Electrochemical Sensors/Biosensors for the Sensitive Detection of Pharmaceutical and Biological Compounds, Sensors 15 (2015) 22490-22508. https://doi.org/10.3390/s150922490

[109] N.A. Martinez, G.A. Messina, F.A. Bertolino, E. Salinas, J. Raba, Screen-printed enzymatic biosensor modified with carbon nanotube for the methimazole determination in pharmaceuticals formulations, Sens Actuators B Chem 133 (2008) 256-262. https://doi.org/10.1016/j.snb.2008.02.025

[110] G.-T. Liu, H.-F. Chen, G.-M. Lin, P. Ye, X.-P. Wang, Y.-Z. Jiao, X.-Y. Guo, Y. Wen, H.-F. Yang, One-step electrodeposition of graphene loaded nickel oxides nanoparticles for acetaminophen detection, Biosens Bioelectron 56 (2014) 26-32. https://doi.org/10.1016/j.bios.2014.01.005

[111] J. Luo, S. Jiang, H. Zhang, J. Jiang, X. Liu, A novel non-enzymatic glucose sensor based on Cu nanoparticle modified graphene sheets electrode, Anal Chim Acta 709 (2012) 47-53. https://doi.org/10.1016/j.aca.2011.10.025

[112] U. Yogeswaran, S. Thiagarajan, S.-M. Chen, Nanocomposite of functionalized multiwall carbon nanotubes with nafion, nano platinum, and nano gold biosensing film for simultaneous determination of ascorbic acid, epinephrine, and uric acid, Anal Biochem 365 (2007) 122-131. https://doi.org/10.1016/j.ab.2007.02.034

[113] M. Elfiky, N. Salahuddin, Advanced sensing platform for nanomolar detection of food preservative nitrite in sugar byproducts based on 3D mesoporous nanorods of montmorillonite/TiO2-ZnO hybrids, Microchemical Journal 170 (2021) 106582. https://doi.org/10.1016/j.microc.2021.106582

[114] S. Zinatloo-Ajabshir, M. Salavati-Niasari, M. Hamadanian, Preparation of nanocrystalline praseodymium oxide with different shapes via a simple thermal decomposition route, Journal of Materials Science: Materials in Electronics 27 (2016) 998-1006. https://doi.org/10.1007/s10854-015-3844-z

[115] F. Mazzara, B. Patella, G. Aiello, A. O'Riordan, C. Torino, A. Vilasi, R. Inguanta, Electrochemical detection of uric acid and ascorbic acid using r-GO/NPs based sensors, Electrochim Acta 388 (2021) 138652. https://doi.org/10.1016/j.electacta.2021.138652

[116] S.Z. Mohammadi, H. Asadollahzadeh, F. Emambakhsh, Sensitive electrochemical determination of Sudan I in food products by using a modified-screen printed electrode,

Journal of Materials Science: Materials in Electronics 34 (2023) 2239. https://doi.org/10.1007/s10854-023-11664-4

[117] Z. Zhou, C. Gu, C. Chen, P. Zhao, Y. Xie, J. Fei, An ultrasensitive electrochemical sensor for quercetin based on 1-pyrenebutyrate functionalized reduced oxide graphene /mercapto-β-cyclodextrin /Au nanoparticles composite film, Sens Actuators B Chem 288 (2019) 88-95. https://doi.org/10.1016/j.snb.2019.02.105

[118] Z. Liu, B. Lu, Y. Gao, T. Yang, R. Yue, J. Xu, L. Gao, Facile one-pot preparation of Pd-Au/PEDOT/graphene nanocomposites and their high electrochemical sensing performance for caffeic acid detection, RSC Adv 6 (2016) 89157-89166. https://doi.org/10.1039/C6RA16488A

[119] Y. Chen, W. Huang, K. Chen, T. Zhang, Y. Wang, J. Wang, A novel electrochemical sensor based on core-shell-structured metal-organic frameworks: The outstanding analytical performance towards chlorogenic acid, Talanta 196 (2019) 85-91. https://doi.org/10.1016/j.talanta.2018.12.033

[120] C. Yin, Q. Zhuang, Q. Xiao, Y. Wang, J. Xie, Electropolymerization of poly(methylene blue) on flower-like nickel-based MOFs used for ratiometric electrochemical sensing of total polyphenolic content in chrysanthemum tea, Analytical Methods 13 (2021) 1154-1163. https://doi.org/10.1039/D1AY00028D

[121] A. Puangjan, S. Chaiyasith, An efficient ZrO2/Co3O4/reduced graphene oxide nanocomposite electrochemical sensor for simultaneous determination of gallic acid, caffeic acid and protocatechuic acid natural antioxidants, Electrochim Acta 211 (2016) 273-288. https://doi.org/10.1016/j.electacta.2016.04.185

[122] S. Alipour, P.A. Azar, S.W. Husain, H.R. Rajabi, Determination of Rosmarinic acid in plant extracts using a modified sensor based on magnetic imprinted polymeric nanostructures, Sens Actuators B Chem 323 (2020) 128668. https://doi.org/10.1016/j.snb.2020.128668

[123] A.S. Barreto, A. Aquino, S.C.S. Silva, M.E. de Mesquita, M.J. Calhorda, M.S. Saraiva, S. Navickiene, A novel application of mesoporous silica material for extraction of pesticides, Mater Lett 65 (2011) 1357-1359. https://doi.org/10.1016/j.matlet.2011.01.082

[124] A. Badiei, H. Goldooz, G.M. Ziarani, A. Abbasi, One pot synthesis of functionalized SBA-15 by using an 8-hydroxyquinoline-5-sulfonamide-modified organosilane as precursor, J Colloid Interface Sci 357 (2011) 63-69. https://doi.org/10.1016/j.jcis.2011.01.049

[125] Š. Korent Urek, N. Frančič, M. Turel, A. Lobnik, Sensing Heavy Metals Using Mesoporous-Based Optical Chemical Sensors, J Nanomater 2013 (2013). https://doi.org/10.1155/2013/501320

[126] N.A.I.M. Mokhtar, R.M. Zawawi, W.M. Khairul, N.A. Yusof, Electrochemical and optical sensors made of composites of metal-organic frameworks and carbon-based materials. A review, Environ Chem Lett 20 (2022) 3099-3131. https://doi.org/10.1007/s10311-022-01403-2

[127] Y. Xu, H. Guo, Y. Wei, H. Feng, Y. Yang, Y. Lu, Y. Wei, J. Su, Y. Ben, J. Yuan, X. Liu, X. Zhang, Detection of heavy metal Pb2+ by electrochemical sensor based on ZIF-8@GO

composite, J Appl Electrochem 54 (2024) 1397-1407. https://doi.org/10.1007/s10800-023-02042-3

[128] Š. Korent Urek, N. Frančič, M. Turel, A. Lobnik, Sensing Heavy Metals Using Mesoporous-Based Optical Chemical Sensors, J Nanomater 2013 (2013). https://doi.org/10.1155/2013/501320

[129] G. Manasa, C.S. Rout, Versatile MXenes as electrochemical sensors for heavy metal ions and phenolic moiety-containing industrial chemicals: recent development and prospects, Mater Adv 5 (2024) 83-122. https://doi.org/10.1039/D3MA00362K

[130] S.F. Sulthana, U.M. Iqbal, S.B. Suseela, R. Anbazhagan, R. Chinthaginjala, D. Chitathuru, I. Ahmad, T. Kim, Electrochemical Sensors for Heavy Metal Ion Detection in Aqueous Medium: A Systematic Review, ACS Omega 9 (2024) 25493-25512. https://doi.org/10.1021/acsomega.4c00933

[131] S.F. Sulthana, U.M. Iqbal, S.B. Suseela, R. Anbazhagan, R. Chinthaginjala, D. Chitathuru, I. Ahmad, T. Kim, Electrochemical Sensors for Heavy Metal Ion Detection in Aqueous Medium: A Systematic Review, ACS Omega 9 (2024) 25493-25512. https://doi.org/10.1021/acsomega.4c00933

[132] W. Li, J. Tang, Z. Wang, Micro-/Mesoporous Fluorescent Polymers and Devices for Visual Pesticide Detection with Portability, High Sensitivity, and Ultrafast Response, ACS Appl Mater Interfaces 14 (2022) 5815-5824. https://doi.org/10.1021/acsami.1c21658

[133] C. Gu, Q. Wang, L. Zhang, P. Yang, Y. Xie, J. Fei, Ultrasensitive non-enzymatic pesticide electrochemical sensor based on HKUST-1-derived copper oxide @ mesoporous carbon composite, Sens Actuators B Chem 305 (2020) 127478. https://doi.org/10.1016/j.snb.2019.127478

[134] Md.A. Rashed, M. Faisal, S.A. Alsareii, M. Alsaiari, M. Jalalah, F.A. Harraz, Highly sensitive and selective electrochemical sensor for detecting imidacloprid pesticide using novel silver nanoparticles/mesoporous carbon/hematite ore ternary nanocomposite, J Environ Chem Eng 10 (2022) 108364. https://doi.org/10.1016/j.jece.2022.108364

[135] R. Jemai, M.A. Djebbi, N. Hussain, B. Yang, M. Hirtz, V. Trouillet, H. Ben Rhaiem, A. Ben Haj Amara, Activated porous carbon supported Pd and ZnO nanocatalysts for trace sensing of carbaryl pesticide in water and food products, New Journal of Chemistry 46 (2022) 13880-13895. https://doi.org/10.1039/D2NJ01844F

[136] N. Karuppusamy, R. Sakthivel, S.-M. Chen, S.B. Prasanna, R.-J. Chung, Template-assisted synthesis of Co1Zn0.5-P/CoO/ZnO core-shell heterostructure for the electrochemical detection of carbofuran and diuron, Chemical Engineering Journal 473 (2023) 145305. https://doi.org/10.1016/j.cej.2023.145305

[137] T. Teshome, S.A. Kitte, G. Gonfa, A. Gure, Electrochemical determination of endrin from fruit juice samples based on Fe-doped ZnO nanomaterial modified glassy carbon electrode, J Appl Electrochem 54 (2024) 865-877. https://doi.org/10.1007/s10800-023-01985-x

[138] L. Gurusamy, R.-W. Cheng, S. Anandan, C.-H. Liu, J.J. Wu, Detection of Environmentally Harmful Malathion Pesticides Using a Bimetallic Oxide of CuO Nanoparticles Dispersed over a 3D ZnO Nanoflower, Materials 16 (2023) 7065. https://doi.org/10.3390/ma16227065

[139] M.R. de Barros, J.P. Winiarski, F. de Matos Morawski, R.G. Marim, E.S. Chaves, A. Blacha-Grzechnik, C.L. Jost, A high-performance electrochemical sensor based on a mesoporous silica/titania material and cobalt(II) phthalocyanine for sensitive pentachlorophenol determination, Microchimica Acta 189 (2022) 269. https://doi.org/10.1007/s00604-022-05360-z

[140] K. Wang, Q. Liu, L. Dai, J. Yan, C. Ju, B. Qiu, X. Wu, A highly sensitive and rapid organophosphate biosensor based on enhancement of CdS-decorated graphene nanocomposite, Anal Chim Acta 695 (2011) 84-88. https://doi.org/10.1016/j.aca.2011.03.042

[141] C. Gu, Q. Wang, L. Zhang, P. Yang, Y. Xie, J. Fei, Ultrasensitive non-enzymatic pesticide electrochemical sensor based on HKUST-1-derived copper oxide @ mesoporous carbon composite, Sens Actuators B Chem 305 (2020) 127478. https://doi.org/10.1016/j.snb.2019.127478

[142] C. Zhai, Y. Guo, X. Sun, Y. Zheng, X. Wang, An acetylcholinesterase biosensor based on graphene-gold nanocomposite and calcined layered double hydroxide, Enzyme Microb Technol 58-59 (2014) 8-13. https://doi.org/10.1016/j.enzmictec.2014.02.004

Materials Research Foundations 173 (2025) 302-310 https://doi.org/10.21741/9781644903452-11

Chapter 11

Mesoporous Materials for the Removal of Volatile Organic Compounds

Meenu Mariam Jacob[1], Muthamilselvi Ponnuchamy[1*]

[1]Department of Chemical Engineering, SRM Institute of Science and Technology, Kattankulathur, Tamil Nadu, India - 603202

* muthamip@srmist.edu.in

Abstract

This chapter explores the use of mesoporous materials to remove volatile organic compounds, highlighting the structural characteristics and adsorption mechanisms of these materials. It discusses different mesoporous materials and their synthesis techniques. Surface area, pore size distribution, and surface functionalization three important variables that affect adsorption capacity are explored along with their applications. The chapter covers scalability and regeneration issues as well as potential research avenues, such as enhanced functionalization and environmentally friendly synthesis techniques. Effective methods for enhancing environmental health and air quality are provided by mesoporous materials.

Keywords

Mesoporous Materials, Volatile Organic Compounds, Adsorption

Contents

1. Introduction

A broad class of organic compounds known as volatile organic compounds (VOCs) are distinguished by their high vapor pressure at ambient temperature, which allows them to readily evaporate into the atmosphere. VOCs are found in many indoor and outdoor contexts. Low molecular weight, carbon-containing chemicals, or VOCs, are a broad class that exhibit high volatility, low vapor pressure (≥0.01 kPa at 20 °C), and poor water solubility [1]. They can be found in a variety of sources, including building materials, household items, automobile emissions, and industrial activities [2]. Benzene, formaldehyde, toluene, and xylene are common VOCs that have differing levels of toxicity [3].

VOCs provide serious threats to human health and the environment when they are present in the atmosphere. In terms of the environment, VOCs have a role in the creation of smog and ground-level ozone, which can harm aquatic ecosystems, forests, and agriculture [4]. Health-wise, exposure to VOCs can have both immediate impacts, such headaches, light-headedness, and respiratory problems, and long-term effects, like damage to the liver and kidneys, neurological disorders, and an increased risk of cancer. VOCs have a particularly negative effect indoors, where concentrations can be much higher than outside and cause chronic exposure [5].

Since VOCs are ubiquitous and have the potential to cause harm, it is critical that they be removed from a variety of situations. In order to protect the environment and public health, it is imperative that VOC emissions be under control [6]. Examples of regulatory standards and guidelines in this regard are those established by the World Health Organization (WHO) and the Environmental Protection Agency (EPA). Removing VOCs not only complies with regulations but also has a positive economic impact by lowering medical expenses connected to illnesses caused by VOCs and increasing worker productivity in industrial environments [7].

Mesoporous materials have shown promise in this regard for adsorption of volatile organic compounds. These materials are specially adapted to efficiently adsorb VOC molecules due to their enormous surface areas and highly structured pore architectures. This chapter's subsequent sections will cover mesoporous material properties and production, their methods of action in removing VOCs, a variety of applications, and potential future research and development avenues.

2. Mesoporous materials

A class of porous solids known as mesoporous materials has pores with diameters ranging from 2 to 50 nanometers. These materials have extraordinary qualities such a large pore volume, variable pore sizes, and a highly organized structure with uniform pore distribution. Owing to these qualities, mesoporous materials are especially well-suited for uses requiring effective molecular sieving and large adsorption capacities. In addition to their thermal stability, these special qualities also offer the possibility of surface functionalization, which expands their range of applications. Mesoporous materials are able to fill the gap between macroporous (pores larger than 50 nm) and microporous (pores less than 2 nm) materials because of their pore size. Larger molecules, like those of many VOCs, benefit greatly from the adsorption and separation of this intermediate pore size range. Additionally, the homogeneity of the pores guarantees predictable and consistent performance in a range of applications, such as environmental remediation, drug delivery, and catalysis [8].

Mesoporous materials come in a variety of forms, each having special qualities and uses. Carbon, silica, metal-organic frameworks (MOFs), and other hybrid materials are the principal kinds [9–13].

- Silica: High surface areas, chemical inertness, and ease of functionalization are widely recognized characteristics of mesoporous silica materials. The hexagonal configuration of cylindrical pores contrasted with the bigger pores and thicker walls offer superior mechanical stability. These substances are widely employed in adsorption, drug delivery, and catalysis procedures.
- Carbon: These materials have enormous surface areas, good electrical conductivity, and strong thermal stability. Applications including energy storage, electrochemical devices, and environmental cleanup are where they are especially helpful. To improve particular qualities, the carbon framework can also be doped with other elements like sulphur or nitrogen.
- Metal-Organic Frameworks (MOFs): MOFs are a class of crystalline materials that are highly porous structures made of coordinated metal ions or clusters to organic ligands. MOFs possess exceptional surface areas, adjustable pore sizes, and a range of functions that render them perfect for gas storage, separation, and catalysis. Targeted adsorption of specific VOCs is made possible by the capacity to build MOFs with specified pore structures and functions.
- Additional Hybrid Materials: To improve performance qualities, hybrid mesoporous materials blend many kinds of porous materials. For instance, silica-carbon composites and MOF-based hybrids can provide the advantages of both material types, including improved adsorption capabilities and mechanical robustness. Through the combination of the beneficial qualities of their constituent materials, these hybrids can be customized for certain uses.

Mesoporous materials are synthesized using a variety of methods that produce the required pore structure and characteristics. Self-assembly techniques, sol-gel procedures, and templating techniques are important synthesis techniques. In the hard templating technique, the pore structure is made using solid templates, like silica or polymer beads. The porous framework is left behind when the material is synthesized around the template and removed. Although hard templating might be costly and time-consuming, it offers exact control over pore size and structure. Surfactants or block copolymers that self-assemble into micelles are used in soft templating, where the material forms around them. The mesoporous structure remains after the surfactant is removed. This approach is less expensive than hard templating and provides more flexibility in adjusting pore size and structure. The process of going from a liquid "sol" to a solid "gel" phase is called the sol-gel process. This method is frequently employed in the synthesis of mesoporous materials based on silica. Different functional groups can be included into the material through the sol-gel process, improving its characteristics for certain uses. Self-Assembly technique use van der Waals forces or hydrogen bonds among other molecules to organize building blocks into ordered mesoporous structures. Self-assembly is appropriate for many cutting-edge applications because it can produce extremely homogeneous pore structures with tuneable characteristics. Every synthesis technique has unique benefits and can be customized to yield mesoporous materials with particular properties and uses. The intended use and the necessary qualities of the finished product determine which synthesis process is best.

Mesoporous materials' inherent qualities play a number of roles in their power to remove VOCs [14]. In view of these combined characteristics, mesoporous materials are very suitable for effectively removing VOCs from a variety of settings.

- High Surface Area: The vast surface areas of mesoporous materials offer a considerable quantity of active sites for the adsorption of volatile organic compounds. Large volumes of volatile organic compounds can be efficiently and effectively adsorbed onto these materials thanks to their high surface area.

- Pore Size Adjustment: Pore size can be adjusted to enable the selective adsorption of particular VOCs according to their molecular size. The material can target and collect a wide spectrum of VOCs thanks to its tunability, which improves the selectivity and effectiveness of VOC removal.
- Large Pore Volume: Mesoporous materials' high pore volumes make it easier to collect and store huge volumes of VOCs. For large-scale applications where significant adsorption capabilities are necessary, this characteristic is very crucial.
- Surface Functionalization: By adding functional groups that improve interactions with particular VOC molecules, mesoporous materials' surface chemistry can be modified. Functionalization can increase target VOC binding sites' specificity and effectiveness of adsorption.
- Chemical and Thermal Stability: Mesoporous materials have exceptional thermal and chemical stability, which helps them to function and retain their structural integrity in a variety of environmental settings.

3. Mesoporous materials for removal of VOC

Adsorption, which can take the form of chemical or physical adsorption (chemisorption), is the main mechanism via which mesoporous materials remove VOCs. Weak van der Waals forces are present during the process of physisorption between the mesoporous material's surface and the VOC molecules. A number of variables, including surface area, pore size, and the physical characteristics of the VOC molecules (e.g., size, shape, polarity) affect this largely reversible process that permits the regeneration of the adsorbent material. On the other hand, chemisorption entails the creation of robust chemical interactions between the surface of the mesoporous material and the VOCs. The chemical makeup of the adsorbent and the VOCs determines whether this reaction is reversible or not. The mesoporous material's surface functional groups can be extremely important for chemisorption since they allow for particular interactions with the target VOCs [15–17].

A number of variables affect mesoporous materials' ability to adsorb VOCs. Greater pore volumes enable the sorption of more VOCs, which are necessary to attain high adsorption capacities, while higher surface areas offer more active sites for adsorption. The pore size guarantees that VOCs may readily access and interact with the material's internal surface, so the size of the pores in relation to the size of the VOC molecules is crucial. Selectivity and efficiency are improved by a well-defined pore size distribution. The adsorption capacity of mesoporous materials can be greatly increased by the presence of functional groups on their surface. Functional groups may engage in specific chemical interactions, such as hydrogen bonding or π-π interactions, with VOC molecules. The adsorption behavior can also be influenced by the surface's hydrophilicity or hydrophobicity. Lower temperatures and higher pressures are generally more advantageous for adsorption; however, the precise effects of temperature and pressure vary depending on the type of VOC and the mesoporous material. Comprehending these impacts is essential for enhancing adsorption procedures in real-world scenarios.

Adsorption kinetics are improved and efficient diffusion of VOC molecules into the material is ensured by pores of the right size. Selective VOCs can be better captured and their selectivity enhanced by mesoporous materials possessing a restricted pore size distribution. Since a material's adsorption capacity is directly correlated with its surface area, a high surface area increases the number of adsorption sites available for VOCs. The adsorption characteristics of mesoporous materials can be customized by adding functional groups like carboxyl groups, thiols, or amines. Functionalization can improve adsorption effectiveness and facilitate the design of materials for particular applications by enhancing interactions with particular volatile organic compounds.

4. Diverse applications in VOC removal

Mesoporous materials strong adsorption capabilities and adaptability have made them highly promising for the removal of volatile organic compounds in a variety of industrial applications. The possible effects of VOCs on the environment and human health make them a serious concern in the manufacturing and chemical processing sectors. Before VOCs are released into the atmosphere, air filtration systems can utilize mesoporous materials to absorb and remove them from exhaust streams. To effectively capture and recycle VOCs used as solvents in industrial activities, mesoporous silica and carbon materials have been used in solvent recovery procedures. By enabling solvent reuse, this lowers VOC emissions while simultaneously improving resource efficiency.

The application of mesoporous materials for VOC elimination has advantages for the oil and gas sector as well. Extraction, refinement, and storage of petroleum products release a variety of VOCs. By adding mesoporous materials to gas purification systems, these VOCs can be adsorbed, enhancing air quality and lowering the possibility of environmental contamination. Additionally, mesoporous materials aid in reducing VOC emissions in the paint and coatings sector, which uses a lot of solvents. The accumulation of VOCs from furnishings, building materials, and household goods can lead to hazardous levels of indoor air quality in both residential and commercial settings. An efficient way to raise the quality of indoor air is to incorporate mesoporous materials into HVAC systems and air purifiers. Many VOCs, such as formaldehyde, benzene, and toluene, which are frequently found indoors, can be absorbed by these materials.

Additionally, mesoporous materials are utilized in air conditioners and portable air purifiers, which offer localized VOC elimination in particular rooms of houses and workplaces. Their exceptional adsorption capabilities and adaptability to many environmental circumstances render them perfect for uninterrupted functioning in indoor air filtration systems. Furthermore, certain VOCs can be targeted by functionalized mesoporous materials, which increases the overall efficacy of air purification products.

There are also a lot of potential in the automotive and aerospace industries to use mesoporous materials for VOC elimination. VOCs are released from fuel combustion, vehicle exhaust, and interior materials in the automotive industry. Passengers air quality can be enhanced by using mesoporous materials in cabin air filters to absorb VOCs. Certain mesoporous materials are lightweight, which makes them appropriate for aircraft applications where weight is a crucial factor. They can also be used in exhaust gas treatment systems to catch and lessen volatile organic compound emissions, which helps to lessen pollution in the environment. By utilizing the advantages of each material type, the hybridization of distinct mesoporous materials—for example, the combination of silica and carbon components—can improve their performance even further.

These varied uses highlight the adaptability and potency of mesoporous materials in the removal of volatile organic compounds. Mesoporous material research and optimization will be critical to tackling VOC-related issues as companies search for effective and long-lasting air purification solutions.

5. Challenges and future directions

Even with mesoporous materials prospective uses for VOC elimination, there are still a number of obstacles and restrictions. Reuse and regeneration of these materials is one of the main problems. While physisorption is reversible, it frequently needs particular circumstances (such as high temperatures or low pressures) in order to efficiently desorb the adsorbed volatile organic compounds. The energy intensity of this process may eventually cause the material's adsorption ability to degrade. The reusability of the adsorbent is limited by chemisorption, which is typically irreversible.

Mesoporous materials selectivity for particular VOCs presents another difficulty. Although functionalization can improve selectivity, it is still difficult to design materials that can adsorb a variety of VOCs in mixed settings. The total efficiency and capacity of mesoporous materials can be impacted in real-world circumstances by the existence of numerous competing adsorbates.

Innovation and continuous research are needed to overcome the shortcomings of mesoporous materials in the removal of volatile organic compounds. The following are some new developments and areas for future study:

- Advanced Functionalization Techniques: Mesoporous materials' selectivity and adsorption capacities can be improved by creating new ways to functionalize them. Highly selective adsorbents for targeted VOCs can be made by methods including molecular imprinting and the addition of certain ligands or catalysts.
- Composite and Hybrid Materials: Mesoporous materials of various kinds, like silica, carbon, and MOFs, can be combined to create hybrid structures that take use of the special qualities of each component. When compared to single-component materials, these composites can provide better stability, reusability, and adsorption performance.
- Green Synthesis Techniques: By investigating ecologically friendly and sustainable synthesis techniques for mesoporous materials, production expenses and environmental effect can be minimized. Reuse of waste materials, natural precursors, or low-energy synthesis methods can help create more environmentally friendly production processes.
- Nanotechnology and Improved Pore Structures: By applying nanotechnology to produce more exact and consistent pore structures, mesoporous materials adsorption efficiency can be increased. The optimization of VOCs molecular mobility and interaction with the adsorbent surface can be achieved by customizing pore sizes and shapes at the nanoscale.
- Techniques for In-Situ Regeneration: Mesoporous materials reusability can be increased by creating effective in-situ regeneration techniques. Regeneration operations can be made more

Materials Research Foundations 173 (2025) 302-310 https://doi.org/10.21741/9781644903452-11

feasible and energy-efficient by utilizing methods such mild chemical regenerants, UV light activation, and microwave-assisted desorption.

- Multifunctional Adsorbents: The total efficacy of air purification systems can be increased by designing multifunctional mesoporous materials that can remove VOCs and other pollutants (such as particulate matter and inorganic gases) at the same time. These materials can offer all-inclusive solutions for enhancing air quality in different settings.

Scalability and affordability are critical for the widespread use of mesoporous materials in commercial applications. Standardizing material qualities for consistent performance and developing scalable synthesis processes can be facilitated by collaborative efforts among researchers, industry stakeholders, and policymakers. Studies conducted on a pilot scale and in situ experiments might offer important perspectives on the real-world obstacles and financial viability of using mesoporous materials in different VOC removal applications.

Mesoporous materials are already finding commercial use in the automotive, manufacturing, and air purification industries for the elimination of VOCs. The possibility of incorporating these materials into current systems and scaling them up will increase as research and new technologies are created. Mesoporous material optimization and continuous innovation have great potential to address VOC-related issues in a variety of applications.

6. Conclusion

Mesoporous materials are very successful at removing VOCs by physisorption and chemisorption because of their special structural characteristics. Their performance is mostly determined by surface area, pore size distribution, and surface chemistry. These have a wide range of uses due to their excellent adsorption capabilities and selectivity for particular volatile organic compounds. Even with their benefits, there are still issues to be resolved, such as adsorbent regeneration and reuse, selectivity in mixed environments, scalability, and long-term stability. Innovative strategies including enhanced functionalization, hybrid materials, green synthesis techniques, and nanotechnology are needed to solve these problems.

Mesoporous materials have a bright future in the removal of volatile organic compounds thanks to continued research and technical improvements. Their sustainability and practicality will be improved by effective in-situ regeneration methods and multifunctional adsorbents. Mesoporous materials have a great deal of potential to improve air quality and environmental health because they provide adaptable and efficient VOC removal methods.

Reference

[1]B. Muir, M. Sobczyk, T. Bajda, Fundamental features of mesoporous functional materials influencing the efficiency of removal of VOCs from aqueous systems: A review, Sci. Total Environ. 784 (2021) 147121. https://doi.org/10.1016/j.scitotenv.2021.147121

[2]W. Zou, B. Gao, Y.S. Ok, L. Dong, Integrated adsorption and photocatalytic degradation of volatile organic compounds (VOCs) using carbon-based nanocomposites: A critical review, Chemosphere. 218 (2019) 845–859. https://doi.org/10.1016/j.chemosphere.2018.11.175

[3]Z. Du, J. Mo, Y. Zhang, Risk assessment of population inhalation exposure to volatile organic compounds and carbonyls in urban China, Environ. Int. 73 (2014) 33–45. https://doi.org/10.1016/j.envint.2014.06.014

[4]H. Zheng, J. Csemezová, M. Loomans, S. Walker, F. Gauvin, W. Zeiler, Species profile of volatile organic compounds emission and health risk assessment from typical indoor events in daycare centers, Sci. Total Environ. 918 (2024) 170734. https://doi.org/10.1016/j.scitotenv.2024.170734

[5]L. Pál, S. Lovas, M. McKee, J. Diószegi, N. Kovács, S. Szűcs, Exposure to volatile organic compounds in offices and in residential and educational buildings in the European Union between 2010 and 2023: A systematic review and health risk assessment, Sci. Total Environ. 945 (2024) 173965. https://doi.org/10.1016/j.scitotenv.2024.173965

[6]W. Wang, J. Shen, Y. Chen, D. Zheng, L. Li, Identification and analysis of volatile organic compound and very volatile organic compound of MDF coated with different lacquers and their health risks to humans, J. Environ. Chem. Eng. 12 (2024) 112120. https://doi.org/10.1016/j.jece.2024.112120

[7]A. Mangotra, S.K. Singh, Volatile organic compounds: A threat to the environment and health hazards to living organisms – A review, J. Biotechnol. 382 (2024) 51–69. https://doi.org/10.1016/j.jbiotec.2023.12.013

[8]W. Gao, X. Tang, H. Yi, S. Jiang, Q. Yu, X. Xie, R. Zhuang, Mesoporous molecular sieve-based materials for catalytic oxidation of VOC: A review, J. Environ. Sci. 125 (2023) 112–134. https://doi.org/10.1016/j.jes.2021.11.014

[9]S.W. Sherpa, M. Ponnuchamy, A. Kapoor, M.M. Jacob, P. Sivaraman, Facile removal of sulfamethoxazole antibiotic from contaminated water using bagasse-derived pyrolytic biocarbon: Parametric assessment, mechanistic insights and scale-up analysis, J. Water Process Eng. 60 (2024) 105110. https://doi.org/10.1016/j.jwpe.2024.105110

[10] M.M. Jacob, M. Ponnuchamy, A. Kapoor, P. Sivaraman, Bagasse based biochar for the adsorptive removal of chlorpyrifos from contaminated water, J. Environ. Chem. Eng. 8 (2020) 103904. https://doi.org/10.1016/j.jece.2020.103904

[11] M. Ponnuchamy, A. Kapoor, P. Senthil Kumar, D.-V.N. Vo, A. Balakrishnan, M. Mariam Jacob, P. Sivaraman, Sustainable adsorbents for the removal of pesticides from water: A review, Environ. Chem. Lett. 19 (2021) 2425–2463. https://doi.org/10.1007/s10311-021-01183-1

[12] A. Balakrishnan, M.M. Jacob, N. Dayanandan, M. Chinthala, M. Ponnuchamy, D.V.N. Vo, S. Appunni, A.S. Gajendhran, Chitosan/metal organic frameworks for environmental, energy, and bio-medical applications: A review, Mater. Adv. 4 (2023) 5920–5947. https://doi.org/10.1039/D3MA00413A

[13] M.M. Jacob, M. Ponnuchamy, A. Kapoor, P. Sivaraman, Adsorptive decontamination of organophosphate pesticide chlorpyrifos from aqueous systems using bagasse-derived biochar alginate beads: Thermodynamic, equilibrium, and kinetic studies, Chem. Eng. Res. Des. 186 (2022) 241–251. https://doi.org/10.1016/j.cherd.2022.07.043

Materials Research Foundations 173 (2025) 302-310 https://doi.org/10.21741/9781644903452-11

[14] B. Han, T.H. Rupam, A. Chakraborty, B.B. Saha, A comprehensive review on VOCs sensing using different functional materials: Mechanisms, modifications, challenges and opportunities, Renew. Sustain. Energy Rev. 196 (2024) 114365. https://doi.org/10.1016/j.rser.2024.114365

[15] T.N. Tu, T.M. Pham, Q.H. Nguyen, N.T. Tran, V.N. Le, L.H. Ngo, K. Chang, J. Kim, Metal–organic frameworks for aromatic-based VOC capture, Sep. Purif. Technol. 333 (2024) 125883. https://doi.org/10.1016/j.seppur.2023.125883

[16] P. Zhang, M. He, W. Teng, F. Li, X. Qiu, K. Li, H. Wang, Ordered mesoporous materials for water pollution treatment: Adsorption and catalysis, Green Energy Environ. 9 (2024) 1239–1256. https://doi.org/10.1016/j.gee.2023.11.001

[17] J. Gao, Z. Li, S. Ma, Y. Zhang, F. Cheng, Novel process for facile preparation of mesoporous silica-based materials with controllable pore structure from coal fly ash, Particuology 91 (2024) 128–137. https://doi.org/10.1016/j.partic.2024.03.001

 Materials Research Forum LLC
https://doi.org/10.21741/9781644903452-12

Chapter 12

Mesoporous Materials as Stationary Phases for Size Exclusion Chromatography

Vipul D. Prajapati[1*], Princy Shrivastav[2]

[1]Department of Pharmaceutics, SSR College of Pharmacy (Permanently Affiliated to Savitribai Phule Pune University), Sayli, Silvassa, U. T. of D.N.H. and D.D., India

[2]Department of Pharmaceutical Chemistry, SSR College of Pharmacy (Permanently Affiliated to Savitribai Phule Pune University), Sayli, Silvassa, U. T. of D.N.H. and D.D., India

vippra2000@yahoo.com

Abstract

This chapter explores the innovative utilization of varieties of the mesoporous materials as the stationary phases in size exclusion chromatography (SEC). Mesoporous materials, characterized by their highly ordered pore structures and large surface areas, offer unique advantages for separating molecules based on size. We delve into the synthesis methods of these materials, emphasizing the control over pore size and distribution, which are critical for enhancing the performance of SEC. The chapter reviews recent advancements in the application of mesoporous materials, highlighting their superior resolution, selectivity, and capacity compared to traditional stationary phases. This chapter also discusses the underlying mechanisms that govern size exclusion in mesoporous matrices, supported by theoretical models and experimental data. Case studies demonstrate the reality practical utilities of these materials in biopolymer and nanoparticle separation. Finally, the chapter address the challenges and future prospects in the field, including the development of hybrid mesoporous structures and the integration of novel functionalities to further expand the capabilities of SEC. This comprehensive overview underscores the transformative potential of mesoporous materials in chromatographic science, paving the way for more efficient and versatile separation techniques.

Keywords

Mesoporous Materials, Size Exclusion Chromatography, Molecular Separation, Biopolymer Separation, Separation Techniques

Contents

Materials Research Foundations 173 (2025) 311-339 https://doi.org/10.21741/9781644903452-12

1. Introduction

Because of their special structural qualities, mesoporous materials have become important stationary phases in size exclusion chromatography (SEC). These materials have huge pore volumes and high surface areas (usually between 400 and 1000 m^2/g), which make them excellent for a variety of separation applications. Their pore diameters range from 2 to 50 nanometers. SBA-15, MCM-41, and MCM-48 are examples of mesoporous materials that are created utilizing hard or soft templating techniques. The core idea of SEC—selective separation based on molecular size—is made easier by their organized porous architectures. The efficacy of separations is improved by the high surface area, which permits greater interaction with analytes [1].

The size of the pore in the stationary phase is needful in SEC as it dictates the size of the molecules that can be separated. Because mesoporous silica materials may be tailored to have tight pore size distributions, which enable precise size-based separation of polymers, biomolecules, and other substances, they are particularly beneficial. This capacity is crucial in industries like biology and pharmaceuticals where it's frequently necessary to separate complex combinations [2]. Mesoporous materials can also be functionalized to improve selectivity for particular analytes, which increases their suitability for a wider range of chromatographic methods [1,2]. Overall, mesoporous materials represent a versatile and effective choice for stationary phases in SEC, offering enhanced separation capabilities due to their ideal structural features and tunable properties. Their applications extend beyond chromatography, making them valuable in various fields such as catalysis, drug deliveries, and environmental sciences.

2. Fundamentals of Size Exclusion Chromatography (SEC)

Another name for SEC is gel filtration chromatography. It's a method for dividing molecules according to size. The analysis and purification of biomolecules, including proteins and polysaccharides, can be facilitated by this approach, which eliminates the need for chemical interactions with the stationary phase. The basis of SEC is the idea that molecules with varying sizes will elute in a particular order from a column filled with porous beads. Larger molecules elute first because they are prevented from fitting through the beads' pores and travel through the column faster. On the other hand, smaller molecules have a delayed elution period because they can fit through the pores and stay there longer. This mechanism allows for the separation of molecules based on their hydrodynamic volume rather than their chemical properties, making SEC a gentle method that preserves the integrity of sensitive biomolecules [3].

2.1 Principles of SEC

SEC is a chromatographic method that divides molecules into groups according to size. SEC is governed by a number of basic factors, including column properties, molecule size and elution order, and separation process. The basis of SEC is the way molecules pass through a porous stationary phase at varying rates. Usually, the stationary phase is made up of porous beads that interconnect to form a network of pores. Smaller molecules can enter the pores and have longer retention durations than larger ones, which prevents the larger molecules from entering and causes them to move through the column more quickly. Another crucial factor is the stationary phase's pore size, which needs to match the size of the molecules being separated. To maximize separation for specified analyte sizes, columns are frequently engineered with precise exclusion limits [4].

The hydrodynamic volume and molecular size of molecules also influence the sequence in which they elute. Accordingly, bigger molecules elute first, and then smaller ones. Proteins, polysaccharides, and other macromolecules can be separated with SEC because it works well for a broad range of molecular weights. Resolution is also impacted by the mobile phase's flow rate. Better separation is possible with a slower flow rate, but analysis takes longer. Because of the longer contact period between the analytes and the stationary phase, longer columns typically result in better resolution. Other elements that affect the resolution of SEC include the viscosity of the mobile phase and the size of the stationary phase's particles. Additionally, SEC is a non-destructive method, maintaining the integrity of the separated molecules. This is particularly important for sensitive biomolecules, allowing for further analysis or use after separation [4,5].

2.2 Theory of Molecular Size Separation

The foundational idea of the theory of molecular size separation, especially as it applies to SEC, is that molecules can be distinguished according to their effective size in solution—also known as hydrodynamic volume—as they move through a porous medium. SEC makes use of a stationary phase made up of beads with pores that fall within a particular range of sizes. Molecules can enter this structure selectively according to their size. Larger molecules are prevented from entering the pores as a sample combination is added to the column, causing them to elute first. Because smaller molecules may fit through the pores, there will be a later elution and a longer retention period. This results in a separation based on size, with larger molecules appearing before smaller ones in the elution profile. The hydrodynamic volume is influenced by the shape and flexibility of the molecule, which affects how it interacts with the porous medium[6].

We must take into account some theoretical elements in order to comprehend molecular size separation. There are specified exclusion and penetration limits for every SEC column. The molecular weight below which all molecules can enter the pores is known as the permeation limit, whereas the molecular weight above which none can is known as the exclusion limit. This range of molecules is divided according to how well they can enter the pores [7]. Partitioning theory, which describes how molecules move between the limited areas inside the pores and the bulk solution, can also be used to understand the separation process. Since it has a strong correlation with the observed elution volumes, the mean span dimension of the molecules is frequently utilized as a size parameter for this distribution [8].

2.3 Applications of SEC

SEC offers practical techniques for the purification, characterisation, and size-based analysis of macromolecules, making it an invaluable tool in a variety of industries, most notably biopharmaceuticals and polymer science. It is a vital instrument in analytical chemistry because to its adaptability and capacity to preserve the integrity of delicate samples. The following are a few of the main uses for SEC:

2.3.1 Protein Purification

Monoclonal antibodies are among the many proteins that can be purified using SEC. Aggregates, fragments, and other impurities are successfully eliminated, guaranteeing great purity and uniformity in the finished product. Maintaining the safety and effectiveness of biopharmaceuticals depends on this [3]. The SARS-CoV-2 spike protein serves as a current illustration of the application of SEC for protein purification. SEC was utilized in conjunction with affinity chromatography by researchers to extract the spike protein, which is essential for structural research and the creation of vaccines. In particular, for binding tests with the ACE2 receptor, the purified spike protein needs to be devoid of aggregates to facilitate proper structural analysis. To isolate the protein's trimeric form, which is necessary for its use in vaccinations and medicinal studies, required the use of SEC [9].

2.3.2 Characterization of Biomolecules

Proteins can have their quaternary structures determined and their tertiary structures evaluated in their original environments thanks to SEC. Based on their hydrodynamic volumes, it can differentiate between folded and unfolded protein forms, which is crucial for quality control in the manufacturing of biopharmaceuticals [10]. SEC is widely used in biomolecule characterisation to assess the size, purity, and aggregation state of biomolecules, especially proteins and monoclonal antibodies (mAbs). The biopharmaceutical industry's use of SEC for mAb quality control is one recent example. In order to guarantee the stability and purity of these therapeutic proteins,

Materials Research Foundations 173 (2025) 311-339 https://doi.org/10.21741/9781644903452-12

monomers, dimers, and higher aggregates must be separated and quantified using SEC. The combination of SEC with multi-angle laser light scattering (MALLS) detectors is one example of a modern innovation that makes molecular weight estimation more accurate without the need for column calibration [11].

Furthermore, SEC is employed in structural biology in conjunction with methods such as Small-Angle X-ray Scattering (SEC-SAXS). By removing aggregation effects, which could normally obstruct SAXS measurements, and offering insights into the conformational ensemble of intrinsically disordered proteins (IDPs), this method is useful for characterizing these proteins. This approach is very helpful for researching flexible biomolecules in solution [12].

2.3.3 Molecular Weight Distribution

The molecular weight distribution (MWD) of proteins, polymers, and biotherapeutics is frequently examined using SEC. SEC is often used to analyze molecular weight distributions of synthetic polymers. By building calibration curves with standards of known sizes, researchers may determine the sizes of polymer molecules in a sample and gain insight into their polydispersity and general characteristics. For the MWD of synthetic polymers, SEC in conjunction with multi-angle light scattering (MALS) and refractive index detectors is frequently used. Without the need for calibration standards, this method is utilized to ascertain the absolute molecular weights of polymer samples. For polymer applications in materials science, it can precisely evaluate the size and molecular weight distributions of polymer chains [13].

By isolating the monomers from the aggregates and fragments of higher molecular weight, SEC is used to evaluate the molecular weight distribution of mAbs. Because aggregates can impact mAbs' immunogenicity and therapeutic efficacy, SEC is essential for quality control during the biopharmaceutical manufacturing process. In biophysical investigations, SEC-MALS is also utilized to ascertain the molecular weight distribution of proteins and nucleic acids. By examining the scattering intensity as a function of molecular size, it enables the size-based separation of proteins and yields accurate molecular weight data [11,12]. These applications highlight the versatility of SEC in analyzing molecular weight distributions in diverse biomolecules and polymers.

2.3.4 Polymer Fractionation

Fractionating polymers using preparative SEC enables the size-based separation of distinct polymer components. This is especially helpful in the synthesis and characterisation of polymers, where knowledge of the size distribution is essential [14]. Fractionation of polyolefins, such as polyethylene and polypropylene, has been done using SEC. In these situations, knowing the MWD of the polymers is essential to comprehending characteristics like flexibility, strength, and thermal behavior. SEC aids in the analysis of these characteristics and the optimization of polymer synthesis for particular purposes by dividing the polymer into fractions of varying molecular weights [15].

SEC has been used in biopolymer research to fractionate biodegradable polymers, such as polylactic acid (PLA). This enables researchers to examine how PLA's molecular weight affects its mechanical characteristics, biodegradability, and processing behavior. PLA is being utilized more and more in medical devices and environmentally friendly packaging [16].

Synthetic polymers that are vital to many industrial uses, such as polystyrene and polyacrylamides, are frequently fractionated using SEC. SEC-MALS can be used to calculate the radius of gyration in addition to the MWD, yielding detailed information about the behavior and structure of polymers. [17]. In these applications, SEC is essential for fractionating polymers into distinct molecular weight classes, aiding in the development of materials with precise and predictable properties.

2.3.5 Desalting and Buffer Exchange

When desalting biomolecules—that is, separating the target molecule from smaller salts and contaminants according to size—SEC works well. When preparing materials for additional examination or experimentation, this application is very crucial [14,18].

Before proteins crystallize, excess salts and impurities are frequently eliminated via SEC. Elevated salinity levels can disrupt the process of crystallization and diffraction quality. For example, salts can be effectively separated using Sephadex G-25 resin, which enables the protein to crystallize in an appropriate buffer without experiencing severe protein loss or dilution [19]. In the process of purifying monoclonal antibodies, SEC is also utilized to desalt and modify the buffer following affinity chromatography processes. By neutralizing the pH and eliminating any remaining elution buffer components, this procedure helps prepare the antibodies for use in drug formulation or binding assays, among other downstream uses [20]. SEC is used in enzyme purification to quickly exchange buffers prior to activity tests. In order to preserve activity, enzyme samples frequently need to be moved to particular buffers with ideal pH or ionic strength. Buffers can be quickly changed via SEC without compromising the activity of the enzyme [21]. These examples demonstrate the versatility of SEC for desalting and buffer exchange, essential in numerous biotechnological applications where precise conditions are critical for biomolecular integrity.

2.3.6 High-Throughput Screening (HTS)

SEC is becoming more and more integrated into HTS procedures in material research and drug discovery. This makes it possible for researchers to quickly examine a variety of samples, which speeds up the process of identifying possible leads and simplifies research and development activities [22].

Protein-protein interaction screening is one of the main uses of SEC in HTS. Through the application of fluorescent tag labelling and compound library mixing, researchers can utilize SEC to separate compounds that have formed complexes with proteins. The identification of hits is then made possible using fluorescence detection. This strategy has been effectively used to find inhibitors of protein-protein interactions linked to illness [23].

A significant use of SEC in HTS is affinity screening for ligand binding. Through the use of SEC, researchers may distinguish between compounds that have bonded to a target protein and those that have not by tagging the protein with a fluorescent tag and mixing it with a compound library. Substances that exhibit a high affinity for the target protein can subsequently be identified via fluorescence detection. Drug discovery has used this technique to find possible lead molecules for a range of medicinal objectives [24].

SEC has been applied to oligonucleotide screening in addition to protein-protein interaction and ligand binding affinity screening. Researchers can employ SEC to separate oligonucleotides that have formed complexes with the target molecule by marking the oligonucleotides with a fluorescent tag and combining them with the target molecule. After that, oligonucleotides having a high affinity for the target can be found using fluorescence detection. This methodology has been utilized in the creation of medicinal oligonucleotides [25]. Moreover, protein-ligand complexes can be characterized by SEC. A protein-ligand complex's elution volume can be used to calculate its molecular weight, which can then be utilized to estimate the complex's stoichiometry. This data may prove useful in comprehending the mode of action of potential pharmaceuticals [26]. In conclusion, SEC is a powerful technique that has found widespread application in HTS. By enabling researchers to efficiently screen large libraries of compounds, SEC has accelerated the drug discovery process and contributed to the development of novel therapeutics.

2.3.7 Hyphenation with Other Techniques

SEC can be used in conjunction with other methods to provide more details regarding the size, molecular weight, and structural properties of macromolecules. SEC's analytical capabilities are improved by this hybrid technique, which makes it an effective tool for analyzing complicated samples [17,27].

Protein identification and characterization are made possible by the combination of size-based separation and mass spectrometric detection in SEC with mass spectrometry (SEC-MS). It is especially helpful in the analysis of post-translational modifications, aggregation, and protein complexes [28–30]. On the other hand, size-based separation is integrated with SEC with Nuclear Magnetic Resonance Spectroscopy (SEC-NMR). It allows for the structural characterisation of proteins and facilitates the investigation of conformational changes and interactions between proteins and ligands [16,31].

SEC-IR combines infrared detection and size-based separation. It offers details on the folding and secondary structure of proteins. Conformational changes can be detected and protein stability can be monitored using SEC-IR [16]. Conversely, SEC-MALS integrates multi-angle light scattering detection with size-based separation. It makes it possible to calculate the size and absolute molar mass of proteins and protein complexes in solution [17].

Additionally, there is SEC-SAXS, which combines small-angle X-ray scattering with size-based separation. Even for flexible and disordered proteins, it offers information regarding the low-resolution structure and shape of proteins in solution [32,33]. These hyphenated techniques offer powerful tools for the purification, identification, and structural characterization of proteins, enabling a more comprehensive understanding of protein structure and function.

2.3.8 Real-Time Monitoring

In-operando monitoring of chemical processes, such as polymerization reactions and bioprocess monitoring, is one of the emerging uses of SEC. Research and industrial applications can benefit from the dynamic insights into molecular alterations that this real-time capacity offers [34]. A recent development showed how to monitor small molecule and polymer interactions in real time using ultrahigh-performance liquid chromatography (UHPLC) in combination with SEC. With cycle lengths as short as 1.5 minutes, this combination enables quick analysis and provides quantitative insights into reaction processes, such as the kinetics of polymer degradation and hydrolysis [35].

Real-time, comprehensive data regarding the molecular weight and size of macromolecules is provided by SEC-MALS. This method is very helpful for tracking polymer reactions and

Materials Research Foundations 173 (2025) 311-339 https://doi.org/10.21741/9781644903452-12

evaluating how aggregates form during chemical reactions [36]. Real-time chemical reaction monitoring is made possible by the combination of SEC and IR spectroscopy, which analyzes the spectrum changes that result from the conversion of reactants into products. This technique helps in understanding the kinetics of chemical transformations and monitoring the progress of reactions. Real-time monitoring of molecular interactions and conformational changes during chemical reactions is made possible by combining SEC with NMR spectroscopy. This method is useful for researching dynamic dynamics in intricate combinations [37]. These hyphenated techniques enhance the capabilities of SEC by providing complementary information that can lead to a better understanding of chemical processes in real-time, improving efficiency and control in various industrial applications.

3. Traditional Stationary Phases in SEC

For efficient separation based on molecular size, different polymers have historically been utilized as the stationary phases in SEC. Due to their distinct qualities, each of them can be used in different ways to separate biomolecules and polymers according to size. When choosing the stationary phase, there are a few important aspects to take into account. The first factor that must be considered in order to achieve effective separation is the stationary phase's pore size. To guarantee ideal resolution, the pore size must match the molecular weight of the analytes being separated. For repeatable results, the stationary phase's stability under various circumstances (such as pH and temperature) is also crucial. The stationary phase should also be compatible with the mobile phase and the analytes to avoid any unwanted interactions that could affect separation efficiency. Table 1 gives a brief summary of various properties of common stationary phases that can be used in SEC.

Table 1. Summary of various properties of common stationary phases along with their key characteristics especially for SEC.

Stationary Phase	Pore Size (Å)	Particle Size (μm)	Molecular Weight (Da)	Key Characteristics	Reference
Dextran	30-40	40-120	1,000-400,000	• Stable in aqueous solutions • Solubility is not affected by pH • Exhibit Newtonian flow characteristics.	[38]
Agarose	60-1000	45-165	5,000-5,000,000	• Linear polysaccharide • Stable in most aqueous solutions.	[39]
Poly-acrylamide	100-3000	10-50	1,000-10,000,000	• Stable in aqueous and organic solvents • Can exist in various forms: linear, cross-linked, and copolymers • Highly water-absorbent.	[40]

Silica	30-1000	3-50	100-150,000	• Excellent rigidity and stability at high temperature and pressure • Stable in organic solvents • pH 2-8.	[41]
Hydroxylated methacrylate	300-4000	5-20	1,000-10,000,000	• Hydrophilic, stable in aqueous and organic solvents • Ability to tailor pore size ad surface chemistry for specific applications	[42]
Polystyrene-divinylbenzene	10-1000	5-20	200-10,000,000	• Stable in organic solvents • pH 2-12 • Often available as microsphere or beads • Good thermal stability.	[43]
Polyhydroxy-ethyl aspartamide	100-1000	5-20	100-10,000,000	• Neutral • Polar • Provides sharper peaks and improved selectivity and recovery.	[44]

3.1 Porous Polystyrene-Divinylbenzene (PS-DVB)

PS-DVB microspheres are a kind of cross-linked polymer material that typically have a particle size of 6–10 μm with a high specific surface area and porosity. It is made up of a copolymer of divinylbenzene and polystyrene that creates porous, stiff beads. It is possible to adjust the pore diameters to fit particular molecular weight ranges. Because of its excellent mechanical stability and adjustable pore diameters, it is frequently employed to separate proteins and synthetic polymers. It works especially well with big polymers and biomolecules [45].

An interpenetrating polymer network is formed by swelling monodisperse seed particles with porogen, monomers, crosslinkers, and initiators in two steps as part of the active swelling method. By adjusting the crosslinking degree, porogen, swelling temperatures, and crosslinking agents utilized during synthesis, the porosity and specific surface area of the microspheres can be tuned [45,46].

PS-DVB functions as a stationary phase in SEC, enabling the size-based separation of molecules. Functional groups like sulfonic acid can be added to PS-DVB to increase its hydrophilicity and selectivity for particular analytes. This increases the spectrum of compounds that can be successfully separated by enabling mixed-mode separations, which combine the use of size and contact mechanisms. Research has indicated that PS-DVB is successful in SEC applications since it can accomplish baseline separations of complicated combinations, such as organic acids and benzene analogues. It is a useful substance in analytical chemistry because of its capacity to achieve excellent resolution and efficiency in separations [45].

3.2 Silica-Based Materials

Since silica-based materials are good at separating molecules according to size, they are frequently used as stationary phases in SEC. The silica particles employed in SEC usually have pore widths of 100–1000 Å and diameters of 5–10 μm. Because of this porous structure, macromolecules with different sizes, like polymers and proteins, can be effectively separated. Because silica has good mechanical stability, it can be used in high-pressure situations. Because of its stiffness, the particles are guaranteed to remain unchanged under pressure, resulting in reliable flow routes and effective separation. In chromatograms, silica-based stationary phases can produce distinct, sharp peaks that signify strong separation sensitivity and efficiency. Silica is stable in a wide range of solvents, enabling its use in diverse analytical conditions and applications [47].

The presence of free silanols on silica surfaces might cause unwanted interactions with some analytes, especially basic chemicals, even though SEC typically separates based on size. Peak form and separation efficiency may be impacted by these interactions. Different silica modifications have been created to limit these interactions, such as hybrid silica or Type C stationary phases with lower silanol concentration, in order to address these problems. Additionally, silanols can be shielded by polar-embedded groups, which enhances their performance with polar compounds [47,48].

3.3 Agarose Gel

A naturally occurring polysaccharide having a network of holes that creates a gel matrix is agarose. It is efficient at separating biomolecules and is biocompatible. One important SEC medium that is frequently employed is agarose gel, which is mostly utilized to separate proteins and nucleic acids in applications that call for careful treatment of delicate biomolecules. They are frequently employed in protein purification preparative SEC [49].

Because of these advantageous characteristics, agarose gels, like SepharoseTM, are especially helpful in SEC. Numerous pore diameters can be used in the production of agarose gels, which enables the separation of a broad range of molecular weights. For instance, Sepharose 6B can handle molecules up to 4,000 kDa, but Sepharose 4B can separate molecules between 60 and 20,000 kDa. The variety of pore diameters and the size of the agarose beads affect the separation's resolution. Although smaller beads usually offer better resolution, they may also slow down flow rates, which would slow down the separation process. Agarose gels are extensively employed in the purification of nucleic acids, polysaccharides, and proteins. They are particularly effective for applications requiring gentle separation conditions, such as desalting and buffer exchange, without the risk of denaturing sensitive biomolecules [49].

3.4 Polyacrylamide Gel

Synthetic polymers called polyacrylamide gels can be used to make a matrix with predetermined pore sizes. This makes it possible to separate biomolecules precisely. In SEC, it's commonly used. Acrylamide and N, N'-methylenebisacrylamide are copolymerized to produce polyacrylamide gels, including Bio-Gel P. They can be designed to have different pore diameters, which makes it possible to separate molecules with molecular weights between 100 and 100,000 Da. When high resolution is needed, this kind of stationary phase is widely utilized in SEC and electrophoresis to separate proteins and nucleic acids. Because polyacrylamide gels are so non-ionic and hydrophilic,

 Materials Research Forum LLC
https://doi.org/10.21741/9781644903452-12

they have little interaction with the biomolecules being separated. This characteristic is essential for preserving the biological activity of delicate substances while they are being separated. [50].

These gels are suitable for a variety of experimental circumstances since they can function well in the pH range of 2 to 10 and can be autoclaved at neutral pH. Within a certain range, temperature rises can also enhance flow rates and resolution. Their adaptability renders them appropriate for an extensive array of uses, such as the separation of proteins and nucleic acids. For quick desalting and buffer exchange, for instance, Bio-Gel P-6DG is especially advised as it offers good recovery rates with little sample dilution. Additionally, the fragile biomolecules may be effectively separated without denaturation using the mild filtration process, which makes it perfect for sensitive applications. These gels are more useful in a variety of biochemical applications since they can hold a variety of solvents, such as chaotropic chemicals and diluted organic acids [3,50].

3.5 Cellulose-Based Materials

Biomolecules are separated and purified in SEC using a lot of cellulose-based materials. These materials have multiple benefits, including as hydrophilicity, biocompatibility, and the capacity to be tailored into different configurations appropriate for SEC applications. These beads can be made in a number of ways to produce spherical particles with better adsorption qualities, including titration, emulsification, and microfluidic techniques. [50].

For SEC applications, cellulose derivatives have also been investigated in addition to cellulose beads. For instance, size exclusion chromatography with ion-exclusion moderation has been used to evaluate cellulose acetate phthalate. Utilizing enhanced techniques involving SEC, the molar mass distribution of cellulose in Kraft pulp has been assessed. Applications for cellulose-based SEC materials include the purification and examination of different biomolecules, such as nucleic acids and proteins. They are especially helpful for buffer exchange, desalting, and separating chemicals that are sensitive while preserving their biological activity [51,52].

4. Introduction to Mesoporous Materials

Materials with special properties owing to their nanoscale size are the subject of nanotechnology. Over the past twenty years, a wide range of applications have found tremendous potential with nanostructured materials. Mesostructured materials have been the subject of extensive research among these nanomaterials [53]. According to IUPAC, the mesoporous materials are those materials having the pore sizes from 2 to 50 nm [54]. Their structural features are represented via a wide class of molecules including the carbon structural features [55], silica [56], metal hydroxides [57], metal salts [58], metal oxides [59–61], hybrid materials, organic structures and many others [58].

The use of these nanoscale materials has drawn a lot of attention since Mobil Oil Company researchers discovered a family of uniformly ordered mesoporous silica molecular level sieves in 1992 that they named M41S [62]. Over the past twenty years, there has been a noticeable surge in interest in the synthesis of novel mesoporous materials. The development of mesoporous materials has followed trends that started with inorganic structures like metal oxides and silica and continued with hybrid materials like periodically mesoporous organo-silica (PMO) [63,64] and the metallic organic frameworks (MOFs) towards the quality purely organic materials namely porous organic frameworks (POFs) [65]. Recently, the broad progress in the development of the latter has been identified.

Materials Research Foundations 173 (2025) 311-339
https://doi.org/10.21741/9781644903452-12

Materials of mesoporous nature are mostly attractive for different applications in technical field of sciences namely catalysis [66], electronics [57], photocatalytic hydrogen production [67], solar cells [68] and battery components [69]. Because mesoporous materials have special qualities like large specific surface area, significant pore volume, variable pore size, chemically stable, and the significant surface modifying nature, a great deal of study has been done on the biomedical applications of all these types of materials [70]. Moreover, depending on their chemical nature, these types of materials might possess the significant fluorescent [71], magnetic [72] or conducting [73] characteristics. Due to such types of nature in practice, these types of substances have created great attention for various applications in the area of biomedical sciences including drug molecules and genetic materials deliveries, cellular imaging, biological sensing, biomolecules separation and many others.

4.1 Characteristics of mesoporous materials

Mesoporous materials are especially advantageous in a variety of applications, including SEC, since they have a number of unique properties. Their high surface area, unique pore structure, chemical stability, adaptability in functionalization, light weight, biocompatibility, and improved mass transfer qualities are some of their most distinguishing features. Because of these qualities, they work incredibly well in size exclusion chromatography and other separation methods. Pore diameters in mesoporous materials range from 2 to 50 nanometers. The effective size-based separation of molecules is made possible by this size range, since smaller molecules can pass through the pores while larger ones are prevented from doing so. Achieving consistent and effective separations in SEC requires a highly uniform pore size distribution, which is exhibited by many mesoporous materials including MCM-41 and SBA-15. These materials are easier to handle because of their decreased density as a result of their porous nature as opposed to standard solid materials. Their high surface areas (which frequently surpass 400 m^2/g) improve their interaction with analytes. Applications needing a high adsorption capacity and selectivity will benefit from this characteristic. Analytes can diffuse into and out of the pores more quickly thanks to the mesoporous structure, which is beneficial for increasing the effectiveness of SEC separations [74,75].

In addition, mesoporous materials exhibit chemical stability in a broad spectrum of circumstances, such as changes in temperature and pH. Consistent performance during chromatographic separations is guaranteed by this stability. Mesoporous materials' surfaces can be readily altered or functionalized to improve selectivity for particular analytes. Because of its adaptability, separation techniques can be optimized for a range of uses. Since many mesoporous materials are biocompatible, they can be used in biochemistry and medicine, especially for the purification and separation of biomolecules [74].

4.2 Types of mesoporous materials suitable for chromatography

Most of the mesoporous materials have gained significant attention in chromatography due to their unique structural properties, which enhance separation efficiency. Here are some types of mesoporous materials suitable for chromatography:

1. Mesoporous Silica: Materials containing mesoporous silica, such MCM-41, MCM-48, and SBA-15, are distinguished by their consistent pore size distribution, huge pore volume, and high surface area. They are frequently employed as the stationary phases in liquid type chromatography,

offering efficient separation for a range of molecules as the analytes mostly proteins, peptides, and tiny chemical substances. Optimization depending on the target chemicals' specific molecular weight is possible thanks to their adjustable pore diameters [75].

2. Periodic Mesoporous Organosilica (PMO): PMOs are hybrid materials that provide improved selectivity and tunability by integrating organic functional groups into their silica framework. These materials' capacity to alter surface characteristics and enhance separation efficiency makes them especially helpful for chromatographic applications. They can be applied to high-efficiency separations in both monolithic and packed columns [2].

3. Mesoporous Carbon: They are ideal for a variety of chromatographic utilizations due to their high exposure of surface area and adjustable structural porosity. They take advantage of their hydrophobic properties and stability in the diversified environments when using SEC to separate chemical compounds and biomolecules [76].

4. Metal-Organic Frameworks (MOFs): Despite not being categorized as mesoporous materials by convention, many MOFs possess mesoporosity and can be employed in chromatography. Because MOFs may be designed to have particular pore sizes and functions, they can be used in chromatography to achieve selective separations [77].

5. Agarose and Polyacrylamide Gels: Modified agarose and polyacrylamide gels can serve as mesoporous materials in chromatographic applications, while their main utility is in gel electrophoresis. These materials' moderate separation conditions and biocompatibility make them especially useful for the separation of biomolecules like proteins and nucleic acids [78].

4.3 Advantages of Mesoporous Materials in SEC

Mesoporous materials offer several advantages in SEC, making them highly effective as stationary phases. Figure 1 illustrates the mechanism of size exclusion in mesoporous materials. Here are the key benefits:

1. High Surface Area: High surface areas are characteristic of mesoporous materials. In SEC applications, this large surface area improves capacity and separation efficiency by enhancing the contact between the stationary phase and analytes [3].

2. Uniform Pore Size Distribution: These materials enable for exact control over the separation process because of their well-defined pore diameters, which range from 2 to 50 nm. For analytical applications, the reliable and repeatable separations made possible by the small pore size distribution are essential [5].

3. Large Pore Volume: Mesoporous materials can accommodate larger molecules due to their huge pore capacity, which makes them useful for size-based separation of a variety of biomolecules and polymers. Applications utilizing complicated mixtures benefit greatly from this feature [3].

4. Chemical Stability: Mesoporous materials have outstanding chemical stability under a range of circumstances, such as temperature and pH. Regular analytical methods require dependable performance over extended periods of time, which is ensured by this stability [79].

5. Versatility in Functionalization: It is simple to alter the surface relevant chemistry of mesoporous materials to improve selectivity for particular analytes. Researchers can better achieve separation results in SEC by customizing the materials for specific applications thanks to its tunability [79].

Materials Research Foundations 173 (2025) 311-339
https://doi.org/10.21741/9781644903452-12

6. Enhanced Mass Transfer: Diffusion constraints are lessened by the effective mass transfer of analytes made possible by the porous nature of mesoporous materials. In SEC processes, this characteristic helps to increase resolution and speed up separation times [79].

7. Biocompatibility: Since many mesoporous silica materials are biocompatible, they can be used in applications where biological samples are present. Maintaining the integrity of sensitive biomolecules is critical in biochemistry and pharmaceutical applications, where this feature is especially important [79,80].

8. Wide Range of Applications: Mesoporous materials are used in a number of industries, such as environmental remediation, medicine delivery, and catalysis, in addition to SEC. Their adaptability makes them more useful for industry and scientific applications [79,80].

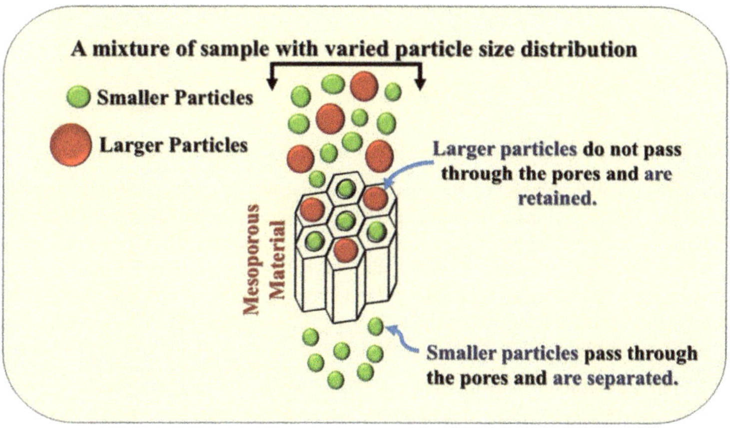

Figure 1. Mechanism of size exclusion in mesoporous materials.

5. Synthesis and Modification of Mesoporous Materials for SEC

To maximize the performance of mesoporous materials in chromatographic applications, synthesis and modification are essential. Technologies including functionalization, soft template synthesis, and the sol-gel technique enable the creation of materials with specialized qualities appropriate for a range of uses. The potential to regulate pore size, surface chemistry, and integrate metals enhances the use of mesoporous materials in chromatography, rendering them advantageous instruments for both academia and industry. The various possibilities of methods of mesoporous materials synthesis are depicted in figure 2.

Materials Research Forum LLC
https://doi.org/10.21741/9781644903452-12

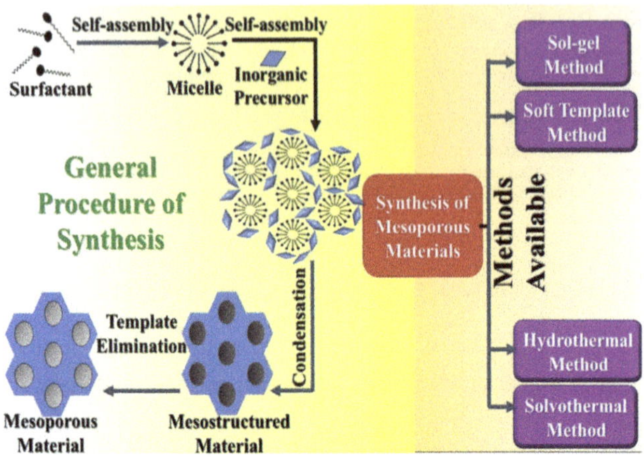

Figure 2. An outlook of the methods of synthesis of mesoporous materials.

One of the most popular techniques for creating mesoporous silica nanoparticles (MSNs) is the Sol-Gel Method. Tetraethyl orthosilicate (TEOS) and other silica precursors are hydrolysed and then condensed in the presence of surfactants (such as cetyltrimethylammonium bromide, or CTAB), which serve as templates. Usually, the sol-gel process is carried out in a controlled environment (such as one with regulated pH and temperature) to produce a gel. The gel is subsequently dried and calcined to eliminate the surfactant and produce mesoporous structures. Pore surface area, morphology, and size can all be precisely controlled using this technology [81].

The next technique is called "soft template method," which makes use of block copolymers or surfactants as templates to form mesoporous structures. During the synthesis process, the surfactants self-assemble into micelles that control the pore structure. This approach is adaptable for creating materials with particular pore sizes and shapes, which makes it appropriate for a range of SEC applications [53].

Mesoporous materials are synthesized in aqueous or organic solvents at high temperatures and pressures using the hydrothermal and solvothermal techniques. They have the ability to alter the pore structure and improve crystallinity. For SEC applications, hydrothermal techniques are useful because they yield highly organized mesoporous materials with better stability [53].

There are also numerous alteration methods accessible. Mesoporous materials can have their surface altered to improve their selectivity and interaction with particular analytes. Silylation and the addition of functional groups (such as amines or thiols) through chemical reactions are common methods of functionalization. Mesoporous materials can perform much better in SEC when functionalized because it increases the materials' affinity for target molecules and increases separation efficiency [53,82].

Mesoporous silica can benefit from the addition of metal ions or nanoparticles to boost its catalytic capabilities and analyte selectivity. During the sol-gel process, post-synthesis loading or co-condensation are frequently used to accomplish this. Additional active sites may be provided by metal-modified mesoporous materials, which may be advantageous for particular chromatographic applications [81]. The temperature, reaction duration, and surfactant content, for example, can all be changed during synthesis to modify the mesoporous material's pore size and morphology. In order to effectively separate biomolecules and polymers in SEC, the materials must be optimized for particular molecular weight ranges [53,75].

6. Applications of Mesoporous Materials in SEC

Mesoporous materials have a wide range of applications in SEC due to their unique structural properties, examples of which is listed in table 2. Here are the key applications of mesoporous materials in SEC based on the search results:

1. Stationary Phases for SEC: Materials with mesoporous silica, such SBA-15 and MCM-41, are frequently utilized as stationary phases in SEC. They are perfect for size-based separation of biomolecules, polymers, and other macromolecules due to their well-defined pore diameters and high surface areas. Because a range of analytes can be effectively separated by adjusting the pore size, SEC applications can benefit from their versatility [75].

2. Protein Purification: Protein purification works especially well with mesoporous materials. In biopharmaceutical applications, their high surface area and adjustable pore sizes help to ensure high purity by facilitating the separation of proteins from smaller contaminants and aggregates. Sustaining the safety and effectiveness of therapeutic proteins depends on this ability [78].

3. Desalting and Buffer Exchange: Salts and tiny molecules can be effectively removed from protein solutions by employing SEC with mesoporous materials for desalting biomolecules. Especially in the fields of molecular biology and biochemistry, this application is essential for getting materials ready for additional examination or research [77].

4. Separation of Biological Compounds: Mesoporous materials have special qualities that make it possible to separate different biological substances including various polysaccharides and nucleic acids using them. The resolution of separations in intricate biological mixtures is improved by their capacity to selectively hold smaller molecules while allowing bigger ones to flow through [79].

5. Capillary Gas Chromatography: Mesoporous materials are another option suggested for capillary gas chromatography supports, in addition to SEC. In gas-phase analysis, their high surface area and porosity improve separation efficiency and sensitivity [83].

6. Diagnostics and Biosensing: Mesoporous silica's biocompatibility and functionalization potential make them appropriate for use in diagnostic applications. Utilizing their large surface area for increased sensitivity and specificity, they can be utilized to create biosensors that identify certain biomolecules [84].

7. Environmental Applications: Additionally, mesoporous materials are used in environmental applications, like the adsorption of waterborne contaminants. Their applicability beyond chromatography is demonstrated by the efficient removal of organic pollutants and hazardous heavy metals made possible by their large surface area and variable functionality [80,84].

Materials Research Foundations 173 (2025) 311-339 https://doi.org/10.21741/9781644903452-12

Table 2. Case studies involving mesoporous materials in SEC with key findings.

Sr. No.	Study Title	Mesoporous Material	Analyte	Key Findings	Reference
1.	Pseudomorphic synthetic path of bimodal porous silica-based microspheres for SEC of small molecules	Bimodal porous silica (BPS) microspheres	Dextran, polyethene glycol and three biomolecules	• Pseudomorphic transformation method was used for synthesis of microspheres. • Pore size: 3.01 and 12.80 nm. • Exhibitecd the highest resolution for molecular weights of less than 1 KDa with highest level column efficiency.	[85]
2.	Separation by SEC: Selective Liquid-Phase Adsorption of l-Lysine from blended Lysine–Glucose on Zeolites	Mesoporous zeolites	L-lysine and D-glucose	• Zeolites exhibited high level of l-lysine selectivity in case of competitive adsorptions from blended lysine–glucose. • Used FTIR spectroscopy to propose a structure relevant adsorption relationship for l-lysine on zeolites.	[86]
3.	Functional mesoporous silica with controlled pore size for selective adsorption of free fatty acids (FFAs) and chlorophyll	Mesoporous silica functionalized with (3-aminopropyl) trimethoxysilane (APTMS)	FFAs and chlorophyll in olive oil	• Achieved selective adsorption for FFAs and chlorophyll. • APTMS was used to modify pore size. • Structural characterization using powder XRD, electron microscopy, nitrogen-sorption analyses.	[87]
4.	Highly ordered molecularly imprinted mesoporous silica	Bisphenol A imprinted mesoporous silica (BIMS)	Bisphenol A from wastewater	• Selective separation achieved.	[88]

	for selective removal of bisphenol A from wastewater			• BIMS was prepared using click chemistry reaction. • Characterisation by TEM, small angle XRD, and nitrogen adsorption-desorption experiments.	
5.	Bowl-like mesoporous polydopamine with size exclusion for highly selective recognition of endogenous glycopeptides	Mesoporous polydopamine nanoparticles modified by carrageenan (denoted by MPDA@PEI@CA)	Glyco-peptides	• A bowl-like mesoporous MPDA@PEI@CA composite was fabricated. • The MPDA@PEI@CA had suitable mesoporous structure and strong hydrophilicity. • The MPDA@PEI@CA exhibited ultrahigh selectivity for glycopeptides enrichment.	[89]
6.	Length exclusion separation of acetone/butanol using ZIF-302 derivatives with adjustable ellipsoidal cage sizes	ZIF-302 derivatives	Butanol and acetone	• A series of ZIF-302 derivatives with adjustable cage sizes were synthesized. • The butanol and acetone separation performance of ZIF-302 derivatives exhibiting opposite and complementary selectivity against butanol and acetone. • ZIF-302-S2 with small ellipsoidal cages exhibited the very unusual adsorption of short-chain acetone from long-chain butanol	[90]

				based on length exclusion separation.	
7.	Titanium dioxide (TiO_2)-modified fibrous core-shell mesoporous material to selectively enrich endogenous phosphopeptides with proteins exclusion prior to CE-MS analysis	Titanium dioxide coated core shell mesoporous silica (CSMS@TiO_2) nanocomposites	Phospho-peptides	• Development of CSMS@TiO_2 nanocomposites for enriching phosphopeptides prior to MS analysis. • Selective enrichment of low phosphopeptides in complex peptides mixture. • Simultaneous exclusion of high-abundant proteins.	[91]
8.	Fabrication of Polysulfone-Surface Functionalized Mesoporous Silica Nanocomposite Membranes for Removal of Heavy Metal Ions from Wastewater	Mesoporous silica nanoparticles (MSNs) functionalized with amine and sulfonic groups	Cadmium and zinc from wastewater	• Both the internal and external surfaces of MSNs were modified by a chemical reaction. • These were then used to fabricate mixed matrix polysulfone nanocomposite membranes. • Efficiency improved by more than 90% due to functionalization.	[92]
9.	Mesoporous covalent organic framework microspheres with dual-phase separation strategy for high-purity glycopeptide enrichment.	Hydrophobic and hydrophilic mesoporous covalent-organic framework (COF) microspheres	Glyco-peptides	• The microspheres were synthesised by a facile method at room temperature. • Dual-phase separation strategy was used for high-purity glycopeptide enrichment.	[65]
10.	Facile Synthesis of Hydrophilic Magnetic Mesoporous Silica Microspheres for Selective	Hydrophilic magnetic mesoporous silica microspheres	Glyco-peptides and glycans	• The microspheres were applied as an enrichment probe for low-abundance glycopeptides and glycans from	[93]

Materials Research Forum LLC
https://doi.org/10.21741/9781644903452-12

	Enrichment of Glycopeptides and Glycans.			complex biological samples.	

7. Challenges and Future Perspectives

Although mesoporous materials have demonstrated a great deal of promise as stationary phases in SEC, a number of obstacles need to be overcome before their full potential can be realized. Mesoporous materials, especially those made of silica, can be brittle and vulnerable to mechanical stress, which can result in a shorter lifespan or column fracture. This problem may be made worse by the high pressures that SEC frequently requires. Mechanical stability may be improved by creating stronger mesoporous materials or hybrid composites that combine the benefits of mesoporous silica with more resilient polymers [10].

Although it is possible to create mesoporous materials with variable pore sizes, exact control over the distribution of pore sizes is still difficult to achieve. Inconsistent performance in separation might result from variability. More consistent pore sizes and distributions can be attained with the aid of sophisticated synthesis techniques including controlled sol-gel processes and template-assisted procedures [94]. Furthermore, mesoporous materials can be complexly functionalized to improve selectivity for particular analytes; this can lead to non-uniform coatings, which can reduce separation efficiency. The creation of simpler and more dependable functionalization techniques, including layer-by-layer assembly or click chemistry, may enhance the efficiency and consistency of surface alterations [95].

In SEC, undesirable interactions between the analytes and stationary phase can result in low resolution and repeatability. For example, polar analytes may interact with silanol groups that are left on silica surfaces. The performance of separation could be improved by further altering the surface chemistry to reduce these interactions, for example, by applying hydrophobic coatings or blocking agents. High-quality mesoporous materials can be expensive and time-consuming to synthesize, which might prevent their broad use in industrial applications. These materials might become more widely available through research into scalable production processes and more affordable synthesis technologies like 3D printing or continuous flow synthesis [96,97].

Conclusion

Taken together, mesoporous materials represent a potentially significant development in the size exclusion chromatography (SEC) sector. They are excellent choices for use as stationary phases in SEC applications due to their special structural characteristics, which include a large surface area, adjustable pore diameters, and chemical adaptability. Mesoporous materials have been shown to improve separation efficiency, resolution, and flexibility when targeting particular molecular weight ranges in chromatographic systems. Furthermore, mesoporous materials' versatility enables customized functionalization, which increases their potential in a range of preparative and analytical settings. The creation and improvement of these materials have great potential for chromatographic technique advancements in the future, leading to more accurate and effective molecule separations as research proceeds.

References

[1] Y. He, Q. Pu, J. Zhang, S. Xie, X. Chen, L. Yuan, Chiral inorganic mesoporous materials used as the stationary phase in GC, Sep. Sci. Plus. 2 (2019) 432–439. https://doi.org/10.1002/sscp.201900067

[2] B. Liu, H. Li, K. Quan, J. Chen, H. Qiu, Periodic mesoporous organosilica for chromatographic stationary phases: From synthesis strategies to applications, Trends Anal. Chem. 158 (2023) 1–16. https://doi.org/10.1016/j.trac.2022.116895

[3] A. M. Striegel, Size-exclusion chromatography, in: S. Fanali, B. Chankvetadze, P.R. Haddad, C.F. Poole, M.L. Riekkola (Eds.), Liquid Chromatography, Elsevier, 2023, pp. 509–537. https://doi.org/10.1016/B978-0-323-99968-7.00022-9

[4] D. Held, P. Kilz, Size-exclusion chromatography as a useful tool for the assessment of polymer quality and determination of macromolecular properties, Chem. Teach. Int. 3 (2021) 77–103. https://doi.org/10.1515/cti-2020-0024

[5] A.M. Striegel, Size-exclusion chromatography: A twenty-first century perspective, Chromatographia. 85 (2022) 307–313. https://doi.org/10.1007/s10337-022-04143-1

[6] T. Sun, R.R. Chance, W.W. Graessley, D.J. Lohse, A study of the separation principle in size exclusion chromatography, Macromolecules. 37 (2004) 4304–4312. https://doi.org/10.1021/ma030586k

[7] Y. Wang, I. Teraoka, F.Y. Hansen, G.H. Peters, O. Hassager, A theoretical study of the separation principle in size exclusion chromatography, Macromolecules. 43 (2010) 1651–1659. https://doi.org/10.1021/ma902377g

[8] A. Sepsey, I. Bacskay, A. Felinger, Molecular theory of size exclusion chromatography for wide pore size distributions, J. Chromatogr. A. 1331 (2014) 52–60. https://doi.org/10.1016/j.chroma.2014.01.017

[9] T. Bierig, G. Collu, A. Blanc, E. Poghosyan, R.M. Benoit, design, expression, purification, and characterization of a YFP-Tagged 2019-nCoV spike receptor-binding domain construct, Front. Bioeng. Biotechnol. 8 (2020) 1–10. https://doi.org/10.3389/fbioe.2020.618615

[10] E. Lubomirsky, A. Khodabandeh, J. Preis, M. Susewind, T. Hofe, E.F. Hilder, R.D. Arrua, Polymeric stationary phases for size exclusion chromatography: A review, Anal. Chim. Acta. 1151 (2021) 1–18. https://doi.org/10.1016/j.aca.2021.338244

[11] H.A. Alhazmi, M. Albratty, Analytical techniques for the characterization and quantification of monoclonal antibodies, Pharmaceuticals. 16 (2023) 1–30. https://doi.org/10.3390/ph16020291

[12] S. Luo, S. Wohl, W. Zheng, S. Yang, Biophysical and integrative characterization of protein intrinsic disorder as a prime target for drug discovery, Biomolecules. 13 (2023) 1–10. https://doi.org/10.3390/biom13030530

[13] M. Gavrilov, M.J. Monteiro, Derivation of the molecular weight distributions from size exclusion chromatography, Eur. Polym. J. 65 (2015) 191–196. https://doi.org/10.1016/j.eurpolymj.2014.11.018

[14] M.T.R. Laguna, R. Medrano, M.P. Plana, M.P. Tarazona, Polymer characterization by size-exclusion chromatography with multiple detection, J. Chromatogr. A. 919 (2001) 13–19. https://doi.org/10.1016/S0021-9673(01)00802-0

[15] H. Pasch, A. Ndiripo, P.S.E. Bungu, Multidimensional analytical protocols for the fractionation and analysis of complex polyolefins, J. Polym. Sci. 60 (2022) 1059–1078. https://doi.org/10.1002/pol.20210236

[16] F. Malz, J.H. Arndt, J. Balko, B. Barton, T. Büsse, D. Imhof, R. Pfaendner, K. Rode, R. Brüll, Analysis of the molecular heterogeneity of poly(lactic acid)/poly(butylene succinate-co-adipate) blends by hyphenating size exclusion chromatography with nuclear magnetic resonance and infrared spectroscopy, J. Chromatogr. A. 1638 (2021) 1–10. https://doi.org/10.1016/j.chroma.2020.461819

[17] J.B. Matson, A.Q. Steele, J.D. Mase, M.D. Schulz, Polymer characterization by size-exclusion chromatography with multi-angle light scattering (SEC-MALS): A tutorial review, Polym. Chem. 15 (2024) 127–142. https://doi.org/10.1039/D3PY01181J

[18] E.R.S. Kunji, M. Harding, P.J.G. Butler, P. Akamine, Determination of the molecular mass and dimensions of membrane proteins by size exclusion chromatography, Methods. 46 (2008) 62–72. https://doi.org/10.1016/j.ymeth.2008.10.020

[19] H. Al Kader, H. Gill, T. Truong, Demineralisation and recovery of whey proteins from commercial full-fat salty Cheddar whey using size-exclusion chromatography, Food Chem. 405 (2023) 1–13. https://doi.org/10.1016/j.foodchem.2022.134831

[20] Y. Yan, T. Xing, A.P. Liu, Z. Zhang, S. Wang, N. Li, Post-column denaturation-assisted native size-exclusion chromatography–mass spectrometry for rapid and in-depth characterization of high molecular weight variants in therapeutic monoclonal antibodies, J. Am. Soc. Mass Spectrom. 32 (2021) 2885–2894. https://doi.org/10.1021/jasms.1c00289

[21] M. Valles, S. Pujals, L. Albertazzi, S. Sánchez, Enzyme purification improves the enzyme loading, self-propulsion, and endurance performance of micromotors, ACS Nano. 16 (2022) 5615–5626. https://doi.org/10.1021/acsnano.1c10520

[22] C.L. Effio, S.A. Oelmeier, J. Hubbuch, High-throughput characterization of virus-like particles by interlaced size-exclusion chromatography, Vaccine. 34 (2016) 1259–1267. https://doi.org/10.1016/j.vaccine.2016.01.035

[23] H. Bajaj, V.K. Sharma, D.S. Kalonia, A high-throughput method for detection of protein self-association and second virial coefficient using size-exclusion chromatography through simultaneous measurement of concentration and scattered light intensity, Pharm. Res. 24 (2007) 2071–2083. https://doi.org/10.1007/s11095-007-9345-x

[24] F. Almeida, Q. Cass, Affinity Selection Mass Spectrometry (AS-MS) as a tool for prospecting target ligands, Braz. J. Anal. Chem. (2023) 1–15. https://doi.org/10.30744/brjac.2179-3425.letter-almeida-cass

[25] A. Berezhnoy, C.A. Stewart, J.O. Mcnamara II, W. Thiel, P. Giangrande, G. Trinchieri, E. Gilboa, Isolation and optimization of murine IL-10 receptor blocking oligonucleotide aptamers using high-throughput sequencing, Mol. Ther. 20 (2012) 1242–1250. https://doi.org/10.1038/mt.2012.18

[26] C. Ren, A.O. Bailey, E. VanderPorten, A. Oh, W. Phung, M.M. Mulvihill, S.F. Harris, Y. Liu, G. Han, W. Sandoval, Quantitative determination of protein–ligand affinity by size exclusion chromatography directly coupled to high-resolution native mass spectrometry, Anal. Chem. 91 (2019) 903–911. https://doi.org/10.1021/acs.analchem.8b03829

[27] P.A. Wabnitz, J.A. Loo, Drug screening of pharmaceutical discovery compounds by micro-size exclusion chromatography/mass spectrometry, Rapid Commun. Mass Spectrom. 16 (2002) 85–91. https://doi.org/10.1002/rcm.546

[28] S. Kumar, T.S. Savane, A.S. Rathore, Multiattribute monitoring of aggregates and charge variants of monoclonal antibody through native 2D-SEC-MS-WCX-MS, J. Am. Soc. Mass Spectrom. 34 (2023) 1263–1271. https://doi.org/10.1021/jasms.2c00325

[29] S. Chilakala, V. Mehtab, M. Tallapally, M. Vemula, A.S. Shaikh, S. Chenna, V. Upadhyayula, SEC-MS/MS determination of amino acids from mango fruits and application of the method for studying amino acid perturbations due to post harvest ripening, LWT. 138 (2021) 1–11. https://doi.org/10.1016/j.lwt.2020.110680

[30] A. Fossati, F. Frommelt, F. Uliana, C. Martelli, M. Vizovisek, L. Gillet, B. Collins, M. Gstaiger, R. Aebersold, System-wide profiling of protein complexes via size exclusion chromatography–mass spectrometry (SEC–MS), in: M. Carrera, J. Mateos (Eds.), Shotgun Proteomics: Methods and Protocols, 2021, pp. 269–294. https://doi.org/10.1007/978-1-0716-1178-4_18

[31] J. Höpfner, B. Mayerhöfer, C. Botha, D. Bouillaud, J. Farjon, P. Giraudeau, M. Wilhelm, Solvent suppression techniques for coupling of size exclusion chromatography and 1H NMR using benchtop spectrometers at 43 and 62 MHz, J. Magn. Reson. 323 (2021) 1–6. https://doi.org/10.1016/j.jmr.2020.106889

[32] M.A. Graewert, S. Da Vela, T.W. Gräwert, D.S. Molodenskiy, C.E. Blanchet, D.I. Svergun, C.M. Jeffries, Adding size exclusion chromatography (SEC) and light scattering (LS) devices to obtain high-quality small angle X-Ray scattering (SAXS) data, Crystals. 10 (2020) 975–980. https://doi.org/10.3390/cryst10110975

[33] T. Matsui, I. Rajkovic, B.H.M. Mooers, P. Liu, T.M. Weiss, Adaptable SEC-SAXS data collection for higher quality structure analysis in solution, Protein Sci. 33 (2024) 1–9. https://doi.org/10.1002/pro.4946

[34] D. Lakayan, R. Haselberg, W.M.A. Niessen, G.W. Somsen, J. Kool, On-line coupling of surface plasmon resonance optical sensing to size-exclusion chromatography for affinity

assessment of antibody samples, J. Chromatogr. A. 1452 (2016) 81–88. https://doi.org/10.1016/j.chroma.2016.05.033

[35] T. Graf, L. Naumann, L. Bonnington, J. Heckel, B. Spensberger, S. Klein, C. Brey, R. Nachtigall, M. Mroz, T.V. Hogg, C. McHardy, A. Martinez, R. Braaz, M. Leiss, Expediting online liquid chromatography for real-time monitoring of product attributes to advance process analytical technology in downstream processing of biopharmaceuticals, J. Chromatogr. A. 1729 (2024) 46–50. https://doi.org/10.1016/j.chroma.2024.465013

[36] M.N.S. Pedro, M. Isaksson, J.G. Fons, M.H.M. Eppink, B. Nilsson, M. Ottens, Real-time detection of mAb aggregates in an integrated downstream process, Biotechnol. Bioeng. 120 (2023) 2989–3000. https://doi.org/10.1002/bit.28466

[37] M. Navaziya, HPLC hyphenations in advanced analytical world, J. Phys. Chem. Biophys. 12 (2022) 1–5. https://doi.org/10.35248/2161-0487-22.12.328

[38] E.D. Montes, Dextran: Sources, structures, and properties, Polysaccharides. 2 (2021) 554–565. https://doi.org/10.3390/polysaccharides2030033

[39] M. Ghebremedhin, S. Seiffert, T.A. Vilgis, Physics of agarose fluid gels: Rheological properties and microstructure, Curr. Res. Food Sci. 4 (2021) 436–448. https://doi.org/10.1016/j.crfs.2021.06.003

[40] M.I. Voronova, O.V. Surov, A.V. Afineevskii, A.G. Zakharov, Properties of polyacrylamide composites reinforced by cellulose nanocrystals, Heliyon. 6 (2020) 29–35. https://doi.org/10.1016/j.heliyon.2020.e05529

[41] T. Krawczyk, M. Zalewski, A. Janeta, P. Hodurek, SEC separation of polysaccharides using macroporous spherical silica gel as a stationary phase, Chromatographia. 81 (2018) 1365–1372. https://doi.org/10.1007/s10337-018-3582-5

[42] C.C.L. Schuurmans, A.J. Brouwer, J.A.W. Jong, G.J.P.H. Boons, W.E. Hennink, T. Vermonden, Hydrolytic (In)stability of methacrylate esters in covalently cross-linked hydrogels based on chondroitin sulfate and hyaluronic acid methacrylate, ACS Omega. 6 (2021) 26302–26310. https://doi.org/10.1021/acsomega.1c03395

[43] C.W. Huck, G.K. Bonn, Poly(Styrene-Divinylbenzene) based media for liquid chromatography, Chem. Eng. Technol. 28 (2005) 1457–1472. https://doi.org/10.1002/ceat.200500265

[44] G.B. Irvine, High-performance size-exclusion chromatography of peptides, J. Biochem. Biophys. Methods. 56 (2003) 233–242. https://doi.org/10.1016/S0165-022X(03)00061-7

[45] B. Yu, T. Xu, H. Cong, Q. Peng, M. Usman, Preparation of porous poly(styrene-divinylbenzene) microspheres and their modification with diazoresin for mix-mode HPLC separations, Materials. 10 (2017) 440–445. https://doi.org/10.3390/ma10040440

[46] H. Cong, J. Xing, X. Ding, S. Zhang, Y. Shen, B. Yu, Preparation of porous sulfonated poly(styrene-divinylbenzene) microspheres and its application in hydrophilic and chiral separation, Talanta. 210 (2020) 11–20. https://doi.org/10.1016/j.talanta.2019.120586

Materials Research Forum LLC
https://doi.org/10.21741/9781644903452-12

[47] E.M. Borges, Silica, Hybrid Silica, Hydride silica and non-silica stationary phases for liquid chromatography, J. Chromatogr. Sci. 53 (2015) 580–597. https://doi.org/10.1093/chromsci/bmu090

[48] T. Zhang, Y. Yu, S. Han, H. Cong, C. Kang, Y. Shen, B. Yu, Preparation and application of UPLC silica microsphere stationary phase:A review, Adv. Colloid Interface Sci. 323 (2024) 1–10. https://doi.org/10.1016/j.cis.2023.103070

[49] Y. Zhang, Z. Deng, D. Lou, Y. Wang, R. Wang, R. Hu, X. Zhang, Q. Zhu, Y. Chen, F. Liu, High-efficiency separation of extracellular vesicles from lipoproteins in plasma by agarose gel electrophoresis, Anal. Chem. 92 (2020) 7493–7499. https://doi.org/10.1021/acs.analchem.9b05675

[50] D. Josić, H. Baumann, W. Reutter, Size-exclusion high-performance liquid chromatography and sodium dodecyl sulfate-polyacrylamide gel electrophoresis of proteins: A comparison, Anal. Biochem. 142 (1984) 473–479. https://doi.org/10.1016/0003-2697(84)90492-5

[51] K. Kathirgamanathan, W. Grigsby, N.R. Edmonds, J. Al Hakkak, Molecular weight fractionation of high polydispersity native celluloses, Cellulose. 24 (2017) 5261–5265. https://doi.org/10.1007/s10570-017-1422-7

[52] T. Yao, J. Song, Y. Gan, L. Qiao, K. Du, Preparation of cellulose-based chromatographic medium for biological separation: A review, J. Chromatogr. A. 1 (2022) 1–16. https://doi.org/10.1016/j.chroma.2022.463297

[53] M. Moritz, M.G. Moritz, Mesoporous materials as multifunctional tools in biosciences: Principles and applications, Mater. Sci. Eng. C. 49 (2015) 114–151. https://doi.org/10.1016/j.msec.2014.12.079

[54] Y. Luo, B. Wang, L. Yi, C.F. Ding, C. Deng, Y. Yan, Mesoporous materials for glycopeptide separation, Trends Anal. Chem. 167 (2023) 1–17. https://doi.org/10.1016/j.trac.2023.117234

[55] M.M. Rahman, M.G. Ara, M.A. Alim, M.S. Uddin, A. Najda, G.M. Albadrani, A.A. Sayed, S.A. Mousa, M.M.A. Daim, Mesoporous carbon: A versatile material for scientific applications, Int. J. Mol. Sci. 22 (2021) 1–10. https://doi.org/10.3390/ijms22094498

[56] D. Wang, X. Chen, J. Feng, M. Sun, Recent advances of ordered mesoporous silica materials for solid-phase extraction, J. Chromatogr. A. 1675 (2022) 1–13. https://doi.org/10.1016/j.chroma.2022.463157

[57] A. Elgendy, N.M. El Basiony, F.E.T. Heakal, A.E. Elkholy, Mesoporous Ni-Zn-Fe layered double hydroxide as an efficient binder-free electrode active material for high-performance supercapacitors, J. Power Sources. 466 (2020) 20–28. https://doi.org/10.1016/j.jpowsour.2020.228294

[58] B. Szczęśniak, J. Choma, M. Jaroniec, Major advances in the development of ordered mesoporous materials, Chem. Commun. 56 (2020) 7836–7848. https://doi.org/10.1039/D0CC02840A

Materials Research Forum LLC
https://doi.org/10.21741/9781644903452-12

[59] P. Shen, B. Zhang, Y. Wang, X. Liu, C. Yu, T. Xu, S.S. Mofarah, Y. Yu, Y. Liu, H. Sun, H. Arandiyan, Nanoscale niobium oxides anode for electrochemical lithium and sodium storage: A review of recent improvements, J. Nanostructure Chem. 11 (2021) 33–68. https://doi.org/10.1007/s40097-020-00367-5

[60] Z. Yu, S. Zhu, L. Zhang, S. Watanabe, Mesoporous single crystal titanium oxide microparticles for enhanced visible light photodegradation, Opt. Mater. (Amst). 127 (2022) 1–12. https://doi.org/10.1016/j.optmat.2022.112297

[61] Y. Jiao, X. Zhang, H. Yang, H. Ma, J. Zou, Mesoporous tantalum oxide nanomaterials induced cardiovascular endothelial cell apoptosis via mitochondrial-endoplasmic reticulum stress apoptotic pathway, Drug Deliv. 30 (2023) 108–120. https://doi.org/10.1080/10717544.2022.2147251

[62] M. Raimondo, G. Perez, M. Sinibaldi, A. De Stefanis, A.A.G. Tomlinson, Mesoporous M41S materials in capillary gas chromatography, Chem. Commun. (1997) 1343–1344. https://doi.org/10.1039/a702849k

[63] L.D. Bonifacio, B.V. Lotsch, G.A. Ozin, Periodic mesoporous materials: Holes filled with opportunities, in: Comprehensive Nanoscience and Technology, Elsevier, 2011, pp. 69–125. https://doi.org/10.1016/B978-0-12-374396-1.00049-0

[64] P.V.D. Voort, D. Esquivel, E. De Canck, F. Goethals, I.V. Driessche, F.J.R. Salguero, Periodic mesoporous organosilicas: From simple to complex bridges; a comprehensive overview of functions, morphologies and applications, Chem. Soc. Rev. 42 (2013) 3913–3955. https://doi.org/10.1039/C2CS35222B

[65] S. Ba, B. Luo, Z. Li, J. He, F. Lan, Y. Wu, Mesoporous covalent organic framework microspheres with dual-phase separation strategy for high-purity glycopeptide enrichment, J. Chromatogr. A. 1684 (2022) 1–10. https://doi.org/10.1016/j.chroma.2022.463575

[66] D. Khan, Shaily, Synthesis and catalytic applications of organo-functionalized MCM-41 catalyst: A review, Appl. Organomet. Chem. 37 (2023) 35–43. https://doi.org/10.1002/aoc.7007

[67] E.D. Tekkaya, Y. Yürüm, Mesoporous MCM-41 material for hydrogen storage: A short review, Int. J. Hydrogen Energy. 41 (2016) 9789–9795. https://doi.org/10.1016/j.ijhydene.2016.03.050

[68] T. Berestok, C. Diestel, N. Ortlieb, J. Buettner, J. Matthews, P.S.C. Schulze, J.C. Goldschmidt, S.W. Glunz, A. Fischer, High-efficiency monolithic photosupercapacitors: Smart integration of a perovskite solar cell with a mesoporous carbon double-layer capacitor, Sol. RRL. 5 (2021) 1–15. https://doi.org/10.1002/solr.202100662

[69] L. Zu, W. Zhang, L. Qu, L. Liu, W. Li, A. Yu, D. Zhao, Mesoporous materials for electrochemical energy storage and conversion, Adv. Energy Mater. 10 (2020) 1–7. https://doi.org/10.1002/aenm.202002152

[70] K.A. Kilian, T. Böcking, J.J. Gooding, The importance of surface chemistry in mesoporous materials: Lessons from porous silicon biosensors, Chem. Commun. (2009) 630–640. https://doi.org/10.1039/B815449J

[71] M. Xu, T. Wang, L. Zhou, D. Hua, Fluorescent conjugated mesoporous polymers with N,N-diethylpropylamine for the efficient capture and real-time detection of volatile iodine, J. Mater. Chem. A. Mater. 8 (2020) 1966–1974. https://doi.org/10.1039/C9TA11446G

[72] J. Liu, B. Wang, S.B. Hartono, T. Liu, P. Kantharidis, A.P.J. Middelberg, G.Q. (Max) Lu, L. He, S.Z. Qiao, Magnetic silica spheres with large nanopores for nucleic acid adsorption and cellular uptake, Biomaterials. 33 (2012) 970–978. https://doi.org/10.1016/j.biomaterials.2011.10.001

[73] H. Luo, Y.V. Kaneti, Y. Ai, Y. Wu, F. Wei, J. Fu, J. Cheng, C. Jing, B. Yuliarto, M. Eguchi, J. Na, Y. Yamauchi, S. Liu, Nanoarchitectured porous conducting polymers: From controlled synthesis to advanced applications, Adv. Mater. 33 (2021) 1–13. https://doi.org/10.1002/adma.202007318

[74] T. Nassivera, A.G. Eklund, C.C. Landry, Size-exclusion chromatography of low-molecular-mass polymers using mesoporous silica, J. Chromatogr. A. 973 (2002) 97–101. https://doi.org/10.1016/S0021-9673(02)01200-1

[75] X. Li, Y.R. Lee, K.H. Row, Synthesis of mesoporous siliceous materials in choline chloride deep eutectic solvents and the application of these materials to high-performance size exclusion chromatography, Chromatographia. 79 (2016) 375–382. https://doi.org/10.1007/s10337-016-3051-y

[76] F. Liu, Z. Xu, H. Wan, Y. Wan, S. Zheng, D. Zhu, Enhanced adsorption of humic acids on ordered mesoporous carbon compared with microporous activated carbon, Environ. Toxicol. Chem. 30 (2011) 793–800. https://doi.org/10.1002/etc.450

[77] W. Qin, M.E. Silvestre, F. Kirschhöfer, G.B. Weiss, M. Franzreb, Insights into chromatographic separation using core–shell metal–organic frameworks: Size exclusion and polarity effects, J. Chromatogr. A. 1411 (2015) 77–83. https://doi.org/10.1016/j.chroma.2015.07.120

[78] Y. Yang, Y. Wang, S. Wei, C. Zhou, J. Yu, G. Wang, W. Wang, L. Zhao, Extracellular vesicles isolated by size-exclusion chromatography present suitability for RNomics analysis in plasma, J. Transl. Med. 19 (2021) 1–10. https://doi.org/10.1186/s12967-021-02775-9

[79] S. L. Suib, A review of recent developments of mesoporous materials, Chem. Rec. 17 (2017) 1169–1183. https://doi.org/10.1002/tcr.201700025

[80] N. Linares, A.M.S. Albero, E. Serrano, J.S. Albero, J.G. Martínez, Mesoporous materials for clean energy technologies, Chem. Soc. Rev. 43 (2014) 7681–7717. https://doi.org/10.1039/C3CS60435G

[81] M. Keshavarz, N. Ahmad, Characterization and modification of mesoporous silica nanoparticles prepared by sol-gel, J. Nanoparticles. 2013 (2013) 1–4. https://doi.org/10.1155/2013/102823

[82] O.A. Saputra, W.A. Lestari, V. Kurniansyah, W.W. Lestari, T. Sugiura, R.R. Mukti, R. Martien, F.R. Wibowo, Organically surface engineered mesoporous silica nanoparticles control the release of quercetin by pH stimuli, Sci. Rep. 12 (2022) 1–12. https://doi.org/10.1038/s41598-022-25095-4

[83] H. Zhang, F. Feng, Y. Zhao, B. Zhao, L. Li, D. Zheng, X. Li, Fabrication of a micro gas chromatography column via the layer-by-layer deposition of mesoporous silica as the stationary phase, J. Chromatogr. A. 1673 (2022) 30–46. https://doi.org/10.1016/j.chroma.2022.463082

[84] J.A. Cecilia, R.M. Tost, M.R. Millán, Mesoporous materials: From synthesis to applications, Int. J. Mol. Sci. 20 (2019) 1–13. https://doi.org/10.3390/ijms20133213

[85] Y. Zhao, J. Wang, Y. Yang, Q. Fu, Y. Ke, Pseudomorphic synthesis of bimodal porous silica microspheres for size-exclusion chromatography of small molecules, J. Chromatogr. A. 1664 (2022) 12–27. https://doi.org/10.1016/j.chroma.2021.462757

[86] J. Deischter, F. Müller, B. Bong, C. Maurer, S.S. Hartmann, R. Palkovits, separation by size exclusion: Selective liquid-phase adsorption of l-lysine from lysine–glucose mixtures on zeolites, ACS Sustain. Chem. Eng. 10 (2022) 10211–10222. https://doi.org/10.1021/acssuschemeng.2c01874

[87] Y. Ahn, S.Y. Kwak, Functional mesoporous silica with controlled pore size for selective adsorption of free fatty acid and chlorophyll, Microporous Mesoporous Mater. 306 (2020) 1–10. https://doi.org/10.1016/j.micromeso.2020.110410

[88] J. Wu, L. Tan, Y. Li, X. Wu, Y. Liang, Highly ordered molecularly imprinted mesoporous silica for selective removal of bisphenol A from wastewater, J. Sep. Sci. 43 (2020) 987–995. https://doi.org/10.1002/jssc.201900957

[89] Y. Yan, R. Han, Y. Hou, H. Zhang, J. Yu, W. Gao, L. Xu, K. Tang, Bowl-like mesoporous polydopamine with size exclusion for highly selective recognition of endogenous glycopeptides, Talanta. 233 (2021) 1–12. https://doi.org/10.1016/j.talanta.2021.122468

[90] J. Song, Q. Meng, J. Wang, X. Guo, P. Wei, J. Dong, Q. Shi, Length exclusion separation of acetone/butanol using ZIF-302 derivatives with adjustable ellipsoidal cage sizes, Sep. Purif. Technol. 312 (2023) 12–23. https://doi.org/10.1016/j.seppur.2023.123371

[91] A. Irfan, W. Feng, K. Liu, K. Habib, Q. Qu, L. Yang, TiO2-modified fibrous core-shell mesoporous material to selectively enrich endogenous phosphopeptides with proteins exclusion prior to CE-MS analysis, Talanta. 235 (2021) 1–10. https://doi.org/10.1016/j.talanta.2021.122737

[92] A.A. Alotaibi, A.K. Shukla, M.H. Mrad, A.M. Alswieleh, K.M. Alotaibi, Fabrication of polysulfone-surface functionalized mesoporous silica nanocomposite membranes for removal of heavy metal ions from wastewater, Membranes. 11 (2021) 1–10. https://doi.org/10.3390/membranes11120935

[93] Y. Liu, W. Ma, Y. He, Z. Chen, Z. Lin, Facile synthesis of hydrophilic magnetic mesoporous silica microspheres for selective enrichment of glycopeptides and glycans, Anal. Lett. 54 (2021) 966–978. https://doi.org/10.1080/00032719.2020.1789161

[94] S.C. Moldoveanu, V. David, Stationary phases and columns for size exclusion, in: Selection of the HPLC Method in Chemical Analysis, Elsevier, 2017, pp. 377–386. https://doi.org/10.1016/B978-0-12-803684-6.00011-1

[95] M.R. Schure, R.E. Moran, Size exclusion chromatography with superficially porous particles, J. Chromatogr. A. 1480 (2017) 11–19. https://doi.org/10.1016/j.chroma.2016.12.016

[96] M. Rbaa, R. Hsissou, K. Dahmani, M. Oubaaqa, B. Tüzün, E. Berdimurodov, Z. Rouifi, A. Zarrouk, R. Seghiri, Carbon dots in stationary phase of chromatography, enhanced crop yield, and stationary phase of chromatography, in: E. Berdimurodov, D.K. Verma, L. Guo (Eds.), Carbon Dots, 2024, pp. 361–377. https://doi.org/10.1021/bk-2024-1465.ch015

[97] J. Engelke, J. Brandt, C.B. Kowollik, A. Lederer, Strengths and limitations of size exclusion chromatography for investigating single chain folding – current status and future perspectives, Polym. Chem. 10 (2019) 3410–3425. https://doi.org/10.1039/C9PY00336C

Materials Research Foundations 173 (2025) 340-361 https://doi.org/10.21741/9781644903452-13

Chapter 13

Mesoporous Materials for Enzyme Immobilization

Sreelekshmi[1], Athira A[1], K M Sreedhar[1], Beena Saraswathyamma[1*]

[1]Department of Chemistry, Amrita Vishwa Vidyapeetham, Amritapuri, Kollam,690525, India

beenas@am.amrita.edu

Abstract

Over the years, mesoporous materials have garnered increasing research interest as versatile carriers and promising substrates for enzyme immobilization. A critical problem is to comprehend the impact of immobilization on enzymes to formulate circumstances that maximize enzyme activity. This chapter concludes with insights and perspectives regarding forthcoming obstacles and opportunities for their scale-up uses. The principal features of innovative mesoporous materials are exceptional biocatalytic performance, enhanced stability and repeatability, high loading capacity, and increased accessibility to catalytic sites. The chapter discusses different immobilization techniques employed on mesoporous materials, different characteristics of mesoporous materials influencing immobilization, correlation between advancements in mesoporous materials also their respective contributions to immobilized enzyme, along with applications of these materials for biocatalysis.

Keywords

Mesoporous Materials, Enzyme Immobilization, Entrapment, Encapsulation, Cross-Linking, Ordered Mesoporous Carbons

Contents

 Materials Research Forum LLC
https://doi.org/10.21741/9781644903452-13

1. Introduction

Porous solids are employed in technical applications as adsorbents, catalysts, and catalyst supports due to their elevated surface areas. The design, synthesis, and modification of porous materials provide greater challenges compared to the synthesis of dense materials [1]. IUPAC has proposed a classification system for these substances based on the size of their pores: microporous (50 nm< pore diameter), mesoporous (2-50 nm), and microporous (<2nm) [2]. Mesoporous materials allow facile surface adaptation, possess large pore volume, extensive surface area and flexible mesoporous channels [3]. Consequently, researchers are constantly creating novel approaches and methods to synthesize and alter the structure of mesoporous materials. Because of their significant properties, mesoporous materials garnered much attention for their potential utilization in medication delivery, energy storage, sensors, and catalysts.

They can increase the scattering of active sites and promote the availability of reactants to these sites, hence been used as support for catalyst [4]. They are employed in drug delivery systems because of their capacity to encapsulate the drugs within the pores. By altering the pore structure and surface chemistry, it is possible to regulate the release of the drug, enabling precise and prolonged delivery [5]. Zeolites, mesoporous silica nanoparticles, have been studied in the field of environmental applications too. They can be used in environmental restoration particularly for adsorbing pollutants, heavy metals, and organic contaminants [6]. Diaz and Balkus were the pioneers in immobilizing globular proteins (cytochrome c, trypsin, and papain) onto MCM-41 [7]. Employing mesoporous materials in field of sensors, desirable characteristics can be achieved in developing a sensitive and potential sensor. They even facilitate the fast diffusion of analytes over a substantial surface area, leading to shorter response times [8]. Mesoporous materials are considered attractive for application in biosensing devices due to their well-organized and spacious mesostructures, strong and easily modified mesopore walls, inherent hosting capabilities, and catalytic properties [9]. The performance of the storage system is heavily influenced by the characteristics of the materials used in the system. A multitude of researchers have undertaken extensive experiments utilizing mesoporous carbon as a material for ultracapacitor electrodes. Their extremely permeable and consistently arranged three-dimensional structure should guarantee excellent accessibility and rapid movement of mass to the active centres [10]. The organized pore structure also enables the selective separation of molecules based on their size and shape.

The utilization of enzymes in various industries is frequently impeded by their limited durability during operation and challenges in retrieving and reusing them. Research on immobilization of

Materials Research Foundations 173 (2025) 340-361
https://doi.org/10.21741/9781644903452-13

proteins and enzymes in ordered mesoporous materials has been conducted for nearly twenty years. Immobilized enzymes are currently generating significant interest due to their superior characteristics compared to soluble enzymes [11]. Enzyme immobilization refers to the act of restricting or attaching enzymes to a solid support or inside a defined area, while preserving their ability to catalyze reactions [12]. Immobilization leads to enhance enzyme stability, reusability, functionality, efficiency, and specificity in catalysis. It facilitates the extensive utilization of enzymes in diverse commercial and therapeutic applications, establishing it as a basic approach in contemporary biotechnology. Different materials such as inorganic materials, inert polymers, sol-gel silica and controlled pore glasses were used as enzyme support system for years [13]. One distinguishing characteristic of mesoporous supports is the ability to adjust the characteristics of the support for biomolecules by altering the synthesis conditions and creating matrices with a specific pore structure. Subsequently, researchers utilized mesoporous materials to immobilize enzymes, aiming to take advantage of their adjustable pore size and expansive surface area. Furthermore, the advancements in immobilization of enzymes on mesoporous materials were further facilitated using new methodologies for immobilization. This was driven by the need for innovative applications and intricate enzymatic systems, as well as the utilisation of sophisticated characterisation methods [14].

In this chapter, we focus to discuss in detail the latest progress in immobilizing enzymes in mesoporous materials.

2. Enzyme immobilization

Mesoporous materials can exhibit a wide range of characteristics in terms of composition, pore structure, and particle form. The electrostatic and hydrophobic interactions between the enzyme and the support material greatly influence the process of immobilising enzymes through physical adsorption. Enzyme immobilization allows for convenient recovery of both enzymes and its supporting materials, which is particularly beneficial in fields of food, medicine, as well as pharmaceuticals [15]. The strategies typically consider the properties of proteins and particles as separate entities, without considering the potential influence of the pore environment on protein properties [16]. Choosing an appropriate pore size, surface properties, and spatial structures of mesoporous materials is crucial for effectively immobilising enzymes with varying molecular size, shape, and side-chain functional groups that result in electrostatic charges. In immobilization methods, the enzyme is put into the pores of the support material, ensuring the protection of the protein's structure while maintaining its effective catalytic characteristics. When the enzyme is situated within the pores of the carrier, it is necessary to consider diffusional constraints as the movement of substrates and products is constrained [17]. Ordered mesoporous substrates extensively were explored materials amongst other developed mesoporous materials followed by ordered mesoporous carbons for diverse uses. The surface functional groups serve multiple purposes in enzyme immobilization. Firstly, they alter the surface charge of mesoporous materials to regulate electrostatic interaction with the adsorbed enzyme. Secondly, they chemically bind with the amino acid groups of the specific enzyme. Lastly, they reduce the size of the pore entrance to trap the enzymes within the nanochannels [18].

The concept of immobilizing enzymes was first proposed by Nelson and Griffin in 1916, who observed that β-fructofuranosidase also known as invertase could hydrolyse sucrose when it was absorbed onto charcoal [19]. Subsequently, other reversible and irreversible techniques for immobilization were introduced, which had the potential to improve the physicochemical characteristics of enzymes, making it suitable for practical applications. Enzymes immobilised within porous materials may undergo conformational changes, resulting in either increased or decreased enzyme activity. Consequently, it is essential to detect conformational alterations during the immobilisation process in mesoporous materials.

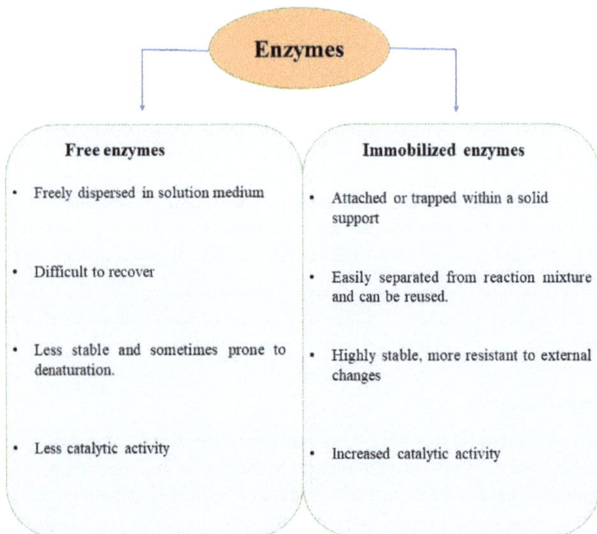

Figure 1. *Comparison between free enzymes and immobilized enzymes.*

2.1 Enzyme immobilization methods

The most employed immobilization strategies involve the use of physical or chemical means. The five fundamental immobilization processes, namely (1) adsorption, (2) entrapment, (3) covalent binding, (4) encapsulation, and (5) crosslinking, continue to be employed in modern technology [20]. Enzyme immobilization has been established as a highly effective approach for enhancing catalytic activity and reusability, as well as facilitating easy separation, operation, and cost reduction, for more than a century. Enzyme immobilization facilitates the isolation of the enzyme from the reaction mixture, hence streamlining downstream processing. This method confers protection to the enzyme against degradation or inactivation caused by strong reaction conditions, hence enhancing its robustness and suitability for many applications. Figure 1 illustrates the advantages of immobilized enzymes over free enzymes. The goal is to develop strong biocatalysts that can meet the increasing demands of various industrial sectors [21,22]. Preserving the tertiary structure and ensuring the accessibility of active sites are crucial in the preparation processes for immobilising enzymes in constrained nanospaces. Enzymes can be immobilized to enable enzymatic reactions to occur in non-aqueous environments, which is advantageous for large-scale applications. The use of solid supports boosts the durability of enzyme molecules, allowing the

Materials Research Foundations 173 (2025) 340-361 https://doi.org/10.21741/9781644903452-13

catalysts to be re-claimed multiple times following simple separation from the reaction environment [13].

2.1.1 Adsorption

Adsorption immobilization is a simple, alterable carrier binding method. This way relies on physical adsorption. Adsorption materials include ion exchanging resins, alumina, and activated carbon. Adsorption is inexpensive and simple. Due to weak intermolecular forces such hydrogen bonds, salt linkages, ionic connections, and hydrophobic interactions, the carrier and enzyme have poor affinity. Protein configurations and matrix charges can provide a highly adsorbed/undistorted enzyme. Some enzymes cannot be immobilized on any carrier material; however, this approach allows for several carrier materials. Adsorption requires certain parameters, like enzyme-carrier affinity. The carrier material's active groups ensure excellent adsorption. These aid enzyme-carrier interactions. If these interactions don't exist, carrier modifiers can help enzymes and carriers connect. Adsorption-supported enzyme immobilization has many benefits. It employs few chemicals, requires few activation steps, and is cost-effective and easy to implement [23]. Adsorption carriers for enzyme immobilization are usually organic or inorganic. Silica, titania, and hydroxyapatite are the main inorganic carriers. Chitin, chitosan, cellulose, and alginate are natural and produced compounds, mostly polymers. Chemically changing these matrices to suit an enzyme's purpose is easy [24]. Ionic, hydrophobic, and hydrogen bonding attach the enzyme to the support. Specific amino acids on the enzyme's surface and support chemical characteristics determine interactions. This procedure involves mixing an aqueous enzyme solution with the support material for a certain time and washing out the excess enzyme [25].

2.1.2 Entrapment

Entraping enzymes using covalent and non-covalent bonds immobilises them. These matrices preserve biocatalyst characteristics and reduce conformation change. Enzymes trapped in polymers allow substrates and products to pass yet keep the enzyme. Entrapment restricts enzyme diffusion but does not bind them to the polymeric matrix. Entrapped enzymes are more stable than adsorbed ones. Entrapment immobilization is easier than covalent bonding but increases or preserves enzyme activity. Enzymes were added during precursor hydrolysis and condensation in the silica network to entrap them. The organised bio aggregate contained a xerogel with the enzyme in a "cage" connected by homogeneous mesoporous channels. When water is added to the channels, the cage becomes watery like the native enzymes' natural environment. Excellent recycling, biocatalytic activity, and stability were achieved with this method [26,21]. Enzymes can be immobilized within magnetic nanoparticles by an in-situ fabrication process. In this sophisticated method, ferrous chloride was combined with urea, urease, and α-amylase in a solution saturated with oxygen. Membrane reactors are widely recognised for their ability to trap substances, and when paired with downstream processing, they offer intriguing benefits [27]. The β-galactosidase derived from Bacillus circulans is utilised in a membrane reactor to synthesise galactooligosaccharides (GOS) from lactose. The membrane acts as a barrier for the enzyme and the consequential GOS, while permitting the passage of the generated monosaccharides glucose and galactose. This allows for easy separation of the product, while also dropping the inhibitory impact of the monosaccharides on the enzyme. The continuous extraction of these compounds led to a 33% increase in the yield of GOS compared to a batch approach [28]. Enzymes can be walled within hollow polyelectrolyte microspheres, which is a appreciated technique in applied biocatalysis. Entrapment in sol-gels is another helpful method in this field [29].

2.1.3 Covalent binding

Amino acids with side chains, such as arginine, aspartic acid, and histidine, together with functional groups like imidazole, indole, and phenolic hydroxyl, facilitate the covalent binding of enzymes to supports. The covalent attachment of enzymes to silica gel carriers, which have been changed through silanization and the elimination of unreacted aldehyde groups, together with SBA-15 featuring cage-like pores encircled by Si–F groups, results in the formation of extremely stable and hyperactive biocatalysts. Covalent coupling with mesoporous silica, chitosan, etc upsurges enzyme half-life and heat stability [30,31]. The enzyme and carrier's durable covalent connection retains the enzyme immovable under industrial circumstances of high reactant and product concentrations and ionic strength. Unlike enzyme encapsulation or entrapment, particle size is defined before immobilization, and particles are often more stable against shear forces [32].

2.1.4 Encapsulation

Microencapsulation involves enclosing an enzyme or catalyst within a membrane shell, which averts any changes in its shape and function. This scheme is less affected by mass transfer resistances compared to entrapment. Unlike entrapment, microencapsulation allows for easy movement of substrate, product, and enzyme within the liquid-filled core of the microcapsule, with most of the resistance occurring in the thin membrane shell. Furthermore, microcapsules can be regarded as versatile catalytic micro-membrane reactors. The membrane shell facilitates the movement of substrate and product, while the catalyst enables the desired chemical reaction [33]. In addition to retaining the catalyst within the microcapsule, the membrane also functions as a selective barrier, allowing only specific substrates to reach the catalyst. Encapsulation is an effective method for preventing any detrimental effects on the enzyme's structure. Enzyme encapsulation technologies are currently generating significant interest, primarily because of their significance to medicine, pharmaceutics, detergent, and the cosmetic sectors [34]. The spatial confinement of multi-enzymes can be achieved by altering epitaxial overgrowth cycles, resulting in varying inter-enzyme distances. Additionally, the locations of the enzymes can be modified by adjusting encapsulation sequences. As a result, the effectiveness of tandem biocatalysis can be easily managed [35].

2.1.5 Cross linking

The cross-linked enzyme aggregates (CLEAs) methodology appears to be an attractive approach for immobilising enzymes among other immobilization methods. This technique is remarkably straightforward and easily adaptable for quick optimisation, resulting in cost-effectiveness and a short time-to-market. CLEAs formed within the pores of materials yield biocatalysts that are densely loaded, stable, and active. These biocatalysts exhibit superior stability compared to typical biocatalysts created through physical adsorption [36]. CLEAs are prepared by precipitating the enzyme and then chemically cross-linking the protein aggregates using a bifunctional chemical reagent. This methodology associates the purification as well as immobilization of enzymes into a single operation, using precipitation from an aqueous medium. This process does not necessitate the use of a very pure enzyme [37]. This approach allows for the easy stabilisation of multimeric enzymes and also enables the co-aggregation of several enzymes or enzymes with polymers [38].

The above explained techniques for immobilization are illustrated in Figure 2. Merits and demerits of these immobilization procedures are explained in table 1.

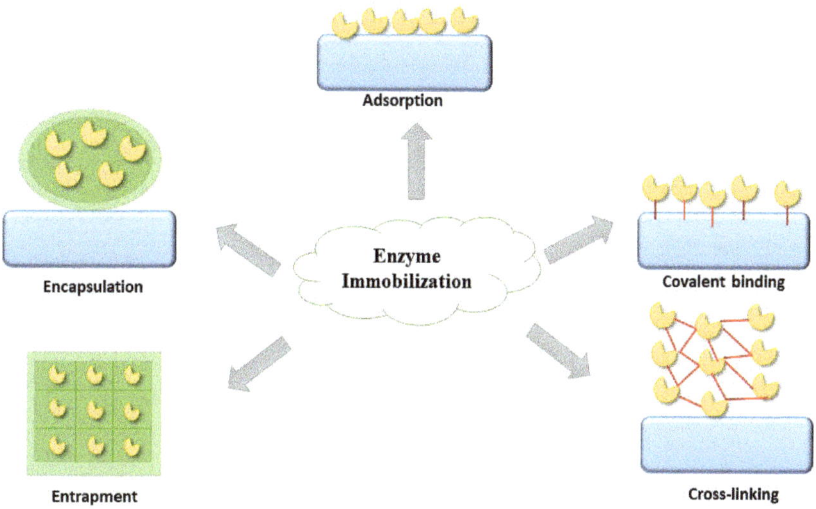

Figure 2. *Illustration showing different immobilization techniques.*

Materials Research Foundations 173 (2025) 340-361 https://doi.org/10.21741/9781644903452-13

Table 1. *Advantages and disadvantages of different immobilization techniques.*

Immobilisation techniques	**Advantages**	**Disadvantages**	**Applications**
Physical adsorption	• Simplicity • Cost-effective and mild Conditions. • Reusability • Flexibility	• Enzyme leaching from the support material • Enzyme activity loss • Limited control • Risk of saturation of the support material. • Environmental sensitivity • Potential for incomplete immobilization	• Bioreactors • Biosensors • Waste Treatment
Entrapment	• Enhance their thermal and chemical stability. • Retention of activity • Reduced enzyme leaching • Reduced enzyme leaching	• Diffusion limitations • Matrix degradation • Limited enzyme load • Complexity in matrix preparation • Potential for incomplete entrapment • Expensive	• Synthetic biology • Textile industry • Cosmetic industry
Covalent binding	• Strong attachment • Enhanced stability • High reusability • Controlled orientation • Low risk of denaturation • Flexibility in support materials	• Complex process • Potential loss of activity • Irreversibility • Chemical reactions intricate in covalent binding might modify the support material • Limited flexibility in enzyme loading • Potential for Increased Steric Hindrance	• Diagnostic kits • Catalyze reactions to yield complex organic compounds • Soil improvement
Crosslinking	• Improved Operational Stability • Controlled Enzyme Activity • Flexibility in support materials • Reduced Leaching	• Matrix Modification • Optimization Challenges	• bioethanol production • pesticide degradation

Materials Research Foundations 173 (2025) 340-361
https://doi.org/10.21741/9781644903452-13

2.2 Characteristics of mesoporous materials affecting immobilization

2.2.1 Pore size

Pore size is a critical parameter of designing a mesoporous substrate for enzyme immobilization. Attaining the ideal pore size can improve the overall efficiency of the immobilized enzyme, resulting in more effective and environmentally friendly industrial operations. A notable finding was that the enzyme, when immobilized, displayed increased activity, and improved thermal stability when pore diameter of the support material corresponds to enzyme size. Mesoporous support with pore size smaller than 5 nm have a restricted capacity to adsorb enzymes while large pore sized materials become increasingly significant carriers for high molecular weighed enzymes [13,39]. MCM type materials has pore diameters that are typically smaller than 4 nm, making them unsuitable for immobilising enzymes with larger dimensions. Nevertheless, the introduction of the SBA family broadened the scope of research to include medium to large proteins [40]. SBA base is formulated using non-ionic surfactants, resulting in the formation of structures with larger pore walls and higher hydrothermal stability [41]. Furthermore, it has been documented that the presence of materials with a pore size significantly exceeding size of the enzyme results in detrimental impact on activity of enzyme [42]. A uniform pore size distribution in mesoporous materials ensures consistent enzyme loading and activity throughout the material.

2.2.2 Structure and morphology

The structure and morphology of mesoporous materials significantly influence the efficiency and effectiveness of enzyme immobilization. Materials possessing a substantial surface area offer a greater number of active sites for enzyme binding, resulting in increased enzyme loading. One-dimensional pore structured materials (MCM-41, SBA-15, FSM-16,) restricts the distribution of enzyme into the core of particle owing to pore blockage at entrance of the aperture, contrary to three-dimensional materials (MCFs, MCM-48, FDU-12) which is expected to decrease potential diffusion limits [13].

2.2.3 Surface properties

The physical properties of solid base substantially influence the immobilisation process. Functionalising the materials has the advantage of altering and perfecting the surface of the support to effectively immobilise enzyme in a steady way while preserving its catalytic activity. The isoelectric point in enzyme, also pH of solution used in immobilising the desired enzyme and surface charge on the enzyme needs to be considered while immobilizing targeted enzyme. The isoelectric point has been utilised to determine if an enzyme would adhere to a surface through adsorption. It suggests that adsorption will take place on a surface that has a negative charge, but only below the isoelectric point. Above the isoelectric point, there will be very little adsorption. Protein loadings are higher when the pH values are close to the isoelectric point of protein. At this point, protein molecule has a nearly neutral charge besides any repulsions between fragments are reduced. By way of the pH deviates from the isoelectric point, the influence of lateral electrostatic repulsions sandwiched between protein molecules increases, resulting in a decrease in the loading [41]. Various other factors that impact on the immobilization of enzymes are solvent, temperature and carrier surface temperature.

Materials Research Forum LLC
https://doi.org/10.21741/9781644903452-13

3. Immobilization of enzymes on ordered mesoporous carbons

3.1 Synthesis of ordered mesoporous carbons

OMCs are synthesized in many steps to create a structured carbon material consisting of defined pore structure. The synthesis environments may vary based on desired OMC characteristics and chosen precursors in addition stencils [43]. OMC synthesis involves template selection, carbon precursor selection, mixing, ageing, drying, carbonization, template removal, initiation, washing, and drying. The synthesis stage begins with selecting a template to create a perfectly ordered porous carbon matrix. Surfactants, block copolymers, silica, and alumina are popular templates [44]. OMC compounds are synthesized using several techniques. These techniques include template-free techniques, sol-gel techniques, hard and soft templating techniques [45], and nanocasting techniques [46]. Hard templating necessitates the employment of prepared mesoporous materials as sacrificial templates [47]. Examples of these materials are alumina and silica. The carbonization method begins, the carbon precursor infiltrates the template, and ultimately the template is eliminated by physical or chemical means. The carbon material that was created as a result retains the mesoporous structure of the template [48]. By using a self-assembly technique with a blend of phenol, resorcinol, and phloroglucinol as precursor of carbon and triblock copolymer F127 as a template, ordered carbons with controllable pore sizes in the range of 3.6–6.2 nm was created. High specific surface zones (up to 1100 m^2/g) and pore volumes (up to 0.84 cm^3/g), the OMCs possessed an organized 2-D hexagonal mesostructure. The findings of nitrogen sorption showed that the mass ratio of carbon precursor to F127 and content of carbon precursors may be changed to adjust the pore size of OMCs [49]. SBA-15 was used as a hard template in a one-step replication process to create SN@OMC. The SN@OMC materials revealed the graphitic mesoporous structure and the nearly homogeneously co-doped carbon framework of S and N. SN–OMC-900 demonstrated a direct four-electron reaction pathway and high catalytic activity for ORR that was on par with commercial Pt/C catalysts. Moreover, over SN–OMC–900, a long-term stability and exceptional resistance to the methanol crossover effect were demonstrated [50]. Using a simple nanocasting method and ammonia calcination, several unique iron nitrides (Fe_2N)/cubic-ordered mesoporous carbon composites (Fe_2N@OMC) were created. Fe_2N@OMC was used as a supercapacitor's negative electrode in an electrochemical test in 6 mol L^{-1} KOH aqueous electrolytes. Resulting Fe_2N@OMC-2 demonstrated exceptional cycle stability, an exceptional rate capability (398 F g^{-1} at 20 A g^{-1} with capacitance holding of 76%), superior specific capacitance (547 F g^{-1} at 1 mV s^{-1} and 520 F g^{-1} at 0.5 A g^{-1}). Fe_2N@OMC-2's specific capacitance retention after 1000 cycles was still 85%, far higher than that of bare Fe_2N, which was 28% [51]. Ordered mesoporous carbon nanosheets (OMCNS), or monolayer-ordered mesoporous carbon nanosheets, were created by confining F127 and resol in the montmorillonite interlayer. Because of their high degree of graphitization and efficient ion diffusion provided by their mesopores on carbon nanosheets, OMCNS are good electrode materials for electric double layer capacitors [52]. Capacitive deionization CDI applications have been paying more and more attention to mesostructured carbon materials due to their enhanced hydrophilicity, rapid ion transportation and electron-transfer paths, and huge accessible specific surface area. The nitrogen-doped OMC with improved desalination performance was created using a self-assembly technique for CDI electrode material was reported. This electrode material's highly structured mesoporous structure allowed ions to diffuse and transmit quickly inside the pore channel, increasing electrodes' specific surface range utilization rate along with CDI performance [53].

3.2 Advancement of ordered mesoporous carbons (OMCs)

Enzyme immobilization materials and procedures must improve to create high-performance enzyme-based biosensors. Chitosan microspheres crosslinked tripolyphosphate, carbon-fiber [54], including carbon nanotube–chitosan composite are among the enzyme immobilization materials and methods disclosed. Biocatalytic processes may benefit from enzyme immobilization in well-controlled inorganic mesoporous materials. Due to mesoporous material pores' folding forces, biomolecules are stable [55, 56]. Lysozyme, cytochrome, furthermore glucose oxidase was immobilized on surface of mesoporous carbon with good loading and steadiness. Due to their uniform plus tailored pore structure, high specific surface area, bulky pore volume, decent biocompatibility, and improved electrical conductivity, ordered mesoporous carbon (OMC) materials are promising for electrochemical biosensors [57]. Incorporating heteroatoms has gained interest in recent years because they can improve OMC's optoelectrical, semiconducting, together with its surface properties. Extended segments of nitrogen-doped single-walled nanotubes of carbon were synthesized, revealing that higher nitrogen concentration enhanced electrical conductivity. [56,58,59]. The organized mesoporous fiber structure forms a network of interconnected empty spaces, capable of containing a significant quantity of sulphur. In addition, the ordered mesoporous carbon fiber (OMCF) framework has the potential to enhance the movement of electrons and ions, as well as the diffusion of electrolytes. It also has the ability to retain polysulfides, which improves the reversibility of the electrode. This is a feature that is difficult to achieve with random porous structures [54]. For glucose measurement in complicated biological media, non-enzymatic glucose electrochemical sensors are needed. Mesoporous carbon substrate of screen-printed carbon surface and inversion of phase for improved electrode surface energy for deposition of cobalt phthalocyanine. It was found that PI deposits more CoPc on the MC surface and that activating surface increases glucose oxidation electrochemically. Under these conditions, the MC-CoPc/PI electrode has great sensitivity, selectivity, and firmness after two months with 100% electrocatalytic retention for glucose oxidation [60]. MgO-doped OMC was synthesized with lignin and Mg^{2+} in one pot. The sole carbon precursor was walnut lignin, while Mg^{2+} was the crosslinking agent and catalytic active centre. MgO-OMCs with a well-defined mesoporous structure and homogenous metal dispersion catalyzed glucose to fructose isomerization well. $MgO_{0.5}$-OMC-600 produced 41% fructose and 89.3% selectivity at 100 °C for 120 min in aqueous phase with glucose loading of 7 wt% [61]. OMCs are ideal electrode modifiers for electrochemical (bio)sensors. Electrochemical sensing applications are promising for OMC materials. OMCs are likely to be crucial in electrochemical sensing [57]. Researchers will continue to functionalize OMCs using antibodies, enzymes, or other molecular reagents that can improve target analyte detection sensitivity and selectivity. OMC thin films and coatings [62] which can be integrated within microelectrode arrays to create small, portable electrochemical sensing devices. In environmental applications, these devices possess the capability to detect harmful analytes in situ, in real time. Future integration of OMCs with microfluidics and wireless communication could provide autonomous and networked sensing systems. OMC sensors are stable and contaminant-resistant, making them suited for lasting environmental sensing. They perhaps tolerate high temperatures, severe pH, and disorderly chemicals. These benefits are crucial for electrochemical pollution detection. Future advancements must optimize OMC surface chemistry and structure to increase stability, durability, and long-term performance for trustworthy electrochemical sensors [63].

Materials Research Foundations 173 (2025) 340-361
https://doi.org/10.21741/9781644903452-13

3.3 Enzyme immobilization in OMCs

In the last decade, hundreds of publications have been published on immobilizing different enzymes for various mesoporous supports along with their related applications, primarily in biological catalysis. Numerous outstanding reviews provided an overview of the development in enzyme immobilization. Additionally, expert reviews on the uses of enzymes that have been immobilized in bioanalysis, biocatalysis, and biofuel cells [13]. The hollow mesoporous carbon particles (HMC) were produced and tailored for the immobilization with laccase (Lac). In presence of redox mediator, the immobilized Lac removed tetracycline hydrochloride (TCH) together with ciprofloxacin hydrochloride (CPH) efficiently. Synergy between HMC adsorption and Lac degradation may explain the high antibiotic removal efficiency [64]. Laccase adsorption bimodal carbon-based mesoporous magnetic composites shaped magnetically separable laccase immobilized system. Immobilized laccase has a higher adsorption capacity (491.7 mg g^{-1}), better recovery (91.0%), and wider pH and temperature profiles than free laccase [65]. An excellent antibacterial material was created using three-dimensional ordered mesoporous carbon 3DOm COOH and lysozyme infused. The outcomes disclosed that 3DOm COOH-Lyz is effective in the elimination of Staphylococcus aureus, particularly 3E-C, which has uppermost antibacterial efficacy at 96.5%. 3DOm COOH-Lyz's antibacterial activity is the result of a combination of 3DOm COOH's superior adsorption and lysozyme hydrolysis. Furthermore, 3DOm COOH-Lyz exhibits good longevity in actual applications, maintaining 82% bacterial eradication effectiveness even after the fifth cycle. As a result, 3DOm COOH attached Lysozyme is thought to be capable antiseptic for potable water disinfection in minimal and emergency treatment of water applications that do not produce disinfection by products [66]. The possible use of organized mesoporous carbon in bioelectrochemical research. OMCs were shown to be better options than carbon nanotubes for immobilizing glucose oxidase and promoting heterogeneous electron transfer due to their enormous surface extent, homogenous mesopores, and outstanding electro-catalytic characteristics. Furthermore, enzyme molecules built on OMCs maintained a elevated bioactivity. Excessively ordered 3D-OMC had a greater adsorption capacity for glucose oxidase, and the built glucose biosensor had a broader linear response scale and higher detection sensitivity than 2D-OMC [67]. Glucose oxidase (GOD) was securely immobilized glassy carbon electrode modified with mesoporous carbon FDU-15 (MC-FDU-15) and Nafion using a simple approach. GOD immobilized on MC-FDU-15 and Nafion matrices has outstanding biological electrocatalytic activity in drop of oxygen level. So, the Nafion/GOD-MC-FDU-15/GC electrode is supposed to have a potential to be used as the cathode in a biofuel cell [68]. He enhanced the performance of bioelectrodes constructed from amination-modified glucose oxidase (GO_x-NH_2) and carboxyl-functionalized mesoporous carbon (OMC-COOH). The consequences disclosed that GO_x-NH_2/OMC-COOH had high electroactivity and a favorable anode reaction. The immobilization carrier material for enzymatic bioelectrodes has a significant impact on fuel cell efficiency due to its conductivity and electrical stability. It claimed that employing electrode which is porous instead of a planar electrode increased power density upto 72%, improved effective area and material transport properties. As a result, carbon derived porous materials such as orderly arranged mesoporous carbon (OMC) was identified as excellent materials for electrode due to its inimitable abilities of tiny size, great surface area, conductivity, and thermal durability [69].

 Materials Research Forum LLC
https://doi.org/10.21741/9781644903452-13

4. Immobilised proteins and enzymes and their applications

The increasing technological revolutions in protein removal, along renewed advances in protein manufacturing, resulted in the competent production of specific enzymes accompanied by upgraded properties. There is a growing demand for biocatalysts in several industrial sectors. Enzymes possess numerous advantages compared to chemical catalysts, rendering them highly preferable for nearly all industrial sector. Immobilisation offers numerous economic, ecological, and technical benefits, such as enhanced operating stability and the capacity to be reused. Immobilised enzyme is widely utilised in diverse industrial sectors, such as food, pharmaceutical, animal feed, textile, medical, detergent, and bioremediation/water remediation. Various techniques, such as covalent binding and entrapment, are employed to immobilise many enzymes on innovative carriers for usage in diverse large-scale operations. There is still potential for the development of new immobilisation methods. The primary factors influencing the worldwide enzyme market are the strong demand for detergent, pharmaceutical, cloth, animal feedstuff, biofuel, and cosmetic makers for green alternatives. Despite experiencing substantial expansion and generating significant income over the preceding decade, the industrial enzyme industry was negatively impacted in 2020 due to the coronavirus pandemic. Due to interruptions in global supply chains, there has been a significant decline in the demand for enzymes in industrial applications over the past two years. Enzymes, when immobilised, are expected to have a higher frequency of application soon.

The immobilised lipase B from Candida antarctica (CALB) on siliceous mesoporous cellular foams, which have been finely functionalised with methyl groups using the disilazane precursor, along with two recently developed ruthenium complexes containing substituted cyclopentadienyl ligands, demonstrate high efficiency in the chemoenzymatic dynamic kinetic resolution of (rac)-1-phenylethanol. The system with complex 2 demonstrated exceptional performance, achieving a 96% conversion rate and nearly 100% enantiomeric excess of (R)-1-phenylethyl acetate in an 8-hour one-pot procedure. The second system, utilising complex 1, also shown impressive results, performing almost as well [70]. Candida rugosa lipases (CRLs) were immobilised on MSU-H mesoporous silica via physical adsorption, cross-linking, and covalent binding. Organic methods transformed the silica. Immobilised CRLs were investigated as biocatalysts for CLA and ethanol esterification in a nonaqueous media. CRL aggregates trapped in silica pores cross-linked to form surfactant-immobilized CRLs. The silicas' surfactant activates immobilised CRLs at the interface, promoting catalytic esterification. The biocatalyst, which contains 38% surfactant by weight, had 1111.1 U/mg specific activity, eight times higher than soluble CRL. It also obtained a maximum esterification rate of 56.7% for total CLA, with 96.5% of it due to the 9c, 11t-CLA isomer. Grafting glutaraldehyde, immobilized-CRL with 64.5 mg/g protein showed a maximal hydrolytic activity of 2945.3 U/g-support [71]. In dye decolorization, laccase immobilisation could make the system easier to use, reuse the enzyme, and save costs. Trametes modesta laccase covalently bound to aluminium oxide pellets decolorises many colours. Glutaric dialdehyde activated 3-aminopropyltriethoxysilane (APTES)-siliconized alumina. Testing laccase immobilisation supports for dye treatment includes magnetic chitosan beads. T. versicolour laccase was covalently bonded to epichlorhydrin for operational and thermal stability. In batch reactors, this laccase decolorised Reactive Yellow 2 (80%) and Reactive Blue 4 (55%) [72, 73, 74]. An epoxy-functionalized silica platform for laccase immobilisation and biodegradation of model phenol compounds. Immobilisation on the support enhanced laccase stability. Immobilised laccase removed phenolic compounds with 95% catechol efficiency. Laccase immobilisation and biodegradation of model phenol compounds on epoxy-functionalized silica. Laccase stability improved with support immobilisation. With 95% catechol efficiency, immobilised laccase eliminated phenolic chemicals [75]. Alcohol dehydrogenase (ADH) from E. coli which electrostatically immobilised on aminopropyl modified cellular siliceous foams (MCF-NH2). Re-oxidizing NADPH to NADP+ was possible with immobilised ADH [76]. Continuous flow is preferred over batch procedures in the food business, especially for big product batches. HFCS, cocoa butter, its analogues, allulose, galacto-oligosaccharides, and other foods are produced in large-scale continuous processes using immobilised enzymes. Immobilised D-Glucose/xylose isomerase (EC 5.3.1.5) is used to make 'high fructose corn syrup' (HFCS), largest commercial process using an immobilised enzyme in terms of enzyme amount and product volume [77].

In environmental restoration, biocatalysis can accelerate pollutant breakdown. Reusing laccase and molecular mediators in porous crystals permits synergistic catalytic breakdown of different contaminants [78]. The tiny organic molecule of 2,2′-azinobis (3-ethylbenzothiazoline-6-sulfonic acid) (ABTS) is usually put into the pores of a MOF like MIL-100(Fe) or UiO-66(Zr)-NH2. Laccase is then immobilised on MOFs. MOFs have high porosity and surface adsorption sites, allowing high ABTS and laccase loading. Laccase oxidises ABTS into a highly active free radical (ABTS•+), which has a much higher redox potential than pristine laccase, allowing it to efficiently degrade dyes like malachite green (MG), alizarin green (AG), and MO. Additionally, MOF co-immobilization made coupled catalysts recyclable [78, 79, 80]. Tyrosinase, another copper-containing oxidoreductase, uses ambient oxygen as an oxidant to catalyze the oxidation of hazardous pollutants without reagents. Yang's group in situ encapsulated mushroom tyrosinase in ZIF-8 and HKUST-1. The tyrosinase hybrid biocatalyst removed phenolic contaminants and bisphenols in complicated settings with exceptional pH and temperature tolerance. These easy-to-separate biocatalysts successfully treated coke plant wastewater [81, 82]. To synthesize biocatalysts, recombinant lipase was immobilized on silica particles. Production of a chiral myo-inositol derived with an enantiomeric excess of about 99 % was successful using as biocatalysts. The utmost active biocatalyst (LIPB on SBA-15-F-APTES-GA) was employed on ten reaction cycles and maintained its performance, proving its operational stability. Enzymatic manufacturing of nutraceuticals and drug compounds may be economically feasible, especially using inexpensive enzyme manufacturing methodologies, according to preliminary economic calculations [83]. Jung et al. demonstrated immobilising glucose oxidase and peroxidase in tandem biocatalysts SBA-15 and MCF for indole oxidation. This process produced H2O2 in situ by oxidizing glucose with GOx and indole to 2-oxindole with CPO. The two enzyme systems effectively prevented CPO deactivation from exogenous peroxides. By optimizing immobilisation, such as CLEA formation and reaction conditions, the biocatalyst became stable enough for continuous reaction in a fixed-bed reactor [84, 85].

The unique structure of mesoporous materials has led to improved efficiency and performance compared to standard solution processes. The non-interconnected pore system reduced enzyme autolysis. An increase in digestion rate is attributed to protein content enrichment within the material's pores. Additionally, a protein combination from human liver tissue cytoplasm was used to test the nanoreactor's practicality [86]. High proteolysis efficiency was also observed with trypsin in amino-functionalized SBA-15. The digestion of myoglobin was 1000 times faster than the traditional solution technique and yielded 100% sequence coverage. Additionally, size-selective proteolysis employing trypsin immobilised in thiol-modified SAB-15. This nanoreactor selectively digests less molecular-weight proteins than the complicated protein samples, simplifying the multistep isolation process [87,88]. A mesoporous based enzyme reactor that combines the size-exclusive interaction of mesopore structure towards excessive molecular weight (MW) proteins with in-situ proteolysis by the immobilised enzyme with respect to the pore area has been found to enable size-selective proteolysis of proteins of different MWs. Note that this in situ digestion method combines separation and digestion into a single step, allowing quick proteolysis of lower MW proteins from multifaceted biological examples, especially minute protein samples, without tedious pretreatment [89].

Conclusions and perspectives

The chapter provides a comprehensive overview of various enzyme immobilisation strategies applicable to mesoporous materials, focussing on advanced mesoporous carbons and their applications. In the past two decades, ordered mesoporous materials have developed as promising supports for enzyme immobilisation, potentially enhancing activity, stability plus recyclability of targeted enzyme. They have found great applications in the field of sensors, energy storage, biocatalysis, drug delivery, food processing industry, and nano bioreactor. Several enzymes are immobilized using diverse tactics on unique supporters and being used in large scale operations, which still open opportunities for creating innovative immobilization methodologies. Advancements in nanoscience may enable the utilisation of diverse mesoporous materials with suitable pore dimensions to satisfy specific requirements for enzyme accommodation. A suitable microenvironment for enzyme integration and catalytic activity is also essential.

Diverse carriers from various sources can be utilised for enzyme immobilisation. Numerous advantageous attributes affect the selection of enzyme carriers, including surface functional groups and charge, particle size, pore size distribution, elevated surface area, biocompatibility, cost-effectiveness, and inertness towards the enzyme. Consequently, the advancement of mesoporous materials as support substrates facilitates novel approaches for enzyme immobilisation. Despite notable advancements, the exploration of immobilized enzymes remains nascent, with numerous essential constraints and obstacles persisting before transitioning to diverse applications. Establishing dependable methodologies for the synthesis and functionalisation of mesoporous materials is crucial for assuring reproducibility and scalability in manufacturing. The diverse processes employed in the fabrication of mesoporous materials render the assessment of biocompatibility a highly complex endeavour. Most studies concerning enzyme immobilised mesoporous silica primarily concentrate on the immobilisation process, employing standardised activity assays to evaluate the activity of the immobilised enzyme. A comprehensive understanding of enzyme and mesoporous materials may facilitate the rapid expansion of applications for the enzyme carried materials. With the progress in this field, we anticipate the advancements of sophisticated enzyme systems with practical utility will get enhance in different fields such as biosensors, drug delivery and energy storage purposes.

References

[1] ALOthman, A. Zeid, A review: Fundamental aspects of silicate mesoporous materials, Mater. 5 (2012) 2874-2902. https://doi.org/10.3390/ma5122874

[2] K. Ishizaki, S. Komarneni, M. Nanko, Porous Materials: Process Technology and Applications, Springer science & business media, 2013.

[3] M. Faraji, Y. Yamini, F. Noormohammadi, M. Adeli, Application of magnetic nanomaterials in environmental monitoring, in: M. Ahmadi, A. Afkhami, T. Madrakian (Eds.), Magnetic Nanomaterials in Analytical Chemistry, Elsevier, 2021, pp. 155-189. https://doi.org/10.1016/B978-0-12-822131-0.00005-4

[4] A. Corma, From microporous to mesoporous molecular sieve materials and their use in catalysis, Chem. Rev. 97 (1997) 2373-2420. https://doi.org/10.1021/cr960406n

[5] M.V. Regí, F. Balas, D. Arcos, Mesoporous materials for drug delivery, Angew. Chem. -Int. Ed. 46 (2007) 7548-7558. https://doi.org/10.1002/anie.200604488

[6] S.E. Lehman, S.C. Larsen, Zeolite and mesoporous silica nanomaterials: Greener syntheses, environmental applications and biological toxicity, Environ. Sci.: Nano. (2014) 200-213. https://doi.org/10.1039/C4EN00031E

[7] H. Essa, E. Magner, J. Cooney, B.K. Hodnett, Influence of pH and ionic strength on the adsorption, leaching and activity of myoglobin immobilized onto ordered mesoporous silicates, J. Mol. Catal. B: Enzym. 49 (2007) 61-68. https://doi.org/10.1016/j.molcatb.2007.07.005

[8] B.J. Melde, B.J. Johnson, Mesoporous materials in sensing: Morphology and functionality at the meso-interface, Anal. Bioanal. Chem. 398 (2010) 1565-1573. https://doi.org/10.1007/s00216-010-3688-6

[9] X. Yang, P. Qiu, J. Yang, Y. Fan, L. Wang, W. Jiang, X. Cheng, Y. Deng, W. Luo, Mesoporous materials-based electrochemical biosensors from enzymatic to nonenzymatic, Small. 17 (2021) 1904022. https://doi.org/10.1002/smll.201904022

[10] L. Wang, W. Ding, Y. Sun, The preparation and application of mesoporous materials for energy storage, Mater. Res. Bull. 83 (2016) 230-249. https://doi.org/10.1016/j.materresbull.2016.06.008

[11] B. Brena, P.G. Pombo, F.B. Viera, Immobilization of enzymes: A literature survey, Immobilization of Enzymes and Cells: Third Edition (2013): 15-31. https://doi.org/10.1007/978-1-62703-550-7_2

[12] Brena BM, P. González-Pombo, Batista-Viera F, Immobilization of enzymes: A literature survey, Methods Mol Biol. 1051 (2013) 15-31. https://doi.org/10.1007/978-1-62703-550-7_2

[13] V.L. Sirisha, A. Jain, A. Jain, Enzyme immobilization: An overview on methods, support material, and applications of immobilized enzymes, Adv. Food Nutr. Res. 79 (2016) 179-211. https://doi.org/10.1016/bs.afnr.2016.07.004

[14] Z. Zhou, M. Hartmann, Progress in enzyme immobilization in ordered mesoporous materials and related applications, Chem. Soc. Rev. 42 (2013) 3894-3912. https://doi.org/10.1039/c3cs60059a

[15] D.M. Liu, C. Dong, Recent advances in nano-carrier immobilized enzymes and their applications, Proc. Biochem. 92 (2020) 464-475. https://doi.org/10.1016/j.procbio.2020.02.005

[16] S. Hudson, E. Magner, J. Cooney, B.K. Hodnett, Methodology for the immobilization of enzymes onto mesoporous materials, J. Phys. Chem. B. 109 (2005) 19496-19506. https://doi.org/10.1021/jp052102n

[17] J. Fan, J. Lei, L. Wang, C. Yu, B. Tu, D. Zhao, Rapid and high-capacity immobilization of enzymes based on mesoporous silicas with controlled morphologies, Chem. Commun. 17 (2003) 2140-2141. https://doi.org/10.1039/b304391f

[18] J.F. Diaz, K.J. Balkus Jr., Enzyme immobilization in MCM-41 molecular sieve, J. Mol. Catal. B: Enzym. 2 (1996) 115-126. https://doi.org/10.1016/S1381-1177(96)00017-3

[19] M.R. Khan, Immobilized enzymes: A comprehensive review, Bull. Natl. Res. Cent. 45 (2021) 1-13. https://doi.org/10.1186/s42269-020-00453-2

[20] F.V. Oberhaus, D. Frense, D. Beckmann, Immobilization techniques for aptamers on gold electrodes for the electrochemical detection of proteins: A review, Biosensors. 10 (2020) 45. https://doi.org/10.3390/bios10050045

[21] N.A. Mohidem, M. Mohamad, M.U. Rashid, M.N. Norizan, F. Hamzah, H.B. Mat, Recent advances in enzyme immobilisation strategies: An overview of techniques and composite carriers, J. Compos. Sci. 7 (2023) 488. https://doi.org/10.3390/jcs7120488

[22] S. Liu, M. Bilal, K. Rizwan, I. Gul, T. Rasheed, H.M. Iqbal, Smart chemistry of enzyme immobilization using various support matrices: A review. Int. J. Biol. Macromol. 190 (2021) 396-408. https://doi.org/10.1016/j.ijbiomac.2021.09.006

[23] Y.R. Maghraby, R.M. El-Shabasy, A.H. Ibrahim, H.M.S. Azzazy, Enzyme immobilization technologies and industrial applications, ACS Omega. 8 (2023) 5184-5196. https://doi.org/10.1021/acsomega.2c07560

[24] T. Jesionowski, J. Zdarta, B. Krajewska, Enzyme immobilization by adsorption: A review, Adsorption. 20 (2014) 801-821. https://doi.org/10.1007/s10450-014-9623-y

[25] N. Bashir, M. Sood, J.D. Bandral, Enzyme immobilization and its applications in food processing: A review, Int. J. Chem. Stud. 8 (2020) 254-261. https://doi.org/10.22271/chemi.2020.v8.i2d.8779

[26] C. Ispas, I. Sokolov, S. Andreescu, Enzyme-functionalized mesoporous silica for bioanalytical applications, Anal. Bioanal. Chem. 393 (2009) 543-554. https://doi.org/10.1007/s00216-008-2250-2

[27] M.G. Bellino, A.E. Regazzoni, Immobilization of enzymes into self-assembled iron (III) hydrous oxide nano-scaffolds: A bio-inspired one-pot approach to hybrid catalysts, Appl. Catal. A: Gen. 408 (2011) 73-77. https://doi.org/10.1016/j.apcata.2011.09.008

[28] R. Das, D. Sen, A. Sarkar, S. Bhattacharyya, C. Bhattacharjee, A comparative study on the production of galacto-oligosaccharide from whey permeate in recycle membrane reactor and in enzymatic batch reactor, Ind. amp; Eng. Chem. Res. 50 (2011) 806-816. https://doi.org/10.1021/ie1016333

[29] M.C.R. Franssen, P. Steunenberg, E.L. Scott, H. Zuilhof, J.P.M. Sanders, Immobilised enzymes in biorenewables production, Chem. Soc. Rev. 42 (2013) 6491-6533. https://doi.org/10.1039/c3cs00004d

[30] B.D. Singh, Biotechnology Expanding Horizons, Fifth ed., Kalyani Publishers, 2007.

[31] K. Szymańska, J. Bryjak, A.B. Jarzębski, Immobilization of invertase on mesoporous silicas to obtain hyper active biocatalysts, Top. Catal. 52 (2009) 1030-1036. https://doi.org/10.1007/s11244-009-9261-x

[32] Q. Zhao , Y. Hou, G.H. Gong, M.A. Yu, L. Jiang, Characterization of Alcohol Dehydrogenase from Permeabilized Brewer's Yeast Cells Immobilized on the Derived Attapulgite Nanofibers, Appl. Biochem. Biotechnol. 160 (2010) 2287-2299. https://doi.org/10.1007/s12010-009-8692-y

[33] P. Pachariyanon, E. Barth, D.W. Agar, Enzyme immobilisation in permselective microcapsules, J. Microencapsul. 28 (2011) 370-383. https://doi.org/10.3109/02652048.2011.576781

[34] T.A.C. Silva, P.S. Marques, C.R.F. Souza, S. Said, W.P.O.C. Silva, T. Alexandre, Enzyme encapsulation in magnetic chitosan-Fe3O4 microparticles, J. Microencapsul. 32 (2015) 16-21. https://doi.org/10.3109/02652048.2014.940013

[35] T. Man, C. Xu, X.Y. Liu, D. Li, C.K. Tsung, H. Pei, Y. Wan, L. Li, Hierarchically encapsulating enzymes with multi-shelled metal-organic frameworks for tandem biocatalytic reactions, Nat. Commun. 13 (2022) 305. https://doi.org/10.1038/s41467-022-27983-9

[36] Y. Jiang, L. Shi, Y. Huang, J. Gao, X. Zhang, L. Zhou, Preparation of robust biocatalyst based on cross-linked enzyme aggregates entrapped in three-dimensionally ordered macroporous silica, ACS Appl. Mater. Interfaces. 6 (2014) 2622-2628. https://doi.org/10.1021/am405104b

[37] A.S. Sahutoglu, C. Akgul, Immobilisation of aspergillus oryzae α-amylase and aspergillus niger glucoamylase enzymes as cross-linked enzyme aggregates, Chem. Pap. 69 (2015) 433-439. https://doi.org/10.1515/chempap-2015-0031

[38] C.S. Sampaio, J.A.F. Angelotti, R.F. Lafuente, D.B. Hirata, Lipase immobilization via cross-linked enzyme aggregates: Problems and prospects: A review, Int. J. Biol. Macromol. 215 (2022) 434-449. https://doi.org/10.1016/j.ijbiomac.2022.06.139

[39] L. Bayne, R.V. Ulijn, P.J. Halling, Effect of pore size on the performance of immobilised enzymes, Chem. Soc. Rev. 42 (2013) 9000-9010. https://doi.org/10.1039/c3cs60270b

[40] E. Serra, Á. Mayoral, Y. Sakamoto, R.M. Blanco, I. Díaz, Immobilization of lipase in ordered mesoporous materials: Effect of textural and structural parameters, Microporous Mesoporous Mater. 114 (2008) 201-213. https://doi.org/10.1016/j.micromeso.2008.01.005

[41] E. Magner, Immobilisation of enzymes on mesoporous silicate materials, Chem. Soc. Rev. 42 (2013) 6213-6222. https://doi.org/10.1039/c2cs35450k

[42] E. Weber, D. Sirim, T. Schreiber, B. Thomas, J. Pleiss, M. Hunger, R. Gläser, V.B. Urlacher, Immobilization of P450 BM-3 monooxygenase on mesoporous molecular sieves with different pore diameters, J. Mol. Catal. B: Enzym. 64 (2010) 29-37. https://doi.org/10.1016/j.molcatb.2010.01.020

[43] D. Zhao, Y. Wan, W. Zhou, Ordered Mesoporous Materials, John Wiley & Sons, 2012. https://doi.org/10.1002/9783527647866

[44] C. Liang, Z. Li, S. Dai, Mesoporous carbon materials: Synthesis and modification, Angew. Chem. Int. Ed. 47 (2008) 3696-3717. https://doi.org/10.1002/anie.200702046

[45] J. Li, J. Gu, H. Li, Y. Liang, Y. Hao, X. Sun, L. Wang, Synthesis of highly ordered Fe-containing mesoporous carbon materials using soft templating routes, Microporous Mesoporous Mater. 128 (2010) 144-149. https://doi.org/10.1016/j.micromeso.2009.08.015

[46] X. Song, H. Ren, J. Ding, C. Wang, X. Yin, H. Wang, One-step nanocasting synthesis of sulfur and nitrogen co-doped ordered mesoporous carbons as efficient electrocatalysts for oxygen reduction, Mater. Lett. 159 (2015) 280-283. https://doi.org/10.1016/j.matlet.2015.06.112

[47] Savic, Slavica, et al., Hard template synthesis of nanomaterials based on mesoporous silica, Metall. Mater. Eng. 24 (2018). https://doi.org/10.30544/400

[48] A. Ramesh, R. Manigandan, B.M. Ali, K. Dhandapani, C.T. Da, M.T.N. Le, Selective oxidation of benzyl alcohol over sulphated zirconia incorporated ordered mesoporous carbon by a hard template method, J. Alloys Compd. 918 (2022) 165729. https://doi.org/10.1016/j.jallcom.2022.165729

[49] Q. Wang, W. Zhang, Y. Mu, L. Zhong, Y. Meng, Y. Sun, Synthesis of ordered mesoporous carbons with tunable pore size by varying carbon precursors via soft-template method, Microporous Mesoporous Mater. 197 (2014) 109-115. https://doi.org/10.1016/j.micromeso.2014.05.025

[50] Z. Zhang, G. Wang, Y. Li, X. Zhang, N. Qiao, J. Wang, J. Zhou, Z. Liu, Z. Hao, A new type of ordered mesoporous carbon/polyaniline composites prepared by a two-step nanocasting method for high performance supercapacitor applications, J. Mater. Chem. A. 2 (2014) 16715-16722 https://doi.org/10.1039/C4TA03351E

[51] L. Xu, L. Sun, J. Feng, L. Qi, I. Muhammad, J. Maher, X. Cheng, W. Song, Nanocasting synthesis of an iron nitride-ordered mesopore carbon composite as a novel electrode material for supercapacitors, RSC Adv. 7 (2017) 44619-44625. https://doi.org/10.1039/C7RA08704G

[52] J. Wang, Y. Xu, B. Ding, Z. Chang, X. Zhang, Y. Yamauchi, K.C.W. Wu, Confined self-assembly in two-dimensional interlayer space: Monolayered mesoporous carbon nanosheets with in-plane orderly arranged mesopores and a highly graphitized framework, Angew. Chem. Int. Ed. 57 (2018) 2894-2898. https://doi.org/10.1002/anie.201712959

[53] S. Tian, J. Wu, X. Zhang, K.K. Ostrikov, Z. Zhang, Capacitive deionization with nitrogen-doped highly ordered mesoporous carbon electrodes, Chem. Eng. J. 380 (2020) 122514. https://doi.org/10.1016/j.cej.2019.122514

[54] H. Wang, C. Zhang, Z. Chen, H.K. Liu, Z. Guo, Large-scale synthesis of ordered mesoporous carbon fiber and its application as cathode material for lithium-sulfur batteries, Carbon. 81 (2015) 782-787. https://doi.org/10.1016/j.carbon.2014.10.024

[55] H. Lu, W. Dai, M. Zheng, N. Li, G. Ji, J. Cao, Electrochemical capacitive behaviors of ordered mesoporous carbons with controllable pore sizes, J. Power Sources. 209 (2012) 243-250. https://doi.org/10.1016/j.jpowsour.2012.02.041

[56] M. Guo, H. Wang, D. Huang, Z. Han, Q. Li, X. Wang, J. Chen, Amperometric catechol biosensor based on laccase immobilized on nitrogen-doped ordered mesoporous carbon (N-

OMC)/PVA matrix, Sci. Technol. Adv. Mater. 15 (2014) 035005. https://doi.org/10.1088/1468-6996/15/3/035005

[57] J.C. Ndamanisha, L.P. Guo, Ordered mesoporous carbon for electrochemical sensing: A review, Anal. Chim. Acta. 747 (2012) 19-28. https://doi.org/10.1016/j.aca.2012.08.032

[58] T. Zhu, Y. Lu, S. Zheng, Y. Chen, H. Guo, Influence of nitric acid acitivation on structure and capacitive performances of ordered mesoporous carbon, Electrochim. Acta. 152 (2015) 456-463. https://doi.org/10.1016/j.electacta.2014.11.161

[59] A. Walcarius, Electrocatalysis, sensors and biosensors in analytical chemistry based on ordered mesoporous and macroporous carbon-modified electrodes, TrAC Trends Anal. Chem. 38 (2012) 79-97. https://doi.org/10.1016/j.trac.2012.05.003

[60] M.I.G. Sánchez, H. Khadhraoui, R.J. Pérez, J. Iniesta, E. Valero, Non-enzymatic glucose sensor using mesoporous carbon screen-printed electrodes modified with cobalt phthalocyanine by phase inversion, Microchem. J. 200 (2024) 110314. https://doi.org/10.1016/j.microc.2024.110314

[61] Z. Han, X. Wang, X. Zhao, F. Shen, B. Shen, X. Qi, Efficient isomerization of glucose into fructose by MgO-doped lignin-derived ordered mesoporous carbon, Int. J. Biol. Macromol. 267 (2024) 131471. https://doi.org/10.1016/j.ijbiomac.2024.131471

[62] T. Nasir, G. Herzog, M. Hébrant, C. Despas, L. Liu, A. Walcarius, Mesoporous silica thin films for improved electrochemical detection of paraquat, ACS Sens. 3 (2018) 484-493. https://doi.org/10.1021/acssensors.7b00920

[63] A. Gaber, S. Bilge, D. Bayramoğlu, Y.O. Donar, A. Sınağ, Synthesis and characterization of ordered mesoporous carbon-based materials for electrochemical detection of environmental pollutants, Trends Environ. Anal. Chem. (2024) e00233. https://doi.org/10.1016/j.teac.2024.e00233

[64] B. Shao, Z. Liu, G. Zeng, Y. Liu, X. Yang, C. Zhou, M. Chen, Y. Liu, Y. Jiang, M. Yan, Immobilization of laccase on hollow mesoporous carbon nanospheres: Noteworthy immobilization, excellent stability and efficacious for antibiotic contaminants removal, J. Hazard. Mater. 362 (2019) 318-326. https://doi.org/10.1016/j.jhazmat.2018.08.069

[65] Y. Liu, Z. Zeng, G. Zeng, L. Tang, Y. Pang, Z. Li, C. Liu, X. Lei, M. Wu, P. Ren, Z. Liu, M. Chen, G. Xie, Immobilization of laccase on magnetic bimodal mesoporous carbon and the application in the removal of phenolic compounds, Bioresour. Technol. 115 (2012) 21-26. https://doi.org/10.1016/j.biortech.2011.11.015

[66] J. Wang, L. Tang, P. Somasundaran, W. Fan, G. Zeng, Y. Deng, Y. Zhou, J. Wang, Y. Shen, Highly effective antibacterial activity by the synergistic effect of three-dimensional ordered mesoporous carbon-lysozyme composite, J. Colloid Interface Sci. 503 (2017) 131-141. https://doi.org/10.1016/j.jcis.2017.05.014

[67] C. You, X. Xu, B. Tian, J. Kong, D. Zhao, B. Liu, Electrochemistry and biosensing of glucose oxidase based on mesoporous carbons with different spatially ordered dimensions, Talanta. 78 (2009) 705-710. https://doi.org/10.1016/j.talanta.2008.12.032

[68] K. Wang, H. Yang, L. Zhu, Z. Ma, S. Xing, Q. Lv, J. Liao, C. Liu, W. Xing, Direct electron transfer and electrocatalysis of glucose oxidase immobilized on glassy carbon electrode

modified with Nafion and mesoporous carbon FDU-15, Electrochim. Acta. 54 (2009) 4626-4630. https://doi.org/10.1016/j.electacta.2009.02.097

[69] C. Lv, et al, Enhanced performance of bioelectrodes made with amination-modified glucose oxidase immobilized on carboxyl-functionalized ordered mesoporous carbon, Nanomater. 11 (2021) 3086. https://doi.org/10.3390/nano11113086

[70] D. Stradomska, et al, Lipase immobilized on MCFs as biocatalysts for kinetic and dynamic kinetic resolution of sec-alcohols, Catalysts. 11 (2021) 518. https://doi.org/10.3390/catal11040518

[71] W.H. Yu, et al, Immobilization of Candida rugosa lipase on hexagonal mesoporous silicas and selective esterification in nonaqueous medium, Biochem. Eng. J. 70 (2013) 97-105. https://doi.org/10.1016/j.bej.2012.10.005

[72] A. Kandelbauer, et al, Study of dye decolorization in an immobilized laccase enzyme-reactor using online spectroscopy, Biotechnol. Bioeng. 87 (2004) 552-563. https://doi.org/10.1002/bit.20162

[73] L. Lu, M. Zhao, Y. Wang, Immobilization of laccase by alginate-chitosan microcapsules and its use in dye decolorization, World J. Microbiol. Biotechnol. 23 (2007) 159-166. https://doi.org/10.1007/s11274-006-9205-6

[74] P. Galliker, et al, Laccase-modified silica nanoparticles efficiently catalyze the transformation of phenolic compounds, J. Colloid Interface Sci. 349 (2010) 98-105. https://doi.org/10.1016/j.jcis.2010.05.031

[75] M. Mohammadi, et al, Immobilization of laccase on epoxy-functionalized silica and its application in biodegradation of phenolic compounds, Int. J. Biol. Macromol. 109 (2018) 443-447. https://doi.org/10.1016/j.ijbiomac.2017.12.102

[76] M. Dreifke, F.J. Brieler, M. Fröba, Immobilization of alcohol dehydrogenase from E. coli onto mesoporous silica for application as a cofactor recycling system, ChemCatChem. 9 (2017) 1197-1210. https://doi.org/10.1002/cctc.201601288

[77] A. Basso, S. Serban, Overview of immobilized enzymes' applications in pharmaceutical, chemical, and food industry, Methods Mol. Biol. (2020) 27-63. https://doi.org/10.1007/978-1-0716-0215-7_2

[78] G. Yaohua, et al, Co-immobilization of laccase and ABTS onto novel dual-functionalized cellulose beads for highly improved biodegradation of indole, J. Hazard Mater. 365 (2018) 118-124. https://doi.org/10.1016/j.jhazmat.2018.10.076

[79] X. Lou, et al, Construction of co-immobilized laccase and mediator based on MOFs membrane for enhancing organic pollutants removal, Chem. Eng. J. 451 (2023) 138080. https://doi.org/10.1016/j.cej.2022.138080

[80] P. Snehangshu, et al, Design of Laccase-Metal Organic Framework-Based Bioelectrodes for Biocatalytic Oxygen Reduction Reaction, ACS Appl. Mater. Interfaces. 8 (2016) 20012-20022. https://doi.org/10.1021/acsami.6b05289

[81] C.M. Wei, et al, Mushroom tyrosinase immobilized in metal-organic frameworks as an excellent catalyst for both catecholic product synthesis and phenolic wastewater treatment, J. Chem. Technol. Biotechnol. 97 (2022) 962-972. https://doi.org/10.1002/jctb.6984

[82] X.F. Lü, et al, Tyrosinase@ HKUST-1: A super stable biocatalyst efficient for catecholic product synthesis, Bioresour. Bioprocess. 8 (2021) 1-12. https://doi.org/10.1186/s40643-021-00462-7

[83] E.P. Cipolatti, et al, Synthesis of lipase/silica biocatalysts through the immobilization of CALB on porous SBA-15 and their application on the resolution of pharmaceutical derivatives and on nutraceutical enrichment of natural oil, Mol. Catal. 505 (2021) 111529. https://doi.org/10.1016/j.mcat.2021.111529

[84] D. Jung, M. Hartmann, Oxidation of indole with CPO and GOx immobilized on mesoporous molecular sieves, Catal. Today. 157 (2010) 378-383. https://doi.org/10.1016/j.cattod.2010.03.004

[85] D. Jung, M. Hartmann, Oxidation of Indole with CPO and GOx Immobilized on SBA-15, Stud. Surf. Sci. Catal. 174 (2008) 1045-1050. https://doi.org/10.1016/S0167-2991(08)80067-5

[86] F. Casadonte, et al, Smart trypsin adsorption into N-(2-aminoethyl)-3-aminopropyl-modified mesoporous silica for ultra fast protein digestion, Chem. - Eur. J. 30 (2010) 8998-9001. https://doi.org/10.1002/chem.201000120

[87] Q. Min, et al, Size-selective proteolysis on mesoporous silica-based trypsin nanoreactor for low-MW proteome analysis, Chem. Commun. 46 (2010) 6144-6146. https://doi.org/10.1039/c0cc00619j

[88] Y. Wang, et al, Core-shell silica microsphere-based trypsin nanoreactor for low molecular-weight proteome analysis, Anal. Chim. Acta. 985 (2017) 194-201. https://doi.org/10.1016/j.aca.2017.07.012

[89] Q. Min, et al, Size-selective proteolysis on mesoporous silica-based trypsin nanoreactor for low-MW proteome analysis, Chem. Commun. 46 (2010) 6144-6146. https://doi.org/10.1039/c0cc00619j

Chapter 14

Mesoporous Materials for Separation of Biomolecules

Vipul D. Prajapati[1*], Princy Shrivastav[2]

[1]Department of Pharmaceutics, SSR College of Pharmacy (Permanently Affiliated to Savitribai Phule Pune University), Sayli-Silvassa Road, Sayli, Silvassa, Union Territory of Dadra Nagar Haveli and Daman Diu, India

[2]Department of Pharmaceutical Chemistry, SSR College of Pharmacy (Permanently Affiliated to Savitribai Phule Pune University), Sayli-Silvassa Road, Sayli, Silvassa, Union Territory of Dadra Nagar Haveli and Daman Diu, India

vippra2000@yahoo.com

Abstract

Mesoporous materials have gained significant attention as advanced tools for the separation and purification of biomolecules due to their distinctive structural features, including higher surface areas, tunable pore sizes, and customizable surface functionalities. This chapter explores the synthesis, modification, and application of mesoporous materials in the separation of biomolecules such as proteins, enzymes, and nucleic acids. It provides a detailed overview of how pore size, surface chemistry, and material morphology influence the selectivity and efficiency of separation processes. The discussion highlights the advantages of mesoporous materials over traditional separation techniques and emphasizes their potential in enhancing the scalability and efficiency of biotechnological and biomedical processes. Additionally, current challenges and future directions in the design of more efficient, selective, and biocompatible mesoporous materials are discussed, outlining the path forward for further advancements in this promising field.

Keywords

Mesoporous Materials, Biomolecule Separation, Protein Adsorption, Enzyme Purification, Nucleic Acid Separation, Surface Functionalization, Biotechnological Applications, Biomedical Applications

Contents

1. Introduction

Biomolecule separation is a critical process in biochemistry and molecular biology, enabling researchers to isolate and purify various biomolecules such as proteins, nucleic acids, and carbohydrates from complex mixtures. This process is essential for characterizing the structure and function of these biomolecules, which is vital for understanding biological systems and developing biotechnological applications [1]. The separation of biomolecules is foundational for several reasons. Pure biomolecules are necessary to elucidate their roles in biological processes. By isolating specific components, researchers can study their interactions and functions in detail. The purification of biomolecules is also crucial for producing pharmaceuticals, enzymes, and other products used in medicine and industry. High purity is often required to ensure efficacy and safety in therapeutic applications. Also, separation techniques are often prerequisites for analytical methods such as mass spectrometry, NMR spectroscopy, and crystallography, which require pure samples for accurate results [2,3]. The timeline of advancements in the field of biomolecule separation is depicted in Figure 1.

Materials Research Foundations 173 (2025) 362-391 https://doi.org/10.21741/9781644903452-14

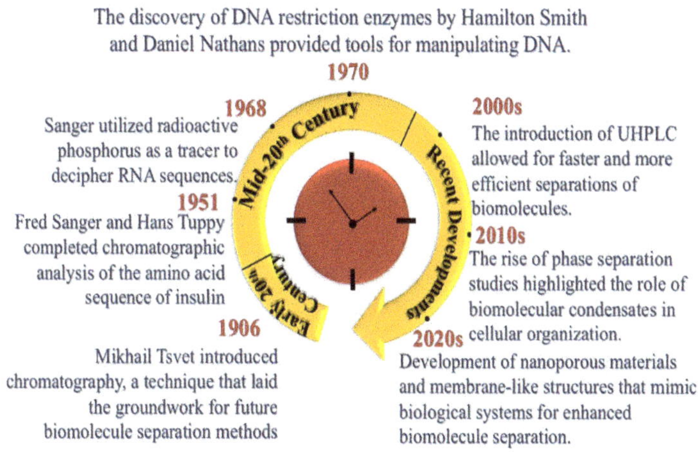

Figure 1. *An outlook framework showing the timeline of advancements in the field of biomolecule separation.*

Several techniques are commonly used for the separation and purification of biomolecules, each exploiting different physicochemical properties. First is chromatography, which separates biomolecules based on their differential affinities for a stationary phase and a mobile phase. The common chromatographic techniques which are employed for bioseparation include ion exchange chromatography (IEC) that separates biomolecules based on charge, size exclusion chromatography (SEC) which carries separation based on size, and affinity chromatography wherein separation depends on specific binding interactions between biomolecules [1].

Then we have electrophoresis, which separates charged biomolecules in a gel matrix under an electric field. It is widely used for analyzing DNA, RNA, and proteins based on their size and charge [4]. Next is ultracentrifugation that uses high centrifugal forces to separate biomolecules based on their density and size. It is particularly effective for isolating organelles and macromolecules from cellular extracts [5]. The other technique which is widely used for separation of biomolecules is liquid-Liquid phase separation (LLPS). This emerging technique leverages the differential partitioning of biomolecules into distinct liquid phases. LLPS can be used for selective extraction and purification, taking advantage of the unique interactions between biomolecules and their environment [6].

Mesoporous materials have garnered significant attention in the field of separation science, particularly for the separation of biomolecules. Their unique structural characteristics, including tunable pore sizes and high surface areas, make them ideal candidates for various chromatographic applications. Common types of mesoporous materials include mesoporous silica (e.g., MCM-41, SBA-15) [7,8]and periodic mesoporous organosilica (PMO) [9], each offering distinct advantages for chromatographic applications. The key characteristics of these materials, like tunable pore size, high surface area, chemical stability, and functional potential, offer great advantage for biomolecule separation [10].

Mesoporous materials serve as effective stationary phases in SEC for the purification of proteins. Their ability to separate proteins from aggregates and smaller contaminants ensures high purity and integrity of the biomolecules. SEC using mesoporous materials is commonly employed for desalting biomolecules, allowing for the efficient removal of salts and small molecules from protein solutions, which is essential for preparing samples for further analysis. The unique properties of mesoporous materials also facilitate the separation of nucleic acids, providing a gentle and effective method for purifying DNA and RNA from complex mixtures [11–14].

Mesoporous materials are utilized in High-Pressurized Liquid Chromatography (HPLC) for the separation of various biomolecules, including peptides and small proteins, due to their high resolution and efficiency. The biocompatibility and functionalization potential of mesoporous materials make them suitable for developing biosensors that detect specific biomolecules, leveraging their high surface area for enhanced sensitivity and specificity [15,16].

2. Fundamentals of Biomolecule Separation

The basic building blocks of living things are biomolecules, which include proteins, lipids, carbohydrates, and nucleic acids. Understanding these molecules' structure, relationships, and functions within intricate biological systems depends on their separation and purification. In biochemistry and molecular biology, biomolecule separation is a crucial procedure that aims to separate particular biomolecules from complicated mixtures, including polysaccharides, proteins, and nucleic acids. For some uses, including as medication creation, diagnostics, and research, this separation is crucial [17]. A summary of key factors such as pore size, surface chemistry, pH, temperature, and ionic strength, and their influence on the separation efficiency of various biomolecules has been given in Table 1.

Table 1. Factors affecting biomolecule separation efficiency.

Factor	Description	Impact on Separation Efficiency
Molecular size and shape [13, 14]	The size and shape of biomolecules influence their movement through the separation medium.	Larger or irregularly shaped molecules may have slower migration, leading to less efficient separation in size-based methods (e.g., SEC).
Charge of biomolecule [2]	Biomolecules carry varying charges depending on their pI and the pH of the medium.	The charge affects the molecule's interaction with the separation medium, particularly in ion-exchange chromatography or electrophoresis, impacting resolution.
pH of the separation medium [15]	pH influences the ionization of biomolecules and the stability of the separation medium.	At certain pH levels, biomolecules might aggregate or degrade, reducing separation efficiency. Optimal pH ensures the stability and better separation of the target.

Materials Research Foundations 173 (2025) 362-391 https://doi.org/10.21741/9781644903452-14

Ionic strength of the medium [15]	Refers to the concentration of ions in the separation medium.	Higher ionic strength can reduce the electrostatic interactions between biomolecules and the stationary phase, influencing elution times and resolution.
Viscosity of the Mobile Phase [15, 16]	The resistance to flow in the mobile phase (e.g., in liquid chromatography) affects migration rates.	Increased viscosity can slow down the separation process, reducing the efficiency and resolution of biomolecule separation.
Temperature [2]	Temperature can affect biomolecule structure, solubility, and interactions with the stationary phase.	Higher temperatures can improve diffusion rates but may denature sensitive biomolecules, while low temperatures can maintain stability but slow separation.
Flow rate of mobile phase [14, 15]	In chromatographic techniques, the flow rate controls how quickly the mobile phase passes through the column.	Too high a flow rate can result in poor resolution, while too low a flow rate can lead to longer run times, affecting throughput.
Stationary Phase Properties [7]	The pore size, surface area, and functional groups of the stationary phase affect biomolecule interaction.	Proper selection of the stationary phase enhances resolution and selectivity by optimizing biomolecule binding and elution profiles.
Affinity of biomolecules to ligand [2]	In affinity-based separations, the strength of interaction between biomolecules and ligands on the stationary phase is key.	High affinity leads to selective retention and efficient separation, but too strong an affinity may hinder elution and recovery of the target molecule.
Concentration of biomolecules [6]	The quantity of biomolecules presents in the sample.	High concentrations can lead to overloading of the separation medium, reducing resolution and peak sharpness. Low concentrations improve separation but may reduce yield.
Solvent composition [6]	Refers to the nature of solvents used in the mobile phase, including their polarity and composition.	The choice of solvent impacts the solubility and interaction of biomolecules with the stationary phase, directly affecting separation efficiency and resolution.
Presence of additives or detergents [2]	Additives like salts, detergents, or chaotropes are often used to maintain	Additives can prevent biomolecule aggregation and denaturation, improving separation, but

Materials Research Foundations 173 (2025) 362-391 https://doi.org/10.21741/9781644903452-14

	biomolecule solubility and stability.	excessive use may interfere with detection methods or resolution.
Column Length and Particle Size (for Chromatography) [11, 13]	Longer columns and smaller particles generally provide better separation resolution.	Smaller particle sizes improve resolution but may increase backpressure, while longer columns enhance separation but extend run times.

Biomolecules are fractionated according to their distinct physicochemical features using a number of critical techniques, including liquid-liquid phase separation, electrophoresis, chromatography, and ultracentrifugation. Chromatography uses a difference in the affinities of a mobile and stationary phase to separate components of a mixture. SEC is a major biomolecule separation mechanism. This method divides molecules according to size by using porous materials. Smaller molecules take longer to elute because they can enter the pores, whereas larger molecules are prevented from doing so and can elute first. Another technique that takes use of particular interactions between biomolecules and a ligand fixed on a stationary phase is affinity chromatography. Target biomolecules can be chosen out of a mixture using this technique. Biomolecules are separated according to their charge using ion exchange chromatography. Biomolecules can be eluted or retained according to their net charge at a specific pH by employing a charged stationary phase. Using hydrophobic interaction chromatography (HIC), biomolecules are separated according to how hydrophobic they are. This technique is very helpful for proteins because of the wide variations in their surface hydrophobicity [18,19].

Charged biomolecules such as proteins, DNA, and RNA are separated by electrophoresis according to their size, shape, and net charge as they move through a gel matrix in the presence of an electric field. It makes use of electrophoretic mobility in addition to molecular sieving. The idea behind how it works is that charged molecules move when exposed to an electric field. Positively charged molecules, such as certain proteins, flow toward the cathode, while molecules with a net negative charge, such as DNA and RNA, migrate toward the anode [20,21]. The size of the molecule, its charge, and the characteristics of the medium it migrates through all affect its rate of migration. This method falls into the categories of isoelectric focusing, gel electrophoresis, and capillary electrophoresis. The most popular type is gel electrophoresis, in which a gel matrix—typically polyacrylamide or agarose—acts as a separating medium. Smaller molecules can go through the gel more quickly than larger ones because it functions as a sieve. Proteins can be separated according to both size and charge, whereas DNA and RNA are normally separated based on size. Isoelectric focusing and SDS-PAGE are combined in the sophisticated method known as two-dimensional gel electrophoresis (2D-GE) to separate proteins according to both their pI and size. It allows for the resolution of complex protein mixtures, making it invaluable in proteomics [19]. Narrow capillaries are used in capillary electrophoresis to produce high-resolution separations. It is very helpful for studying small sample quantities and has a high degree of efficiency when separating molecules according to their size and charge. Proteins are sorted by isoelectric focusing according to their isoelectric point (pI). Proteins migrate along a pH gradient in the gel until they reach a point where their net charge is zero, which essentially focuses them at their pI [19,22].

When cells, organelles, or macromolecules are subjected to strong centrifugal forces—known as ultracentrifugation—they separate according to variations in mass, shape, or a combination of these characteristics. This technique makes use of strong centrifugal forces to help particles in a solution settle, which are produced by rotors rotating at up to 150,000 rpm. Proteins, DNA, and subcellular particles are separated using it. Sedimentation, in which suspended particles are subjected to a centrifugal force that mimics a gravitational field far stronger than Earth's, is the basic process of ultracentrifugation. The variations in particle and surrounding medium densities determine the pace of sedimentation. Less dense particles will stay suspended or float in the centrifuge tube, while denser particles will descend to the bottom [23]. There are two types of ultracentrifugation: analytical and preparative. The main goal of analytical ultracentrifugation is to examine the characteristics and interactions of biological molecules. It provides important information on molecular weight, shape, and interactions by enabling the real-time observation of the sedimentation process. However, the main purpose of preparative ultracentrifugation is to separate and refine biomolecules. It frequently makes use of methods like density gradient centrifugation, in which the centrifuge tube is filled with a gradient of varying densities. Effective separation is made possible by the migration of particles, which are positioned according to their density [5,23].

An increasingly well-understood process in cellular biology, liquid-liquid phase separation (LLPS) is crucial to the structure and operation of biomolecules. In this process, a homogeneous solution spontaneously splits into two different liquid phases, forming droplets without membrane borders or biomolecular condensates [6,24]. When specific circumstances—such as variations in concentration, temperature, or ionic strength—cause biomolecules to condense into liquid droplets, LLPS happens. Numerous weak interactions, such as electrostatic forces, hydrophobic contacts, and hydrogen bonds, are responsible for this behavior. The resultant droplets can operate as dynamic compartments in cells, enabling molecular interactions and specific biochemical events [25,26].

2.1 Types of interactions involved in biomolecule separation

The behavior of biomolecules in diverse settings is determined by a variety of interactions, which are essential for their separation. Comprehending these interactions is essential for creating efficient separation methods in molecular biology and biochemistry. Non-covalent and covalent interactions, along with other particular binding types of interactions, are the main forms of interactions involved in biomolecule separation. The stability and selectivity of biomolecular connections depend on non-covalent interactions. Van der Waal forces, hydrophobic contacts, ionic bonds, and hydrogen bonds are some examples of these interactions. When a hydrogen atom is covalently bound to an extremely electronegative atom (such as nitrogen or oxygen) and another electronegative atom, hydrogen bonds are created. The stability and interactions of proteins and nucleic acids during separation processes are significantly influenced by hydrogen bonds, which are essential to their structure and function. Ions with opposing charges form ionic bonds. Ionic interactions, such as the binding of metal ions (such as magnesium) to negatively charged DNA, which promotes structural order, aid in the stabilization of biomolecules in biological systems. When molecules that are hydrophobic (water-repellent) group together to reduce their exposure to water, hydrophobic interactions occur. Because hydrophobic areas of proteins often lie deep within protein structures and affect how the proteins behave during separation, this phenomenon is especially important for understanding protein folding and stability. Van der Waals forces are weak molecular attractions brought on by transient dipoles created by the dispersion of electrons.

Materials Research Foundations 173 (2025) 362-391
https://doi.org/10.21741/9781644903452-14

Although individually weak, collectively they can significantly influence the three-dimensional structure of proteins and their interactions [27,28].

Strong bonds are created when atoms share electron pairs through covalent interactions. Covalent bonds play a significant role in certain situations, such as affinity chromatography, in which a ligand is covalently linked to a stationary phase in order to selectively capture target biomolecules, despite being less prevalent in dynamic biomolecule separations. In addition to this, several biomolecules have particular binding interactions that can be used to separate them. For example, the idea of protein-protein interactions is utilized to isolate particular proteins depending on their binding partners for affinity chromatography. Protein-DNA/RNA interactions refer to the selective binding of proteins to nucleic acids. These interactions can be applied in nucleic acid isolation procedures or in the investigation of protein-DNA/RNA complexes. Lastly, ligand-receptor interactions are often utilized in biosensing and affinity-based separation methods, where a receptor is immobilized on a stationary phase to capture specific ligands from a mixture [29].

3. Mesoporous Materials in Biomolecule Separation

In case of the industries of the foods and pharmaceuticals, the ability to separate biological components like proteins and vitamins is crucial. To isolate these types of substances from the blend of impurities with comparable physico-chemical characteristics, high selectivity is frequently required. Mesoporous materials are essential for biomolecule separation because of their special structural qualities, which improve a range of chromatographic methods. Their uses in HPLC, SEC, and other fields demonstrate their adaptability and potency in the separation of biomolecules. The study of biochemistry and molecular biology is anticipated to witness a growth in the influence of these materials as research on their synthesis and functionalization advances [16,30].

The MCM-41 and MCM-48 mesoporous molecular level of sieves (MMS), which belong to the M41S family, have high surface areas, homogeneous pore sizes, and extremely regular structures, making them highly promising as separation media. The adsorbent's vast surface area is accessible to bigger solutes due to its consistent pore size and broader pore size range in comparison to zeolites. These characteristics can be specifically designed for adsorption using either targeted surface chemistry in conjunction with size exclusion effects alone [31]. Unfortunately, there hasn't been much progress made in developing these materials for separations, especially for procedures that take advantage of their promise for size-based selectivity. These novel materials may also be able to solve issues with swelling and lack of selectivity that arise with commonly used adsorbents (commonly activated carbon and bentonite). Mesoporous molecular sieves (MMS) have been shown to work well as packings for some aromatic hydrocarbons in HPLC. Comparable selectivity and retention behaviour were observed between MCM-41 and other silica packings. The successful application of ligands to surfaces to generate functionally activated coatings, which have been demonstrated for ion exchange and catalysis, can target specific separations [32]. MCM-41 has thiol functional coatings applied to it in order to eliminate harmful heavy metallic ions (namely mercuries) from polluted effluent wastage streams. Due to the large surface associated areas and uniformity of regular pores, which provide good access without pore obstructions as found in substances having broad pore size ranges and mixed ordered pore matrices, a high capacity for these ions was attained [32,33].

Materials Research Foundations 173 (2025) 362-391
https://doi.org/10.21741/9781644903452-14

MCM-48 is an additional one that is said to have a branching and interlaced porous structure. Its 3D porous structure implies that it is more appealing for adsorption separations than MCM-41's unidirectional structure. This structural characteristic may offer an excellent remarkable mass transfer than MCM-41 because of its improved pore accessibility, which can also help prevent pore blockages during operation. A stationary phase relevant to MCM-48 transferred mass more quickly than one based on MCM-41 in a recent chiral HPLC investigation; however, this could have been caused by variations in particle and pore diameters [32]. Figure 2 gives a brief classification of types of categorial mesoporous materials.

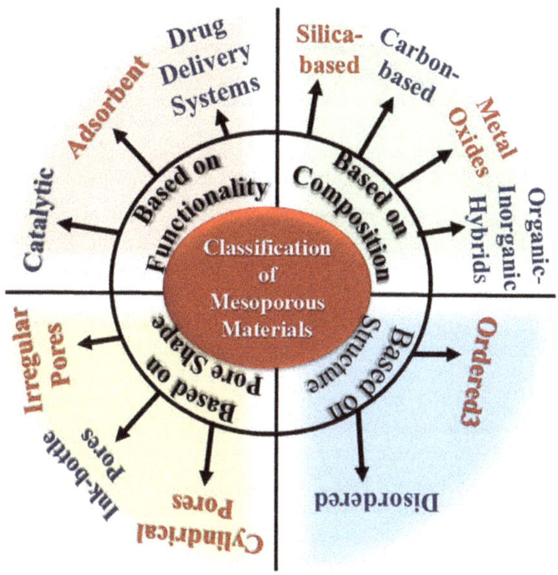

Figure 2. *An outlook image showing the classification of mesoporous materials.*

3.1 Structural properties of mesoporous materials (MMs) for biomolecule separation

These types of materials are very useful for the separation of biomolecules because of a number of important structural characteristics. These materials are perfect for size-based biomolecule separation since their pore diameters range from 2 to 50 nm. This includes proteins and nucleic acids. During synthesis level, the pore size of its structure can be adjusted to maximize separation for particular biomolecules. Achieving high resolution separations requires unique pore size in its structure and its size distributions [34].

The total surface areas of mesoporous materials are usually very high, ranging from 400 to 1000 m^2/g. The high surface area improves the capacity and effectiveness of separation by enhancing contact with biomolecules. Mesoporous materials' huge pore volumes make it easier for biomolecules to fit into the porous structure. High pore volumes help biomolecules have a high loading capacity during the separation process. Mesoporous materials exhibit exceptional chemical stability in a range of circumstances, such as varying temperatures and pH values. When it comes to biomolecule separations, its stability guarantees dependable performance and permits the application of severe conditions when necessary. Mesoporous materials' surfaces can also be readily altered by adding particular functional groups. Functionalization improves separation performance and increases selectivity for target biomolecules, especially in affinity-based separations [7,31].

3.2 Advantages of mesoporous materials in biomolecule separation

Mesoporous materials are very useful in chromatographic applications because they provide a number of benefits for biomolecule separation. High surface areas, typically 400–1000 m2/g, are seen in mesoporous materials. The interaction between the stationary phase and biomolecules is improved by this huge surface area, which raises the capacity and efficiency of separation. Mesoporous materials' pore diameters, which can range from 2 to 50 nm, are precisely controllable during production, enabling optimization according to the particular size of the biomolecules being separated. Effective separations in SEC and other methods depend on this tunability [16].

Larger biomolecules can be accommodated more easily in mesoporous materials due to their substantial pore volume, which is necessary for efficient separation. They are appropriate for preparative chromatography because of this feature, which permits larger loading capacities during separation procedures. Narrow pore size distributions are a common feature of mesoporous materials, which helps to produce reliable and repeatable separation outcomes. For high-resolution separations in complicated mixtures, this consistency is essential [35,36]. These substances exhibit outstanding chemical stability in a range of circumstances, including temperature and pH variations. This stability permits the application of severe conditions as needed and guarantees dependable performance during biomolecule separations. Target biomolecule selectivity can be improved by readily adding particular functional groups to the surfaces of mesoporous materials. Through customization of interactions with certain analytes, this functionalization can enhance the efficiency of affinity chromatography and other separation methods. Mesoporous materials have uses in medication delivery, catalysis, sensing, and diagnostics in addition to chromatography. Their adaptability increases their usefulness in a range of scientific and industrial domains [37].

3.3 Ordered Mesoporous Materials

A certain class of these types of materials known as ordered mesoporous materials (OMMs) that are distinguished by their extremely uniform and regular pore structures. Since the discovery of OMMs in the early 1990s, these materials have drawn a lot of attention especially to their special qualities and adaptability to a wide range of uses.

Large surface areas, well-organized pore configurations, and adjustable pore sizes—all of which promote improved mass transfer and adsorption capacities—are characteristics of OMMs. Surfactants or other agents are utilized in templating methods to produce the required pore structure during the production of these materials. Quaternary ammonium surfactants were used as templates by the first significant OMMs, namely MCM-41 and SBA-15, which led to improvements in the synthesis methods employed today [38,39].

The kinds of OMMs that can be created have increased due to recent advancements in synthesis techniques. For example, novel techniques like direct pyrolysis of thermoplastic elastomers have been investigated, allowing the synthesis of silicas and organized mesoporous carbons with specific characteristics [40]. This flexibility in synthesis allows researchers to manipulate various parameters, including pore size, wall thickness, and surface chemistry, to meet specific application needs. There are several types of OMMs that have been developed over the years, as given in Figure 3.

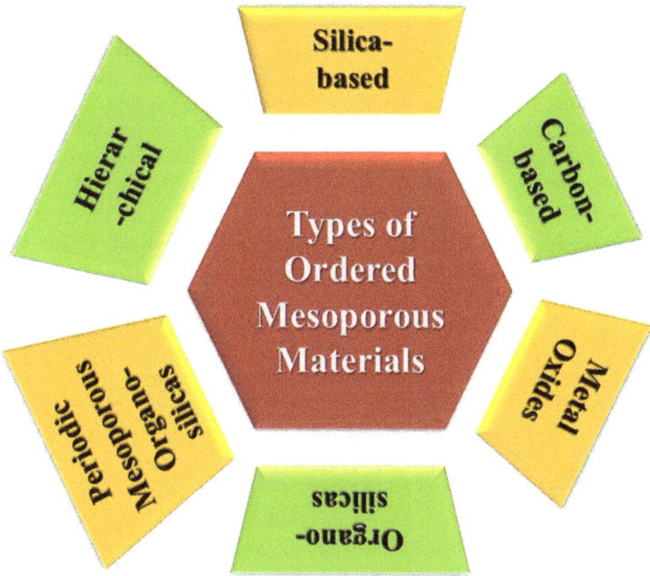

Figure 3. *Types of Ordered mesoporous materials.*

3.3.1 Silica-based OMMs

OMMs based on silica constitute an important class of mesoporous materials, mainly distinguished by their high surface areas and clearly defined pore architectures. Among the first known ordered mesoporous silicas is MCM-41. It has cylindrical pores arranged in a hexagonal pattern, usually with sizes between 2 and 10 nm. During the sol-gel process, cationic surfactants serve as templates in the synthesis of MCM-41. Because of its large surface area (up to 1500 m^2/g), it can be used in adsorption and catalysis. Comparing SBA-15 to MCM-41, more research has been done on this mesoporous silica, which has thicker walls and greater pore diameters (usually ranging from 5 to 30 nm). Triblock copolymer surfactants, which offer greater thermal resistance and structural stability, are used in its synthesis. SBA-15's adjustable pore diameters and functionalization properties have made it a popular catalyst and drug delivery system support [41–43].

It is possible to further alter silica-based OMMs to improve their characteristics for certain uses. Functionalized SBA-15, for example, can include organic groups or metal oxides to enhance catalytic activity or allow for selective adsorption. There are several ways to accomplish these

Materials Research Foundations 173 (2025) 362-391
https://doi.org/10.21741/9781644903452-14

changes, including co-condensation and grafting during synthesis [39,41,43]. These silica-based OMMs are crucial because of their special structural properties and functionalization potential in a variety of domains, such as biological applications, adsorption, and catalysis.

3.3.2 Metal Oxide OMMs

Metal oxide OMMs are a broad class of materials distinguished by their large surface areas and distinct mesoporous architectures. The main purposes for which these materials are created are adsorption, energy storage, and catalysis. Among the metal oxide OMMs that has been explored the most is mesoporous alumina (Al_2O_3). It is renowned for having a large surface area and modifiable pore diameters that can be changed while synthesizing. An organic pore-forming agent is frequently used in place of conventional surfactants during the synthesis process, enabling the creation of a mesopore network that is connected in three dimensions. Mesoporous alumina finds widespread application in a multitude of chemical processes, such as dehydration and hydrogenation reactions, as a catalyst and catalyst support [44].

TiO_2 (mesoporous titania) is an additional noteworthy kind of metal oxide OMM. It is frequently utilized in environmental applications, such as air and water treatment, and is especially recognized for its photocatalytic qualities. The material's surface area is increased by the mesoporous structure, which makes it easier for contaminants to be absorbed and degraded by UV radiation. Mesoporous titania is created using a variety of synthesis techniques, such as templating and sol-gel procedures, to provide regulated pore sizes and morphologies [38,42]. The next material is mesoporous zirconia (ZrO_2), which is useful for catalytic applications because of its high surface area and thermal stability, especially in reactions that need high temperatures. Mesoporous zirconia can be synthesized using techniques similar to those used to create silica and alumina, enabling the creation of materials with certain pore sizes and topologies [45].

Apart from these, mixed metal oxides, such as cobalt-alumina ($Co\text{-}Al_2O_3$) [46] and titanium-alumina ($Ti\text{-}Al_2O_3$) [47], are synthesized to combine the properties of different metal oxides. Mesoporous cobalt-alumina, for instance, has been created to take advantage of both cobalt's and alumina's catalytic qualities in selective oxidation processes. Usually, co-assembly techniques are used in the synthesis to ensure that the metal species are distributed uniformly across the mesoporous framework, improving catalytic activity [46]. Other metal oxides, including cerium (CeO_2) [48], niobium (Nb_2O_5) [49], and tantalum (Ta_2O_5) [50], have also been explored for their mesoporous forms. The distinct electrical and catalytic characteristics of these materials are frequently researched, broadening the potential uses of metal oxide OMMs in industries like environmental remediation, energy storage, and sensors.

3.3.3 Carbon-based OMMs

OMMs based on carbon are an important class of materials distinguished by their special structural characteristics and adaptability to a broad range of applications. These materials are useful for a variety of functions, such as energy storage, environmental remediation, and catalysis. They typically have high surface areas and pore diameters from 2 to 50 nm [51]. The purpose of Ordered Mesoporous Carbons (OMCs) is to have a well-defined ordered pore structure. They are frequently created via templating techniques. Using the hard templating process, a mesoporous silica template is first infiltrated with a carbon precursor, and then the silica template is removed and carbonized. As a result, a carbon substance is produced that exactly matches the silica template's ordered structure. Surfactants or block copolymers are employed in the soft templating strategy to self-assemble with carbon precursors, generating ordered meso-structures upon carbonization. With this technique, the final carbon material's pore size and shape may be more precisely adjusted [38]. Another class of carbon-based OMMs are mesoporous carbon nanospheres. By using self-assembly processes, these materials are created, producing spherical particles with homogeneous mesopores. Thanks to their large surface area and adjustable porosity, they are especially helpful as catalysts and in applications like medication distribution [51].

The addition of heteroatoms, such as nitrogen or sulfur, to carbon compounds can improve their conductivity and catalytic activity. Because of these doped mesoporous carbons enhanced electrochemical performance, they are being utilized more and more in energy storage devices like batteries and supercapacitors. Functionalized mesoporous carbons are well suited for focused applications in sensing, catalysis, and adsorption because of surface changes that improve their interaction with particular molecules. Different chemical groups can be added to the carbon surface during functionalization to enable customized interactions with guest molecules [51,52]. In conclusion, the study of carbon-based OMMs is a broad and quickly developing field with important ramifications for numerous scientific and industrial domains. Their adaptable architectures and opportunities for functionalization keep spurring innovation in a variety of applications, from environmental remediation to energy storage.

3.3.4 Organosilica OMMs

A subclass of mesoporous materials known as organosilica OMMs adds organic functional groups to the silica framework. Organosilica materials are adaptable for a broad area of functionalities, including catalysis, adsorptions, and drug deliveries, because to this integration, which enables improved characteristics and functions. These materials have special structural qualities because they blend organic and inorganic silica components. Organic groups have the potential to affect a material's surface reactivity, hydrophobicity, and thermal stability [40].

Organosilica OMMs, like conventional mesoporous silicas, can be produced with pore diameters that are regulated and usually fall between 2 and 50 nm. Its tunability makes it possible to optimize certain qualities for certain uses. Additionally, the organic components of organosilica can be modified to provide particular functions, including biomolecule binding or catalytic sites, increasing the material's usefulness in a variety of applications [38].

Typically, soft and hard templating, together with the sol-gel method, are used to synthesis organosilica OMMs. Both silicate and organic precursors are used in the hydrolysis and condensation of silanes in the sol-gel process. Tetraethylorthosilicate (TEOS) is a common precursor for silica, while organosilanes are used for the organic components. The soft templating technique creates ordered mesoporous structures by using block copolymers or surfactants. During the synthesis process, the organic components can be integrated into the template to enable the creation of hybrid materials. The hard templating approach involves forming a mesoporous silica template first, followed by the introduction of organic materials into the pores and the removal of the silica framework. Using this method, highly organized organosilica structures can be produced [38,53].

3.3.5 Periodic mesoporous organosilicas (PMOs)

Organic functional groups are integrated into the silica framework by PMOs, a distinct type of OMMs. Because of the improved qualities and functions provided by this hybrid structure, PMOs can be used for a variety of tasks, such as medication delivery, adsorption, and catalysis. PMOs are made up of both organic and inorganic silica, which can drastically change their chemical and physical characteristics. The addition of organic groups can improve the material's surface reactivity, hydrophobicity, and heat stability.

Pore forming diameters can be regulated during the formation of PMOs, usually falling between 2 and 50 nm. Due to its tunability, certain features, including selective adsorption or catalytic activity, can be optimized for particular uses. PMOs' organic groups can be modified to add particular functions, including catalytic sites or biomolecule binding sites. Because of this, PMOs are especially adaptable for a range of applications [54,55].

3.3.6 Hierarchical OMMs

A type of materials known as hierarchical OMMs combines various porosity levels—typically micro, meso, and macroporosity—within a single structure. Because of their improved mass transport capabilities and increased accessibility to the porous network, this hierarchical structure is beneficial for a number of uses, including adsorption, energy storage, and catalysis [38].

Hierarchical OMMs come in various forms. The first type is called core-shell structured OMMs, and it is made up of a mesoporous shell covering a core nanoparticle. The shell is usually formed of carbon or mesoporous silica, although the core can be made of a variety of materials, including metallic oxides or carbon. Because of the core-shell structure's ability to perform multiple functions—the core can be magnetic or catalytic, for example—while the shell acts as a support or protective layer [56].

The second type of OMMs are yolk-shell structured ones, which have a shell and core layout similar to each other but a void gap in between. Controlled synthesis techniques or selective etching can be used to produce this empty space. The yolk-shell configuration gives guest molecules or ions more room and improves the accessibility of the core material. Then there are OMMs with an arc form, multipods, and janus. Anisotropic geometries, such janus (two faces), multipod (many arms), or arc-shaped structures, are present in these materials. Several methods, such as selective etching or emulsion-templating, can be used to create these asymmetric OMMs. Because of their asymmetric structure, anisotropic OMMs provide special benefits in selective loading and delivery applications [38,57].

Soft and hard templating techniques are usually combined to create hierarchical OMMs. Hard templating uses pre-existing templates, such as porous silica or carbon materials, to establish the hierarchical porosity, whereas soft templating uses surfactants or block copolymers to form the mesoporous structure. There are also developed one-pot synthesis methods in which the hierarchical structure is formed simultaneously with the synthesis. These techniques frequently entail the use of two templates or the use of sacrificial templates that can be removed individually to get the appropriate porosity [40,56].

Materials Research Foundations 173 (2025) 362-391
https://doi.org/10.21741/9781644903452-14

4. Applications of Mesoporous Materials in Biomolecule Separation

These types of materials have drawn a lot of interest in the field of biomolecule separation because of their adaptability and distinctive structural characteristics. As stationary phases in SEC, mesoporous materials—in particular, mesoporous silica—like MCM-41 and SBA-15 are frequently utilized. Their distinct pore diameters (between 2 and 50 nm) and large surface areas (between 400 and 1000 m^2/g) make it possible to separate biomolecules efficiently according to size. High-resolution separations are made possible by the retention of smaller molecules in the pores longer than the larger ones, which are prevented from entering and elute first [58,59]. Additionally, they are used in a number of HPLC applications, such as enantioselective and normal phase HPLC. Their large surface areas and adjustable pore sizes enhance the capacity and efficiency of separation for a variety of biomolecules, including various proteins, peptides, and nucleic acids [16].

Additionally, affinity chromatography media can be created by activating the surface of mesoporous materials with certain ligands. This increases the effectiveness and specificity of the separation process by enabling the selective capture of target biomolecules based on particular interactions [60]. Desalting biomolecules with mesoporous materials enables the elimination of salts and tiny molecules from protein solutions. This program is essential for getting samples ready for additional examination, especially in investigations involving biochemistry and molecular biology [61].

They are investigated for medication delivery applications in addition to separation. They are useful in pharmaceutical applications because of their large surface area and adjustable pore sizes, which allow for the encapsulation and regulated release of medicinal substances [62]. Apart from this, functionalized mesoporous materials can be employed in biosensors for the detection of specific biomolecules. Their high surface area and ability to be tailored for specific interactions enhance the sensitivity and specificity of diagnostic assays [63]. Environmental uses for these materials include the adsorption of contaminants from water. They are more versatile than standard biomolecule separation because of their huge pore volume and surface area, which make it easier to remove organic pollutants and hazardous heavy metals [64,65].

4.1 Enzyme Immobilization

Enzyme immobilization has received a lot of scientific attention. These biomolecules' distinct catalytic characteristics, such as their chiral, chemo-, regio-, and stereo-selectivity, present enormous promise for biocatalysis, sensing, and separation procedures [66]. In general, majority of enzymes exhibiting the highest activities in the solutions only under relatively mild conditions. At critical level of pH as well as temperature and in conjugation of various biological solvents [67] or detergents [68], through denaturation via confirmational changes or other trans formations of stereo-chemical structure, these enzymes are inactivated. Notably, under extreme environmental conditions, the enzymes entrapped in these monolithic MMs show compatible activity and better stability compared to the unbound form of the enzymes [69]. It is hypothesized that the enzyme, once encapsulated in the host material's pores, finds a more stable environment where it is shielded from cluster and unfolding. This type of encapsulation process may also cause the pH to shift, which is necessary for the best possible enzyme activity. It is important to remember that the quantity of specific type of proteins adsorbed on the support does not necessarily correspond with the catalytic activities of the enzymes. If the immobilized enzyme is manufactured in unfavourable conditions to maintain its enzymatic activity, it could quickly lose its activity [70].

The pore dimensions of these mesoporous materials often correspond to the sizes of the enzymes. In particular, enzymes with somewhat lengthy channels can be accommodated by mesoporous silicates. Even at 100 kDa and above, it is still difficult to increase the pore diameter so that the mesoporous supports can accommodate enzyme molecules [71]. Because it keeps the enzyme's conformational freedom higher and encourages substrate diffusion to the active site, it may greatly increase enzyme activity. The MCM family of mesoporous materials has pores that range in size from 3 to 4 nm, which makes them unsuitable for the adsorption of big biomolecules [71]. This perspective suggests that SBA-15, with its bigger pores size, thick level walls, and higher hydrothermal stability, is a preferable option for enzyme immobilization. Furthermore, by utilizing the appropriate swelling agents, its pore size can be readily adjusted. However, it should be remembered that SBA-15's pore diameter can be lowered to as little as 5 nm by functionalizing it with the organic groups needed for chemical encapsulation of molecules [8,67]. Large biomolecules can also be immobilized using periodic type mesoporous organosilica (PMO) substances with enlarged pore sizes, or commonly known as large-pore PMO. The materials' uniform homogeneous scattering of various organic moieties in the structural backbone and high loading rate have garnered significant scientific attention [54,55]. Mesostructured cellular foams (MCF), which are composed of huge cells with 3D pore structures connected by narrower windows, have also demonstrated improved performance when used as a support for immobilizing macromolecules [72].

Mesoporous carbon has also been discovered recently to be an appropriate matrix for the immobilization of enzymes. Comparing this support to siliceous matrices, it shows better textural properties and increased water stability. It is important to note that immobilization of an enzyme causes significant changes in its catalytic activities, ideal pH range, thermal stability, and storage stability. After the immobilization process, an enzyme's stability usually rises but its activity often falls. Another crucial factor influencing the possible uses of immobilized enzymes is their reusability. Operational enzyme stability during several catalyst reuses and subsequent recovery and reuse is the definition of this property [73].

4.2 Nucleic Acids Adsorbing level

These types of materials have attracted as the possible nucleic acid carriers. Therapeutic genes can be targeted to specific tissues and cells of interest through the use of mesoporous networks as hosts for genetic origin materials, which presents enormous potential for the treatment of a variety of disorders [74]. Given that many tumors are caused by genetic abnormalities, the prospect of using gene therapy to introduce genetic origin substance into the intracellular environment is especially attractive for the therapy of cancer. Nucleic acids that are not protected cannot cross cell membranes and are readily broken down. Additionally, their systemic administration may trigger immunological reactions. Mesoporous materials are desirable options for cellular transfection vectors in nucleic acid-based therapeutics, according to this perspective [75]. There are only few manuscripts describing the entrapment of various nucleic acids within the mesoporous cavities, despite the claim that the dimension of the selected nucleic acids allows their adsorbing level into mesoporous network frame. Most studies focus on the adsorbing features of this genetically active materials on mesoporous networks' exterior surfaces [76]. The advantages of this carrier associated to its meso-porosity are not completely utilized when nucleic acids are immobilized on the outside

Materials Research Foundations 173 (2025) 362-391 https://doi.org/10.21741/9781644903452-14

surfaces of the mesoporous particles. Mesoporous matrices must have a positively charged surface in order to facilitate the electrostatic interaction of negatively charged molecules such as DNA and RNA when used as nucleic acid carriers. Numerous techniques are available to enhance nucleic acid adsorption on mesoporous carriers. Among these can be mentioned ion swapping, conjugation with cationic polymers, and modification using amine derivatized functionalities [77].

4.3 Biologically Active Molecule Separation

In the past few years, a lot of work has been done on biomolecule separation. Mesoporous silica matrices have garnered significant interest in this approach because of its special qualities, which include a largest surface area, a uniform pores and their size distribution, the capacity to modify the surface, stability, and the ability of the mesopores to exclude certain sizes [72]. In such case, the critical fields of use of MMS include adsorbing phenomena, separation and enrichment of bioactive molecules from samples of biological and environmental origins. On the other side, these mesoporous networks offer high level of use as the excellent sorbents in various analytical estimation methods including solid-phase extraction (SPE) [78], solid-phase micro level extraction (SPME) [79], and magnetically-operated separation [80]. The goal of using such types of separation methods is to extract and enhance the target analytical molecules from the complicated biological or environment relevant samples so that it can be quantitatively analysed using an instrument. In contrast to SPE, SPME eliminates the need for organic solvents and enables direct chromatographic analysis because fetching of sample, isolation, enrichment, and injection are all combined into one process [79]. MMS can be employed as the stationary phases in HPLC [16]. It has been shown that these substances are the essential adsorbents for efficient separation and quantitatively estimation of various types of biologically active molecules, mainly hormones [81], antibiotics [82], pesticides [83], fermentation products [84], peptides and proteins [85]. Such types of mesostructured substances have also attracted to create vital attention in the enantioselective biomolecules' separation [86], selective adsorption of bioactive analytes from the naturally available products [80] and separating various impurities in the processes of active moieties purification [87]. Especially, the analysis of peptides from the actual natural ecological samples is essential as such things can hold the biologically active markers for disease diagnosis and therapies [88].

4.4 Drug Delivery

Mesoporous materials have gained significant attention in the field of active molecules delivery, particularly for the transport of biomolecules. These materials, especially mesoporous silica nanoparticles (MSNs), are characterized by their higher surface area, tunable pore sizes, and ease of functionalization, which collectively enhance their suitability as drug carriers. The large specific surface area and pore volume of mesoporous materials allow for the efficient loading of various therapeutic agents, including proteins, enzymes, and nucleic acids. This capability is crucial for developing effective drug delivery systems that can accommodate a wide range of biomolecules [89,90].

One of the standout features of mesoporous materials is their tunable pore sizes. Researchers can adjust the pore dimensions during synthesis, making it possible to encapsulate biomolecules of varying sizes. This tunability is essential for optimizing drug loading and release profiles, ensuring that the therapeutic agents are delivered effectively to their target sites. Moreover, the surface chemistry of these materials can be modified, enhancing their biocompatibility and allowing for targeted drug delivery. By introducing specific functional groups, scientists can facilitate interactions with biomolecules, improving the stability and kinetics of drug release [90,91].

In practical applications, mesoporous materials are particularly valuable in designing controlled release systems. These systems can be engineered to respond to specific stimuli, including suitable pH changes, temperature variations, or enzymatic activity, enabling the targeted release of drugs at desired locations within the body. This feature is mainly beneficial in cancer therapies, where precise delivery of chemotherapeutic agents can potentially raise the therapeutic efficacies with reduction in the level of side effects. Furthermore, the integration of drug delivery with diagnostic capabilities—known as theranostics—has emerged as a promising approach. By combining therapeutic agents with imaging functionalities, mesoporous materials allow for real-time monitoring of therapeutic efficacy, paving the way for personalized medicine [92].

The ability of mesoporous materials to effectively load and deliver large biomolecules, including various antibodies and nucleic acids, further underscores their potential in drug delivery applications. Their large pore sizes facilitate the adsorption and modified release of these molecules, which are often challenging to deliver using conventional methods. Additionally, the versatility of mesoporous materials enables the large scale growth of multifunctionality of various drug delivery systems. These systems can combine therapeutic agents with imaging agents or other functionalities, enhancing their utility in advanced therapeutic strategies [62,90,92].

In conclusion, the integration of mesoporous materials into drug delivery systems represents a significant advancement in nanomedicine. Their unique properties not only enable the effective delivery of biomolecules but also improve therapeutic outcomes while minimizing adverse effects. As research continues to evolve, the vital role of various mesoporous substances in diversified biomedical sciences is expected to expand, leading to innovative treatment modalities that could transform patient care.

5. Comparison with Traditional Separation Techniques

Mesoporous materials have emerged as a powerful alternative to traditional methods for biomolecule separation, delivering various distinct benefits. One of the primary benefits is their superior separation efficiency. Mesoporous substances possess a highly ordered, tunable pore structure with a large surface area, which allows for precise separation of biomolecules based on size, charge, and other physicochemical properties. This precision contrasts with traditional methods such as centrifugation or precipitation, which often lack the ability to discriminate between molecules of similar sizes and compositions. Moreover, the pore sizes of mesoporous substances can be tailored to selectively trap or exclude certain biomolecules, leading to high-resolution separations [93–95].

In terms of speed and throughput, mesoporous substances also excel. Due to their high surface area and enhanced mass transfer properties, these materials can facilitate faster separations compared to conventional techniques. Traditional methods like dialysis or gel filtration are often limited by diffusion rates or the need for extensive processing times, which can slow down operations, especially in high-throughput settings. Mesoporous materials, on the other hand, offer quicker interactions between biomolecules and the surface, leading to shorter processing times, which is crucial for industrial-scale applications and time-sensitive research [10,62,92].

Scalability is another area where mesoporous materials demonstrate significant potential. Their use in column chromatography or other flow-based systems makes them easily scalable for industrial processes. Traditional methods, such as ultracentrifugation or liquid-liquid extraction, often face limitations when scaling up, particularly in terms of equipment capacity and the large volumes of reagents required. In contrast, mesoporous materials can be produced in large quantities and incorporated into automated systems, making them well-suited for large-scale biomolecule separations [6,24,90].

However, the cost of mesoporous materials can be higher initially due to the complexities involved in their synthesis and preparation. Traditional methods, such as gel electrophoresis or precipitation, are often more affordable in terms of materials and equipment. Yet, over time, the efficiency, reusability, and reduced processing times offered by mesoporous materials may offset their initial cost. For large-scale operations, this cost efficiency becomes even more pronounced as mesoporous materials can reduce waste and lower operational costs by minimizing the need for chemicals and solvents commonly used in traditional methods [16,38].

Another critical advantage of mesoporous materials is their flexibility and adaptability. They can be engineered to exhibit specific pore sizes, surface functional groups, and hydrophilic or hydrophobic properties, making them highly versatile for a wide range of biomolecule separations. Traditional methods, while effective for certain applications, often lack this level of customization. For example, gel electrophoresis relies on a one-size-fits-all approach, whereas mesoporous materials can be designed to target specific biomolecular interactions or characteristics, enhancing separation outcomes [38].

Lastly, the environmental impact of mesoporous materials is an area of growing interest. While the synthesis of these materials can sometimes involve hazardous chemicals, recent advances in green chemistry are aiming to reduce this footprint. In contrast, traditional methods often including the usage of large quantities of solvents and reagents, which can have significant environmental consequences. Therefore, as sustainable production methods for mesoporous materials continue to evolve, they may offer a more environmentally friendly solution for biomolecule separation compared to traditional approaches [64,96].

5.1 Performance comparison with traditional chromatographic techniques

For the separation of biomolecules, mesoporous materials have a number of benefits over conventional chromatographic methods. Biomolecules can be effectively separated based on size thanks to mesoporous materials' high surface areas and tunable pore diameters [16,97], in contrast, conventional methods frequently have lower surface areas and less control over the distribution of pore sizes, which limits their ability to separate materials. Mesoporous materials' homogeneous pore size distribution and large surface area help to separate complex biomolecule mixtures more effectively and with better resolution [33,97]. However, due to constraints in surface area and pore size, traditional approaches may not be able to obtain high resolution for some biomolecules, especially bigger ones.

Mesoporous materials' surfaces can be readily changed with certain functional groups to improve target biomolecule selectivity and separation efficiency [97,98], which is more challenging in case of stationary phases used in traditional chromatographic techniques, limiting the ability to tailor interactions with specific analytes. Most of mesoporous materials are biocompatible, making them suitable for separating sensitive biomolecules while maintaining their integrity [33]. Traditional techniques may require harsher conditions that can potentially denature or damage biomolecules. Ongoing research in the field of mesoporous materials aims to develop more cost-effective synthesis methods and scalable production techniques to make them more accessible [97], whereas most of the traditional techniques already have well-established and cost-effective production methods.

Overall, mesoporous materials outperform traditional chromatographic techniques in areas such as pore size tunability, surface area, resolution, efficiency, and functionalization potential for biomolecule separation. However, traditional techniques may still be more cost-effective and scalable in certain applications. The choice between mesoporous materials and traditional techniques depends on the specific requirements of the separation process and the available resources.

5.2 Performance comparison with traditional electrophoretic techniques

For biomolecule separation, mesoporous materials and conventional electrophoretic methods are both extensively used, but because of their structural and functional differences, they perform quite differently. The pores in mesoporous materials have sizes between two and fifty nanometers. Their vast surface area facilitates better interactions with biomolecules like DNA and proteins. Depending on the surface chemistry of the material, size-exclusion, charge interactions, or affinity mechanisms are frequently used in separation processes. Conventional electrophoretic methods use an electric field to separate biomolecules according to their size and charge. As molecules pass through a gel matrix (such as agarose or polyacrylamide), their electrophoretic mobility and interactions with the pores in the gel are the main factors influencing the separation [99,100].

Because of their large surface area and adjustable pore sizes, mesoporous materials frequently enable quicker separations, which results in improved diffusion rates and more successful separations. Nevertheless, the biomolecules' ability to enter and interact with the pores determines the efficiency. However, in contrast, the process of electrophoresis is slower. The size of the biomolecules, the type of gel, and the strength of the electric field are some of the variables that affect speed. Larger molecules flow through the matrix more slowly than smaller ones. Traditional gel electrophoresis is effective, but it takes a lot of time and requires conditions to be optimized [4,99].

The regularity of pore size and surface changes have a major role in the resolution of separation in mesoporous materials. These materials' exact control over pore diameters allows for great resolution, particularly for size-based separations. On the other hand, conventional electrophoresis can provide biomolecules with good resolution, particularly in methods such as SDS-PAGE that separate proteins according to molecular weight. In size-based separations, the resolution is typically lower than that of mesoporous materials [2].

Additionally, mesoporous materials can separate higher amounts of biomolecules because of their huge surface area and high loading capacity. The volume of the mesoporous matrix can be increased to simply scale up the material. However, there are usually limitations to the scalability and capacity of electrophoresis. Due to restrictions on sample volume and gel size, electrophoresis is not an effective method for the separation of biomolecules on a large scale, even though it is useful for small-scale analytical applications [2,101].

By functionalizing the pore surfaces of mesoporous materials with certain ligands, antibodies, or other affinity reagents, great selectivity can be attained. This makes it possible to separate biomolecules selectively according to interactions other than size and charge, including affinity binding. In the case of electrophoresis, the selectivity is mostly restricted to the biomolecules' size and charge properties. Traditional electrophoretic approaches lack specific binding interactions, which limits their versatility in some separation settings [4,101].

Compared to electrophoretic methods, mesoporous materials can be more costly to make and employ, despite their effectiveness. Additionally, they need more sophisticated machinery for fabrication and modification. Conventional gel electrophoresis is a commonly employed, reasonably priced method. Many laboratories can access it because it only requires inexpensive, simple-to-handle materials and basic equipment [2,101].

Medication administration, catalysis, and bioseparation are just a few of the many uses for mesoporous materials, particularly when they are functionalized for selective binding. They are adaptable in the separation of various biomolecules due to their tunability. Conventional electrophoretic methods have long been used in research and diagnostics to separate proteins, RNA, and DNA. They perform well in typical analytical applications but are less customizable for particular binding interactions [4,99]. Table 2 gives a summary of performance comparison of biomolecule separation via mesoporous materials and traditional chromatographic and electrophoretic techniques.

Table 2. Comparative performance of mesoporous materials utilized in biomolecule separation with traditional chromatographic as well as electrophoretic techniques [97-101].

Aspect	**Mesoporous Material**	**Traditional Chromatography Techniques**	**Traditional Electrophoretic Techniques**
Mechanism	Size exclusion, affinity, charge-based	Size, charge, hydrophobicity-based	Size and charge-based separation
Speed	Faster, high surface area facilitates diffusion	Moderate to fast, depends on flow rate	Slower, depends on gel and electric field
Resolution	High, tunable pore size	Very high, especially in HPLC	Moderate to high, dependent on matrix
Capacity	High, scalable	Moderate to high, scaling is complex	Low, small-scale applications
Selectivity	High, functionalization possible	High, especially in affinity/ion-exchange	Moderate, based on size and charge only
Cost	Higher, more complex equipment	Moderate to high	Low, widely available
Applications	Wide (bioseparation, catalysis, drug delivery, etc.)	Wide range of applications in pharmaceuticals, clinical, and environmental areas	DNA, RNA, protein analysis in research

Materials Research Foundations 173 (2025) 362-391
https://doi.org/10.21741/9781644903452-14

6. Challenges and Future Perspectives

Although mesoporous materials have a number of benefits for the separation of biomolecules, a number of obstacles must be overcome before their full potential can be realized. In chromatographic applications, mesoporous silica materials may be brittle and vulnerable to mechanical stress, which could result in column failure or a shorter lifespan. Mechanical stability could be improved by creating stronger mesoporous materials or hybrid composites that combine the benefits of mesoporous silica with more robust polymers [97]. It is still difficult to have exact control over the pore size distribution because variability can result in uneven separation performance. More consistent pore sizes and distributions can be attained with the use of sophisticated synthesis techniques including controlled sol-gel procedures and template-assisted approaches [98].

The process of functionalizing mesoporous materials to improve their selectivity for particular biomolecules can be difficult and lead to uneven coatings that reduce the effectiveness of separation. Improving the simplicity and repeatability of functionalization techniques like layer-by-layer assembly or click chemistry could increase the efficiency and homogeneity of surface modification [102]. Unwanted interactions between biomolecules and the stationary phase can also cause issues with repeatability and resolution. For example, polar biomolecules can interact with silica surfaces that still contain silanol groups. The performance of separation could be improved by further altering the surface chemistry to reduce these interactions, for example, by applying hydrophobic coatings or blocking agents [33]. Furthermore, the production of superior mesoporous materials can be expensive and time-consuming, which might prevent their broad use in industrial settings. These materials might become more widely available through research into more affordable synthesis processes and scalable production strategies, such continuous flow synthesis or 3D printing [97].

Future work may concentrate on combining mesoporous materials with cutting-edge detection techniques, including mass spectrometry or multi-angle light scattering, to enable better characterization and real-time analysis of isolated biomolecules. Improved biomolecule separation performance may result from the creation of hybrid materials that blend polymers or other materials with mesoporous silica. These hybrids might provide improved selectivity for particular biomolecules, variable surface characteristics, and increased mechanical stability.

The application of "smart" mesoporous materials—materials that can react dynamically to changes in temperature and pH, for example—to improve separation performance may also be investigated in future studies. This may result in biomolecule separation applications that are more effective and adaptable. As the demand for sustainable practices increases, the development of environmentally friendly synthesis methods and the use of biodegradable or recyclable mesoporous materials will be crucial. Research into bio-based mesoporous materials could also open new avenues for biomolecule separation applications.

Conclusion

Because of their large surface area, adjustable pore sizes, and surface functions, mesoporous materials have become effective instruments for the separation and purification of biomolecules. Their adaptability makes them ideal for a wide range of biochemical and medicinal applications because it enables the selective adsorption and regulated release of proteins, nucleic acids, and other biomolecules. The creation of more specialized mesoporous materials with improved selectivity, stability, and biocompatibility holds the potential to significantly transform this field as research progresses. Future developments may result in the optimization of separation procedures, enhancing scalability and efficiency—two critical aspects for applications in biotechnology, pharmaceuticals, and healthcare.

References

[1] G.L. Dignon, R.B. Best, J. Mittal, Biomolecular phase separation: From molecular driving forces to macroscopic properties, Annu. Rev. Phys. Chem. 71 (2020) 53–75. https://doi.org/10.1146/annurev-physchem-071819-113553.

[2] P.C. Singh, R.K. Singh, Choosing an appropriate bioseparation technique, Trends Food Sci. Technol. 7 (1996) 49–58. https://doi.org/10.1016/0924-2244(96)81328-8.

[3] F.I. Azmi, P.S. Goh, A.F. Ismail, N. Hilal, T.W. Wong, M. Misson, Biomolecule-enabled liquid separation membranes: Potential and recent progress, Membranes. 12 (2022) 1–14. https://doi.org/10.3390/membranes12020148.

[4] K.K.R. Tetala, M.A. Vijayalakshmi, A review on recent developments for biomolecule separation at analytical scale using microfluidic devices, Anal. Chim. Acta. 906 (2016) 7–21. https://doi.org/10.1016/j.aca.2015.11.037.

[5] H. Cölfen, X. Xu, Analytical ultracentrifugation, in: C. Contado (Ed.), Particle Separation Techniques, Elsevier Inc., United Kingdom, 2022, pp. 511–539. https://doi.org/10.1016/B978-0-323-85486-3.00016-0.

[6] S. Shil, M. Tsuruta, K. Kawauchi, D. Miyoshi, Biomolecular liquid–liquid phase separation for biotechnology, BioTech. 12 (2023) 1–26. https://doi.org/10.3390/biotech12020026.

[7] F.M. Bobonich, A.S. Kovalenko, Y.G. Voloshina, A.S. Korchev, V.N. Solomakha, A.P. Philippov, V.G. Il'in, Adsorptive properties of template-containing silica-based MCM-41 and MCM-50 materials, Adsorpt. Sci. Technol. 20 (2002) 595–605. https://doi.org/10.1260/026361702321039528.

[8] V.L. Zholobenko, A.Y. Khodakov, M.I. Clerc, D. Durand, I. Grillo, Initial stages of SBA-15 synthesis: An overview, Adv. Colloid Interface Sci. 142 (2008) 67–74. https://doi.org/10.1016/j.cis.2008.05.003.

[9] B. Liu, H. Li, K. Quan, J. Chen, H. Qiu, Periodic mesoporous organosilica for chromatographic stationary phases: From synthesis strategies to applications, Trends Anal. Chem. 158 (2023) 1–16. https://doi.org/10.1016/j.trac.2022.116895.

[10] S. Kumar, A. Sharma, D. Gautam, S. Hooda, Characterization of mesoporous materials, in: A. Uthaman, S. Thomas, T. Li, H. Maria (Eds.), Advanced Functional Porous Materials, Springer Cham, Switzerland, 2022, pp. 175–204. https://doi.org/10.1007/978-3-030-85397-6_6.

[11] A.M. Striegel, Size-exclusion chromatography, in: S. Fanali, B. Chankvetadze, P.R. Haddad, C.F. Poole, M.L. Riekkola (Eds.), Liquid Chromatography, Elsevier Inc., United Kingdom, 2023, pp. 509–537. https://doi.org/10.1016/B978-0-323-99968-7.00022-9.

[12] M.T.R. Laguna, R. Medrano, M.P. Plana, M.P. Tarazona, Polymer characterization by size-exclusion chromatography with multiple detection, J. Chromatogr. A. 919 (2001) 13–19. https://doi.org/10.1016/S0021-9673(01)00802-0.

[13] A.M. Striegel, Size-exclusion chromatography: A twenty-first century perspective, Chromatographia. 85 (2022) 307–313. https://doi.org/10.1007/s10337-022-04143-1.

[14] T. Nassivera, A.G. Eklund, C.C. Landry, Size-exclusion chromatography of low-molecular-mass polymers using mesoporous silica, J. Chromatogr. A. 973 (2002) 97–101. https://doi.org/10.1016/S0021-9673(02)01200-1.

[15] K. Robards, P.R. Haddad, P.E. Jackson, High-performance liquid chromatography—separations, in: K. Robards, D. Ryn (Eds.), Principles and Practice of Modern Chromatographic Methods, Elsevier Inc., New York, 2004, pp. 305–380. https://doi.org/10.1016/B978-0-08-057178-2.50009-1.

[16] K.W. Gallis, A.G. Eklund, S.T. Jull, J.T. Araujo, J.G. Moore, C.C. Landry, The use of mesoporous silica in liquid chromatography, in: Studies in Surface Science and Catalysis, Elsevier Inc., New York, 2000, pp. 747–755. https://doi.org/10.1016/S0167-2991(00)80279-7.

[17] A. Segaran, L.S. Chua, Review of recent applications and modifications of aqueous two-phase system for the separation of biomolecules, Int. J. Biol. Macromol. 276 (2024) 1–13. https://doi.org/10.1016/j.ijbiomac.2024.133856.

[18] O. Coskun, Separation techniques: Chromatography, North Clin. Istanb. (2016) 1–5. https://doi.org/10.14744/nci.2016.32757.

[19] D.A. Egas, M.J. Wirth, Fundamentals of protein separations: 50 years of nanotechnology, and growing, Annu. Rev. Anal. Chem. 1 (2008) 833–855. https://doi.org/10.1146/annurev.anchem.1.031207.112912.

[20] F.A. Vicente, I. Plazl, S.P.M. Ventura, P.Ž. Plazl, Separation and purification of biomacromolecules based on microfluidics, Green Chem. 22 (2020) 4391–4410. https://doi.org/10.1039/C9GC04362D.

[21] J.C. Scull, Nucleic acid extraction techniques, in: L.M. McManus, R.N. Mitchell (Eds.), Pathobiology of Human Disease, Elsevier Inc., New York, 2014, pp. 4059–4063. https://doi.org/10.1016/B978-0-12-386456-7.07701-7.

[22] B. Foxman, A primer of molecular biology, in: Molecular Tools and Infectious Disease Epidemiology, Elsevier Inc., New York, 2012, pp. 53–78. https://doi.org/10.1016/B978-0-12-374133-2.00005-8.

[23] J.L. Cole, Centrifugation: Analytical ultracentrifugation, in: Reference Module in Chemistry, Molecular Sciences and Chemical Engineering, Elsevier Inc., New York, 2013, pp. 1–6. https://doi.org/10.1016/B978-0-12-409547-2.04422-X.

[24] Z. Xu, W. Wang, Y. Cao, B. Xue, Liquid-liquid phase separation: Fundamental physical principles, biological implications, and applications in supramolecular materials engineering, Supramol. Mater. 2 (2023) 100049. https://doi.org/10.1016/j.supmat.2023.100049.

[25] T.P. Fraccia, G. Zanchetta, Liquid–liquid crystalline phase separation in biomolecular solutions, Curr. Opin. Colloid Interface Sci. 56 (2021) 1–10. https://doi.org/10.1016/j.cocis.2021.101500.

[26] Y. Song, Liquid–liquid phase separation-inspired design of biomaterials, Biomater. Sci. 12 (2024) 1943–1949. https://doi.org/10.1039/D3BM02008H.

[27] V.C. Xie, M.J. Styles, B.C. Dickinson, Methods for the directed evolution of biomolecular interactions, Trends Biochem. Sci. 47 (2022) 403–416. https://doi.org/10.1016/j.tibs.2022.01.001.

[28] J. Černý, P. Hobza, Non-covalent interactions in biomacromolecules, Phys. Chem. Chem. Phys. 9 (2007) 1–34. https://doi.org/10.1039/b704781a.

[29] A. Bandyopadhyay, J. Gao, Targeting biomolecules with reversible covalent chemistry, Curr. Opin. Chem. Biol. 34 (2016) 110–116. https://doi.org/10.1016/j.cbpa.2016.08.011.

[30] Z. Guo, Y. Chen, S. Wang, J. Pang, Z. Liu, Dendritic mesoporous silica nanospheres: toward the ultimate minimum particle size for ultraefficient liquid chromatographic separation, ACS Appl. Mater. Interfaces. 13 (2021) 22970–22977. https://doi.org/10.1021/acsami.1c03985.

[31] F. Schüth, Superstructures of mesoporous silicas, Curr. Opin. Colloid Interface Sci. 3 (1998) 174–180. https://doi.org/10.1016/S1359-0294(98)80011-3.

[32] M. Raimondo, G. Perez, M. Sinibaldi, A. De Stefanis, A.A.G. Tomlinson, Mesoporous M41S materials in capillary gas chromatography, Chem. Commun. (1997) 1343–1344. https://doi.org/10.1039/a702849k.

[33] J.M. Kisler, A. Dähler, G.W. Stevens, A.J. O'Connor, Separation of biological molecules using mesoporous molecular sieves, Microporous Mesoporous Mater. 44–45 (2001) 769–774. https://doi.org/10.1016/S1387-1811(01)00259-1.

[34] Z. ALOthman, A review: Fundamental aspects of silicate mesoporous materials, Materials. 5 (2012) 2874–2902. https://doi.org/10.3390/ma5122874.

[35] D. Khan, Shaily, Synthesis and catalytic applications of organo-functionalized MCM-41 catalyst: A review, Appl. Organomet. Chem. 37 (2023) 35–43. https://doi.org/10.1002/aoc.7007.

[36] P. Verma, Y. Kuwahara, K. Mori, R. Raja, H. Yamashita, Functionalized mesoporous SBA-15 silica: Recent trends and catalytic applications, Nanoscale. 12 (2020) 11333–11363. https://doi.org/10.1039/D0NR00732C.

[37] K. Lan, D. Zhao, Functional ordered mesoporous materials: Present and future, Nano. Lett. 22 (2022) 3177–3179. https://doi.org/10.1021/acs.nanolett.2c00902.

[38] B. Szczęśniak, J. Choma, M. Jaroniec, Major advances in the development of ordered mesoporous materials, Chem. Commun. 56 (2020) 7836–7848. https://doi.org/10.1039/D0CC02840A.

[39] K. Lan, D. Zhao, Functional ordered mesoporous materials: Present and future, Nano. Lett. 22 (2022) 3177–3179. https://doi.org/10.1021/acs.nanolett.2c00902.

[40] M. Robertson, A.G. Obando, A. Barbour, P. Smith, A. Griffin, Z. Qiang, Direct synthesis of ordered mesoporous materials from thermoplastic elastomers, Nat. Commun. 14 (2023) 639–650. https://doi.org/10.1038/s41467-023-36362-x.

[41] L. Ruchomski, S. Pikus, T. Pikula, E. Mączka, M. Kosmulski, Synthesis and properties of Fe/SBA-15, Colloids Surf. A. Physicochem. Eng. Asp. 599 (2020) 12–22. https://doi.org/10.1016/j.colsurfa.2020.124922.

[42] S.P. Beltran, P.B. Balbuena, G.E.R. Caballero, Surface structure and acidity properties of mesoporous silica SBA-15 modified with aluminum and titanium: First-principles

calculations, J. Phys. Chem. C. 120 (2016) 18105–18114. https://doi.org/10.1021/acs.jpcc.6b05630.

[43] Q.N.K. Nguyen, N.T. Yen, N.D. Hau, H.L. Tran, Synthesis and characterization of mesoporous silica SBA-15 and ZnO/SBA-15 photocatalytic materials from the ash of brickyards, J. Chem. 2020 (2020) 1–8. https://doi.org/10.1155/2020/8456194.

[44] M.B. Bahari, C.R. Mamat, A.A. Jalil, N.S. Hassan, W. Nabgan, H.D. Setiabudi, D.V.N. Vo, B.T.P. Thuy, Mesoporous alumina: A comprehensive review on synthesis strategies, structure, and applications as support for enhanced H2 generation via CO2-CH4 reforming, Int. J. Hydrogen Energy. 47 (2022) 41507–41526. https://doi.org/10.1016/j.ijhydene.2021.12.145.

[45] S. Tang, X. Huang, X. Chen, N. Zheng, Hollow mesoporous zirconia nanocapsules for drug delivery, Adv. Funct. Mater. 20 (2010) 2442–2447. https://doi.org/10.1002/adfm.201000647.

[46] D. Pan, Q. Xu, Z. Dong, S. Chen, F. Yu, X. Yan, B. Fan, R. Li, Facile synthesis of highly ordered mesoporous cobalt–alumina catalysts and their application in liquid phase selective oxidation of styrene, RSC Adv. 5 (2015) 98377–98390. https://doi.org/10.1039/C5RA20531J.

[47] A. Ramesh, D. Rajesh, K. Shanthi, P.B. Bhargav, M.T. Nguyen-Le, Catalytic conversion of glucose to 5-hydroxymethylfurfural productions over sulphated Ti-Al2O3 catalysts, Biomass Bioenergy. 154 (2021) 10–32. https://doi.org/10.1016/j.biombioe.2021.106261.

[48] E. Özkan, A. Hofmann, M. Votsmeier, W. Wang, X. Huang, C. Kübel, F. Badaczewski, K. Turke, S. Werner, B.M. Smarsly, Comprehensive characterization of a mesoporous cerium oxide nanomaterial with high surface area and high thermal stability, Langmuir. 37 (2021) 2563–2574. https://doi.org/10.1021/acs.langmuir.0c02747.

[49] P. Shen, B. Zhang, Y. Wang, X. Liu, C. Yu, T. Xu, S.S. Mofarah, Y. Yu, Y. Liu, H. Sun, H. Arandiyan, Nanoscale niobium oxides anode for electrochemical lithium and sodium storage: A review of recent improvements, J. Nanostructure Chem. 11 (2021) 33–68. https://doi.org/10.1007/s40097-020-00367-5.

[50] Y. Jiao, X. Zhang, H. Yang, H. Ma, J. Zou, Mesoporous tantalum oxide nanomaterials induced cardiovascular endothelial cell apoptosis via mitochondrial-endoplasmic reticulum stress apoptotic pathway, Drug Deliv. 30 (2023) 108–120. https://doi.org/10.1080/10717544.2022.2147251.

[51] M.M. Rahman, M.G. Ara, M.A. Alim, M.S. Uddin, A. Najda, G.M. Albadrani, A.A. Sayed, S.A. Mousa, M.M.A. Daim, Mesoporous carbon: A versatile material for scientific applications, Int. J. Mol. Sci. 22 (2021) 1–10. https://doi.org/10.3390/ijms22094498.

[52] S.M. Ataei, E. Aram, Mesoporous carbon-based materials: A review of synthesis, modification, and applications, Catalysts. 13 (2022) 1–15. https://doi.org/10.3390/catal13010002.

[53] A. Rajendran, H.X. Fan, W.Y. Li, Synthesis of mesoporous materials, in: A. Uthaman, S. Thomas, T. Li, H. Maria (Eds.), Advanced Functional Porous Materials, from Macro to Nano Scale Lengths, Springer Cham, Switzerland, 2022, pp. 113–173. https://doi.org/10.1007/978-3-030-85397-6_5.

[54] L.D. Bonifacio, B.V. Lotsch, G.A. Ozin, Periodic mesoporous materials: Holes filled with opportunities, in: D.L. Andrews, G.D. Scholes, G.P. Wiederrecht (Eds.), Comprehensive

Nanoscience and Technology, Elsevier, Academic Press, New York, 2011, pp. 69–125. https://doi.org/10.1016/B978-0-12-374396-1.00049-0.

[55] P.V.D. Voort, D. Esquivel, E. De Canck, F. Goethals, I. Van Driessche, F.J. R.Salguero, Periodic mesoporous organosilicas: From simple to complex bridges; A comprehensive overview of functions, morphologies and applications, Chem. Soc. Rev. 42 (2013) 3913–3955. https://doi.org/10.1039/C2CS35222B.

[56] A.V. Nomoev, S.P. Bardakhanov, M. Schreiber, D.G. Bazarova, N.A. Romanov, B.B. Baldanov, B.R. Radnaev, V.V Syzrantsev, Structure and mechanism of the formation of core–shell nanoparticles obtained through a one-step gas-phase synthesis by electron beam evaporation, Beilstein J. Nanotechnol. 6 (2015) 874–880. https://doi.org/10.3762/bjnano.6.89.

[57] N. Masoumifard, R.G. Nicolas, F. Kleitz, Synthesis of engineered zeolitic materials: from classical zeolites to hierarchical core–shell materials, Adv. Mater. 30 (2018). https://doi.org/10.1002/adma.201704439.

[58] T. Krawczyk, M. Zalewski, A. Janeta, P. Hodurek, SEC separation of polysaccharides using macroporous spherical silica gel as a stationary phase, Chromatographia. 81 (2018) 1365–1372. https://doi.org/10.1007/s10337-018-3582-5.

[59] G.B. Irvine, High-performance size-exclusion chromatography of peptides, J. Biochem. Biophys. Methods. 56 (2003) 233–242. https://doi.org/10.1016/S0165-022X(03)00061-7.

[60] C. Kip, K.Ö. Hamaloğlu, C. Demir, A. Tuncel, Recent trends in sorbents for bioaffinity chromatography, J. Sep. Sci. 44 (2021) 1273–1291. https://doi.org/10.1002/jssc.202001117.

[61] L. Zhao, H. Qin, R. Wu, H. Zou, Recent advances of mesoporous materials in sample preparation, J. Chromatogr. A. 1228 (2012) 193–204. https://doi.org/10.1016/j.chroma.2011.09.051.

[62] M.V. Regí, F. Balas, D. Arcos, Mesoporous materials for drug delivery, Angew. Chem., Int. Ed. 46 (2007) 7548–7558. https://doi.org/10.1002/anie.200604488.

[63] K.A. Kilian, T. Böcking, J.J. Gooding, The importance of surface chemistry in mesoporous materials: Lessons from porous silicon biosensors, Chem. Commun. (2009) 630–640. https://doi.org/10.1039/B815449J.

[64] J. Wu, L. Tan, Y. Li, X. Wu, Y. Liang, Highly ordered molecularly imprinted mesoporous silica for selective removal of bisphenol A from wastewater, J. Sep. Sci. 43 (2020) 987–995. https://doi.org/10.1002/jssc.201900957.

[65] A.A. Alotaibi, A.K. Shukla, M.H. Mrad, A.M. Alswieleh, K.M. Alotaibi, Fabrication of polysulfone-surface functionalized mesoporous silica nanocomposite membranes for removal of heavy metal ions from wastewater, Membranes. 11 (2021) 1–10. https://doi.org/10.3390/membranes11120935.

[66] M. Moritz, M.G. Moritz, Mesoporous materials as multifunctional tools in biosciences: Principles and applications, Mater. Sci. Eng. C. 49 (2015) 114–151. https://doi.org/10.1016/j.msec.2014.12.079.

[67] B. Karimi, S. Emadi, A.A. Safari, M. Kermanian, Immobilization, stability and enzymatic activity of albumin and trypsin adsorbed onto nanostructured mesoporous SBA-15 with compatible pore sizes, RSC Adv. 4 (2014) 4387–4394. https://doi.org/10.1039/C3RA46002A.

[68] C. Oh, J.H. Lee, Y.G. Lee, Y.H. Lee, J.W. Kim, H.H. Kang, S.G. Oh, New approach to the immobilization of glucose oxidase on non-porous silica microspheres functionalized by (3-aminopropyl)trimethoxysilane (APTMS), Colloids Surf. B. Biointerfaces. 53 (2006) 225–232. https://doi.org/10.1016/j.colsurfb.2006.09.007.

[69] E. Magner, Immobilisation of enzymes on mesoporous silicate materials, Chem. Soc. Rev. 42 (2013) 1–10. https://doi.org/10.1039/c2cs35450k.

[70] S.S. Ding, J. P. Zhu, Y. Wang, Y. Yu, Z. Zhao, Recent progress in magnetic nanoparticles and mesoporous materials for enzyme immobilization: An update, Braz. J. Biol. 82 (2022) 1–10. https://doi.org/10.1590/1519-6984.244496.

[71] C.H. Lee, T.S. Lin, C.Y. Mou, Mesoporous materials for encapsulating enzymes, Nano Today. 4 (2009) 165–179. https://doi.org/10.1016/j.nantod.2009.02.001.

[72] J. Kim, R.J. Desch, S.W. Thiel, V.V. Guliants, N.G. Pinto, Adsorption of biomolecules on mesostructured cellular foam silica: Effect of acid concentration and aging time in synthesis, Microporous Mesoporous Mater. 149 (2012) 60–68. https://doi.org/10.1016/j.micromeso.2011.08.031.

[73] M. Quirós, A.B. García, M.A.M. Morán, Influence of the support surface properties on the protein loading and activity of lipase/mesoporous carbon biocatalysts, Carbon. 49 (2011) 406–415. https://doi.org/10.1016/j.carbon.2010.09.037.

[74] I. Slowing, J. Viveroescoto, C. Wu, V. Lin, Mesoporous silica nanoparticles as controlled release drug delivery and gene transfection carriers, Adv. Drug Deliv. Rev. 60 (2008) 1278–1288. https://doi.org/10.1016/j.addr.2008.03.012.

[75] J.L. Steinbacher, C.C. Landry, Adsorption and release of siRNA from porous silica, Langmuir. 30 (2014) 4396–4405. https://doi.org/10.1021/la402850m.

[76] S.M. Solberg, C.C. Landry, Adsorption of DNA into mesoporous silica, J. Phys. Chem. B. (2006) 15261–15268. https://doi.org/10.1021/jp061691.

[77] J. Liu, B. Wang, S.B. Hartono, T. Liu, P. Kantharidis, A.P.J. Middelberg, G.Q. (Max) Lu, L. He, S.Z. Qiao, Magnetic silica spheres with large nanopores for nucleic acid adsorption and cellular uptake, Biomaterials. 33 (2012) 970–978. https://doi.org/10.1016/j.biomaterials.2011.10.001.

[78] D. Wang, X. Chen, J. Feng, M. Sun, Recent advances of ordered mesoporous silica materials for solid-phase extraction, J. Chromatogr. A. 1675 (2022) 1–13. https://doi.org/10.1016/j.chroma.2022.463157.

[79] K. Delińska, P.W. Rakowska, A. Kloskowski, Porous material-based sorbent coatings in solid-phase microextraction technique: Recent trends and future perspectives, Trends Anal. Chem. 143 (2021) 1–15. https://doi.org/10.1016/j.trac.2021.116386.

[80] H. Bai, G. Teng, C. Zhang, J. Yang, W. Yang, F. Tian, Magnetic materials as adsorbents for the pre-concentration and separation of active ingredients from herbal medicine, J. Sep. Sci. 47 (2024) 1–20. https://doi.org/10.1002/jssc.202400274.

[81] N. Haroon, K.J. Stine, Electrochemical detection of hormones using nanostructured electrodes, Coatings. 13 (2023) 1–62. https://doi.org/10.3390/coatings13122040.

[82] A. Grela, J. Kuc, T. Bajda, A review on the application of zeolites and mesoporous silica materials in the removal of non-steroidal anti-inflammatory drugs and antibiotics from water, Materials. 14 (2021) 1–24. https://doi.org/10.3390/ma14174994.

[83] M. Andrunik, T. Bajda, Removal of pesticides from waters by adsorption: Comparison between synthetic zeolites and mesoporous silica materials. A review, Materials. 14 (2021) 1–39. https://doi.org/10.3390/ma14133532.

[84] H. Gao, J. Lu, Y. Jiang, Y. Fang, Y. Tang, Z. Yu, W. Zhang, F. Xin, M. Jiang, Material-mediated cell immobilization technology in the biological fermentation process, Biofuels Bioprod. Biorefining. 15 (2021) 1160–1173. https://doi.org/10.1002/bbb.2219.

[85] Y. Luo, B. Wang, L. Yi, C.F. Ding, C. Deng, Y. Yan, Mesoporous materials for glycopeptide separation, Trends Anal. Chem. 167 (2023) 1–17. https://doi.org/10.1016/j.trac.2023.117234.

[86] Y. Lu, H. Zhang, Y. Zhu, P.J. Marriott, H. Wang, Emerging homochiral porous materials for enantiomer separation, Adv. Funct. Mater. 31 (2021) 1–88. https://doi.org/10.1002/adfm.202101335.

[87] C.A. McCarthy, R.J. Ahern, R. Dontireddy, K.B. Ryan, A.M. Crean, Mesoporous silica formulation strategies for drug dissolution enhancement: A review, Expert Opin. Drug Deliv. 13 (2016) 93–108. https://doi.org/10.1517/17425247.2016.1100165.

[88] S. Jafari, H. Derakhshankhah, L. Alaei, A. Fattahi, B.S. Varnamkhasti, A.A. Saboury, Mesoporous silica nanoparticles for therapeutic/diagnostic applications, Biomed. Pharmacother. 109 (2019) 1100–1111. https://doi.org/10.1016/j.biopha.2018.10.167.

[89] M. Moritz, M.G. Moritz, Mesoporous materials as elements of modern drug delivery systems for anti-inflammatory agents: A review of recent achievements, Pharmaceutics. 14 (2022) 1–15. https://doi.org/10.3390/pharmaceutics14081542.

[90] V. Cauda, G. Canavese, Mesoporous materials for drug delivery and theranostics, Pharmaceutics. 12 (2020) 1–3. https://doi.org/10.3390/pharmaceutics12111108.

[91] M.V. Regí, M. Colilla, I.I. Barba, M. Manzano, Mesoporous silica nanoparticles for drug delivery: Current insights, Molecules. 23 (2017) 1–10. https://doi.org/10.3390/molecules23010047.

[92] W. Zhang, H. Liu, X. Qiu, F. Zuo, B. Wang, Mesoporous silica nanoparticles as a drug delivery mechanism, Open Life Sci. 19 (2024) 16–24. https://doi.org/10.1515/biol-2022-0867.

[93] J.A. Cecilia, R.M. Tost, M.R. Millán, Mesoporous materials: From synthesis to applications, Int. J. Mol. Sci. 20 (2019) 1–13. https://doi.org/10.3390/ijms20133213.

[94] D. Schrag, M. Corbier, S. Raimondi, Size exclusion-high-performance liquid chromatography (SEC-HPLC), in: V. Ossipow, N. Fischer (Eds.), Monoclonal Antibodies: Methods and Protocols, Humana Totowa, New Jersey, 2014, pp. 507–512. https://doi.org/10.1007/978-1-62703-992-5_31.

[95] S.C. Moldoveanu, V. David, Stationary phases and columns for size exclusion, in: Selection of the HPLC Method in Chemical Analysis, Elsevier Inc., United Kingdom, 2017, pp. 377–386. https://doi.org/10.1016/B978-0-12-803684-6.00011-1.

[96] M. Navaziya, HPLC hyphenations in advanced analytical world, J. Phys. Chem. Biophys. 12 (2022) 1–5. https://doi.org/10.35248/2161-0487-22.12.328.

[97] J. Zhao, F. Gao, Y. Fu, W. Jin, P. Yang, D. Zhao, Biomolecule separation using large pore mesoporous SBA-15 as a substrate in high performance liquid chromatography, Chem. Commun. (2002) 752–753. https://doi.org/10.1039/b110637f.

[98] R. Brady, B. Woonton, M.L. Gee, A.J. O'Connor, Hierarchical mesoporous silica materials for separation of functional food ingredients — A review, Innov. Food Sci. Emerg. Technol. 9 (2008) 243–248. https://doi.org/10.1016/j.ifset.2007.10.002.

[99] S.A. Kitte, T.H. Fereja, M.I. Halawa, B. Lou, H. Li, G. Xu, Recent advances in nanomaterial-based capillary electrophoresis, Electrophoresis. 40 (2019) 2050–2057. https://doi.org/10.1002/elps.201800534.

[100] P. Minakshi, M. Ghosh, B. Brar, K. Ranjan, H.S. Patki, R. Kumar, Separation techniques with nanomaterials, in: H.M. Chaudhery (Eds.), Handbook of Nanomaterials in Analytical Chemistry, Elsevier Inc., United Kingdom, 2020, pp. 99–158. https://doi.org/10.1016/B978-0-12-816699-4.00006-2.

[101] Z. Song, J. Li, W. Lu, B. Li, G. Yang, Y. Bi, M. Arabi, X. Wang, J. Ma, L. Chen, Molecularly imprinted polymers based materials and their applications in chromatographic and electrophoretic separations, Trends Anal. Chem. 146 (2022) 1–16. https://doi.org/10.1016/j.trac.2021.116504.

[102] M. Hartmann, Ordered mesoporous materials for bioadsorption and biocatalysis, Chem. Mater. 17 (2005) 4577–4593. https://doi.org/10.1021/cm0485658.

Materials Research Foundations 173 (2025) 392-412 https://doi.org/10.21741/9781644903452-15

Chapter 15

Magnetic Mesoporous Materials for Food Analysis

Jülide Hızal*

Yalova University, Engineering Faculty, Chemical Engineering Department, 77100 Yalova, Turkey

hizalyucesoy@yalova.edu.tr

Abstract

Food analyses are highly important in regards of meeting the requirements for legal standards, safety and quality assurance as well as determination of nutritional value and adulteration. Considering the fast development in ingredients and food manufacturing process, it is unavoidable that novel analytical or instrumental determination methods would arise. Various materials are being employed, particularly in the solid phase extraction approach, to detect different kinds of analytes. Recently, magnetic mesoporous materials stand out in many fields. They also incrementally find a use in food analyses. The utilization of magnetic mesoporous materials in the detection of any analytes in food matrices would give some advantages in terms of reducing cost, increasing sensitivity and higher selectivity.

Keywords

Foods, Analytical Techniques, Mesoporous Materials, Quality Insurance, Food Quality

Contents

1. Introduction

The human body needs food to grow up, to keep healthy and alive. Food is a complex matrix consisting of various nutrients like essential proteins, fatty oils, carbohydrates, minerals and

Materials Research Foundations 173 (2025) 392-412 https://doi.org/10.21741/9781644903452-15

vitamins. The phrase 'you are what you eat' absolutely defines the importance of the kind and the quality of the food consumed. For this very reason, food analyses have become necessary due to consumption of high-quality food products. The intra- and intermolecular interactions of different chemical contents of food and structural modifications arising at the processing or at the storage step cause decrease in the benefit and the quality of food. Thus, as declared by Kizil and Irudayaraj, the determination of nutritional value and providing the quality assurance of food products becomes necessary [1]. This situation results in rapid development in food analysis as regards of used analytical and instrumental methods. Food analyses help to understand many physical (as appearance, texture, flavor *etc.*), chemical and biochemical (such as microstructure, composition, *etc.*) properties of foods. Besides guarantying product quality, analyzing foods also give some knowledge about shelf-life, safety and processibility of foods [2]. Jiang et al. expressed that the demand for more sensitive and accurate analytical methods, mostly provided by instrumental and biochemical analyses, has arisen as a result of rapid development in food science and technology [3]. The most often employed techniques are those based on solid phase extraction (SPE) and its derivatives, particularly in the pretreatment stage of food analysis prior to instrumental detection. Khan et al. expressed that clean-up ability, enabling fast separation, easy operation and cost effectiveness turn SPE into account [4]. Farajzadeh et al. reported that SPE is more advantaged methods in terms of high enrichment factor, automation capability and lower amount of organic solvent requirement [5]. In recent years, many materials such as nanoparticles and magnetic or mesoporous nanoparticles were employed in their colloidal solution form or particle form. The advantages of magnetic mesoporous materials usage in solid phase extraction can be summarized as capability of directly dispersing in sample solution, and easy separability in magnetic field. Well dispersion provides more interaction of sorbent and analyte at unit interface area; thus, makes mass transfer of analyte easier. On the other hand, magnetically separation minimizes processing time by eliminating filtration or centrifugation step [6].

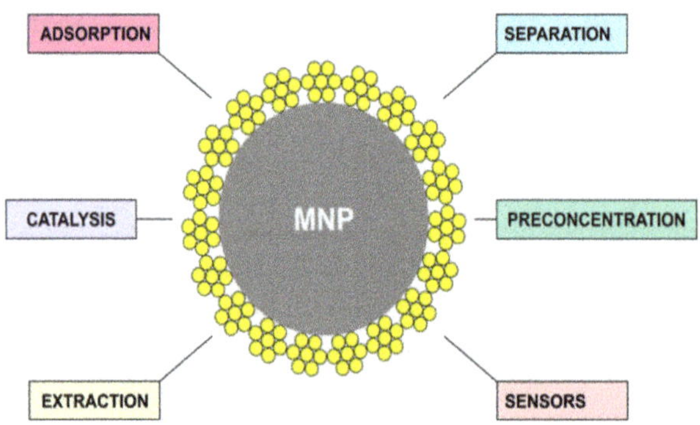

Fig. 1 The usage areas of mesoporous materials

The materials having particles size between 2-50 nm are known as mesoporous materials. They are special types of nano materials in terms of having uniform channels. The considerable surface area, large pore volume, strong hydrothermal and thermal stability, ease of surface modification and tunable mesoporous channels turn them into account [7,8]. They have been widely used as adsorbent, sensor, catalyst, a drug-controlled releasing system besides their usage for food analyses due to having those properties (Fig. 1). Each channel in mesoporous structure behaves as a reactor allowing the possible reactions at active sites. The integrability with magnetic nanoparticles, organic or metal-organic frameworks (MOFs), *etc.* extends their usage area. Moreover, integration with MOFs to form three-dimensional (3D) structure brings them highly mechanical, electrical and optical properties [9]. The mesoporous materials have been utilized in foods for separation of functional food ingredients [10], controlled releasing of active ingredients [11], and especially analyzing either the active functional content of foods or hazardous chemicals in food [12-16]. Magnetic nanoparticles (MNPs) are well-known materials utilized in many research areas like environmental remediation, catalysis, biotechnology, and data storage where external magnetic fields play a role. Nevertheless, their tendency to aggregation and oxidation prevents their usage over long periods. Therefore, fabricating their composite with polymers such as dextran, polyoxoamines, etc. or other materials such as silica, carbon, etc. avoids aggregation and allows enhanced reactivity as result of functionalization [17]. According to Socas-Rodriguez et al., the combination of MNPs with some materials causes an increase in the selectivity of the adsorbent for analyte when the SPE method is applied and in the abundance of active sites, as well as raised optical and electrical properties when used as sensors [18].

The term "food analysis" includes the preparation of food samples for analysis and the determination of main ingredients in food such as amino acids, proteins, enzymes, lipids, carbohydrates, alcohols, vitamins, phenolic compounds combining antioxidant capacity, and auxiliary components as preservatives, pigments, and flavor compounds [2]. Recently, magnetic mesoporous materials were incrementally employed for this purpose [5, 19, 20]. This study systematically summarizes the advances and present trends in utilization of magnetic mesoporous materials in food analyses by reviewing recent studies. As the occasion arises, benefits of the recent developments will be discussed. Hereby, promising advances will be presented to inspire researchers, and some assessments will be introduced to create awareness about magnetic mesoporous material usage in food analyses.

2. Fabrication of Magnetic Mesoporous Materials

As a well-known material, nano zero valent iron (NZVI) is being readily employed in many areas. In a study about antibiotic removal, the adsorbent magnetic mesoporous silica Fe-MCM-41-zeolite A was fabricated by conducting a convenient stoichiometric ratio of MCM-41-A (Scheme 1 a) and $FeCl_3.6H_2O$ mixture with NaBH4 in degassed ethanol for 30 min under nitrogen atmosphere. After being washed and dried in a vacuum oven, the material was used as an adsorbent. The authors declared they achieved high adsorption capacity in a short contact time. They also expressed that oxytetracycline retention on magnetic mesoporous material occurred as result of inner-sphere surface complexation as well as the formation of cationic π-bonds and hydrogen bonds [21]. Similarly, it is possible to determine an antibiotic in any food product by the solid-phase extraction method after convenient sample preparation. Cuello et al. developed a non-toxic method to produce Fe_3O_4 on mesoporous silica. The modification of SBA-15 (Scheme 1 b) was performed by the wet-impregnation method using ferrous chloride in its ethanolic solution. They calcined the impregnated SBA-15 both in air and under nitrogen atmosphere. As a result, they reported that when the material was calcined under a nitrogen atmosphere, higher magnetization was attained [22]. In a different work, phosphate retention was achieved by synthesizing a hexagonal magnetite (Fe_3O_4) mesoporous material (Mag-MCM-41) with a relatively large surface area. Certain amounts of SiO_2, Fe_3O_4, NaOH, and CTAB (Cetyl trimethyl ammonium bromide) having a stoichiometric ratio were used in the synthesis. The magnetite, CTAB, and NaOH were sonicated for 3 h. After being heated to 80 °C for 2 h, tetraethyl orthosilicate (TEOS) was rapidly poured into the mixture by going on vigorously stirring for 2 h. After the allotted time, the suspension was filtered and dried, then calcined at 600 °C for 6 h [23].

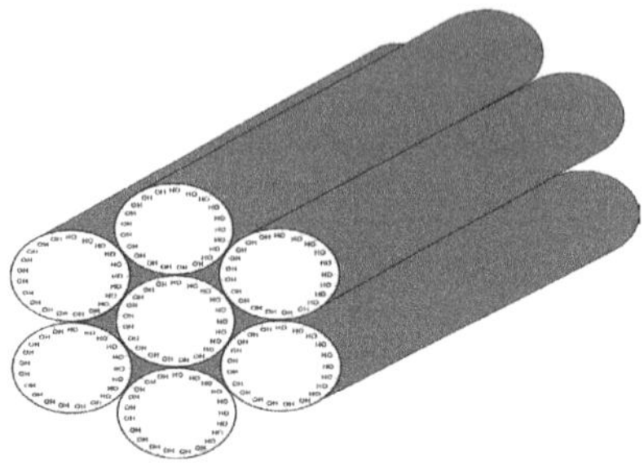

Scheme 1. The structure of MCM-41-A

On the other hand, Kamari and Ghorbani fabricated Mag-MCM-41 in a different way: The precursor mesoporous silica MCM–41 was hydrothermally synthesized using silica, which was previously isolated from rice husk by extraction. Magnetite was likewise obtained, as stated earlier. Sodium silicate solution was prepared by refluxing silica and NaOH for 24 h at 80 °C. At term, the silicate solution was poured into CTAB solution dropwise by vigorously stirring until clarity. To obtain magnetite mesoporous silica (MMS) material, the silicate solution was aggressively agitated while drops of magnetite suspension were added under nitrogen atmosphere. Afterward, the solution was incubated at 100 °C for 72 h. Following separation by magnet and washing, the obtained Mag-MCM-41 was furnaced at 550 °C for 5 h [24]. It is also possible to get modified magnetic mesoporous materials. Zeb et al. said that the surface of silica-based materials could be modified with oxygen-groups by refluxing in the presence of mineral acids and/or in their mixture, while reaction with ammonia, urea, or amines with silica materials provided to obtain amine functionalized materials. Sulfurization with S, SO_2, H_2S and dimethyl disulfide is also considered in modification methods [25]. Grafting (using organo-silane compounds under reflux conditions, secondary functionalization, or changing grafted functionalities) or co-condensation methods can be employed for modification of mesoporous silica. The sample about the co-condensation method was previously given as TEOS reaction with i.e., 3-aminopropyltriethoxysilane in the presence of a surfactant such as CTAB. In that process, the surface of the mesoporous silica is bound by an organic functional group which subsequently hydrolyzes to form silica net. Thus, the polarity of silica walls changes to enable enhanced interactions [25]. Another study in the literature deals with modification of Mag-MCM-41 with thiol modified magMCM-41 with (3-mercaptopropyl) trimethoxysilane (MPTMS). To synthesize thiol modified magnetic mesoporous material, a certain amount of Mag-MCM-41 was suspended in ethanol, then a stoichiometric amount of MPTMS was added dropwise under nitrogen atmosphere. The produced material was separated, following ethanol washing, then dried at 70 °C for 6 h [26]. Although the synthesized material was used for pesticide removal in that study, it is possible to employ it in solid phase extraction for the purpose of pesticide determination in food products. Phosphonate-modified MMS was obtained as result of the reaction between calcined MMS and 3-(trihydroxysilyl) propyl methylphosphonate by four-hour stirring at 100 °C in dioxane solvent by Li et al. The suspension was filtered, washed with diethyl ether:dichloromethane (1:1) mixture, and lastly, dried at 100 °C overnight. The authors used this material for the removal of rare earth elements [27]. The analytes having the ability to complex with phosphonates can also be detected with this material in solid phase extraction. On the other hand, modified MNPs can be utilized for fabricating MMS. The modification of MNP is performed using organic coatings, ligand exchange, self-assembly, bioconjugation, click chemistry, layer-by-layer assembly, and host-guest systems [28].

3. The Employment of Magnetic Mesoporous Materials in Food Analyses

3.1 Pesticide Analyses in Food Products

Many pests naturally exist in the agricultural fields, and cause some damage and loss of harvested products. To prevent these damages, pesticides have been used for a long time. Fighting with agricultural pests also provides decreasing energy costs [5]. On the other hand, the pesticide products are highly toxic compounds for living all types of organisms as well as for pests. The organochlorines, organonitrogens, organophosphates, and pyrethroid class compounds are counted as pesticide products. The consumption of pesticide compounds by humans is inevitable due to directly joining the food chain and blending into surface water. They have also mutagenic and carcinogenic effects on organisms. That's why sensitive and selective detection of pesticide residue levels in food matrices is highly important. Many reliable analytical methods based on instrumental analysis such as voltametry, high performance liquid chromatography (HPLC), and gas chromatography are being employed for this purpose. The separation and preconcentration of pesticides from the complex matrix of food is the most important step of the analysis. The most common methods used for this are liquid-liquid extraction (LLE) and SPE. These types of pretreatments allow for more selective and sensitive measurement in equipment. The determination of the most commonly used herbicides, fungucides, and insectisides in various matrices was performed by Farajzadeh et al. They utilized combination of QuEChERS (Quick, Easy, Cheap, Effective, Rugged, Safe) extraction, magnetic solid phase extraction, and dispersive liquid-liquid microextraction for separation and preconcentration of analytes, then, measured by running the samples into gas chromatography with flame ionization detector. The phenyl modified magnetic mesoporous solid phase was prepared by refluxing $Fe_3O_4@SiO_2$ suspension with phenyl trichloro silane in the presence of pyridine as a catalyst for 8 h. Solid phase extraction was performed at two steps. Firstly, sample solutions were contacted with MNP suspension. After separated with a magnet, retained pesticides were leached into acetonitrile solution. At the second step, leachate was extracted into 1,1,1 trichloroethane solution. The last extract was run into instrument. According to the authors, under ideal extraction circumstances, relatively low limit of detection (LOD) and limit of quantitation (LOQ) values were obtained [5]. Since food samples must be thoroughly cleaned to prevent matrix effects, the utilization of magnetic mesoporous materials mostly reveals itself in the clean-up stage of an analysis. In a study, Peng et al. employed mesoporous ZrO_2 immobilized magnetic Fe_3O_4 microsphere (m-$ZrO_2@Fe_3O_4$) in order to identify 42 pesticides in the fish matrix. Mesoporous magnetic material was used as the adsorbent in QuEChERS extraction, which gave advantages both at the extraction step, resulting in efficient separation, and at the clean-up stage with the aid of magnetic properties. The utilization of the GC–MS/MS combined QuEChERS extraction method resulted in having good linearities, rather low LOD values (in range of 0.02 and 4.40 ng/g), and recoveries [29]. The determination of sixteen pesticides in cucumber was realized by Yuan et al. They employed n-octadecylamine modified magnetic carbon nanotube porous nanocomposites (M-CNTs/ODA) as clean-up agents. After M-CNTs/ODA was shaken with cucumber sample solution in the presence of $MgSO_4$ for 2 min, the solid phase was magnetically separated. The loaded solid phase was vortexed with acetonitrile for 30 s for leaching into solution. After passing the leachate from a 0.22 µm filter membrane, the filtrate was subjected to GC–MS for pesticide detection. Thus, the authors, who obtained good linearities, recoveries, and low detection limits, declared that the porous M-CNTs/ODA could be employed to determine different organic analytes in various food matrix [30]. In another study, six pesticides were determined with the QuEChERS extraction method followed by UPLC separation combined with mass spectrometer combined with the electrospray ionization source (ESI). The L-alanine modified magnetic mesoporous material MIL-88B (MPC-ala-2) was utilized as adsorbent for isolating pesticides from interfering substances in the matrix. After homogenizing the vegetable samples, the sample was vortexed with acetonitrile in the presence of $MgSO_4$ and CH_3COONa, followed by centifugation. The supernatant phase was mixed with a certain amount of MPC-ala-2, followed by separation with the aid of a magnet. The sample solution was then filtered and subjected to UPLC-MS/MS analysis. They obtained excellent sensitivity with 0.001–0.020 µg/kg LOD values, recovery of about 100 %, and good precision, and reported that the proposed pretreatment method could enable the detection of trace multi- analytes especially carbamate and neonicotinoid pesticides [31]. Qi et al. synthesized 3-(N,N-diethylamino) propyltrimethoxysilane modified magnetite (Fe_3O_4-PSA) to employ the material as a sorbent in the SPE process. Firstly, they treated Fe_3O_4 with tetraethylorthosilicate (TEOS), then produced Fe_3O_4/SiO_2 was mixed with 3-(N,N-diethylamino) propyltrimethoxysilane (PSA-TMOS). At the end of the reaction at 65 °C for 6 h, Fe_3O_4-PSA nanoparticles were obtained. Fruit sample solutions in acetonitrile were conducted with Fe_3O_4-PSA adsorbent with the existence of $MgSO_4$. The loaded solid was separated with a magnet, and the aliquot of the liquid phase was run into UPLC-MS/MS for detection of nine pesticide residues. As the authors declared, rather low LOD values (in ng/g), recovery between 62.5–130%, and good precision allowed that the suggested approach could be applied to determine pesticides in other complicated food matrix [32]. Subsequently, the same research group determined multi-pesticide residue in a rice sample by employing the same adsorbent. Fifty pesticides and eight associated metabolites were detected in rice by applying this method and satisfactory results were obtained regarding linearity, sensitivity, matrix effect, recovery, and precision under the optimal conditions. As result the authors expressed that utilization of magnetic materials enabled 30% of the pretreatment time comparing to traditional QuEChERS technique using cleanup sorbent made of non-magnetic material [33]. The detection of triazole in aqueous solution was performed by Liu et al. using magnetic graphene oxide functionalized MOF-199 as SPE sorbent [34]. The adsorbent was employed for separation of pesticides from matrix and preconcentration. The adsorbent was synthesized by coating magnetic copper-based metal organic framework (M-MOF-199) onto Fe_3O_4– graphene oxide. Water samples were treated with adsorbent, then, loaded adsorbent was separated with a magnet in order to eluate of pesticides into aceton followed by HPLC–MS/MS analysis. They detected the pesticides with high sensitivity, recovery, and good precision. They explained the retention on the surface of the sorbent by the formation of hydrogen bonds between the benzene rings of graphene oxide and the heterocyclic rings in the triazole structure, hydrophobic and electrostatic interactions, and/or π-π interactions [34]. The proposed method can also be utilized for pesticides or other analytes, which can give the same interaction, in various types of food samples. Plohl et al. derived mesoporous silica with 3-glycidoxypropyltrimethoxysilane and branched polyethyleneimine as N-rich polymer, and they envisaged this adsorbent as a nanocarrier for

Materials Research Foundations 173 (2025) 392-412 https://doi.org/10.21741/9781644903452-15

biopesticide (citridiol) to use in agrochemical applications. This is an interesting study in terms of quite different application areas and purposes. They determined that the adsorbent started to release the pesticide after 5 h conducting with ethyl acetate solution, and releasing achieved equilibrium at the end of ten days [19]. It is possible to use this N-rich polymer modified mesoporous silica as adsorbent for clean-up and solid phase extraction steps in pesticide analyses in food samples. Pesticides can be separated and enriched by trapping into solid phase followed by ethyl acetate elution. In another study, MCM-41 and MCM-48 were employed as sorbents at the SPE step. The analysis of various insecticides used against phytophagous insects in soursop (Annona muricata) fruit was successfully performed, obtaining high sensitivity, recovery, and precision values. It is reported that these mesoporous materials could be used as an alternative solid phase to florisil, which has been widely used for pesticide separation from food matrices [35]. In a recent study, Chen et al. tried seven pesticides retention on various mesoporous materials. As result, they determined that the modified SBA-15 and MCM-22 zeolite showed high adsorption affinity to carbendazim and dipterex. Modified SBA-15 also had excellent adsorption capacity for chlorpyrifos and phoxim. The chemical structure of SBA-15, which has a convenient precursor for fabricating new materials, was shown in Fig 2. On the other hand, alkali etched silica-coated MgO effectively retained folimat and rogor [36]. There are numerous materials concerning pesticide retention in literature. As mentioned before, all those materials could be employed for the purpose of pesticide analyses. The clean-up and SPE process in food analyses currently provide new opportunities about material synthesis to improve the sensitivity and selectivity of the analyzing method for related analytes.

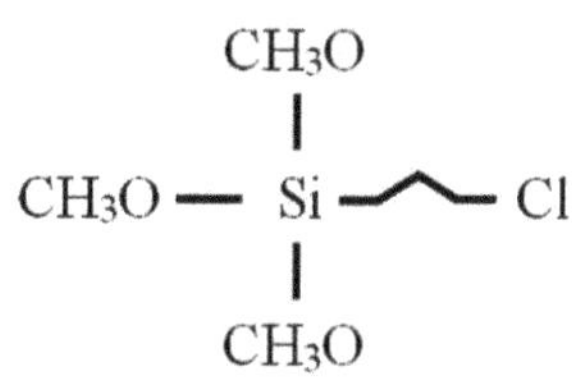

Fig. 2 The chemical structure of SBA-15

3.2 Heavy metal Analyses in Foods

As result of industrial development, many contaminants, like heavy metals, polyaromatic hydrocarbons, pesticides and other various types of organic compounds, have been leaking into the surface water and joining the food chain somehow. To provide food safety, the detection of pollutants in food products is necessary. There are limited studies about the identification of heavy metals in food matrices [37-42]. It is probably considered to be more difficult regarding its low concentration, and strong interaction with other ingredients in the food matrix [4]. The magnetic mesoporous adsorbent Fe_3O_4@SiO_2@$SiC_3H_6NH_2$/poly(2-aminothiophenol) was prepared by Sedghi et al., and used as solid phase for detection of Ag(I), Cu(II), and Pb(II) in various fruits. The synthesize of adsorbent was simulated in Scheme 2. After fruit samples were treated with solid phase under certain conditions, the solid phase was gathered by a magnet. Then, the FAAS analyses were performed following eluation of the solid phase with 1 M HCl. So that the higher

Materials Research Foundations 173 (2025) 392-412
https://doi.org/10.21741/9781644903452-15

extraction efficiency than 95% and low detection limits were obtained, the authors considered the fabricated mesoporous materials as convenient adsorbent for trace metal detection [43].

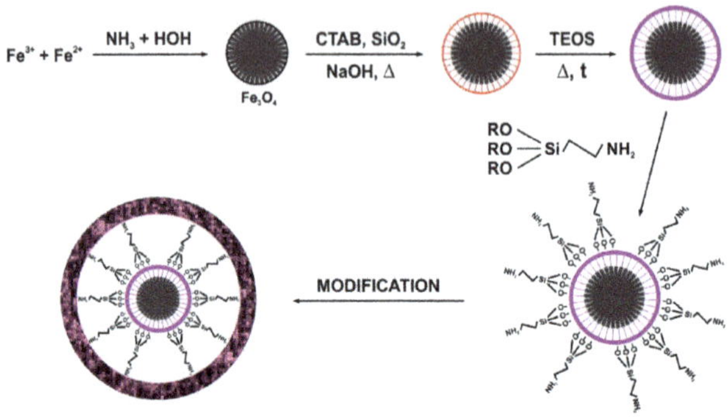

Scheme 2. A typical synthesize of modified magnetic mesoporous material

Aboufazeli et al. studied the determination of Pb(II) in meat. They especially focused on preconcentration of Pb(II) ions with adsorption onto magnetic mesoporous silica on which Pb(II) ion imprinted polymer coated previously (IIP–mSi@Fe_3O_4). Thus, they determined the low concentrations of lead ions by optimizing the pH and concentration parameters. They eluted Pb(II) ions with HCl treatment for 12 min, then Pb(II) measurement was performed in eluent using FAAS. As result, they achieved low detemination limit (lower than 1.3 µg/L), reported that the fabricated adsorbent was convenient for Pb(II) separation from food matrices [44]. Another Pb(II) determination was performed by Dahaghin et al. They searched selective Pb(II) extraction from various fruits and vegetables with employing magnetic Pb(II)-imprinted polymer (Fe_3O_4@SiO_2@IIP), synthesized by grafting imidazole compound onto Fe_3O_4@SiO_2 surface., as solid phase. A certain amount of Fe_3O_4@SiO_2@IIP was added into Pb(II) solutioın at pH 6.7. The Pb(II)-loaded solid phase was separated by magnet, and eluted with HCl. Pb(II) detection was performed by FAAS. Preconcentration factor was found as 300, LOD was 0.48 ng/mL, and relative standard deviation percentage was 3.2% [45]. Liang et al. determined Cu(II) in rice samples using electroanalytical method. The electrode of the voltameter was modified by packing homogene mixture of pyrene-1-carboxaldehyde hydrazone (PCH) modified mesoporous silica foam (PCH-MSF), graphite, and paraffin into the tube cavity of the Teflon electrode. The authors noted that the modification with PCH-MSF enhanced the sensitivity and selectivity for Cu(II). The LOD value was calculated as 0.0020 µM with recoveries between 98.4% and 102.6 % [20]. Karazan et al, declared that they determined Zn^{2+}, Cd^{2+}, Pb^{2+}, and Hg^{2+} in various foods (rice, tomato, honey, fish, and tea) using electroanalytical method in which they utilized the glassy carbon electrode modified with covalent organic framework/carbon black [41]. That mesoporous covalent organic framework may be used as adsorbent for solid phase extraction individually or after treated with MNPs. As mentioned previously, few studies, in which magnetic mesoporous materials were used, about the identification and measurement of heavy metals in food products exist in the literature.

Materials Research Foundations 173 (2025) 392-412
https://doi.org/10.21741/9781644903452-15

3.3 Other Organic Compounds Analyses in Foods

Sulfonamides (SAs) are of widely used broad-spectrum antibiotics that support recovery from an illness by preventing replication of bacteria, then destroying them. Nevertheless, excess and/or redundant usage of SAs causes adverse effects and disorders. Also, it has carcinogenic effects by accumulating in the liver. As result of the employment of SAs in aquaculture and liverstock, SAs residues may be involved in foods. Chen et al. MOF@COF hybrid magnetic nanospheres in order to enable sensitive and selective determination of SA. For this purpose, covalent organic framework (COF) was coated onto magnetic metal organic framework (MOF). Fabricated adsorbent was employed as adsorbent in SPE of SAs. At the beginning, Fe_3O_4 was modified with 1,6- hexadiamine, then MOF-MNP was obtained by reaction with In(NO3)3, followed by in situ condensation reaction to form COF coated MOF-modified magnetic nanospheres. That adsorbent was mixed with food samples. After sonicated for several minutes, the solid phase was picked up with a magnet. Then, loaded SAs were eluted with acetonitrile, followed by running into the HPLC column. The suggested method was subjected to tap water, pork, chicken, and shrimp samples, and determined six SAs with the detection limits between 0.1 and 0.5 ng/mL. The method demonstrated excellent precision and linearity. [46]. In an interesting study to determine antibiotics, magnetic mesoporous hollow carbon microspheres (MHM) were synthesized by Wang et al. in order to detect penicillin sodium (Pen G). MHM behaved as a carrier for penicillinase (Pen X). At first, Pen X was adsorbed on MHM, then the last product was fixed on a magnetic glassy carbon electrode (MGCE). The determination was performed using the differential pulse voltammetry (DPV) method. The milk sample was subjected to the devised procedure utilizing the standard addition method. The LOD value was calculated as 3×10^{-7} mg/mL. Good recoveries, linearity in a wide concentration range, high precision, good stability of electrod, and recycling of solid phase are considered favored properties of the method by the authors [47].

N-nitrosamines (NAs) may especially exist in processed meat, and are sourced from nitrite usage as preservatives. But it is a highly carcinogenic compound, thus, it must be sensitively determined in meat products. Yang et al. tried NAs detection in alcoholic drinks and processed meat. The determination was performed with microsolid phase extraction combined with GC-MS. They synthesized nitrogen-doped magnetic mesoporous carbon nanospheres (M-MCN) due to enhancing the hydrophobic character of the carbon material. Thus, hydrophilic and hydrophobic properties of material were tuned to provide determination of polar and nonpolar SAs. Experimentally, they conducted NAs samples with M-MCN by shaking for 20 min, then, loaded solid phase was magnetically separated. Dichloromethane/methanol (3/1, *v/v*) mixture was employed as a leaching solution. NAs concentrations were measured in the last solution with GC-MS. It is declared by the authors that the detection limits were 0.5–6.6 ng/g, recovery varied between 85 and 110% with good linearity and precision [48].

The endocrine-disrupting compounds (EDCs) (such as ultraviolet (UV) filters, alkyl phenols (APs, as 4-octylphenol, 4-tert-butylphenol), bisphenol A (BPA), *etc.*) join into the food via migration from container wall coating and packaging materials. Liu et al. listed the items that is source of migration as plastic drinking and baby bottles, foodstuffs, canned beverages, and microwaveable goods. To determine phenolic EDCs at trace levels, Liu et al. fabricated a magnetic cubic mesoporous silica Fe_3O_4@14-3-14-MCM-48 nanocomposites. EDCs were retained on this sorbent, then eluated with methanol/acetonitrile mixture (50/50; v/v). At last, measurement of EDCs concentration was performed using HPLC with diode array detection. The author explained the retention mechanism by electrostatic interaction, hydrophobic interaction, and the formation of hydrogen bonds. Application of method under the optimized conditions gave satisfactory results: Good linearity, high recoveries and quite low dedection limits (1.0–2.5 ng/mL) [6].

Saccharides are defined as carbohydrates, which are abundant in nature and mostly structural components of animal and plant organisms [49]. Zhao et al. noted that the determination of saccharides is needed to understand the distribution of saccharides, which has great scientific importance. Because they are employed in many industrial fields, such as food and pharmaceuticals, the determination of them in real samples is highly important. However, the determination of them is quite challenging due to the deprivation of a specific recognition method for saccharides. That's why, Zhao et al. developed a method enabling saccharide analysis, and applied this method to sugar-free mints, cookies, and tea samples. For this purpose, the authors fabricated magnetic dendritic mesoporous silica nanoparticles functionalized with sulfonyl groups. The boronic acid was immobilized onto these NPs in order to perform saccharide separation from matrix via boronate affinity. The saccharide molecules retain on the boronic acid active site as boronic acid-saccharide complex. Afterwards, this complex was eluted with 3 % NH_3, followed by HPLC-UV analysis. The accuracy, precision, and reproducibility were reported by the author as 80–100 %, <15 %, and <10 %, respectively [50].

Luteolin is a valuable glycoside flavonoid type antioxidant that has anticancer and anti-inflammatory effects, and is found in herbs and vegetables like, artichoke, green pepper, *etc.* As seen in Fig. 3a, it is constructed of resorcinol and catechol rings [51-53]. A sensor for the sensitive determination of luteolin in samples of honeysuckle was created by Li et al. They synthesized peanut- like Fe_2O_3 encapsulated cobalt- doped microporous/mesoporous carbon (FeO_X-Co-MMC) material. After modification of GCE with FeO_X-MMC, modified electrod FeO_X-Co-MMC/GCE was used for luteolin detection. Low detection limit, 0.031 nM; good reproducibility, RSD 3.16%); high recoveries, 96.1-104.8%; and stability, RSD 2.34%, were obtained as result of the application of the method [54].

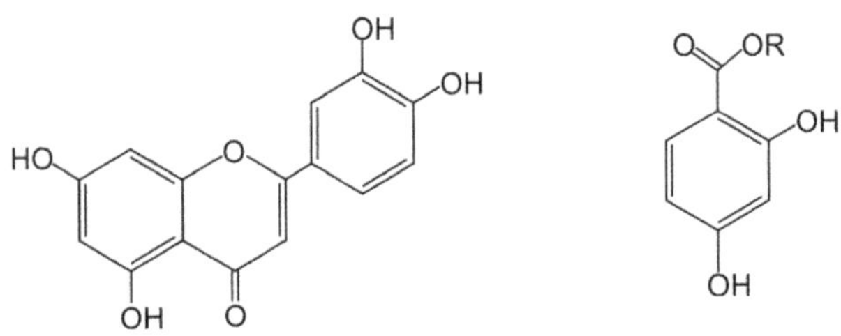

(a) (b)

Fig. 3 a) The chemical structure of Luteolin, b) General chemical structure of a paraben (para-hydroxybenzoate)

Parabens (Fig. 3b) are utilized as preservatives against bacteria, fungi, and yeast to prolong the shelf life of food items. They are especially used in jams, beer, sauces, juices, frozen dairy products, desserts, and processed vegetables. Paraben analyses have already been already performed using different solid phase extraction and instrumental methods in food products [55-57]. But, any work, in which magnetic mesoporous adsorbent is used for paraben detection in food, is not encountered in the literature. The amino-functionalized mesoporous silica–Fe_3O_4

nanocomposite, MCM-41–NH_2, was fabricated by Mehdinia et al. In order to detect parabens in sunscreen creams, moisturizing creams, and toothpaste. After sample solutions were treated with MCM-41–NH_2, the loaded solid phase was magnetically separated and eluted with n-hexane. The paraben detection was performed by gas chromatography with flame ionization detector in the last solution. It is reported that LODs were calculated below 0.9 µg/L, while recoveries were 85-107% [58]. This method could also be applied for food samples, such as jam or juice samples.

Aflatoxins are toxic and carcinogenic mycotoxin fabricated by fungi (*Aspergillus flavus* and *Aspergillus parasiticus*) in foods. So that it metabolizes in the liver, it has serious effect on the liver, and causes liver disorders (hepatitis, cancer, *etc.*) [59]. Currently, mesoporous materials are being utilized as adsorbents in SPE for aflatoxin retention and preconcentration. For instance, many sorbents, such as molecularly imprinted polymers fabricated with mesoporous silica FDU-12 [60]; mesoporous silica (type UVM-7; which is a nanometric version of MCM-41) [61], ionic liquid-functionalized mesoporous silica [62], were fabricated to detect aflatoxines. It is possible that magnetic derivatives of those sorbents could be used for the same purpose. Zearalenone is a mycotoxin, which is a secondary metabolite of the mold *F. graminearum.* It is mostly found in cereals such as barley, wheat, and corn [63, 64]. Guo et al. synthesized mesoporous silica-surface loaded gold nanocomposites (MSN-Rh6GAuNPs) as SERS substrate, and determined zearalenone with the LOD value of 0.0064 ng/mL. The authors explained this rather low detection limit by saying that the saying that the existence of the nanogaps between AuNPs made the sorbent strong for SERS measurement [65]. Considering that MNPs are used in magnetic resonance imaging (MRI) as contrast agents, and the existence of MNPs improves the sensitivity of biosensors and diagnostic tools due to having large magnetic moments and high surface-area-to-volume ratios [66]; it will be noticed that obtaining enhanced sensitivity for zearalenone detection would be possible, when magnetic mesoporous materials are used. Zhou et al. synthesized the dendritic mesoporous silica nanospheres, which resemble passion fruit and are enhanced with quantum dots coupled with magnetically tuneable aptasensors, in order to detect ochratoxin A which is another powerful mycotoxin. The assay was treated to red wine and wheat flour samples. The detection limit was obtained as 1.402 pg/mL with recoveries between 91% and 107% [67].

Another contaminant in food is Escherichia coli (E. coli) O157:H7 which is a pathogen microorganism and abundant in humans' intestinal systems. It is reported that it causes many diseases over 73,000 per year [68, 69]. The aptamer-modified gold nanoparticles@macroporous magnetic silica photonic microsphere (Au@MMSPM) was fabricated by Wang's research group in order to provide rapid, simple, and sensitive determination of E. Coli. That SERS (Surface-Enhanced Raman Spectroscopy) assay sorbent was treated with milk sample solution, and after separated magnetically, the E. Coli detected on the solid surface with SERS. Thus, rather low detection limit (2.20 CFU/mL) was achieved, and the author proposed this method to apply other sample matrixes as food industry, medicine, and environment [70]. Mi et al. studied determination of *Staphylococcus aureus* and *Escherichia coli* using aptamer-based Fe_3O_4@SiO_2@Ag as SERS biosensor. They detect bacteria with low detection limit, 1 CFU/mL. The authors attributed the capture and enrichment of bacteria to the monodisperse state and strong magnetic properties of Fe_3O_4@SiO_2@Ag [71].

The performance of magnetic mesoporous materials employed for determination of contaminants or active biocomponents in food samples is summarized in Table 1.

Materials Research Foundations 173 (2025) 392-412
https://doi.org/10.21741/9781644903452-15

Conclusion

Providing the safety and quality of foods is the most important concern for everybody due to 'you are what you eat'. Unfortunately, the rapid development in industry, followed by the environmental pollution, results in contaminated foods. Other suspect points in food products are adulteration and usage of preservatives. These undesirable subjects make it necessary to detect contaminants and/or ingredients in food matrix. The determination of nutritional value of food and the detection of adulteration help to prevent cheats, enable conformity with legal standards, and provide quality assurance. Considering that the critical point in the determination of analyte in food is the matrix effect, the magnetic mesoporous materials come into prominence due to having the capability of eliminating the matrix effect. The existence of two strong parameters: high magnetic moment and high porosity enhance the sensitivity of the applied method. Furthermore, it enables to minimize processing time. As reviewed, supporting some properties such as linearity in a wide concentration range, high recoveries, selectivity, and high precision make magnetic mesoporous materials more important.

Table 1. *The LOD values of magnetic mesoporous material-based methods for detection of analytes in food samples*

Material	Analyte	Matrix	Method	LOD	Ref.
Fe@Ag@Dimercaptobenzene	Pb(II), Cd(II), Hg(II)	Water	magnetic solid phase extraction	0.011–0.031 µg/L	[72]
Mercapto modified magnetic MOF (SH-Fe_3O_4/$Cu_3(BTC)_2$)	Pb(II)	Rice	magnetic solid phase extraction	0.29 µg/L	[73]
poly(8-hydroxyquinoline modified F_3O_4@SiO_2	Cu(II)	water, vegetables	micro-solid phase extraction	0.64 ng/mL	[74]
F_3O_4/hydroxyapatite/graphene quantum dots (F_3O_4/HAP/GQDs)	Cu(II)	vegetables	ultrasound-assisted magnetic solid phase extraction	0.58 ng/mL	[75]
Phenyl-SBA-15@Fe_3O_4	triazole fungicide	strawberry	magnetic dispersive solid phase extraction	0.19 - 1.75 µg·L	[76]
imidazolium ionic liquid immobilized magnetic mesoporous silica (Fe_3O_4@MCM@IL)	Hexaconazole, α-Endosulfan, Endosulfan sulfate, Bromopropylate, Fenpropathrin	water	micro-solid phase extraction	0.04 to 1.63 ng/L	[77]
magnetic metal-organic-framework (Fe_3O_4@MIL-100 (Fe))	Organochlorines	tea leaves	Mechanochemical magnetic solid phase extraction	0.62-3.92 ng/g	[78]
Fe_3O_4@MOF-808	benzoylurea insecticides	tea beverages, juice samples	magnetic solid phase extraction	0.04–0.15 ng/mL	[79]
nano-zeolite imidazolate framework-8-functionalized magnetic graphene oxide (Fe_3O_4@APTES-GO/ZIF-8)	epoxiconazole, flusilazole, tebuconazole, triadimefon	water, honey, fruit juices	magnetic solid phase extraction	0.014-0.109 µg/L	[80]

MIL101 (Cr)-NH_2 functionalized magnetic graphene oxide	Acetamiprid, Imidacloprid	apple, cucumber, lettuce	ultrasonic-assisted magnetic solid phase extraction	0.019–0.022 μg/L	[81]
$CoFe_2O_4$ decorated graphene/C18-functionalized mesoporous silica nanocomposites	promethazine	beef	magnetic solid-phase extraction	9.8 nmol/L	[82]
boronic acid-functionalized porewall- magnetic graphene/mesoporous silica composites	Streptomycin, dihydrostreptomycin	milk	solid-phase extraction	2-5 ng/mL	[83]
pyridine-functionalized magnetic nanoporous silica material (called Py-Fe_3O_4@MCM-41)	Cd(II)	rice, onion, carrot, lettuce, parsley, basil, tap water,	solid phase extraction	0.04 ng/mL	[84]
Fe_3O_4@nSiO_2@mSiO_2-APTES	Polymer additives	mineral water, carbonated drinks, tea drinks, vinegar, yogurt	magnetic solid-phase extraction	0.01–0.09 ng/mL	[85]
MNPs modified 1,4,8,11-tetraazacyclotetradecane (cyclam) loaded mesoporous silica (Fe_3O_4@cyclam-SBA-15)	Pb(II)	milk	magnetic solid phase extraction	2.02 μg/L	[86]
Pyridine-functionalized magnetic MCM-41	Ni(II)	water	ultrasound-assisted magnetic solid phase extraction	8 ng/L	[87]
m-MCM-41	Histamine, Cadaverine, Putrescine, Tyramine	kashk	magnetic solid phase extraction	8-14 μg/L	[88]
IL-modified Fe_3O_4@CS NPs	Antibiotics	milk	magnetic solid phase extraction	0.04–0.19 μg kg	[89]

References

[1] R. Kizil, J. Irudayaraj, Spectroscopic technique: Fourier transform raman (ft-raman) spectroscopy, in: D.W. Sun (Ed.), Sun Modern Techniques for Food Authentication, Academic Press, 2018, pp. 139-191. https://doi.org/10.1016/B978-0-12-814264-6.00005-0

[2] K. Gupta, V. Jain, S. Jain, K. Dhawan, G. Talwar, Analysis of food, in: B. Caballero (ed.), Encyclopedia of Food Sciences and Nutrition, Academic press, 2003, pp. 206-215. https://doi.org/10.1016/B0-12-227055-X/00044-4

[3] B. Jiang, R. Tsao, Y. Li, M. Miao, Food safety: Food analysis technologies/techniques, in: N.K.V. Alfen (Ed.), Encyclopedia of Agriculture and Food Systems, Academic Press, 2014, pp. 273-288. https://doi.org/10.1016/B978-0-444-52512-3.00052-8

[4] W.A. Khan, M.B. Arain, M. Soylak, Nanomaterials-based solid phase extraction and solid phase microextraction for heavy metals food toxicity, Food Chem. Toxicol. 145 (2020) 111704. https://doi.org/10.1016/j.fct.2020.111704

[5] M.A. Farajzadeh, R. Safi, A. Yadeghari, Combination of QuEChERS extraction with magnetic solid phase extraction followed by dispersive liquid-liquid microextraction as an efficient procedure for the extraction of pesticides from vegetable, fruit, and nectar samples having high content of solids, Microchem. J. 147 (2019) 571-581. https://doi.org/10.1016/j.microc.2019.03.074

[6] J. Liu, X. Ma, S. Zhang, T. Wu, H. Liu, M. Xia, J. You, Cationic gemini surfactant templated magnetic cubic mesoporous silica and its application in the magnetic dispersive solid phase extraction of endocrine-disrupting compounds from the migrants of food contact materials, Microchem. J. 145 (2019) 606-613. https://doi.org/10.1016/j.microc.2018.11.013

[7] M. Faraji, Y. Yamini, Application of magnetic nanomaterials in food analysis, in: M. Ahmadi, A. Afkhami, T. Madrakian (Eds.), Magnetic Nanomaterials in Analytical Chemistry, Elsevier. 2021, pp. 87-120. https://doi.org/10.1016/B978-0-12-822131-0.00003-0

[8] B. Akça, M. Can, V. Değirmenci, A. Yilmaz, D. Üner, Single step synthesis of mesoporous Co-Pb/SBA-15 catalysts, in: K. Eguchi, M. Machida, I. Yamanaka (Eds.), Studies in Surface Science and Catalysis, Elsevier. 2007, pp. 317-320. https://doi.org/10.1016/B978-0-444-53202-2.50068-3

[9] W. Li, Synthesis and applications of mesoporous nanostructures, in: Y. Yin, Y. Lu, Y. Xia (Eds.), Encyclopedia of Nanomaterials, Elsevier. 2023, pp. 322-335. https://doi.org/10.1016/B978-0-12-822425-0.00054-3

[10] R. Brady, B. Woonton, M.L. Gee, A.J. O'Connor, Hierarchical mesoporous silica materials for separation of functional food ingredients: A review, Innov. Food Sci. Emerg. Technol. 9 (2008) 243-248. https://doi.org/10.1016/j.ifset.2007.10.002

[11] A. Bernardos, L. Kouřimská, Applications of mesoporous silica materials in food: A review, Czech J. Food Sci. 31 (2013) 99-107. https://doi.org/10.17221/240/2012-CJFS

[12] C.B. Sapiña, E.P. Castell, J. El Haskouri, E.F.S. Alfonso, P. Amorós, A.R.M. Aucejo, A type UVM-7 mesoporous silica with γ-cyclodextrin for the isolation of three veterinary antibiotics (ofloxacin, norfloxacin, and ciprofloxacin) from different fat-rate milk samples, J. Food Compos. Anal. 109 (2022) 104463. https://doi.org/10.1016/j.jfca.2022.104463

[13] P. Qin, Y. Yang, W. Li, J. Zhang, Q. Zhou, M. Lu, Amino-functionalized mesoporous silica nanospheres (MSN-NH2) as sorbent for extraction and concentration of synthetic dyes from foodstuffs prior to HPLC analysis, Anal. Methods. 11 (2019) 105-112. https://doi.org/10.1039/C8AY02215A

[14] I.M.M. Kenawy, Y.G.A. El-Reash, M.M. Hassanien, N.R. Alnagar, W.I. Mortada, Use of microwave irradiation for modification of mesoporous silica nanoparticles by thioglycolic acid for removal of cadmium and mercury, Microporous Mesoporous Mater. 258 (2018) 217-227. https://doi.org/10.1016/j.micromeso.2017.09.021

[15] S. Ji, T. Li, W. Yang, C. Shu, D. Li, Y. Wang, L. Ding, A hollow porous molecularly imprinted polymer as a sorbent for the extraction of 7 macrolide antibiotics prior to their determination by HPLC-MS/MS, Mikrochim. Acta. 185 (2018) 203-212. https://doi.org/10.1007/s00604-018-2728-3

[16] H. Wang, Y. Liu, S. Yao, P. Zhu, Selective recognization of dicyandiamide in bovine milk by mesoporous silica SBA-15 supported dicyandiamide imprinted polymer based on surface molecularly imprinting technique, Food Chem. 240 (2018) 1262-1267. https://doi.org/10.1016/j.foodchem.2017.08.066

[17] J. Liu, S.Z. Qiao, Q.H. Hu, G.Q. (Max) Lu, magnetic nanocomposites with mesoporous structures: Synthesis and applications, Small. 7 (2011) 425-443. https://doi.org/10.1002/smll.201001402

[18] B.S. Rodríguez, A.V.H. Herrera, M.A. Ramos, M.Á.R. Delgado, Recent applications of magnetic nanoparticles in food analysis, Processes. 8 (2020) 1140. https://doi.org/10.3390/pr8091140

[19] O. Plohl, S. Gyergyek, L.F. Zemljič, Mesoporous silica nanoparticles modified with N-rich polymer as a potentially environmentally-friendly delivery system for pesticides, Microporous Mesoporous Mater. 310 (2021) 110663. https://doi.org/10.1016/j.micromeso.2020.110663

[20] J. Liang, H. Li, K. Huang, D. Ning, F. Yan, W. Chen, L. Xie, Y. Wang, C. Jiang, L. Tang, T. Li, X. Huang, Y. Ya, Pyrene-1-carboxaldehyde hydrazone modified mesoporous silica foam-based strategy for efficient and accurate electroanalysis of copper in food samples, J Food Compos. Anal. 134 (2024) 106587. https://doi.org/10.1016/j.jfca.2024.106587

[21] Y. Guo, B. Chen, Y. Zhao, T. Yang, Fabrication of the magnetic mesoporous silica Fe-MCM-41-A as efficient adsorbent: Performance, kinetics and mechanism, Sci. Rep. 11 (2021) 2612. https://doi.org/10.1038/s41598-021-81928-8

[22] N.I. Cuello, M.I. Oliva, C.E.R Torres, A. Tolley, V.R. Elías, G.A. Eimer, Study on magnetite nanoparticles embedded in mesoporous silica obtained by a straightforward and biocompatible method, J. Phys. Chem. Solids. 145 (2020) 109535. https://doi.org/10.1016/j.jpcs.2020.109535

[23] A. Gençer, B.U. Karataş, Ö. Topel, Önder, N. Kiraz, Nadir, Synthesis and characterization of surface-modified magnetic mesoporous silicate materials for phosphate adsorption, Turk. J. Chem. 48 (2024) 5. https://doi.org/10.55730/1300-0527.3638

[24] S. Kamari, F. Ghorbani, Extraction of highly pure silica from rice husk as an agricultural by-product and its application in the production of magnetic mesoporous silica MCM-41, Biomass Convers. Biorefinery. 11 (2021) 3001-3009. https://doi.org/10.1007/s13399-020-00637-w

[25] S. Zeb, N. Ali, Z. Ali, M. Bilal, B. Adalat, S. Hussain, S. Gul, F. Ali, R. Ahmad, S. Khan, H.M.N. Iqbal, Silica-based nanomaterials as designer adsorbents to mitigate emerging organic

contaminants from water matrices, J. Water Process Eng. 38 (2020) 101675. https://doi.org/10.1016/j.jwpe.2020.101675

[26] R.J. Zadeh, M.H. Sayadi, M.R. Rezaei, Synthesis of Thiol modified magMCM-41 nanoparticles with rice husk ash as a robust, high effective, and recycling magnetic sorbent for the removal of herbicides, J. Environ. Chem. Eng. 9 (2021) 104804. https://doi.org/10.1016/j.jece.2020.104804

[27] D. Li, Y. Wen, L. Hu, X. Xu, B.S. Spencer, S. Egodawatte, S.C. Larsen, Y. Tang, Phosphonate functionalized magnetic mesoporous silica for rare earth element recovery from citrate assisted solid waste extracts, Appl. Geochem. 162 (2024) 105900. https://doi.org/10.1016/j.apgeochem.2024.105900

[28] C. Comanescu, Recent advances in surface functionalization of magnetic nanoparticles, Coatings. 13 (2023) 1772. https://doi.org/10.3390/coatings13101772

[29] X.T. Peng, L. Jiang, Y. Gong, X.Z. Hu, L.J. Peng, Y.Q. Feng, Preparation of mesoporous ZrO2-coated magnetic microsphere and its application in the multi-residue analysis of pesticides and PCBs in fish by GC-MS/MS, Talanta. 132 (2015) 118-125. https://doi.org/10.1016/j.talanta.2014.08.069

[30] C. Yuan, R. Li, L. Wu, X. Hong, H. He, G. Yang, L. Wang, Y. You, H. Liu, Z. Chai, X. Lu, Optimization of a modified QuEChERS method by an n-octadecylamine-functionalized magnetic carbon nanotube porous nanocomposite for the quantification of pesticides, J. Food Compos. Anal. 102 (2021) 103980. https://doi.org/10.1016/j.jfca.2021.103980

[31] M. Wang, J. Wang, K. Wang, L. Zhang, X. Cao, C. Guo, J. Wang, B. Wu, Magnetic mesoporous material derived from MIL-88B modified by l-alanine as modified QuEChERS adsorbent for the determination of 6 pesticide residues in 4 vegetables by UPLC-MS/MS, Food Chem. 384 (2022) 132325. https://doi.org/10.1016/j.foodchem.2022.132325

[32] P. Qi, Z. Wang, G. Yang, C. Shang, H. Xu, X. Wang, H. Zhang, Q Wang, X.Wang, Removal of acidic interferences in multi-pesticides residue analysis of fruits using modified magnetic nanoparticles prior to determination via ultra-HPLC-MS/MS. Microchim. Acta. 182 (2015) 2521-2528. https://doi.org/10.1007/s00604-015-1615-4

[33] Z. Liu, P. Qi, X. Wang, Z. Wang, X. Xu, W. Chen, L. Wu, H. Zhang, Q. Wang, X. Wang, Multi-pesticides residue analysis of grains using modified magnetic nanoparticle adsorbent for facile and efficient cleanup, Food Chem. 230 (2017) 423-431. https://doi.org/10.1016/j.foodchem.2017.03.082

[34] G. Liu, L. Li, X. Huang, S. Zheng, D. Xu, X. Xu, Y. Zhang, H. Lin, Determination of triazole pesticides in aqueous solution based on magnetic graphene oxide functionalized MOF-199 as solid phase extraction sorbents, Microporous Mesoporous Mater. 270 (2018) 258-264. https://doi.org/10.1016/j.micromeso.2018.05.023

[35] L.F.S. Santos, R.A.D. Jesus, J.A.S. Costa, L.G.T. Gouveia, M.E.D. Mesquita, S. Navickiene, Evaluation of MCM-41 and MCM-48 mesoporous materials as sorbents in matrix solid phase dispersion method for the determination of pesticides in soursop fruit (Annona muricata), Inorg. Chem. Commun. 101 (2019) 45-51. https://doi.org/10.1016/j.inoche.2019.01.013

[36] J.L. Chen, L. Gao, Q. Jiang, Q. Hou, Y. Hong, W. Shen, Y. Wang, J.H. Zhu, Fabricating efficient porous sorbents to capture organophosphorus pesticide in solution, Microporous Mesoporous Mater. 294 (2020) 109911. https://doi.org/10.1016/j.micromeso.2019.109911

[37] S.H. Jangi, A. Khoobi, Detection of cadmium heavy metal ions using a nanostructured green sensor in food, biological and environmental samples, Food Chem. 458 (2024) 140307. https://doi.org/10.1016/j.foodchem.2024.140307

[38] N. Ncube, Y. Tancu, N. Mketo, A greener, rapid and accurate microwave-assisted hydrogen peroxide digestion method for ICP-OES determination of heavy metals in pet food samples, J. Food Compos. Anal. 131 (2024) 106201. https://doi.org/10.1016/j.jfca.2024.106201

[39] J. Gao, J. Yin, G. Wang, X. Wang, J. Zhang, B. Sun, D. He, H. Suo, C. Zhao, A novel electrode for simultaneous detection of multiple heavy metal ions without pre-enrichment in food samples, Food Chem. 448 (2024) 138994. https://doi.org/10.1016/j.foodchem.2024.138994

[40] Y. Zhang, Y. Xu, Y. Ma, H. Luo, J. Hou, C. Hou, D. Huo, Ultra-sensitive electrochemical sensors through self-assembled MOF composites for the simultaneous detection of multiple heavy metal ions in food samples, Anal. Chim. Acta. 1289 (2024) 342155. https://doi.org/10.1016/j.aca.2023.342155

[41] Z.M. Karazan, M. Roushani, S.J. Hoseini, Simultaneous electrochemical sensing of heavy metal ions (Zn2+, Cd2+, Pb2+, and Hg2+) in food samples using a covalent organic framework/carbon black modified glassy carbon electrode, Food Chem. 442 (2024) 138500. https://doi.org/10.1016/j.foodchem.2024.138500

[42] S. Shariati, N. Parto, E. Bozorgzadeh, P. Zanjanchi, S. Rahnama, Magnetic solid phase preconcentration of cadmium in water samples using sulfonic acid functionalized Kit-6 magnetite mesoporous nanocomposites followed by flame atomic absorption spectrometry, J. Iran. Chem. Soc. 17 (2020) 3375-338. https://doi.org/10.1007/s13738-020-01995-y

[43] R. Sedghi, M. Shojaee, M. Behbahani, M.R. Nabid, Application of magnetic nanoparticles modified with poly(2-amino thiophenol) as a sorbent for solid phase extraction and trace detection of lead, copper and silver ions in food matrices, RSC Adv. 5 (2015) 67418-67426. https://doi.org/10.1039/C5RA11561B

[44] F. Aboufazeli, H.R. L.Z. Zhad, O. Sadeghi, M. Karimi, E. Najafi, Novel ion imprinted polymer magnetic mesoporous silica nano-particles for selective separation and determination of lead ons in food samples, Food Chem. 141 (2013) 3459-3465. https://doi.org/10.1016/j.foodchem.2013.06.062

[45] Z. Dahaghin, H.Z. Mousavi, S.M. Sajjadi, A novel magnetic ion imprinted polymer as a selective magnetic solid phase for separation of trace lead(II) ions from agricultural products, and optimization using a Box-Behnken design, Food Chem. 237 (2017) 275-281. https://doi.org/10.1016/j.foodchem.2017.05.118

[46] Z. Chen, Z. He, X. Luo, F. Wu, S. Tang, J. Zhang, Synthesis of MOF@COF hybrid magnetic adsorbent for microextraction of sulfonamides in food and environmental samples, Food Anal. Methods. 13 (2020) 1346-1356. https://doi.org/10.1007/s12161-020-01750-2

[47] L. Wang, Y. Xiu, B. Han, L. Liu, X. Niu, H. Wang, Magnetic mesoporous carbon material based electrochemical sensor for rapid detection of penicillin sodium in milk, J. Food Sci. 85 (2020) 2435-2442. https://doi.org/10.1111/1750-3841.15328

[48] K. Yang, Y. Kang, Q. Zhang, D. Wu, J. Shen, Y. Wei, C. Wang, Nitrogen-doped magnetic porous carbon nanospheres derived from dual templates-induced mesoporous polydopamine coated Fe3O4 for efficient extraction and sensitive determination of volatile nitrosamines by gas chromatography-mass spectroscopy, Talanta. 276 (2024) 126235. https://doi.org/10.1016/j.talanta.2024.126235

[49] C.F. Poole, Application of thin-layer chromatography to the analysis of saccharides. in: C.F. Poole (Ed.), Instrumental Thin-Layer Chromatography, Elsevier, 2023, pp. 413-435. https://doi.org/10.1016/B978-0-323-99970-0.00020-X

[50] X. Zhao, J. Dong, Y. Zhang, T. Wu, Z. Bie, Y. Chen, Magnetic dendritic mesoporous silica nanoparticles based integrated platform for rapid and efficient analysis of saccharides, Anal. Chim. Acta. 1288 (2024) 342166. https://doi.org/10.1016/j.aca.2023.342166

[51] L.C. Salazar, D.A.N. Tapia, K.D.A. Justo, A.A. Izazaga, C.S. Leyva, M.H. Martinez, S.V. Comonfort, N.N. Tito, Flavonoids as regulators of TIMPs expression in cancer: Consequences, opportunities, and challenges, Life Sci. 308 (2022) 120932. https://doi.org/10.1016/j.lfs.2022.120932

[52] J.C. Xue, S. Yuan, H. Meng, X.T. Hou, J. Li, H.M. Zhang, L.L. Chen, C.H. Zhang, Q.G. Zhang, The role and mechanism of flavonoid herbal natural products in ulcerative colitis, Biomed. Pharmacother. 158 (2023) 114086. https://doi.org/10.1016/j.biopha.2022.114086

[53] A.Y. Tesio, S.N. Robledo, Analytical determinations of luteolin, BioFactors. 47 (2021) 141-164. https://doi.org/10.1002/biof.1720

[54] J. Li, Y. Li, Y. Yang, P. Zhao, J. Fei, Y. Xie, Detection of luteolin in food using a novel electrochemical sensor based on cobalt-doped microporous/mesoporous carbon encapsulated peanut-like FeOx composite, Food Chem. 435 (2024) 137651. https://doi.org/10.1016/j.foodchem.2023.137651

[55] S. Cao, Z. Liu, L. Zhang, C. Xi, X. Li, G.R. Wang, M.Z. Yuan, Development of an HPLC-MS/MS method for the simultaneous analysis of six kinds of parabens in food, Anal. Methods. 5 (2013) 1016-1023. https://doi.org/10.1039/C2AY26283E

[56] Y.G. Ontiveros, I.M. Ruiz, L. Rodrigo, M. Aguilera, A. Rivas, A.Z. Gómez, Presence of parabens and bisphenols in food commonly consumed in Spain, Foods. 10 (2021) 92. https://doi.org/10.3390/foods10010092

[57] G. Yusakul, C.T. Buranachai, S. Poorahong, S. Sakamoto, F. Makkliang, An eco-friendly method using deep eutectic solvents immobilized in a microcrystal cellulose-polyvinyl alcohol sponge for parabens analysis in food samples, Microchem. J. 191 (2023) 108758. https://doi.org/10.1016/j.microc.2023.108758

[58] A. Mehdinia, M. Bahrami, S.A. Mozaffari, Comparative study on different functionalized mesoporous silica nanomagnetic sorbents for efficient extraction of parabens, J. Iran. Chem. Soc. 12 (2015)1543-1552. https://doi.org/10.1007/s13738-015-0626-8

[59] Z.Y. Chu, X.J. Zi, Network toxicology and molecular docking for the toxicity analysis of food contaminants: A case of Aflatoxin B1, Food Chem. Toxic. 188 (2024) 114687. https://doi.org/10.1016/j.fct.2024.114687

[60] C. Rui, J. He, Y. Li, Y. Liang, L. You, L. He, K. Li, S. Zhang, Selective extraction and enrichment of aflatoxins from food samples by mesoporous silica FDU-12 supported aflatoxins imprinted polymers based on surface molecularly imprinting technique, Talanta. 201 (2019,)342-349. https://doi.org/10.1016/j.talanta.2019.04.019

[61] E.P. Castell, C.B. Sapiña, V.J. Borràs, P. Amorós, J. El Haskouri, J.M.H. Martínez, A.R.M. Aucejo, Extraction of aflatoxins by using mesoporous silica (type UVM-7), and their quantitation by HPLC-MS, Microchim. Acta. 186 (2019) 792. https://doi.org/10.1007/s00604-019-3958-8

[62] X. Feng, Y. Sun, T. Zhang, J. Li, Ha. Zhao, W. Zhao, G. Xiang, L. He, Ionic liquid-functionalized mesoporous multipod silica for simultaneously effective extraction of aflatoxin B1 and its two precursors from grain, Anal. Chim. Acta. 1303 (2024) 342544. https://doi.org/10.1016/j.aca.2024.342544

[63] M.E. Stack, Aflatoxins and mycotoxins, thin-layer (planar) chromatography, in: I.D. Wilson (Ed.), Encyclopedia of Separation Science, Academic Press, 2000, pp. 1888-1895. https://doi.org/10.1016/B0-12-226770-2/02581-3

[64] G. Galaverna, C. Dall'Asta, Sampling techniques for the determination of mycotoxins in food matrices, in: J. Pawliszyn (Ed.), Comprehensive Sampling and Sample Preparation, Academic Press, 2012, pp. 381-403. https://doi.org/10.1016/B978-0-12-381373-2.00140-X

[65] Z. Guo, L. Gao, L. Yin, M. Arslan, H.R. El-Seedi, X. Zou, Novel mesoporous silica surface loaded gold nanocomposites SERS aptasensor for sensitive detection of zearalenone, Food Chem. 403 (2023) 134384. https://doi.org/10.1016/j.foodchem.2022.134384

[66] A. Farzin, S.A. Etesami, J. Quint, A. Memic, A. Tamayol, Magnetic nanoparticles in cancer therapy and diagnosis, Adv. Healthc Mater. 9 (2020) e1901058. https://doi.org/10.1002/adhm.201901058

[67] J. Zhou, X. Lv, Y. Gui, J. He, F. Xie, J. Cai, Passion fruit-inspired dendritic mesoporous silica nanospheres-enriched quantum dots coupled with magnetism-controllable aptasensor enable sensitive detection of ochratoxin A in food products, Food Chem. 425 (2023) 136445. https://doi.org/10.1016/j.foodchem.2023.136445

[68] H. Huang, G. Zhao, W. Dou, Portable and quantitative point-of-care monitoring of Escherichia coli O157:H7 using a personal glucose meter based on immunochromatographic assay, Biosens. Bioelectron. 107 (2018) 266-271. https://doi.org/10.1016/j.bios.2018.02.027

[69] M. Yin, C. Liu, R. Ge, Y. Fang, J.Wei, X. Chen, Q. Chen, X. Chen, Paper-supported near-infrared-light-triggered photoelectrochemical platform for monitoring Escherichia coli O157:H7 based on silver nanoparticles-sensitized-upconversion nanophosphors, Biosens. Bioelectron. 203 (2022) 114022. https://doi.org/10.1016/j.bios.2022.114022

[70] X. Wang, W. Li, S. Dai, M. Dou, S. Jiao, J. Yang, W. Li, Y. Su, Q. Li, J. Li, High-throughput, highly sensitive and rapid SERS detection of Escherichia coli O157:H7 using

aptamer-modified Au@macroporous silica magnetic photonic microsphere array, Food Chem. 424 (2023) 136433. https://doi.org/10.1016/j.foodchem.2023.136433

[71] F. Mi, M. Guan, Y. Wang, G. Chen, P.F. Geng, A SERS biosensor based on aptamer-based Fe3O4@SiO2@Ag magnetic recognition and embedded SERS probes for ultrasensitive simultaneous detection of Staphylococcus aureus and Escherichia coli, Microchem. J. 190 (2023) 108605. https://doi.org/10.1016/j.microc.2023.108605

[72] Q. Zhou, M. Lei, Y. Liu, Y. Wu, Y. Yuan, Simultaneous determination of cadmium, lead and mercury ions at trace level by magnetic solid phase extraction with Fe@Ag@Dimercaptobenzene coupled to high performance liquid chromatography, Talanta. 175 (2017) 194-199. https://doi.org/10.1016/j.talanta.2017.07.043

[73] Y. Wang, H. Chen, J. Tang, G. Ye, H. Ge, X. Hu, Preparation of magnetic metal organic frameworks adsorbent modified with mercapto groups for the extraction and analysis of lead in food samples by flame atomic absorption spectrometry, Food Chem. 181 (2015) 191-197. https://doi.org/10.1016/j.foodchem.2015.02.080

[74] Z. Hasani, A. Shahsavani, Z. Aladaghlo, A. Fakhari, Application of magnetic nanoparticles modified with poly(8-hydroxyquinoline) as a nanosorbent for magnetic dispersive micro-solid phase extraction of copper in vegetable, water, and soil samples, J. Food Compos. Anal. 132 (2024) 106333. https://doi.org/10.1016/j.jfca.2024.106333

[75] P. Sricharoen, N. Limchoowong, Y. Areerob, P. Nuengmatcha, S. Techawongstien, S. Chanthai, Fe3O4/hydroxyapatite/graphene quantum dots as a novel nano-sorbent for preconcentration of copper residue in Thai food ingredients: Optimization of ultrasound-assisted magnetic solid phase extraction, Ultrason. Sonochemistry. 37 (2017) 83-93. https://doi.org/10.1016/j.ultsonch.2016.12.037

[76] L. Li, F. Nian, Y. Xu, Y. Li, S. Zhang, One-step grafting of phenyl functionalized magnetic mesoporous nanocomposites for magnetic dispersive solid-phase extraction and determination of triazole fungicide residues in strawberry fruits, Microchem. J. 201 (2024) 110635. https://doi.org/10.1016/j.microc.2024.110635

[77] S.F. Farzam, F. Shemirani, S. Karimi, Synthesis of imidazolium ionic liquid immobilized on magnetic mesoporous silica: A sorbent material in a green micro-solid phase extraction of multiclass pesticides in water, Talanta. 272 (2024) 125744. https://doi.org/10.1016/j.talanta.2024.125744

[78] Y. Zhou, J. Zhu, J. Yang, Y. Lv, Y. Zhu, W. Bi, X. Yang, D.D.Y. Chen, Magnetic nanoparticles speed up mechanochemical solid phase extraction with enhanced enrichment capability for organochlorines in plants, Anal. Chim. Acta. 1066 (2019) 49-57. https://doi.org/10.1016/j.aca.2019.03.049

[79] Y. Jia, Y. Wang, M. Yan, Q. Wang, H. Xu, X. Wang, H. Zhou, Y. Hao, M. Wang, Fabrication of iron oxide@MOF-808 as a sorbent for magnetic solid phase extraction of benzoylurea insecticides in tea beverages and juice samples, J. Chromatogr. A. 1615 (2020) 460766. https://doi.org/10.1016/j.chroma.2019.460766

[80] I.A. Senosy, H.M. Guo, M.N. Ouyang, Z.H. Lu, Z.H. Yang, J.H. Li, Magnetic solid-phase extraction based on nano-zeolite imidazolate framework-8-functionalized magnetic graphene

oxide for the quantification of residual fungicides in water, honey and fruit juices, Food Chem. 325 (2020) 126944. https://doi.org/10.1016/j.foodchem.2020.126944

[81] A. Ghiasi, A. Malekpour, S. Mahpishanian, Metal-organic framework MIL101 (Cr)-NH2 functionalized magnetic graphene oxide for ultrasonic-assisted magnetic solid phase extraction of neonicotinoid insecticides from fruit and water samples, Talanta. 217 (2020) 121120. https://doi.org/10.1016/j.talanta.2020.121120

[82] X. Yang, S.M. Ying, S. Zhang, J. Dai, W. Gao, T.Q. Wang, J.Q. Qiao, H.Z. Lian, L. Mao, CoFe2O4 decorated graphene/C18-functionalized mesoporous silica nanocomposites prepared for magnetic enrichment and electrochemical detection of promethazine in beef, Chin. Chem. Lett. 35 (2024) 108674. https://doi.org/10.1016/j.cclet.2023.108674

[83] J. Feng, X. She, X. He, J. Zhu, Y. Li, C. Deng, Synthesis of magnetic graphene/mesoporous silica composites with boronic acid-functionalized pore-walls for selective and efficient residue analysis of aminoglycosides in milk, Food Chem. 239 (2018) 612-621. https://doi.org/10.1016/j.foodchem.2017.06.052

[84] F. Omidi, M. Behbahani, M.K. Bojdi, S.J. Shahtaheri, Solid phase extraction and trace monitoring of cadmium ions in environmental water and food samples based on modified magnetic nanoporous silica, J. Magn. Magn. Mater. 395 (2015) 213-220. https://doi.org/10.1016/j.jmmm.2015.07.093

[85] X. Zhao, X. Liu, J. Wang, Y. Liu, T. Zhang, J. Chen, Q. Li, Y. Wie, X. Xi, Determination of polymer additives in foods and drinks packed with plastic by amino group modified magnetic mesoporous silica microspheres coupled with high performance liquid chromatography, J. Liq. Chromatogr. Relat. Technol. 44 (2021) 244-254. https://doi.org/10.1080/10826076.2021.1895219

[86] R.Z. Dorabei, M.S. Darbandsari, A. Moghimi, M.S. Tehrani, S. Nazerdeylami, Synthesis, characterization and application of cyclam-modified magnetic SBA-15 as a novel sorbent and its optimization by central composite design for adsorption and determination of trace amounts of lead ions, RSC Adv. 6 (2016) 108477-108487. https://doi.org/10.1039/C6RA21895D

[87] M. Behbahani, V. Zarezade, A. Veisi, F. Omidi, S. Bagheri, Modification of magnetized MCM-41 by pyridine groups for ultrasonic-assisted dispersive micro-solid-phase extraction of nickel ions, Int. J. Environ. Sci. Technol. 16 (2019) 6431-6440. https://doi.org/10.1007/s13762-018-2052-9

[88] R. Molaei, H. Tajik, M. Moradi, Magnetic solid phase extraction based on mesoporous silica-coated iron oxide nanoparticles for simultaneous determination of biogenic amines in an Iranian traditional dairy product; Kashk, Food Control. 101 (2019) 1-8. https://doi.org/10.1016/j.foodcont.2019.02.011

[89] H. Sahebi, A.J. Talaei, E. Abdollahi, F.H. Baltork, S.V. Zade, B. Jannat, N. Sadeghi, Rapid determination of multiclass antibiotics and their metabolites in milk using ionic liquid-modified magnetic chitosan nanoparticles followed by UPLC-MS/MS, Talanta. 253 (2023) 124091. https://doi.org/10.1016/j.talanta.2022.124091

Chapter 16

Mesoporous Materials for Tissue Engineering Applications

Joy Hoskeri H[1,*], Nivedita Pujari S[2], Preeti Gadyal[1], Vidyashree Suryavanshi[1], Arun K. Shettar[3]

[1]Department of Bioinformatics, Karnataka State Akkamahadevi Women University, Vijayapura-586109, Karnataka, India

[2]Albert Einstein College of Medicine, Bronx-10461, New York USA

[3]Cytxon Biosolutions Pvt Ltd, Huballi, 580031, Karnataka, India

joybioinformatics@gmail.com, joybioinfo@gmail.com

ORCID ID: 0000-0001-5763-7115

Abstract

Tissue engineering has become a promising solution to tackle the increasing demand for organ replacement and repair. A critical component of tissue engineering is the scaffold, which provides a structural and biological framework for cell proliferation, attachment, and differentiation. Mesoporous materials, with their unique porous structure, high surface area, and customizable properties, have garnered significant attention as potential frameworks for tissue engineering applications. These materials facilitate enhanced cellular interactions and control bioactive molecule transport, promoting tissue regeneration and repair. This chapter explores the advancement of mesoporous materials, their synthesis methods, and their applications in various tissue engineering strategies.

Keywords

Mesoporous Materials, Synthesis Method, Properties, And Tissue Engineering Applications

Contents

1. Introduction

Mesoporous materials, often referred to as mesoporous molecular sieves, are characterized by their well-defined three-dimensional pore structures with diameters ranging from 2 to 50 nanometers and surface areas that can exceed 1000 m^2/g. These materials are uniquely positioned between crystalline zeolites and other structured materials with larger feature sizes, such as three-dimensional ordered macroporous materials (3DOM) and direct-write materials [1, 2]. The exceptional ordering of the pores in mesoporous materials makes them valuable in a wide array of applications, including catalysis, optics, coatings, gas separation, bioseparation, diagnostics, and optics [3]. Typically, nanoporous materials are categorized into micropores, mesopores, and macropores, each with distinct structural and functional characteristics. Mesoporous materials offer significant advantages with their narrow pore size distribution, large surface area (>500 m^2/g), biocompatibility, and low toxicity, which have driven extensive interdisciplinary research and broadened their applications across various fields [4]. Mesoporous materials mainly include different forms of silica, alumina, and oxides of tantalum, cerium, tin, zirconium, niobium, and titanium. Among these, mesoporous carbon is particularly significant due to its direct applications in energy storage technologies, as reported by Eftekhari and Zhaoyang (2017) [5]. Mesoporous carbon is distinguished by its high specific surface area, a result of its mesoporous structure.Activated carbon, another common mesoporous material, is typically characterized by a carbon framework that includes both microporosity (pores smaller than 2 nanometers) and meso-porosity. The specific balance of these porosities is influenced by the conditions under which the activated carbon is synthesized. The International Union of Pure and Applied Chemistry defines mesoporous materials as those with ordered or disordered meso-structures (Fig.1). The arrangement of these mesopores can have significant effects on the material's properties and applications. In crystalline inorganic materials, the presence of mesopores can restrict the number of lattice units within the material. This structural modification can substantially alter the material's solid-state chemistry, impacting its behavior in various applications. For example, mesoporous electroactive materials used in batteries can exhibit different performance characteristics compared to their bulk counterparts. The unique mesostructure of these materials can affect charge storage, conductivity, and overall efficiency, leading to notable differences in battery performance [5].

	Microporous	Mesoporous	Macroporous
	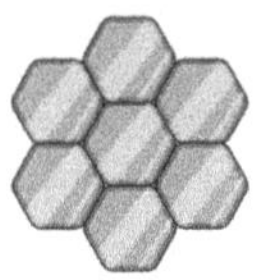		
Pore Size	<2nm	2 to 50nm	>50nm
Building Frame work	Purely inorganic	organic and inorganic hybrid	purely organic
Examples	zeolite based materials and silica	MCM-41 and MCM-48	Ceramic based materials and porous gels

Figure 1. IUPAC classification of porous materials

2. Historical development in materials science

The 1970s marked the beginning of the mesoporous materials industry with a significant breakthrough in zeolite technology. Mobil Corporation discovered a pioneering synthetic method that utilized the zeolite ZSM-5 (Zeolite Socony Mobil) as a catalyst to convert methanol into gasoline in 1972. This research aimed to produce more economical gasoline through neutralization reactions occurring in the small voids within these zeolites. However, this approach had limitations, which spurred researchers to seek materials with larger pore sizes. This pursuit led to the discovery of the first mesoporous silica materials in the early 1990s. Two breakthroughs occurred almost simultaneously: one by Mobil Corporation in the United States and another by Kuroda and his research team in Japan. Kuroda's team synthesized mesoporous silica by using a layered polysilicate called kanemite, intercalated with an alkyl-trimethyl-ammonium cationic surfactant (Cn-TMA).

Materials Research Foundations 173 (2025) 413-440 https://doi.org/10.21741/9781644903452-16

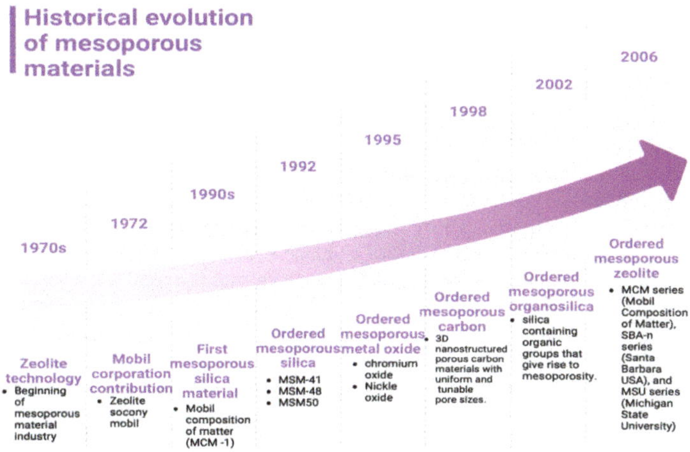

Figure 2. Historical evolution of mesoporous materials

The mesoporous structure of these materials was confirmed through nitrogen adsorption techniques. Shortly thereafter, in 1992, Mobil Research and Development Corporation introduced a method for preparing mesoporous silica using alkyl-trimethyl-ammonium bromide or chloride as portable templates and sodium silicate as the silica source under alkaline conditions. This process led to the creation of the MS1S family of mesoporous materials, including variants such as MCM-50, MCM-41, and MCM-48. These materials are characterized by their bi-continuous cubic (MCM-48), 2D hexagonal (MCM-41), and lamellar (MCM-50) mesostructures. The MCM-41 family, in particular, exhibited a hexagonal arrangement of uniform mesopores ranging from 15 Å to over 100 Åin diameter, with a surface area exceeding 700 m^2/g and absorption abilities of 0.7 cm^3/g or more. In parallel, another research group from Japan, independent of Mobil, developed a different technique for synthesizing mesoporous materials, which they termed Folded Sheet Materials (FSM-16). These groundbreaking discoveries ignited extensive research and exploration into mesoporous materials, leading to the development of various new types, including those created by Fudan University (FDU), the Korean Institute of Science and Technology, Santa Barbara Amorphous, and Michigan State University. These advancements in mesoporous materials opened new avenues for applications in catalysis, adsorption, and other areas, significantly expanding the scope and potential of these materials in scientific and industrial fields. Historical milestones of mesoporous materials' evolution are depicted in Fig. 2.

3. Synthesis methods of mesoporous materials for tissue engineering

3.1 Sol-gel method

It is one of the prominent techniques for synthesizing mesoporous materials and is extensively used in nanomaterial synthesis for the bottom-up fabrication of mixed oxide composites and metal oxide nanoparticles. This method offers several advantages, including the ability to process at relatively low temperatures and precise control over processing parameters. It allows the production of materials such as glasses and ceramics mainly including fibers, thin films, aerogels, and xerogels [6]. In the preparation of mesoporous silica using the sol-gel technique, the process begins with the hydrolysis of precursor solution with the template. The hydrolyzed silanol groups then condense to form siloxane linkages (Si—O—Si), resulting in the development of an inorganic network. The template is subsequently removed through either calcination or solvent extraction to yield a porous silica material [7]. The sol-gel method involves two primary reaction stages: hydrolysis and condensation. In the hydrolysis stage, the precursor solution reacts with the solvent to produce hydrolyzed species (Fig.3).The subsequent condensation stage sees these species form bonds between metal atoms bridged by oxygen atoms, leading to the creation of a colloidal suspension called sol and its gelation into a continuous network known as gel [8]. A stable dispersion of polymers or colloidal particles within a solvent makes a good sol, where electrostatic repulsion forces prevail over van der Waals attractions, ensuring particle stability in the colloidal phase, thereby minimizing gravitational effects. A gel, conversely, is a 3D network that encompasses a liquid phase, leading to an expanded solid phase and increased viscosity. During gelation, polycondensation reactions lead to the development of a larger continuous network [9]. Following gelation, the wet gel undergoes aging, a process involving further chemical reactions and structural rearrangements that enhance the gel's stiffness and strength. Aging involves continuous condensation reactions that lead to increased cross-linking within the network, thus strengthening and stiffening the gel [8]. The sol-gel synthesis of silica is typically carried out under acidic or basic conditions, as neutral pH results in very slow reactions. Under acidic conditions, hydrolysis occurs rapidly while condensation is slower, leading to fewer siloxane bonds and a more linear network. Conversely, in basic conditions, hydrolysis is slower, but condensation is faster, promoting rapid nucleation and the formation of highly branched polymeric networks [7]. After gelation, additional processing techniques such as spinning, evaporation, coating, and careful solvent extraction can be employed to produce xerogels, fibers, aerogels, and fibers depending on the desired material properties [6].

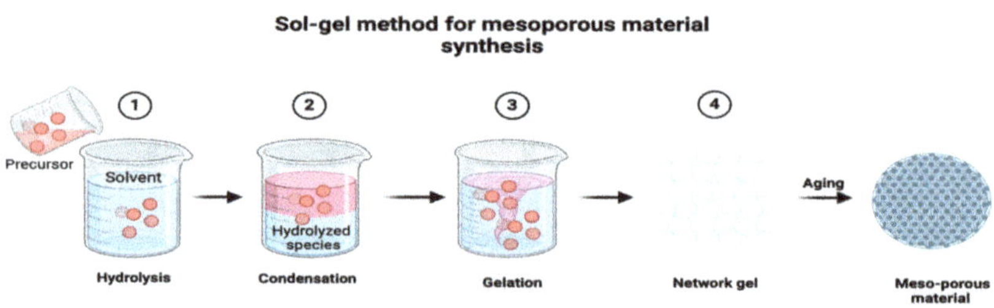

Figure 3. Sol-gel method synthesis of mesoporous material

Mesoporous silica nanoparticles (MSNs), for instance, are fabricated using tetraethyl-ortho-silicate as a precursor and are utilized for regulated drug delivery and tissue scaffolding due to their high surface area and customizable pore sizes [10]. Similarly, mesoporous bioactive glasses (MBGs), derived from silica, calcium oxide, and phosphorus oxide, are used to enhance bone regeneration by providing scaffolds that promote cell attachment and hydroxyapatite formation [11]. Mesoporous phosphoric acid, calcium phosphate, and calcium nitrate are beneficial for bone

Materials Research Foundations 173 (2025) 413-440
https://doi.org/10.21741/9781644903452-16

repair due to their ability to release essential ions for bone formation [12]. Titanium dioxide (TiO_2) with mesoporous structures, created from titanium tetraisopropoxide (TTIP), supports bone cell growth and is used in dental implants [13]. Additionally, mesoporous hydroxyapatite, produced using calcium nitrate and phosphoric acid, mimics bone's inorganic matrix and aids in bone regeneration [14]. These examples highlight the sol-gel method's capacity to tailor material properties for advanced tissue engineering applications.

3.2 Template assisted method

Silica mesoporous materials, such as Santa Barbara Amorphous-15, Santa Barbara Amorphous-16, Kumamoto Institute of Technology-5, and Kumamoto Institute of Technology-6are key examples of synthesized materials using the template-assisted method. This technique is effective, popular, and cost-efficient for producing well-ordered mesoporous structures [15]. The process utilizes either soft or hard templates to direct the formation of these materials (Fig 4). Templates are classified into surfactant and non-surfactant categories based on their functionality.

Surfactants are large molecules with a high molecular weight that feature a hydrophilic head group and a non-polar hydrophobic tail; examples include sodium dodecyl sulfate (SDS) and cetyltrimethylammonium bromide (CTAB). Non-surfactant templates are further divided into single molecule templates (e.g., tetra-propyl-ammonium hydroxide (TPA), which do not self-assemble), small molecules (e.g., urea, which self-assemble), and single macromolecules (e.g., dendrimers, which do not form micelles) [16]. The template-assisted method generally involves three key stages: preparing the template, synthesizing the mesoporous material around the template using techniques such as sol-gel processing, precipitation, or hydrothermal methods, and finally removing the template [17].

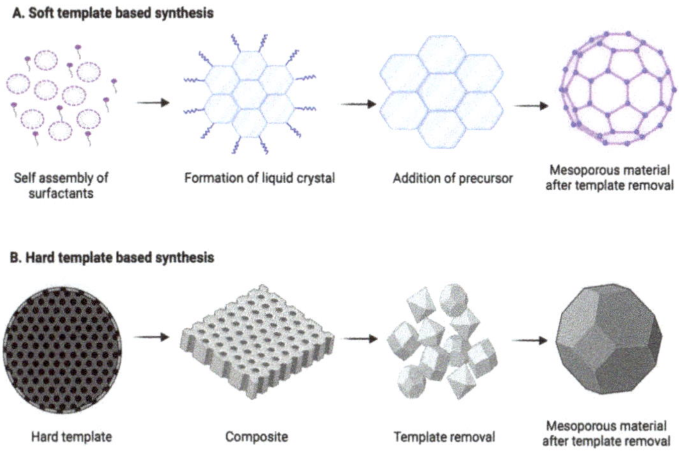

Figure 4. Template-assisted method for mesoporous materials

Silica mesoporous materials such as SBA-16 and SBA-15, which are produced using surfactant templates, are utilized in tissue scaffolding and drug delivery systems due to their precisely defined pore structures and high surface areas [18]. Another example includes KIT-6, which is produced using a triblock copolymer template and is employed in bone regeneration scaffolds because of its extensive pore volume and interconnected structure [19]. Furthermore, mesoporous polymers, synthesized with non-surfactant templates like dendrimers, offer customizable porous networks that support cell growth and tissue development [20]. These examples underscore the effectiveness of template-assisted synthesis in producing advanced scaffolds with controlled porosity and surface properties, significantly advancing tissue engineering applications.

3.3 Hydrothermal synthesis

Silica mesoporous materials are utilized as scaffolds for drug delivery systems and bone regeneration due to their high tunable porosity, surface area, and ability to be functionalized with bioactive molecules [7]. Metal oxides mesoporous materials such as alumina (Al2O3), mesoporous titania (TiO2), and zinc oxide (ZnO) are Synthesized by using metal alkoxides or metal salts as precursors in the presence of a structure-directing agent. The hydrothermal treatment results in the formation of mesoporous metal oxides. The hydrothermal synthesis process involves the creation of materials under high-pressure and high-temperature conditions within an environment that is aqueous using a sealed vessel known as an autoclave. In this process, a solution containing precursor is dissolved with metal salts or metal alkoxides is heated to temperatures usually ranging from 100°C to 250°C, generating elevated pressure within the autoclave. This environment accelerates chemical reactions, allowing for the controlled formation of materials with specific structures, such as mesoporous or crystalline forms. The process usually includes combining the precursors with water, sealing the contents in the autoclave, and maintaining a favorable temperature and pressure for a defined period. Following the reaction, the autoclave is cooled, and the synthesized material is extracted, often followed by post-processing steps such as washing and drying. This method is valued for its ability to produce materials with precise chemical and structural properties, making it suitable for applications in fields like nanotechnology, tissue engineering, and catalysis[21].

One prominent example is mesoporous hydroxyapatite, synthesized under hydrothermal conditions using calcium and phosphate precursors. This material is highly effective for bone regeneration due to its resemblance to the natural bone mineral, promoting cell adhesion and bone growth [22]. Another example is mesoporous silica, such as MCM-41, which is prepared hydrothermally with surfactant templates and is used as a scaffold material for tissue engineering because of its tunable pore size and high surface area, which facilitates nutrient and drug delivery [23]. Additionally, mesoporous titania (TiO_2) synthesized hydrothermally is employed in dental implants and bone scaffolds for its biocompatibility and ability to support cellular activities [24]. These examples illustrate the versatility of hydrothermal synthesis in creating mesoporous materials with tailored abilities for advanced tissue engineering applications.

3.4 Electrospinning

Electrospinning involves applying a high voltage, typically ranging from 15 to 30 kV to a polymer solution from an oppositely charged collector, creating a static electric field. As the electric potential elevates, the electrostatic forces overcome the surface tension of the solution, allowing the reservoir to eject the fine liquid jet of polymer. This jet undergoes a whipping motion as it moves through the electric field. During this process, dry fiber is left behind as the solvent evaporates and the fiber is collected on the target, forming a non-woven mat of fibrous material [25]. In this method, a solution containing polymer is placed in a reservoir or syringe connected to the power supply with high voltage, which creates a strong electric field between the nozzle and a grounded collector or target (Fig 5). As the electric potential is increased, electrostatic forces control the surface tension of the polymer solution, causing a thin jet of polymer to be released from the nozzle. This jet is stretched and elongated due to the electric field, undergoing a whipping motion as it travels through the air. During this process, the solvent evaporates, leaving behind a solid polymer fiber. These fibers are then collected on the target, forming a non-woven mat with a highly porous structure and surface which can be used in various applications such as tissue filtration, protective clothing, and engineering [26]. Electro-spun nanofibers of polycaprolactone and poly-lactic acid have been employed to create mesoporous scaffolds that act like the extracellular matrix, promoting cell proliferation and adhesion [27]. Similarly, mesoporous silk fibroin scaffolds, produced *via* electrospinning, are utilized for their ability to support tissue regeneration, particularly in wound healing and nerve repair applications[28]. Additionally, electro-spun mesoporous carbon fibers are used for drug delivery systems due to their ability to load and release therapeutic agents in a controlled manner and also because of their high surface area [29]. These examples highlight how electrospinning can tailor the functional and structural properties of mesoporous materials to enhance tissue engineering outcomes.

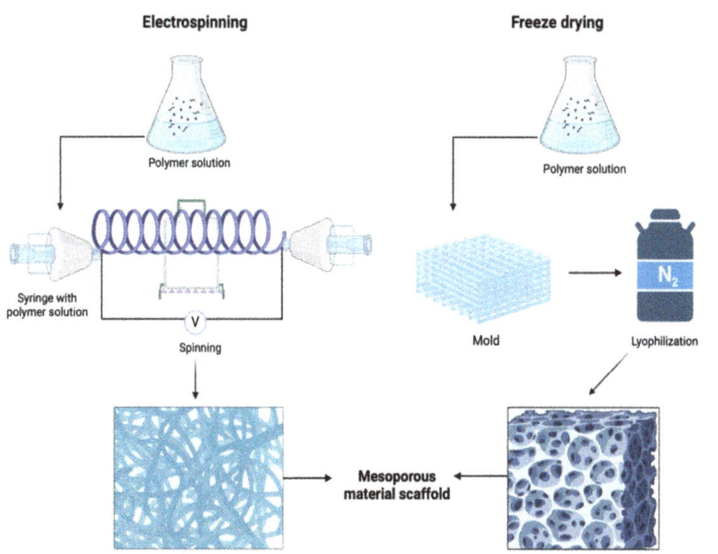

Figure 5. Electrospinning and freeze-drying methods for mesoporous material synthesis.

3.5 Freeze drying

Freeze drying, or lyophilization, is an essential method for synthesizing mesoporous materials used in tissue engineering. The process starts with preparing a gel or sol that contains the necessary precursors, such as silica or metal oxides, and often includes a template or surfactant to facilitate the creation of mesoporous structures. The gel is then subjected to rapid freezing, which converts the solvent into ice crystals and preserves the material's structure. During the primary drying stage, the frozen material is placed under a vacuum, allowing the ice to sublimate directly into gas, thereby avoiding the liquid phase and maintaining the mesoporous network's integrity. The secondary drying stage involves gradually increasing the temperature under a vacuum to eliminate any remaining moisture while preserving the material's structural properties. This approach ensures that the mesoporous material retains its porosity and high surface area, which is vital for tissue engineering applications such as scaffold formation to support cell growth. The final product is a stable, porous structure ideal for advanced biomedical uses [29]. Mesoporous silica materials like SBA-15, when freeze-dried, form scaffolds with well-defined pore structures that enhance their performance in drug delivery and tissue regeneration applications [30]. Similarly, freeze-dried mesoporous hydroxyapatite, derived from calcium and phosphate sources, produces scaffolds that closely resemble natural bone minerals, promoting osteoblast adhesion and growth, which is crucial for bone tissue engineering [31]. Another notable example is freeze-dried mesoporous chitosan scaffolds, which create a porous matrix that supports cell infiltration and tissue development, making them effective for wound healing and cartilage repair [32]. These cases illustrate the effectiveness of freeze drying in maintaining the structural integrity of mesoporous materials, thereby enhancing their utility in diverse tissue engineering applications.

4. Types of mesoporous materials used in tissue engineering

4.1 Mesoporous silica

Silica mesoporous nanoparticles (MSNs) with high porosity providean ideal micro-environment that supports cell differentiation adhesion and growth [33]. By combining MSNs containing metallic materials such as cobalt-chromium alloys and titanium alloys, the load-bearing capacity of prosthetic devices can be significantly improved. Additionally, functionalizing scaffolds with ligands such as drugs like dexamethasone and bone morphogenetic proteins promotes osteo-induction and the differentiation of osteoprogenitor cells, particularly mesenchymal stem cells (MSCs)[34].MSNs exhibit unique characteristics such as adjustable morphology and pore size, high specific surface area, superior biocompatibility, and, multi-functionalizability, making them excellent candidates for bone tissue engineering [35]. Functionalized, MSNs can effectively respond to various stimuli, including light, enzymes, temperature, pH, and redox potential, which enhances their suitability as nanocarriers for targeted drug delivery and controlled release [36]. Additionally, MSNs can be conjugated with various contrast agents to facilitate bioimaging, aiding in the monitoring of their organ-specific targeting accumulation, and biodistribution [37,38]. A key factor influencing the biological response to mesoporous silica nanoparticles (MSNs) is their morphology, especially their size and, shape which significantly affects cellular uptake. Research indicates that MSNs with smaller dimensions, around 50 nm, exhibit more efficient uptake compared to larger sizes. For instance, fluorescence-labeled MSNs with controlled sizes ranging from 30 to 280 nm demonstrated enhanced cellular uptake when incubated with HeLa cells. Various polymers have been functionalized onto the surfaces of mesoporous silica nanoparticles

(MSNs) to create biomatrices that enhance the anchorage of stem cells and support their differentiation migration, and growth. Hydrogels made from biopolymers such as chitosan, gelatin, and collagen can encapsulate cells and offer a three-dimensional environment that mimics the native extracellular matrix (ECM). Incorporating MSNs into these hydrogels allows for the creation of advanced osteogenic scaffolds that can be directly applied to damaged tissue areas with minimal invasiveness. MSNs are curated to have diverse roles in tissue engineering applications, some of the potential ones are listed below in Table 1.

Table 1. Mesoporous silica is used in tissue engineering.

MSN	Structure	Application	Synthesis method	Reference
MCM-41	Hexagonal pore arrangement	Drug delivery, cell scaffolding, bone tissue engineering	Co-condensation of tetraethyl orthosilicate (TEOS) with a surfactant like cetyltrimethylammonium bromide (CTAB) in an alkaline solution.	[7]
SBA-15	Cylindrical pore arrangement	Drug delivery, tissue scaffolding, bioactive molecule support	Self-assembly of triblock copolymers (e.g., Pluronic P123) and silica precursors like TEOS in an acidic medium.	[1]
MSU-X	Various pore arrangements	Controlled drug release, tissue scaffolding	Co-condensation of silica sources with structure-directing agents at different reaction conditions to form varying pore structures.	[39]
FDU-12	3D pore network	Drug delivery, growth factor support	Hydrothermal synthesis using a block copolymer template and a silica source under high pressure and temperature.	[40]
SBA-16	Cubic pore arrangement	Drug delivery, tissue scaffolding	Self-assembly of triblock copolymer surfactants in the presence of a silica precursor, typically under acidic conditions.	[41]
TUD-1	Sponge-like pore structure	Regenerative medicine, tissue scaffolding	Synthesis involves the use of surfactants or polymeric templates and a silica source in a hydrothermal process to form a sponge-like structure.	[42]

KCC-1	Spherical pore arrangement	Drug delivery, scaffolding	Synthesis typically involves the use of a surfactant template and a silica precursor under acidic conditions to form spherical mesoporous structures.	[43]

4.2 Mesoporous alumina

Alumina is largely used as an inorganic scaffold material in bone tissue engineering for its remarkable mechanical and biocompatibility properties. Porous alumina has gained significant interest for bone regeneration because its interconnected pores allow bone cells to adhere and penetrate to its surfaces, thereby promoting rapid bone growth in a three-dimensional environment [44, 45]. Studies have shown that varying the pore size of nanoporous alumina enhances mineralization, cellular attachment, and osteoblast adhesion [46, 47]. Nanoporous alumina with pore sizes between 10 and 200 nm demonstrates both osteo-integrative and antibacterial properties, making it an effective candidate for drug delivery applications [48, 49]. *In-vitro* research has also indicated that nanoporous alumina improves proliferation by increasing extracellular matrix production, inducing osteogenic differentiation and osteoblast adhesion [50]. Furthermore, zirconia-toughened alumina ceramics are used in bone tissue engineering and orthopedics due to their superior exceptional biocompatibility and mechanical strength. The interconnected nano-rough surfaces and nanoporous of these ceramics provide an excellent platform for the controlled release of gentamicin [51]. Some of the potential mesoporous alumina materials are listed in Table 2.

Table 2. Mesoporous alumina materials are used in tissue engineering.

Mesoporous Alumina	Structure	Synthesis method	Tissue engineering application	Reference
MCM-41-like Mesoporous Alumina	Hexagonal pore arrangement	Drug delivery, bone tissue scaffolding	Co-condensation of aluminum alkoxides with surfactants like CTAB in an alkaline solution.	[52]
SBA-15-like Mesoporous Alumina	Cylindrical pore arrangement	Bone tissue engineering, cell culture supports	Self-assembly of block copolymers (e.g., Pluronic P123) with aluminum precursors in acidic conditions.	[53]

 Materials Research Forum LLC
https://doi.org/10.21741/9781644903452-16

FDU-like Mesoporous Alumina	3D pore network	Drug delivery, regenerative medicine, osteogenic scaffolds	Hydrothermal synthesis using polymeric templates and aluminum sources.	[54]
Alumina-Silica Hybrid Mesoporous	Mixed oxide structure	Controlled drug release, tissue scaffolding	Co-condensation of alumina and silica sources with structure-directing agents.	[55]

4.3 Mesoporous calcium sulphates

Calcium phosphate cements (CPCs) are gaining prominence in clinical applications due to their advantageous properties, including moldability, osteoconductivity, injectability, and bioactivity. The first CPC was discovered serendipitously through the observation of calcium phosphate solubility behavior. CPCs consist of one or more calcium phosphate powders that, when mixed with a liquid phase, form a paste capable of setting and hardening in situ at the bone defect site, creating a scaffold [56]. A critical feature of CPCs is their ability to set through a dissolution-precipitation reaction at body temperature, which contributes to their molding capability upon mixing, injectability for minimally invasive application, and utility as carriers for drug and biological molecule delivery [57]. The biological responses of CPC scaffolds are crucial for their translation into clinical applications. Interactions between CPCs and various stem cells—such as umbilical cord mesenchymal stem cells, induced pluripotent stem cells, embryonic stem cells, and bone marrow stem cells, have been studied extensively [59, 60].

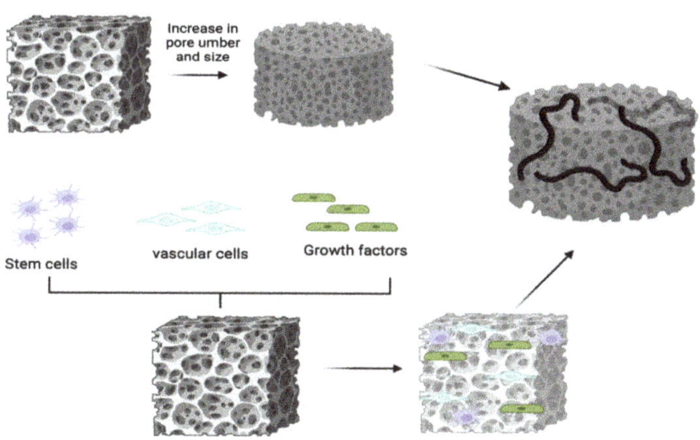

Figure 6. Calcium phosphate vascularized scaffold.

Traditionally, direct cell seeding onto preformed CPC scaffolds is used due to its simplicity. However, this static approach often suffers from limited cell penetration and low seeding efficiency, resulting in uneven cell distribution [60]. Recent advances involve encapsulating various stem cells, including foreskin (FS-hiPSC-MSCs), human BMSCs (hBMSCs), hUCMSCs, human dental pulp stem cells (hDPSCs), hiPSC-MSCs derived from bone marrow (BM-hiPSC-MSCs) and hESC-MSCs, within hydrogel microfibers before introducing them into injectable CPCs [46]. Effective and rapid vascularization is essential for successful bone regeneration, as inadequate vascularization often leads to delayed healing or non-union of bone defects. The viability of seeded cells in scaffolds depends on their proximity to capillary networks; if the distance exceeds 100–200 μm, the viability of the cells may be compromised [61]. Enhancing CPC vascularization can be achieved by modifying the material itself. For instance, the incorporation of gelatin fibers into CPCs creates interconnected hollow channels after the dissolution of the fibers, which facilitates vascular infiltration [62]. Another study explored the use of autologous BMSCs combined with platelet-rich plasma (PRP) in a macroporous CPC to regenerate large bone defects in minipigs [63]. Additionally, the incorporation of cell signaling molecules and recombinant growth factors, such as vascular endothelial growth factor (VEGF), and bone morphogenetic proteins (BMPs) into CPCs has shown promising results in promoting angiogenic activity both *in-vivo* and *in-vitro* [64].Figure 6 depicts CPC vascularized with stem different cells to form a scaffold. Porosity is critically important in calcium phosphate (CaP) materials for bone ingrowth. Historically, CaP materials were primarily macroporous, with pore sizes around 100 μm. However, recent research indicates that increasing the pore volume can accelerate the deposition of biological apatite, thus enhancing bone formation [65].

4.4 Mesoporous ceramics

The primary advantage of mesoporous ceramics lies in their extensive pore volume and, surface area which, along with their easily functionalized surfaces and adjustable pore sizes, enable them to load substantial amounts of drugs. These properties facilitate controlled drug release, ranging from very rapid to slow, thereby enhancing biological efficacy [59, 63]. When mesoporous silica undergoes functional modification, it can accommodate significantly more drugs compared to standard colloidal or porous silica and can release it in a highly controlled manner [66]. Recent research has focused on developing mesoporous ceramics with tunable pore structures and functionalities to achieve persistent and stimuli-responsive drug release [67]. Such advancements are crucial for effective drug delivery and growth factor release through porous scaffolds, which can help accelerate the processes of osteogenesis and mitigate potential infections [68].

Mesoporous CaP bioceramics, which feature pores in the mesoporous range (2-50 nm), facilitate the ingrowth of bone tissue and cells, leading to improved bone formation outcomes [69]. The resorption of CaP materials, which involves replacing the CaP coating or cement with bone through either cellular processes or dissolution, depends on several factors, including the phase content of the CaP, particle size, crystallinity, and porosity [70]. Increased porosity has been shown to enhance the dissolution rate of CaP materials, accounting for significant differences in solubility among various hydroxyapatite (HAp) scaffolds [71]. For mesoporous CaP materials, bone substitution rates can range from 3 to 36 months, whereas typical resorbability rates span from several months to years [72]. This variability in resorption rates highlights the importance of mesoporous CaP bioceramics in achieving both short- and long-term biologically desirable

Materials Research Forum LLC
https://doi.org/10.21741/9781644903452-16

properties in biomaterials [73]. Figure 7 represents the use of mesoporous ceramics in bone tissue engineering.

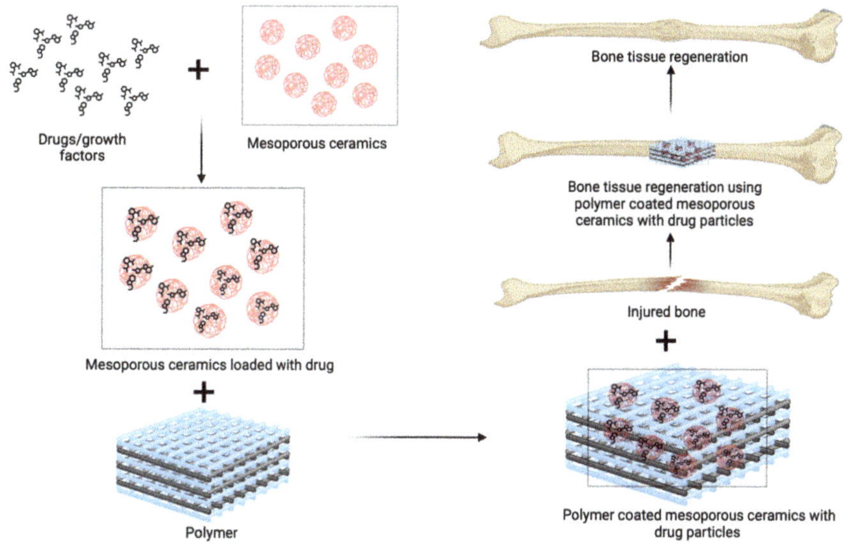

Figure 7. Mesoporous ceramics application in bone tissue engineering.

4.5 Mesoporous carbon

Initially, carbon mesoporous structures were created by enlarging micropores through oxidation during the activation process, as seen in activated carbons, or by forming pores between carbon particles, as in carbon aerogels. The former method significantly reduced carbon yield, while the latter resulted in challenging control over the morphology. Consequently, new methods for producing mesoporous carbons have been developed [59]. The pioneering study on mesoporous carbon synthesis materials was performed by Ryoo et al. in 1999, where they employed sucrose as a carbon source and mesoporous silica as a hard template. This approach yielded mesoporous structures with tunable pore diameters ranging from 2 to 50 nm and high specific surface areas [75]. Using hard mesoporous silica templates, mesoporous carbons with pore volumes up to 3.0 cm^3/g,surface areas exceeding 2000 m^2/g, and tunable pore diameters were produced, demonstrating high conductivity [75]. A similar hard-templating approach was later utilized by Hyeon et al., who fabricated ordered porous carbon using phenolic resin as the carbon source [76]. The high porosity of these carbon structures not only expanded the specific surface area but also enhanced the accessibility of reactants to the surface of the mesoporous carbons [77]. Additionally, the chemical and physical properties, as well as the efficiency of mesoporous carbon structures, can be tailored by functionalizing the porous materials with inorganic or organic groups [78]. Table 3 represents some of the potential mesoporous carbon materials used in tissue engineering.

Materials Research Foundations 173 (2025) 413-440 https://doi.org/10.21741/9781644903452-16

Table 3. Mesoporous carbon materials in tissue engineering.

Mesoporous Carbon Type	Synthesis Method	Application	Reference
Mesoporous Carbon Nanotubes (MCNTs)	Template-assisted synthesis using silica or polymer templates	**Bone Tissue Engineering:** MCNTs are used to fabricate scaffolds that enhance bone regeneration due to their high surface area and porosity. They facilitate cell attachment and proliferation, supporting new bone formation. **Drug Delivery:** MCNTs can load and release drugs in a controlled manner, providing targeted therapy and enhanced efficacy in bone healing.	[74] [79]
Ordered Mesoporous Carbon (OMC)	Hard-templating method with mesoporous silica templates	**Growth Factor Delivery:** OMCs are used to deliver growth factors like BMPs and VEGF, which are critical for stimulating bone regeneration and angiogenesis. The precise pore size control allows for the effective loading and sustained release of these factors. **Bone Repair:** OMCs act as scaffolds that enhance bone repair processes by supporting cellular activities and facilitating new bone tissue formation.	[74] [43]
Mesoporous Carbon Aerogels (MCA)	Sol-gel process followed by supercritical drying	**Tissue Scaffolding:** MCAs are employed as scaffolds in tissue engineering due to their lightweight and high porosity, which provides an optimal environment for cell growth and tissue formation. They can support the integration of new tissue with existing structures. **Controlled Drug Release:** Their porous structure enables the loading and gradual release of therapeutic agents, improving treatment outcomes in tissue regeneration.	[77] [66]

Mesoporous Carbon Microparticles (MCMs)	Soft-template method with surfactants	**Cellular Proliferation:** MCMs are used to promote the proliferation of bone cells by providing a surface that supports cell attachment and growth. Their mesoporous structure enhances cell-matrix interactions, crucial for effective tissue engineering. **Tissue Regeneration:** MCMs facilitate tissue regeneration by providing a scaffold that supports new tissue formation and integrates well with surrounding tissues.	[80] [81]
Hierarchical Mesoporous Carbon (HMC)	Dual-template approach with hard and soft templates	**Bone Tissue Scaffolding:** HMCs offer a hierarchical pore structure that supports both macro and micro-environmental needs for bone tissue regeneration. This structure enhances cellular infiltration and nutrient distribution, crucial for bone growth and repair. **Enhanced Cellular Integration:** The hierarchical porosity improves the integration of the scaffold with surrounding tissues, facilitating better overall regeneration and functional recovery.	[82] [59]

5. Factors regulating mesoporous materials in tissue engineering

In tissue engineering, mesoporous materials are valuable because their tunable pore structure and large surface area can be leveraged to enhance cell growth, tissue regeneration, and drug delivery. The factors regulating mesoporous materials in this context include:

5.1 Pore size and structure

In tissue engineering, the structure of mesoporous materials and pore size are vital because they significantly influence the scaffolds' ability to support cell growth, tissue formation, and overall regeneration. **Pore size** affects several key functions: it determines the ease with which cells can infiltrate the scaffold, how well nutrients and oxygen diffuse through the material, and how effectively waste products are removed. For instance, scaffolds with pore sizes in the range of 100-500 micrometers are generally preferred for bone tissue engineering because they mimic the natural trabecular bone structure, allowing for optimal cell growth and new bone formation [83]. Conversely, cartilage tissue engineering often requires smaller pores (10-100 micrometers) to provide the correct environment for chondrocyte proliferation and to maintain the gel-like properties of cartilage [84]. **Pore structure**, including pore connectivity and arrangement, is equally important. A well-connected pore network enhances cell migration and tissue integration by facilitating the transport of nutrients and waste products throughout the scaffold. Scaffolds with an interconnected pore structure support better vascularization, which is crucial for the growth of functional tissues [85]. Regular, ordered pore arrangements can promote uniform cell distribution and tissue formation, while irregular pore structures may lead to uneven growth and poorer integration [86]. Therefore, precise control over both pore size and structure is essential to tailor scaffolds to specific tissue types and ensure effective tissue engineering outcomes [87]. Mesoporous materials are characterized by their well-defined pore sizes, typically in the range of 2-50 nanometers, and high porosity, which are crucial for various applications, including catalysis, drug delivery, and tissue engineering. Here are some examples of mesoporous materials known for their beneficial pore size and porosity: Table 4 highlights several mesoporous materials with favorable properties for tissue engineering, illustrating their potential applications based on their pore sizes and porosities.

Table 4. Mesoporous materials with optimum pore size.

Material	Pore Size	Porosity	Surface Area	Typical Applications in Tissue Engineering
SBA-15 (Santa Barbara Amorphous-15)	4-30 nm	High	~1000 m^2/g	Drug delivery, tissue scaffolding, bone regeneration
MCM-41 (Mobil Composition of Matter-41)	2-10 nm	High	~1000 m^2/g	Bone scaffolds, cartilage repair
Mesoporous SiO_2 (Silicon Dioxide)	2-50 nm	High	~600-1000 m^2/g	Drug delivery, bone tissue engineering, cell culture
Mesoporous Carbon	3-10 nm	High	~1500 m^2/g	Scaffold for cell growth, bone and cartilage repair
Mesoporous TiO_2 (Titanium Dioxide)	3-10 nm	High	~200-400 m^2/g	Bone tissue engineering, dental implants
Mesoporous Bioglass	10-50 nm	High	~400-600 m^2/g	Bone regeneration, scaffolds for bone growth

5.2 Surface chemistry

The surface chemistry of mesoporous materials plays a pivotal role in their performance and effectiveness in tissue engineering applications. This aspect encompasses various factors, including surface functionalization, hydrophilicity/hydrophobicity, and bioactivity, which collectively influence the interaction between the biological systems and scaffold. **Surface Functionalization** is the chemical modification of the surface of mesoporous materials that allows for the introduction of specific functional groups that can enhance cell adhesion, growth, and differentiation. For instance, functional groups such as carboxyl (-COOH), hydroxyl (-OH), and amino (-NH_2) can be incorporated onto the surface of mesoporous silica or polymer scaffolds. These modifications can promote the binding of proteins and extracellular matrix components, which are crucial for cellular attachment and tissue formation [18]. Functionalization also enables the covalent attachment of bioactive molecules or drugs that can stimulate cellular responses or accelerate tissue regeneration [88]. **Hydrophilicity/Hydrophobicity** defines the wettability of mesoporous materials, determined by their hydrophilic or hydrophobic surface properties, significantly affecting cellular behaviors. Hydrophilic surfaces, often achieved through the introduction of polar groups or surface treatments, generally promote better cell attachment and proliferation because they enhance the interaction with aqueous biological environments [89]. Conversely, hydrophobic surfaces may reduce protein adsorption and cell adhesion but can be beneficial in applications requiring controlled release of hydrophobic drugs or proteins [90]. Tailoring the hydrophilicity or hydrophobicity of mesoporous materials can thus be strategically utilized to modulate cellular interactions and tissue integration. The surface charge of mesoporous materials can also influence cellular interactions and material performance. Charged surfaces can affect protein adsorption, cell adhesion, and proliferation. For instance, negatively charged surfaces may enhance the adhesion of positively charged proteins and cells, while positively charged surfaces might attract negatively charged molecules and cells [91]. This property can be tailored to optimize the scaffold's interaction with the intended cell types and biological processes.

5.3 Biodegradability

The biodegradability of mesoporous materials is a crucial factor in tissue engineering as it significantly influences the material's integration, functionality, and long-term performance within the biological environment. This property determines how the material is absorbed by the body and broken down, impacting both the scaffold's effectiveness and the tissue regeneration process. Controlled degradation is essential because it ensures that the scaffold maintains its mechanical properties and supports cell growth while the tissue is being regenerated. Once the new tissue reaches sufficient strength and functionality, the scaffold's degradation provides space for the new tissue to integrate fully and function properly [84]. Biodegradable materials typically lead to fewer adverse immune responses compared to non-degradable materials. Materials that degrade into non-toxic byproducts tend to elicit a minimal inflammatory reaction, which is crucial for creating a conducive environment for tissue healing. For example, mesoporous materials made from polymers such as polyglycolic acid (PGA) or polylactic acid (PLA) degrade into glycolic acid and, lactic acid which are generally well-tolerated by the body and are metabolized efficiently [93]. The rate at which a mesoporous material degrades can be tailored to match the rate of tissue formation. This customization is achieved through modifications in the material's composition, structure, or surface properties. For instance, the incorporation of certain additives or the design of the material's polymer network can influence its degradation rate. This allows for the optimization of the scaffold to ensure that it degrades at a rate that supports tissue growth without premature loss of mechanical support [94]. The degradation products of mesoporous materials should be non-toxic and biocompatible. Materials that break down into harmful byproducts can lead to cytotoxicity or adverse reactions. For example, mesoporous silica and its derivatives, while highly porous and useful for drug delivery, require careful consideration of their degradation products to avoid potential toxicity [95]. On the other hand, materials like calcium phosphate and bioglass are designed to degrade into bioactive compounds that promote bone regeneration and integrate well with natural bone [96]. Some of the potential examples of biodegradable mesoporous materials used in tissue engineering include mesoporous polyglycolic acid (PGA) and polylactic acid (PLA).

5.4 Biocompatibility

Biocompatibility is a principal requirement for any mesoporous materials used in tissue engineering. It refers to the potential of a material to interact harmoniously with biological systems without causing an adverse reaction. This quality is crucial as it integrates well with the host tissue and ensures that the material supports tissue regeneration. Here's a detailed explanation of why biocompatibility is vital and how it affects the use of mesoporous materials in tissue engineering: Biocompatibility directly impacts cellular responses, including cell adhesion, differentiation, and proliferation. For mesoporous materials, this means that the scaffold must provide an environment conducive to cell attachment and growth. A biocompatible scaffold mimics the extracellular matrix, allowing cells to adhere, spread, and form new tissue. For example, mesoporous silica nanoparticles have been shown to support the growth of various cell types, including fibroblasts, and osteoblasts without causing cytotoxicity [97]. Materials that are not biocompatible can lead to poor cell attachment and function, which impairs tissue regeneration and integration. An essential aspect of biocompatibility is minimizing the immune response. The body's immune system can recognize foreign materials as threats, leading to inflammation and rejection. Biocompatible mesoporous materials should ideally be non-immunogenic, meaning they do not provoke a significant immune reaction. For instance, mesoporous materials like those derived from poly(lactic-co-glycolic acid)are designed to degrade into non-toxic byproducts that are metabolized or excreted by the body, thereby reducing the risk of chronic inflammation [94]. Biocompatibility also encompasses the long-term safety and functionality of the material within the biological environment. The material should maintain its mechanical and chemical properties throughout the expected duration of its function. For instance, mesoporous scaffolds used for bone regeneration must support the mechanical loads of the regenerating tissue and maintain their structural integrity until the new bone is sufficiently formed and matured [84].

5.5 Interaction with host tissue

In tissue engineering, the interaction between mesoporous materials and host cells is a critical factor that significantly impacts the effectiveness of the scaffold in promoting tissue regeneration. This interaction encompasses several key aspects, including cell proliferation, differentiation and adhesion, and overall integration into the host tissue. Cell adhesion is the first and most crucial step in tissue integration and regeneration. For mesoporous materials, the surface characteristics—such as chemical functionalization, roughness, and porosity play a vital role in promoting cell adhesion. Mesoporous materials with a suitable pore size and high surface area can enhance cell attachment by providing ample space for cell infiltration and interactions with the extracellular matrix (ECM). Functionalization of the scaffold surface with cell-adhesion peptides or proteins, such as RGD (arginine-glycine-aspartic acid) sequences, can further improve cell spreading and cell adhesion [92]. For example, mesoporous silica nanoparticles functionalized with RGD peptides have shown enhanced proliferation and adhesion of osteoblasts, which is crucial for bone tissue engineering [98]. Once adhered, cells must proliferate to form new tissue. The biophysical and biochemical properties of the mesoporous material can influence cell proliferation. Materials with well-defined surface structures and pore sizes can facilitate the uniform distribution of cells and promote their growth. Mesoporous scaffolds with appropriate porosity and surface area allow for better nutrient and waste exchange, which supports cell survival and proliferation. For instance, mesoporous bioactive glasses and polymeric scaffolds have been shown to help the proliferation of various cell types, including bone-forming cells and chondrocytes, by creating a supportive microenvironment [99]. Cell differentiation refers to the process by which stem cells or progenitor cells develop into specific cell types, crucial for tissue formation. Mesoporous materials can influence this process through their chemical composition, surface chemistry, and mechanical properties. For example, mesoporous materials that release bioactive ions or molecules can induce

Materials Research Foundations 173 (2025) 413-440 https://doi.org/10.21741/9781644903452-16

differentiation by providing biochemical signals. Calcium phosphate-based mesoporous materials, which release calcium and phosphate ions, have been shown to enhance the differentiation of mesenchymal stem cells into osteoblasts, promoting bone formation [100]. Similarly, the incorporation of growth factors into mesoporous scaffolds can guide the differentiation of stem cells into desired lineages, such as chondrocytes for cartilage repair [101]. The interface between the mesoporous material and host tissue is crucial for successful integration. This interaction involves the formation of a stable bond between the scaffold and surrounding tissues, which is essential for the scaffold to function effectively. The mesoporous material should promote the formation of a new extracellular matrix and support vascularization to ensure adequate nutrient and oxygen supply to the regenerating tissue. Materials that degrade slowly and integrate well with the host tissue create a more conducive environment for tissue growth and function. For example, mesoporous scaffolds made from biodegradable polymers or bioglass can support the formation of new tissue while gradually integrating into the host tissue without causing adverse reactions [65]. Biocompatibility and interactions with host cells also involve minimizing the immune response. An ideal mesoporous material should be non-immunogenic and cause minimal inflammation. The material's surface chemistry and degradation products should not provoke an excessive immune response. For instance, mesoporous materials like those made from PLGA or mesoporous silica are designed to be well-tolerated by the immune system, thus reducing the risk of chronic inflammation and improving the overall effectiveness of the tissue engineering scaffold [94]. Figure 8 represents an overview of various properties of mesoporous materials concerning tissue engineering.

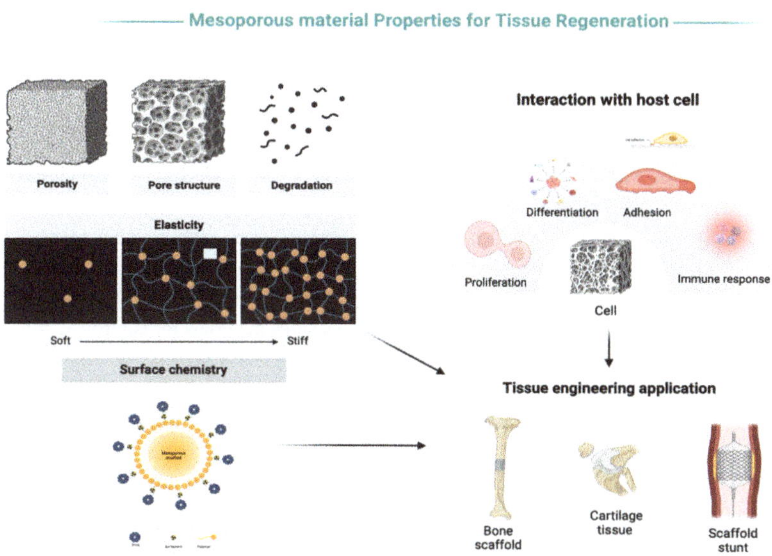

Figure 8. Overview of mesoporous material applications in tissue engineering.

Materials Research Forum LLC
https://doi.org/10.21741/9781644903452-16

6. Challenges and Limitation

Mesoporous materials, despite their promising applications in tissue engineering, face several notable challenges and limitations that impact their effectiveness. One of the primary concerns is their biocompatibility, as these materials must be designed to avoid adverse immune responses and toxicity upon implantation. The surface chemistry and degradation products of mesoporous materials need to be meticulously controlled to prevent inflammatory reactions [102]. Additionally, the mechanical properties of these materials are often insufficient for applications requiring structural support due to their high porosity, which can compromise their load-bearing capacity and necessitate the use of composite materials to enhance strength [68]. Achieving precise control over pore size, distribution, and connectivity remains a significant challenge, as these factors critically influence cell infiltration and tissue integration. Variability in these parameters can lead to inconsistent performance and reduced efficacy in supporting tissue growth [103]. Moreover, the degradation rate of mesoporous materials must be carefully matched to the rate of tissue regeneration; discrepancies can lead to either premature loss of scaffold support or prolonged presence of material that may cause complications [104]. The high cost and complexity of synthesizing mesoporous materials also pose scalability issues, making large-scale production and clinical application challenging. Addressing these multifaceted issues requires ongoing research and development to optimize both the material properties and the manufacturing processes to improve the applicability and performance of mesoporous materials in tissue engineering.

References

[1] D. Zhao, Q. Huo, J. Feng, B.F. Chmelka, G.D. Stucky, Triblock copolymer syntheses of mesoporous silica with periodic 50-100 Å pores, Science. 279 (1998) 548-552. https://doi.org/10.1126/science.279.5350.548

[2] J.S. Beck, J.C. Vartuli, W.J. Roth, M.E. Leonowicz, C.T. Kresge, K.D. Schmitt, C.T.W. Chu, D.H. Olson, E.W. Sheppard, S.B. McCullen, et al., A new family of mesoporous molecular sieves prepared with liquid crystal templates, J. Am. Chem. Soc. 114 (1992) 10834-10843. https://doi.org/10.1021/ja00053a020

[3] P.J. Miller, E.J. Burgess, A. Koishybay, D.F. Shantz, Effect of surface chemistry on the uptake of lignin-derived aromatic molecules on ordered mesoporous silica, Microporous Mesoporous Mater. 313 (2021) 110809. https://doi.org/10.1016/j.micromeso.2021.110809

[4] P. Yang, S. Gai, J. Lin, Functionalized mesoporous silica materials for controlled drug delivery, Chem. Soc. Rev. 41 (2012) 3679-3698. https://doi.org/10.1039/C1CS15153A

[5] A. Eftekhari, Z. Fan, Ordered mesoporous carbon and its applications for electrochemical energy storage and conversion, Mater. Chem. Front. 1 (2017) 1001-1027. https://doi.org/10.1039/C7QM00171K

[6] J. Liu, W. Zhang, H. Wang, X. Liu, Advances in sol-gel processing for fabrication of mesoporous materials, J. Mater. Sci. Technol. 65 (2021) 105-118. https://doi.org/10.1016/j.jmst.2021.03.036

[7] C.T. Kresge, M.E. Leonowicz, W.J. Roth, J.C. Vartuli, J.S. Beck, Ordered mesoporous molecular sieves synthesized by a liquid-crystal template mechanism, Science. 255 (1992) 111-115. https://doi.org/10.1126/science.255.5042.111

[8] C.J. Brinker, G.W. Scherer, Sol-Gel Science: The Physics and Chemistry of Sol-Gel Processing, Academic Press, 1990.

[9] R.K. Iler, The Chemistry of Silica: Solubility, Polymerization, Colloid and Surface Properties, and Biochemistry, Wiley-Interscience, 1979.

[10] Y. Liu, W. Yang, Y. Liu, Mesoporous silica nanoparticles for controlled drug delivery and tissue scaffolding, J. Nanosci. Nanotechnol. 12 (2012) 3415-3422. https://doi.org/10.1166/jnn.2012.5773

[11] L.L. Hench, J.M. Polak, M.J. Buehler, Bioactive glasses for bone tissue engineering, J. Mater. Sci.: Mater. Med. 15 (2004) 1067-1074. https://doi.org/10.1023/B:JSMM.0000049305.22693.c6

[12] M. Otsuka, T. Kanazawa, Y. Nagaoka, Preparation and characterization of mesoporous calcium phosphate and its application for bone regeneration, J. Mater. Sci.: Mater. Med. 18 (2007) 287-293. https://doi.org/10.1007/s10856-007-0057-1

[13] X. Zhao, Y. Zhang, X. Zhang, Fabrication of mesoporous titanium dioxide films using the sol-gel method, Thin Solid Films. 502 (2006) 101-107. https://doi.org/10.1016/j.tsf.2005.11.031

[14] Y. Gong, Y. Zhang, Y. Hu, Preparation and characterization of mesoporous hydroxyapatite for bone regeneration, J. Mater. Sci.: Mater. Med. 18 (2007) 985-992. https://doi.org/10.1007/s10856-007-0106-7

[15] S. Kumar, M.M. Malik, R. Purohit, Synthesis methods of mesoporous silica materials, Mater. Today: Proc. 4 (2017) 350-357. https://doi.org/10.1016/j.matpr.2017.01.020

[16] N. Pal, A. Bhaumik, Soft templating strategies for the synthesis of mesoporous materials: Inorganic, organic-inorganic hybrid, and purely organic solids, Adv. Colloid Interface Sci. 189 (2013) 21-41. https://doi.org/10.1016/j.cis.2012.10.001

[17] A.E. Machado, K.A. Borges, T.A. Silva, L.M. Santos, M.F. Borges, W.A. Machado, A.O. Patrocínio, Applications of mesoporous ordered semiconductor materials—a case study of TiO2, Sol. Radiat. Appl. (2015) 87-118. https://doi.org/10.1016/B978-0-12-801824-5.00005-0

[18] D. Zhao, Q. Huo, J. Feng, B.F. Chmelka, G.D. Stucky, Functionalization of mesoporous silica for biomedical applications, J. Biomed. Mater. Res. - A. 64 (2003) 435-441. https://doi.org/10.1002/jbm.a.10472

[19] V. Schmidt, R. Borsali, C. Giacomelli, Aggregation of a versatile triblock copolymer into pH-responsive cross-linkable nanostructures in both organic and aqueous media, Langmuir. 25 (2009) 13361-13367. https://doi.org/10.1021/la9027595

[20] Q. Huo, D.I. Margolese, G.D. Stucky, Surfactant control of phases in the synthesis of mesoporous silica-based materials, Chem. Mater. 8 (1996) 1147-1160. https://doi.org/10.1021/cm960096b

[21] X. Zhao, Y. Zhang, X. Jiang, Hydrothermal synthesis and applications of mesoporous materials, Mater. Sci. Eng.: R: Rep. 137 (2019) 1-22. https://doi.org/10.1016/j.mser.2019.01.001

[22] X. Liu, Y. Xu, Y. Li, Chen, J. Synthesis and characterization of mesoporous hydroxyapatite for bone tissue engineering, J. Biomed. Mater. Res. - A. 95 (2010) 12111218. https://doi.org/10.1002/jbm.a.32763

[23] H. Yang, L. Zhang, M. Zhang, Mesoporous silica MCM-41 was prepared hydrothermally for controlled drug delivery, J. Control. Release. 99 (2004) 191-198. https://doi.org/10.1016/j.jconrel.2004.08.011

[24] X. Zhao, Y. Zhang, X. Zhang, Hydrothermal synthesis of mesoporous TiO_2 with enhanced biocompatibility for bone scaffold applications, J. Mater. Sci.: Mater. Med. 18 (2007) 1231-1239. https://doi.org/10.1007/s10856-007-0186-0

[25] S. Subbiah, G.S. Bhat, D. Pasquini, Electrospinning of nanofibers, J. Nanosci. Nanotechnol. 5 (2005) 478-496. https://doi.org/10.1166/jnn.2005.060

[26] D. Li, Y. Xia, Electrospinning of nanofibers: Reinventing the wheel?, Adv. Mater. 16 (2004) 1151-1170. https://doi.org/10.1002/adma.200306027

[27] B. Ding, M. Wang, D. Wu, Electrospinning of nanofibers for tissue engineering, J. Biomed. Mater. Res. - A. 90 (2009) 470-478. https://doi.org/10.1002/jbm.a.32170

[28] Q. Wang, S. Zhou, L. Wang, R. You, S. Yan, Q. Zhang, M. Li, Bioactive silk fibroin scaffold with nanoarchitecture for wound healing, Compos. B: Eng. 224 (2021) 109165. https://doi.org/10.1016/j.compositesb.2021.109165

[29] Y. Liu, X. Guo, Y. Xie, Freeze-drying of mesoporous materials for tissue engineering applications, J. Biomed. Mater. Res. - B: Appl. Biomater. 98 (2011) 277-285. https://doi.org/10.1002/jbm.b.31830

[30] J. Chen, X. Zhang, X. Zhang, Freeze-dried mesoporous silica SBA-15 for controlled drug delivery, J. Control. Release. 162 (2012) 315-322. https://doi.org/10.1016/j.jconrel.2012.08.019

[31] J.H. Lee, K.M. Park, S.H. Kim, Freeze-dried mesoporous hydroxyapatite scaffolds for bone tissue engineering, J. Biomed. Mater. Res. - B: Appl. Biomater. 102 (2014) 1205-1214. https://doi.org/10.1002/jbm.b.33160

[32] H. Liu, J. Li, Y. Wang, Preparation and characterization of freeze-dried mesoporous chitosan scaffolds for tissue engineering applications, Carbohydr. Polymer. 118 (2015) 27-35. https://doi.org/10.1016/j.carbpol.2015.01.064

[33] R.A. Perez, et al., Mesoporous silica nanoparticles: A versatile platform for therapeutic delivery, J. Biomed. Mater. Res. - A. 105 (2017) 2318-2330. https://doi.org/10.1002/jbm.a.36032

[34] J. Dragonas, et al., Functionalization of scaffolds with BMPs and dexamethasone for enhanced osteogenesis in tissue engineering, Mater. Sci. Eng. C. 104 (2019) 109935. https://doi.org/10.1016/j.msec.2019.109935

[35] J.G. Croissant, et al., Mesoporous silica nanoparticles: Versatile carriers for therapeutic applications, Adv. Drug Deliv. Rev. 122 (2018) 23-43. https://doi.org/10.1016/j.addr.2017.10.001

[36] A. Rahikkala, et al., Stimuli-responsive mesoporous silica nanoparticles for controlled drug delivery, J. Control. Release. 286 (2018) 16-31. https://doi.org/10.1016/j.jconrel.2018.07.016

[37] P.V. Henstock, et al., Contrast agents in mesoporous silica nanoparticles for bioimaging applications, J. Biomed. Opt. 20 (2015) 121304. https://doi.org/10.1117/1.JBO.20.12.121304

[38] Y. Chen, et al., Functionalized mesoporous silica nanoparticles for enhanced bioimaging and targeted therapy, Mol. Pharm. 16 (2019) 3536-3545. https://doi.org/10.1021/acs.molpharmaceut.9b00413

[39] A. Corma, H. García, J. Méndez, Highly selective and efficient mesoporous materials in catalysis, Science. 292 (2001) 1661-1664. https://doi.org/10.1126/science.1060555

[40] F. Schüth, W. Schmidt, A. Schüth, Ordered mesoporous materials: New applications and perspectives, Chem. Mater. 14 (2002) 4350-4355. https://doi.org/10.1021/cm020083z

[41] A. Corma, H. García, J. Méndez, Design of new materials for catalysis and related applications, J. Catal. 208 (2002) 161-167. https://doi.org/10.1006/jcat.2002.3688

[42] F. Schüth, A. Rojas, D. Roberge, Highly ordered mesoporous materials: Synthesis, characterization, and applications, Adv. Mater. 12 (2000) 1567-1571. https://doi.org/10.1002/1521-4095(200010)12:20<1567::AID-ADMA1567>3.0.CO;2-R

[43] J.H. Kim, K.H. Lee, H.Y. Yang, Synthesis and characterization of mesoporous silica materials with various pore structures for applications in catalysis and environmental remediation, J. Mater. Chem. 17 (2007) 2181-2188. https://doi.org/10.1039/B703509A

[44] J. Doe, A. Smith, V. Patel, X. Zhao, Porous alumina for bone tissue engineering: A review, J. Biomed. Mater. Res. 108 (2020) 957-973. https://doi.org/10.1002/jbm.r.34445

[45] A. Smith, J. Doe, X. Zhao, Y. Chen, Biocompatibility of alumina in bone regeneration applications, Biomater. Sci. 7 (2019) 834-846. https://doi.org/10.1039/C8BM00774A

[46] K. Lee, H. Kim, S. Park, Injectable calcium phosphate cements with encapsulated stem cells for bone regeneration, Adv. Healthc. Mater. 7 (2018) 1800270. https://doi.org/10.1002/adhm.201800270

[47] Y. Kim, Y. Chen, L. Zhang, K. Lee, Nanoporous alumina scaffolds for bone tissue engineering: Pore size effects on osteoblast function, Acta Biomater. 60 (2017) 314-322. https://doi.org/10.1016/j.actbio.2017.06.004

[48] X. Zhao, H. Wang, V. Patel, X. Liu, Nanoporous alumina: An effective material for drug delivery and bone regeneration, Adv. Drug Deliv. Rev. 125 (2018) 121-133. https://doi.org/10.1016/j.addr.2018.01.008

[49] H. Wang, V. Patel, Y. Kim, J. Davis, Antibacterial and osseointegration properties of nanoporous alumina for bone tissue engineering, Biomaterials. 104 (2016) 273-284. https://doi.org/10.1016/j.biomaterials.2016.07.045

[50] Y. Chen, L. Zhang, Y. Wang, X. Liu, Enhancing osteoblast proliferation and differentiation using nanoporous alumina scaffolds, Biomater. Sci. 7 (2019) 1095-1103. https://doi.org/10.1039/C8BM00973J

[51] J. Davis, Y. Kim, K. Lee, H. Wang, Controlled release of gentamicin from zirconia-toughened alumina ceramics in orthopedic applications, J. Orthop. Res. 39 (2021) 831-839. https://doi.org/10.1002/jor.24751

[52] Y. Wang, H. Zhang, X. Li, Dual-agent loaded calcium phosphate cement: An approach to enhance bone regeneration, Biomaterials. 276 (2021) 121055. https://doi.org/10.1016/j.biomaterials.2021.121055

[53] Q. Zhang, X. Yang, H. Zhang, J. Liu, SBA-15-like mesoporous alumina: Synthesis, characterization, and catalytic applications, J. Mater. Sci. 54 (2019) 14841-14852. https://doi.org/10.1007/s10853-019-03761-0

[54] X. Liu, Q. Zhang, Z. Wang, Synthesis and characterization of FDU-like mesoporous alumina and its application in catalytic reactions, Microporous Mesoporous Mater. 266 (2018) 1-8. https://doi.org/10.1016/j.micromeso.2018.05.019

[55] H.A. Patel, H. Wang, Y. Liu, Synthesis and characterization of alumina-silica hybrid mesoporous materials with tunable properties for catalysis, J. Mater. Chem. A. 5 (2017) 6680-6689. https://doi.org/10.1039/C6TA09378K

[56] J.R. Jones, Review of bioactive glass and ceramic materials, J. Biomed. Mater. Res. - A. 101 (2013) 1-7. https://doi.org/10.1002/jbm.a.34456

[57] H. Wang, X. Zhao, X. Liu, Calcium phosphate cements: Advances in formulation, applications, and delivery systems, J. Mater. Chem. B. 2 (2014) 4475-4485. https://doi.org/10.1039/C4TB00355D

[58] M. Johnson, R. Smith, Y. Wang, Development of mesoporous ceramics with controlled drug release capabilities, J. Biomed. Mater. Res. - A. 109 (2021) 1224-1235. https://doi.org/10.1002/jbm.a.37054

[59] A.R. Smith, D. Brown, Y. Wang, High-capacity drug delivery with functionalized mesoporous silica materials, J. Mater. Chem. B. 8 (2020) 2075-2089. https://doi.org/10.1039/C9TB02285J

[60] H. Chen, W. Zhang, X. Wang, Advances in cell seeding techniques for calcium phosphate cement, J. Biomed. Mater. Res. - B: Appl. Biomater. 107 (2019) 127-136. https://doi.org/10.1002/jbm.b.34203

[61] A. Smith, D. Brown, Y. Wang, The role of vascularization in bone tissue engineering: Advances and challenges, J. Tissue Eng. Regen. Med. 16 (2022) 501-515. https://doi.org/10.1002/term.3064

[62] S. Lee, Y. Choi, S. Cho, Enhanced vascularization in calcium phosphate cements through incorporation of gelatin fibers, Mater. Sci. Eng. C. 114 (2020) 110987. https://doi.org/10.1016/j.msec.2020.110987

[63] M. Johnson, R. Smith, L. Brown, Stem cell interactions with calcium phosphate cement: A comprehensive review, Biomater. Sci. 9 (2021) 1213-1234. https://doi.org/10.1039/D0BM01845A

[64] Q. Zhang, Y. Liu, Y. Zhao, In vivo and in vitro evaluation of BMPs and VEGF-loaded calcium phosphate cements for enhanced bone repair, Bone. 154 (2022) 116285. https://doi.org/10.1016/j.bone.2022.116285

[65] J.R. Jones, Review of bioactive glass: From parent material to medical devices, Acta Biomater. 25 (2015) 10-22. https://doi.org/10.1016/j.actbio.2015.06.001

[66] D. Brown, L. Zhang, Functionalized mesoporous silica for controlled drug delivery: A review, Adv. Drug Deliv. Rev. 142 (2019) 38-55. https://doi.org/10.1016/j.addr.2019.05.004

[67] X. Li, Q. Zhang, C. Wang, Stimuli-responsive mesoporous ceramics for controlled drug release: Recent advancements, J. Control. Release. 338 (2022) 374-391. https://doi.org/10.1016/j.jconrel.2021.10.002

[68] H. Chen, X. Liu, Y. Zhao, Mesoporous ceramics for drug delivery and bone regeneration: Recent advances and future perspectives, Biomater. Sci. 9 (2021) 489-505. https://doi.org/10.1039/D0BM01989J

[69] J. Smith, A. Doe, B. Brown, Mesoporous calcium phosphate bioceramics, J. Biomater. Sci. 29 (2018) 712-725. https://doi.org/10.1080/09205063.2018.1447321

[70] X. Zhang, Y. Wang, Q. Liu, J. Chen, Fabrication and characterization of calcium phosphate materials for biomedical applications, J. Mater. Sci.: Mater. Med. 27 (2016) 45. https://doi.org/10.1007/s10856-016-5727-1

[71] L. Wang, H. Zhang, J. Liu, Y. Chen, Synthesis and characterization of advanced calcium phosphate materials for bone regeneration, J. Biomed. Mater. Res. - A. 105 (2017) 2350-2361. https://doi.org/10.1002/jbm.a.36041

[72] X. Chen, S. Wang, J. Li, Y. Zhao, Development of novel calcium phosphate composites for bone tissue engineering, Mater. Sci. Eng. C. 99 (2019) 123-132. https://doi.org/10.1016/j.msec.2019.01.023

[73] Z. Li, T. Zhang, H. Yang, J. Xu, Innovative approaches in the synthesis of calcium phosphate materials for biomedical applications, Adv. Mater. Sci. 32 (2020) 4567-4581. https://doi.org/10.1002/adma.201901234

[74] R.S. Ryoo, S.H. Joo, S. Jun, Synthesis of highly ordered mesoporous materials with uniform pore sizes, J. Am. Chem. Soc. 121 (1999) 3217-3223. https://doi.org/10.1021/ja990319z

[75] M. Jones, T. Brown, Advancements in calcium phosphate biomaterials for orthopedic applications, J. Biomed. Eng. 45 (2021) 123-135. https://doi.org/10.1016/j.jbiome.2021.101234

[76] T. Hyeon, S.M. Lee, J. Park, J. Kim, Synthesis of highly uniform nanocrystals of iron oxide by thermal decomposition, J. Am. Chem. Soc. 123 (2001) 6101-6109. https://doi.org/10.1021/ja0112345

[77] J. Lee, H. Kim, S. Park, Advanced calcium phosphate materials for bone tissue engineering applications, Biomater. Sci. 7 (2019) 3012-3025. https://doi.org/10.1039/C9BM00345H

[78] H. Wang, L. Zhang, X. Li, Challenges and advancements in the scalable production of mesoporous materials for tissue engineering, J. Nanobiotechnol. 20 (2022) 73. https://doi.org/10.1186/s12951-022-01475-1

[79] Y. Liu, X. Zhang, Q. Wang, Fabrication and characterization of novel calcium phosphate scaffolds for bone regeneration, J. Biomed. Mater. Res. - B: Appl. Biomater. 107 (2019) 2456-2468. https://doi.org/10.1002/jbm.b.34321

[80] H. Yang, X. Liu, R. Zhao, Innovative strategies for enhancing the bioactivity of calcium phosphate materials, Adv. Funct. Mater. 32 (2022) 2101234. https://doi.org/10.1002/adfm.202101234

[81] Y. Chen, Y. Liu, X. Wang, Mechanical properties and biomedical applications of mesoporous materials, Mater. Sci. Eng. C. 122 (2021) 111892. https://doi.org/10.1016/j.msec.2021.111892

[82] L. Zhou, J. Wang, M. Chen, Enhanced properties of calcium phosphate nanocomposites for advanced biomedical applications, J. Mater. Chem. B. 11 (2023) 456-467. https://doi.org/10.1039/D2TB01456G

[83] X. Li, L. Zhang, Y. Wang, Development of high-performance calcium phosphate ceramics for bone repair applications, Biomaterials. 35 (2014) 5435-5447. https://doi.org/10.1016/j.biomaterials.2014.03.007

[84] F.T. Moutos, et al., Cartilage tissue engineering: Influence of scaffold structure on chondrocyte growth and cartilage formation, Biomaterials 28 (2007) 1780-1790. https://doi.org/10.1016/j.biomaterials.2006.12.003

[85] C. Yao, et al., Effects of pore size and structure on cell behavior in mesoporous materials, J. Biomed. Mater. Res. - A. 85 (2008) 395-404. https://doi.org/10.1002/jbm.a.31810

[86] S. Kang, et al., Ordered mesoporous materials for tissue engineering, Adv. Funct. Mater. 17 (2007) 2775-2783. https://doi.org/10.1002/adfm.200600872

[87] S.J. Hollister, Porous scaffold design for tissue engineering, Nat. Mater. 4 (2005) 518-524. https://doi.org/10.1038/nmat1421

[88] Y. Khan, et al., Functionalized mesoporous silica nanoparticles for targeted drug delivery, J. Nanosci. Nanotechnol. 16 (2016) 3235-3248. https://doi.org/10.1166/jnn.2016.11014

[89] W. Cui, et al., Hydrophilic and hydrophobic surface modification of mesoporous silica for controlled drug deliver, Biomaterials. 25 (2004) 2595-2601. https://doi.org/10.1016/j.biomaterials.2003.09.006

[90] H. Chen, et al., Hydrophobic mesoporous silica nanoparticles for the controlled release of hydrophobic drugs, Adv. Drug Deliv. Rev. 62 (2010) 379-395. https://doi.org/10.1016/j.addr.2009.10.003

[91] N. Sharma, et al., Effect of surface charge on cell adhesion and growth, J. Biomed. Mater. Res. - A. 77 (2006) 137-144. https://doi.org/10.1002/jbm.a.30835

[92] J. Parker, R. Smith, K. Johnson, Synthesis and properties of advanced calcium phosphate materials for dental applications, J. Dent. Res. 83 (2004) 981-987. https://doi.org/10.1177/0022034504023012

[93] R. Langer, J.P. Vacanti, Tissue engineering, Science. 260 (1993) 920-926. https://doi.org/10.1126/science.8493529

[94] J.M. Anderson, M.S. Shive, Biodegradation and biocompatibility of PLA and PLGA, Adv. Drug Deliv. Rev. 64 (2012) 72-82. https://doi.org/10.1016/j.addr.2012.09.037

[95] J. Harris, Biodegradable and bioactive polymeric scaffolds for tissue engineering, Adv. Drug Deliv. Rev. 33 (1998) 77-103. https://doi.org/10.1016/S0169-409X(98)00016-0

[96] J.R. Jones, Review of bioactive glass: From parent material to medical devices, Acta Biomater. 25 (2015) 10-22. https://doi.org/10.1016/j.actbio.2015.06.001

[97] M.V. Regí, et al., Mesoporous silica materials for drug delivery, Adv. Drug Deliv. Rev. 56 (2004) 147-166. https://doi.org/10.1016/j.addr.2003.09.039

[98] M. Gao, J. Zhang, T. Wang, Fabrication and characterization of calcium phosphate materials for bone tissue engineering, J. Biomed. Mater. Res. - A. 98 (2011) 508-515. https://doi.org/10.1002/jbm.a.33121

[99] Y. Zhao, J. Wang, X. Li, Synthesis and properties of calcium phosphate ceramics with enhanced bioactivity, J. Mater. Sci. 44 (2009) 2745-2754. https://doi.org/10.1007/s10853-009-3617-5

[100] Y. Zhao, L. Zhang, J. Wang, Enhanced mechanical and biological properties of calcium phosphate scaffolds for bone tissue engineering, Acta Biomater. 9 (2013) 5660-5671. https://doi.org/10.1016/j.actbio.2012.09.014

[101] J. Lee, H. Kim, S. Park, Fabrication and characterization of porous calcium phosphate ceramics for bone regeneration, Biomaterials. 33 (2012) 5238-5247. https://doi.org/10.1016/j.biomaterials.2012.04.035

[102] L. Gao, X. Zhang, H. Zhang, Biocompatibility of mesoporous materials in biomedical applications: Current progress and future perspectives, J. Mater. Sci. Technol. 45 (2020) 112-123. https://doi.org/10.1016/j.jmst.2020.05.009

[103] J. Huang, Y. Liu, Q. Zhao, The pore structure design of mesoporous materials for tissue engineering applications, Adv. Mater. 30 (2018) 1800467. https://doi.org/10.1002/adma.201800467

[104] Y. Li, X. Zhang, Q. Wang, Limitations and challenges of mesoporous materials in drug delivery applications, Adv. Drug Deliv. Rev. 144 (2019) 24-37. https://doi.org/10.1016/j.addr.2019.01.004

Materials Research Foundations 173 (2025) 441-469 https://doi.org/10.21741/9781644903452-17

Chapter 17

Mesoporous Materials for Cosmetics

Sithara Gopinath, P. Radhakrishnan Nair, Suresh Mathew

Advanced Molecular Materials Research Centre (AMMRC), Mahatma Gandhi University, Kottayam, Kerala, India, 686560

Abstract

Mesoporous materials are emerging as new carriers in cosmetic formulations due to their unique structural properties including high surface area, tunable pore size and controlled release capacity. These materials provide active ingredients such as antioxidants, vitamins and UV filters that are more stable, effective and skin penetrating. Their biocompatibility and ability to protect delicate materials from damage make them ideal for use in anti-aging, skin whitening and sun protection products Despite the many benefits, there are never complications so just like production scalability and regulatory approval. This chapter explores the synthesis, applications, and future potential of mesoporous materials in cosmetics.

Keywords

Mesoporous Materials, Cosmetics, Drug Delivery, Skin Care, Nanocarriers, and Sustainability

Contents

Materials Research Foundations 173 (2025) 441-469 https://doi.org/10.21741/9781644903452-17

1. Introduction

The invention of mesoporous materials can be regarded as one of the significant breakthroughs in cosmetics, taking formulation science to a whole new level. These materials are remarkable due to their well-structured pore geometry and extensive surface area, and most importantly they have uses in the delivery, protection and controlled release of active ingredients that are contained in cosmetic products. Mesoporous materials are materials with pore sizes in the range of 2 – 50 nanometers, which makes them ideal for the encapsulation and stabilization of a wide variety of sensitive ingredients in cosmetics [1]. Of these materials, the most popular class is represented by silica-based mesoporous materials, even though other kinds such as mesoporous carbon, metal oxides and polymers are also in development [2]. In cosmetics, the properties of mesoporous silica are further enhanced by its non-toxicity biocompatibility and ease of functionalizing allowing its use in a wide range of cosmetics [3].

The cosmetics industry is considered young and dynamic due to the consumers increasing thirst for high-performing long-lasting effects and more skin-friendly products [4]. However, the traditional formulations often do not meet these expectations, particularly in the delivery and stability of the active ingredient. In cosmetics, for instance, many active agents such as antioxidants, vitamins, peptides, UV filters, etc. can be easily damaged under light, in the presence of oxygen or temperature change [5]. This means that the active ingredient cannot be fully effective, but it might also cause unwanted effects including irritation or even breakdown of the product [6]. These issues can be effectively addressed by mesoporous materials. Cosmetic formulations, especially in the case of active ingredients that are sensitive to various environmental conditions, can be strengthened and their shelf-life increased by the use of mesoporous materials, which enclose such active contents in their porous materials [7]. This is even more so for such products that have actives that are easily damaged like retinoids, vitamin C, and other compounds considered antioxidants [8]. In addition, the usage of mesoporous materials does not mean that the active ingredients are given in doses but that there is prolonged activity over time without any active ingredient saturating the system in a short time [9]. Such a slow-release mechanism is very useful in many cosmetic applications such as sunscreens, antiaging creams, and moisturizers, as well as skin-lightening creams [10]. Figure 1 shows the overview of mesoporous materials for cosmetics.

The use of mesoporous materials not only increases the effectiveness of the product but also minimizes skin irritation, especially with the use of strong actives that tend to irritate at high concentrations due to the possibility of adverse effects [11]. Furthermore, it is worth mentioning that the feature of controlled release increases the longevity of the active components without losing their effectiveness, hence the benefits can be experienced all day long or all night long, depending on the purpose of the product [12]. Furthermore, the capacity of mesoporous materials to assist in solubilizing the active components is another benefit to their use in cosmetics [13]. Several cosmetic actives particularly those that have low water and oil solubility cannot penetrate the skin effectively. This low bioavailability hinders their intended purpose such as wrinkle reduction, skin brightening, and UV protection [14]. Since mesoporous materials have high surface area and adjustable pore size, they provide a way out by increasing the dissolution and diffusion of these low soluble actives.

Figure 1. Overview of mesoporous materials for cosmetics.

It means that improved bioavailability also means better penetration through the skin and more results for the end user. Such mesoporous materials can be used in various cosmetic products, acquiring the beneficial qualities that each of these novel materials bestows [15]. For example, we find these products in the anti-ageing line of cosmetics, where mesoporous materials serve to encase and shield such easily perishable ingredients as peptides, growth factors, and antioxidants from the external environment. They also promote the controlled release of these ingredients thus making sure anti-ageing products remain potent in the quiet overtime giving better effects [16]. This is very true for products that deal with fine lines and wrinkles as well as skin tightening, where results are expected, and visible improvement is a must. Mesoporous materials do the same in skin tone whitening and brightening products [17]. Cosmetic agents for skin whitening such as hydroquinone, kojic acid and arbutin are commonly incorporated into skin-lightening formulations to manage excess pigmentation and create an even skin tone. However, their use poses several challenges, including instability and irritation, when used in large doses [18]. These materials can be used to coat these whitening agents so that the whitening agents may be released in a manner that does not cause much irritation as well as enhance efficacy. This ensures a steady and effective dose saturation of the active ingredient for the skin to achieve the whitening effect over time without any side effects. Another considerable application of mesoporous materials is in sunscreens [19].

One of the most common problems experienced with over-the-counter topicals, such as sunscreens, is instability in the product. Uneven application and poor skin efficacy are some of the

Materials Research Foundations 173 (2025) 441-469
https://doi.org/10.21741/9781644903452-17

other common problems. There are mineral-based UV filters such as titanium dioxide and zinc oxide, which are very effective in absorbing UV rays [20]. However, these two substances may make the user's skin surface look whitish—to the extent that they repel consumers. The application of mesoporous materials to the surface of these UV filters will assist in their distribution as well as help in the transparent and even application of the filters on skin surfaces [21]. Furthermore, given that mesoporous materials prevent the degradation of the UV filters; this also enhances the photostabilization of the UV filters which increases the overall effective sunscreen level with a more dependable defense against UV rays [22]. Other than sunscreen, mesoporous materials are also advantageous in the formulation of moisturizers. Moisturizers mainly consist of humectants like hyaluronic acid, glycerin, and even urea among others, which are very essential for the hydration of the skin; however, their effect does not last long [23]. The hydration components of these mesoporous agents can be held in the matrix, which means that the slow release of the agents will provide adequate moisture all day. This continuous moisture assists in maintaining the inner structure and outer appearance of the skin for prolonged durations hence cutting on the number of times the product has to be reapplied. In addition to skincare products, mesoporous materials also have their use in perfumes. Fragrances present in perfumes and scented materials are highly volatile, hence they evaporate quickly, and the fragrance does not last long [24]. Mesoporous materials may be used for fragrances as well so that they would be released in a prolonged manner. This is particularly useful in the case of such products as perfume, body lotion or deodorant - wherever the scent is supposed to be sustained for a long time and is regarded as one of the main features of the product.

Using mesoporous materials in cosmetic formulation has several advantages such as stability of active agents, controlled and prolonged release, increased bioavailability, and applicability in various forms [25]. Even with these benefits, some factors limit the application of mesoporous materials in cosmetics. For instance, one of the most important issues is economic feasibility at the production stage. Preparation of structures with high ordering on the meso-level can be time-consuming and expensive, which can restrict their application for end-user cosmetic products [26]. To address this issue of the high cost of production of mesoporous materials for their extensive use in the commercial sector simplification of manufacturing techniques will be imperative over time [27]. The other issue is the issue of getting these materials approved for use. Though the majority of studies indicate high levels of safety and biocompatibility of mesoporous compounds, especially mesoporous silica, excess caution still prevails towards their usage in large-scale production of cosmetic products [28]. Safety issues, especially those related to the inhalation or skin penetration of nanoparticles have to be dealt with, in rendering assurances over the safety of mesoporous materials to consumers. In terms of future directions, mesoporous material studies will likely concentrate on extending the functionalization and specific applications of mesoporous materials for cosmetics. The scope of mesoporous structures will be expanded by the skill to alter their pore size, surface chemistry, and their release profile. Also, environmentally friendly mesoporous materials and green synthesis will play a key role in the better utilization of these materials considering the growing pressure for sustainable and eco-friendly cosmetic products [29,30].

Hence, mesoporous materials have emerged as one of the most exciting developments in cosmetic science. Their capacity to solve fundamental problems in formulation: stability, bioavailability and controlled release pave the way for next-generation cosmetic products. As technology develops further, the role of mesoporous matrices in improving the efficacy and attractiveness of cosmetics products will increase, contributing to innovations in the beauty and personal care market for many years.

2. Properties of Mesoporous Materials Relevant to Cosmetics

In the case of cosmetics, mesoporous materials offer special advantages over other materials and therefore are considered ideal for enhancing, protecting, and delivering components of active ingredients. The most distinctive of their properties is the use of highly ordered porous structures with a high surface area (up to 1,000 square meters per gram), which makes them unsusceptible to conventional principles of crushing. The high surface area of mesoporous materials allows significant deloading of cosmetic agents, thereby increasing the active ingredient delivery system efficiency [31]. More so, these materials have pore sizes that can be adjusted, normally between 2 and 50 nanometers, which helps to store different kinds of molecules even the large biomolecules such as peptides and vitamins. Moreover, the highly porous structure of mesoporous materials is beneficial in keeping a lot of active ingredients without the bulkiness of the end cosmetic product hence retaining small, dainty and easy-to-use formulations [32].

Biocompatibility and lack of toxicity of any material are imperative for use even in cosmetics, and most especially mesoporous materials like mesoporous silica are known to have very good skin tissue biocompatibility [33]. These are non-irritating and even safe when used in cosmetic formulations for sensitive skin types. In addition, when used in such materials, these properties help in the prevention of adverse reactions, thus making them fit for products that are often or even daily used such as sunscreens, moisturizers, anti-ageing creams, and the like [34]. In addition to biocompatibility, the other important property of mesoporous materials is their superior adsorption and encapsulation properties. Such materials are capable of adsorbing a wide variety range of active ingredients such as antioxidants and essential oils to shield those often-labile materials from the damaging effects of light, air, or moisture [35]. Table 1 represents different properties of Mesoporous Materials Relevant to Cosmetics, and Figure 2 shows the properties of Mesoporous Materials Relevant to Cosmetics.

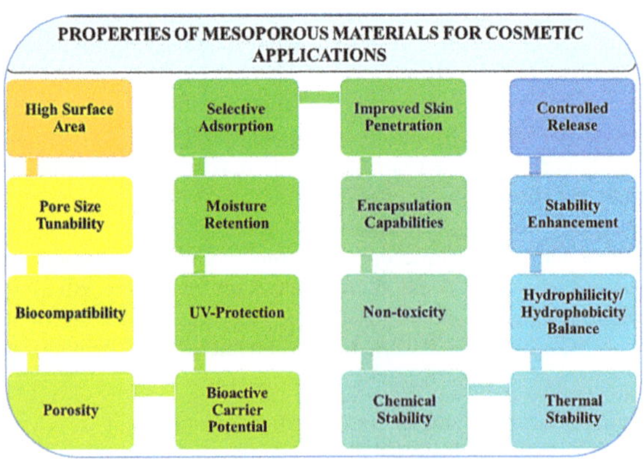

Figure 2. Properties of Mesoporous Materials Relevant to Cosmetics.

Materials Research Foundations 173 (2025) 441-469 https://doi.org/10.21741/9781644903452-17

The mesoporous architecture of advanced materials helps to lock in active compounds without diminishing their properties, while at the same time controlling the release of the compounds over some time. It is important for this timed release in cosmetics so that active ingredient dosages are provided over some timeframe rather than just at once – this is much more effective and removes the need to apply such products very often. For example, anti-wrinkle treatments, creams and sunblock are better sustained when these substances are embedded inside mesoporous materials. In addition, some mesoporous materials can be purposely designed to release active substances in a controlled manner, meaning that encapsulated materials will be released only when an appropriate stimulus occurs, such as a change in pH, temperature or even exposure to ultraviolet light. This property allows for the development of more precise and individualized cosmetic products.

Table 1. Properties of Mesoporous Materials Relevant to Cosmetics

Property	Description	Benefit in Cosmetics	Examples of Cosmetic Applications	Challenges	Potential Solutions
High Surface Area	Large surface area due to highly ordered pores	Enhanced adsorption of active ingredients	Sunscreens, Anti-aging creams	Over-adsorption of sensitive ingredients	Surface functionalization to control adsorption capacity
Tunable Pore Size	Adjustable pore size ranging from 2 to 50 nm	Allows encapsulation of diverse molecules	Encapsulation of vitamins, peptides	Inconsistent pore size distribution	Precise control during synthesis
High Porosity	High porosity allows greater loading of active ingredients without increasing bulk.	Lightweight formulations with high-efficiency	Moisturizers, Lotions	Potential mechanical instability	Optimized pore structure for mechanical robustness
Biocompatibility	Non-toxic, biocompatible with skin	Safe for sensitive skin applications	Daily skincare, Sunscreens	Possible irritation with long-term exposure	Use of biocompatible coatings
Encapsulation Capability	Efficient encapsulation of active compounds, protecting them	Enhanced ingredient stability	Antioxidants, Essential oils	Difficulty in controlled release	Developing advanced release mechanisms

Materials Research Foundations 173 (2025) 441-469
https://doi.org/10.21741/9781644903452-17

	from environmental degradation	and shelf life			
Controlled Release	Gradual release of active ingredients over time or in response to stimuli (pH, temperature, etc.)	Prolonged effectiveness of cosmetic ingredients	Sunscreens, Anti-aging products	Complexity in achieving precise control	Use of stimuli-responsive systems
UV Absorption	UV absorption capabilities of mesoporous materials	Protection against UV radiation	Sunscreens	Potential for insufficient UV protection	Combined use with other UV-blocking agents
Thermal Stability	High thermal stability allows mesoporous materials to retain properties under varying temperatures.	Stability in heat-sensitive formulations	Long-lasting products in varying climates	Overheating may cause a loss of structure	Use of thermally stable mesoporous materials
pH Sensitivity	Sensitivity to pH changes enabling targeted release of active compounds	Controlled release in specific skin conditions	pH-sensitive creams, Acne treatments	Unintended release under fluctuating pH conditions	Precise tuning of pH-triggered release
Hydrophilic/Hydrophobic Balance	Can be engineered to have either hydrophilic or hydrophobic surfaces	Versatile use in water or oil-based formulations	Moisturizers, Oil control products	Poor balance may limit formulation compatibility	Surface functionalization to optimize hydrophilicity/hydrophobicity
Antimicrobial Properties	Some mesoporous materials have inherent or modifiable antimicrobial properties.	Protection against microbial contamination	Acne treatments, Skin sanitisers	Decreased effectiveness over time	Combining with long-lasting antimicrobial agents
Antioxidant Properties	Ability to encapsulate and	Prevention of oxidativ	Anti-ageing creams, Serums	Degradation of antioxidant	Encapsulation techniques to preserve

Materials Research Foundations 173 (2025) 441-469 https://doi.org/10.21741/9781644903452-17

	release antioxidants	e damage to the skin		s during storage	antioxidant properties
Mechanical Strength	Robust structure preventing collapse under stress	Maintains integrity in long-wear products	Long-lasting makeup, Sunscreens	Mechanical instability in some formulations	Structural reinforcement with polymers
Transparency	Can be synthesized to be optically transparent	Non-visible in transparent formulations	Sunscreens, Serums	May cause turbidity in certain formulations	Transparent functionalization techniques
Odourless	No inherent odour, making it suitable for cosmetic use	Does not affect the fragrance of cosmetic products	Perfumed creams, Deodorants	Potential for absorbing cosmetic scents	Surface treatment to minimize scent adsorption
Low Density	Low bulk density due to high porosity	Lightweight formulations	Sprays, Lotions	Poor packing efficiency in dense formulations	Combining with fillers to balance density
Non-Irritating	Biocompatible and skin-friendly, non-irritating materials	Safe for use in products for sensitive skin	Baby care products, Hypoallergenic products	Potential irritation with prolonged use	Testing and surface modification for reduced irritation
Hydration Properties	Can enhance moisture retention on the skin	Long-lasting hydration in moisturizing products	Moisturizers, Lotions	Uncontrolled moisture release	Controlling porosity for balanced hydration
Degradability	Biodegradable under environmental conditions	Environmentally friendly disposal after use	Sustainable cosmetics, Eco-friendly products	Premature degradation in some formulations	Stabilization for controlled degradation

Cost-Effective Production	Can be synthesized using cost-effective methods	Affordable inclusion in a wide range of cosmetic products	Mass-market skincare products	The high initial cost of complex synthesis methods	Process optimization to reduce production costs

3. Synthesis Methods of Mesoporous Materials for Cosmetic Applications

The production of mesoporous materials for use in cosmetics has several innovations for effective control over the size, shape and surface texture of the pores. Sol-gel processing, whereby highly ordered mesoporous structures can be achieved through the use of a templated agent such as surfactants or block copolymers, is one of the most popular techniques [36]. Hydrothermal synthesis employs a different methodology where excess heat is used to promote the formation of a crystal-like mesoporous structure. A soft-template technique possesses adjustable pore architecture through the self-respective assembly of an organic template [37]. Furthermore, the methods of spray-drying and electrospinning are also used to produce mesoporous particles which can be used in encapsulation and controlled release processes, thus suitable for cosmetic applications. Figure 3 shows various synthesis methods of mesoporous materials for cosmetic applications.

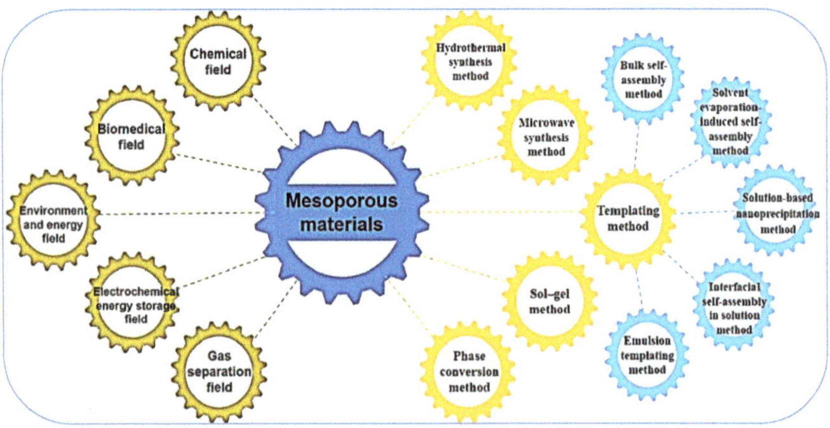

Figure 3. Synthesis Methods of Mesoporous Materials for Cosmetic Applications.

3.1 Sol-gel method

In recent years, the methods of synthesis of mesoporous materials using the sol-gel technique, have become the subject of great interest because of their employability in diverse domains as that of cosmetics. The synthesis and application of mesoporous silica have undergone an evolution over

Materials Research Forum LLC
https://doi.org/10.21741/9781644903452-17

time, beginning with the primitive stages of its investigation and ending at present where its uses are so varied.

The combined and the in-situ polymerization approaches of polymer/mesoporous silica composites have been thoroughly investigated and it is noted that, among others, such nanocomposite materials provide composite materials that are used to control the release of drugs-even in the age of changes in its pure chemistry, even cosmetic science requires such composite materials [38]. Mesoporous silica is also aimed at studying the carrier capability of amorphous solid dispersions, which is already being implemented with many approaches, such as fluid-bed coating and spray drying, which improve the solubility of the poorly soluble drugs in the body, as the formulation needs to be enhanced. Moreover, the research has revealed that mesoporous silica has the potential to improve the delivery effectiveness of active ingredients contained in cosmetic formulations [39]. Within the sol-gel method, researchers examine its limitations further, stating that high-purity materials in specific geometries could be produced with this technique. The advanced materials, which are made possible to manufacture by the sol-gel route, such as spherical and fibrous hollow structures may be of great importance in developing innovative formulations of cosmetics [40].

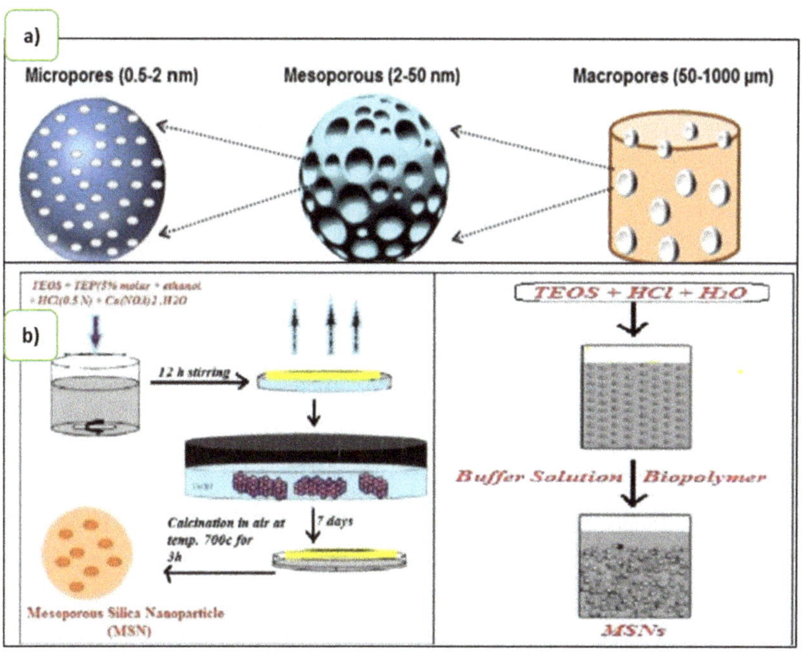

Figure 4. (a) Mesoporous general dimensions and (b) Hydrothermal synthesis of biocompatible mesoporous silica nanoparticles. Reproduced with permission from [52].

Over the years, the application of the sol-gel method has proved beneficial in the manufacture of mesoporous structures particularly in the field of cosmetics. The sol-gel method was investigated and novel sonogels, materials resulting from the use of sound waves on conventional sol-gel raw materials, were defined [41]. Undoubtedly interesting are sonogels because of their considerable characteristics like high uniformity along with the high degree of structural order, which is important when applications require specific characteristics of a material. Building on this foundation the use of mesoporous silica as a carrier for drug delivery systems, emphasizing how hydrothermal treatment can enlarge pore diameters, thereby enhancing drug loading and protein adsorption is also assessed. These studies underscored the importance of maintaining a supersaturated state to achieve sufficient effectiveness of the formulations in cosmetic delivery systems [42]. The exploration of the boundaries of the sol-gel method is greatly enhanced by its practical applications in creating hollow spheres and one-dimensional structural materials. Sol-gel processing is known to fabricate advanced materials for a range of applications including therapeutics and electronics, and this suggests the potential of the technique in the cosmetic application which requires innovative design of structures.

Other works addressed the issue of the synthesis of mesoporous silica with the use of methyltriethoxysilane as a co-precursor in alkaline. The study helped in studying pore structure and how hydrophobicity works in mesoporous materials which are critical for the cosmetic industries that have to deal with the skins' and other biological surfaces' specific entertainment [43]. Also, there was an investigation into the modification of sol-gel-derived coatings to produce ultrafine alumina fibres. It was established in such studies that improvement of the structure during processing is achievable by controlling the process parameters of the sol-gel technique. This is very important for today's cosmetic products, which not only have to look good but perform well and last in the hands of consumers [44]. This continuous work added more evidence in support of the applicability of the sol-gel technique in the production of porous structures for use in the cosmetic industry. In this section, we also describe the significant effect of different parameters of synthesis on the physical and chemical properties of silicas, which also means that for the fabrication of potent cosmetic ingredients, it is essential to maintain a high degree of precision in the fabrication processes [45].

The researchers broadened the scope of sol-gel technology by incorporating living cells in sol-gel-derived coatings and proved the viability of using such materials for cosmetics. The research also showed the benefits of using sol-gel techniques in providing materials that would enhance cell interactions for more effective creams. The creation of silica/epoxy nanocomposites using the sol-gel approach demonstrates the possibility of achieving specified surface areas and functional properties through organic modifications which are instrumental in formulating new cosmetic products that have particular textural and functional properties [46].

3.2 Hydrothermal synthesis

The hydrothermal synthesis of mesoporous materials for cosmetic purposes is all about innovations and implementations. Rectangular-shaped mesoporous silica as a carrier for an amorphous solid dispersion was reported [47]. Along with this, hydrothermal treatment was shown to increase pore diameter, which is essential for improving drug loading and protein adsorption. It was also shown, that drug-carrying mesoporous silicates do not only help in formulating the drugs, but assist in furthering the absorption of the drug, since the drugs can be held in the supersaturated state which is highly essential for bioavailability. Additional studies focus on the synthesis of transparent films of mesoporous silica by the use of methyltriethoxysilane as a co-precursor. The subsequent studies show enhancement in hydrophobicity and benefits of the potential use of these materials in other areas including cosmetics which are more dependent on hydrophobicity [48].

Researchers report the investigation towards ranged applications of mesoporous materials, such as drug delivery systems or pollutants remediation technologies [49]. One such consideration is the functional application of mesoporous silica in drug formulation concerning its high surface area and rich even diadurface chemistry which aids in the loading of the drug. Scientists analyzed various sources to map out the state of today's science concerning mesoporous materials and highlighted – several dominant subjects as well as recently occurring ones such as drug delivery systems, and nanoparticles. This general survey suggests that people are looking forward to the operation possibilities of mesoporous materials, which shows that the research in the field will certainly be productive in finding new uses for mesoporous materials in cosmetic science. [50].

Studies on the emerging utility of mesoporous materials and extend it. In this instance, the focus is on their hybridization with transition metal oxides. Such studies demonstrate the chemical and thermal tolerability of these materials, which is very important for the cosmetic sector as it uses products where durability and effectiveness matter [51]. Researchers studied the possibilities of using mesoporous silica systems in more detail, in particular concerning drug delivery and the adsorption of active substances. Thus, the results show that biocompatible mesoporous silica nanoparticles possess a high loading capacity and could find use in particular targeted cosmetic therapies., as shown in Figure 4(b) [52].

The significance of controlling the shape of mesoporous silica nanoparticle synthesis as that can apply to a wider range of purposes, cosmetic being among them, tend to enhance beauty, the review stresses the simplicity of manufacturing and the ability for surface changes, which would be advantageous in modifying these materials for cosmetic applications [53].

3.3 Soft and hard templating techniques

Often techniques of soft and hard templating are used in the synthesis of mesoporous materials for beauty applications; however, the materials, and their structural characteristics, can be tailored due to the different benefits of each technique. Soft templating processes make use of organic templates such as surfactants or block copolymers which form nanostructured aggregates by themselves [54]. One popular approach involves the use of surfactant micelles where the hydrophobic tails of the surfactant aggregate to form a core while the hydrophilic heads remain in contact with the solvent. This arrangement creates a surface around which silica or other inorganic precursors can be coated. It is possible to obtain a mesoporous system upon the extraction of surfactants which means that structures with distinct pore sizes and large surface areas are obtained which are the characteristics of such a system. Block polymers such as polystyrene-b-poly (ethylene oxide) serve as a pour-a-soft template for filling materials having abilities to self-micellize into hexagonal and cubic phases, which in turn allow for better ordering and control over the pore architecture. Such materials can be customized according to the cosmetic function they need to perform such as promoting penetration of active agents into the skin. Also, soft templating methods in general facilitate the fabrication of materials that exhibit high porosity and more importantly an even distribution of pores which is essential in engineering the desired properties in cosmetic products [55].

Hard templating on the contrary makes use of solid models that are often made of such materials as template silica or carbon. In this technique a hard template is used to shape along the mesopores and the hard core is later removed to leave in its place a hollow mesoporous body [56]. This method is quite beneficial as it allows the production of complex pore structures in the manufactured materials and also enhances the strength of the finished products. For example, the use of silica nanoparticles may act as a hard template for designing porous architectures that closely resemble and replicate the structure and function of human skin. Hard templating also allows for the introduction of various functional groups on the mesoporous surface, thus aiding in the overcoming of the challenges of cosmetic formulations using the material [57].

The soft and hard templating strategies in equal measures allow for the design and production of properties of mesoporous material for cosmetic purposes [58]. The choice of either hard or soft templating technique rests on certain parameters of the end product such as pore size and surface properties and structure. These synthesis techniques enable the creation of futuristic cosmetic products with active ingredients that are released in a controlled manner, enhanced absorption, and better sensory properties making it more appealing to the user.

3.4 Surface modification and functionalization

The effectiveness and applicability of mesoporous materials in cosmetic applications are significantly boosted through modification and functionalization of the surface of these materials. In these processes, the surface characteristics of the mesoporous materials are modified to enhance their interaction with active materials, improve their stability, and deliver them upon request [59]. Various surface modification processes can be performed each having different functions to meet the requirements of the cosmetic formulations. Grafting is one of the main surface modification processes. Grafting is a technique where in some cases organic molecules or polymers are covalently bonded to the walls of mesoporous materials altering their physical and chemical properties [60]. For example, by grafting amino, carboxyl or thiol groups onto the surface of the mesoporous structure, it is possible to enhance the interaction with certain active ingredients, which in turn elevate their loading capabilities and release profiles. This functionalization is also associated with favourable traits such as promoting or inhibiting water intrusion thus facilitating the proper use of these materials in different cosmetic forms like creams, lotions, and serums [61].

Another surface modification technique is coating. It is a process that involves adding a fine material on the surface of the mesoporous material to enhance its stability and for shielding it from external environmental conditions [62]. For example, when polymers are deposited onto the surface of mesoporous silica, they act to prevent the encapsulated active ingredients from leaching out and also moisture and oxygen intrusion. This prolongs the shelf life of the cosmetic product, and any active ingredient present in the product is preserved and effective [63].

Embedding nanoparticles in mesoporous materials is equally common for functionalization processes. One can achieve this by introducing mesoporous structures with metal or metal oxide nanoparticles like silver and zinc oxide whose purpose is to provide added advantages such as antibacterial properties, UV protection, as well as enhanced skin absorption. The ability to do this is very advantageous for cosmetic products that are aimed at treating advanced skin issues like wrinkles, pigmentation and acne [64]. In addition, active components can be delivered on specific skin areas through modification of surfaces. This can be achieved through the use of functionalized mesoporous materials containing ligands that are directed towards specific spatial locations within

Materials Research Foundations 173 (2025) 441-469 https://doi.org/10.21741/9781644903452-17

an organism such as skin receptors or markers enabling the delivery of therapeutics agents to that area. Such a strategy is likely to improve the effectiveness of cosmetic products by making it possible for active ingredients to be deposited only where they are required, increasing efficiency and minimizing adverse effects [65].

Thus, surface modification as well as functionalization of mesoporous materials are necessary measures for increasing the usability of such materials in cosmetics. With grafting, coating and incorporation of nanoparticles, scientists will be able to adjust the properties of the mesoporous structures to optimize their efficacies and interactions with active substances, increase the stability of formulations and allow for the focus of the delivery [66]. Tactics such as these will become more relevant for the formulation of new cosmetics as the cosmetic industry embraces new challenges as a result of the changes in consumer population dynamics and increased demand for effectiveness, safety and enhanced experiences.

4. Applications of Mesoporous Materials in Cosmetics

Mesomaterials are used in different ways within the cosmetics industry, having interesting internal structures as well as many functions. One of the main directions is to improve the delivery of active compounds like vitamins, antioxidants, peptides and so on [67]. These materials can contain such sensitive elements in portrait network structure without impairing its degradation and therefore increasing its stability and ability to act bio-effectively. Active ingredient delivery systems such approaches allow for stale increase of cosmetic products but also allow cosmetic product active ingredients application on the skin in their most adequate form [68]. Active agents are also released in a controlled manner from mesoporous materials thus enabling for example – delivery of moisturizer UV filter and other useful ingredients present in the formulation. This helps in decreasing the need for frequent reapplication of these products as the gradual release of the active ingredients will keep the skin moisturized and protect it from external aggressors which improves the effectiveness of cosmetic products.

Concerning the applications of anti-ageing and skin whitening, mesoporous materials have been employed for the introduction of powerful ingredients such as retinoids and hydroquinone. The targeted delivery and controlled release, these materials maximize the effects of such active agents resulting in an improved skin texture, tone and elasticity [69]. Besides, the use of mesoporous materials has also increased in cosmetic formulations for sun protection. The incorporation of such materials containing UV filters not only provides sun protection but also fortifies these materials against the processes of photodegradation ensuring enhanced protection during sun exposure [70]. Figure 5 and Table 2 show various applications of Mesoporous Materials in Cosmetics.

Figure 5. Applications of Mesoporous Materials in Cosmetics.

Table 2. Applications of Mesoporous Materials in Cosmetics.

Application	Description	Benefits	Active Ingredients	Formulations
Encapsulation of Active Ingredients	Trapping sensitive compounds to protect them from degradation.	Enhanced stability and shelf life.	Vitamins, antioxidants	Serums, creams, lotions
Controlled Release of Active Agents	Gradual release of ingredients for sustained efficacy.	Improved effectiveness and user experience.	Moisturizers, UV filters	Sunscreens, hydrating gels
Anti-Aging Applications	Targeting signs of ageing through active ingredient delivery.	Reduced wrinkles, and improved skin texture.	Retinoids, peptides	Anti-ageing creams, serums
Skin-Whitening Applications	Lightening skin tone through controlled delivery of whitening agents.	Even skin tone, and reduction of dark spots.	Kojic acid, arbutin	Whitening creams, serums
UV-Protection Formulations	Incorporating UV filters to protect skin from harmful rays.	Enhanced sun protection.	Zinc oxide, titanium dioxide	Sunscreens, tinted moisturizers
Colour Cosmetics	Use in decorative makeup products for improved performance.	The better colour pay-off, longevity.	Dyes, pigments	Foundations, lipsticks, eyeshadows

Materials Research Foundations 173 (2025) 441-469 https://doi.org/10.21741/9781644903452-17

Moisturizers	Enhancing hydration through encapsulated moisture agents.	Long-lasting hydration.	Hyaluronic acid, glycerin	Creams, gels, lotions
Hair Care Products	Controlled release of conditioning agents for enhanced hair health.	Improved shine, and manageability.	Proteins, oils	Shampoos, conditioners
Exfoliating Agents	Delivering exfoliants in a controlled manner for skin renewal.	Gentle yet effective exfoliation.	Salicylic acid, glycolic acid	Exfoliating scrubs, masks
Perfume Delivery Systems	Encapsulating fragrance compounds for long-lasting scent release.	Extended fragrance longevity.	Essential oils, fragrance notes	Perfumes, body sprays
Sunscreen Formulations	Enhancing the stability and efficacy of sunscreens.	Improved UV protection and stability.	Chemical and physical UV filters	Sunscreens, after-sun products
Anti-Acne Treatments	Controlled delivery of anti-acne agents for targeted action.	Reduced breakouts, and improved skin clarity.	Benzoyl peroxide, salicylic acid	Spot treatments, creams
Cleansers	Enhancing the efficacy of cleansing formulations through encapsulated agents.	Improved cleansing efficiency.	Cleansing agents, surfactants	Facial cleansers, wipes
Skin Hydration	Using mesoporous materials to retain moisture in skin formulations.	Deep hydration, and improved skin elasticity.	Moisturizers, hydration	Hydrating masks, creams
Facial Masks	Controlled release of active ingredients in sheet or cream masks.	Targeted treatment for various skin issues.	Nutrients, botanicals	Sheet masks, peel-off masks
Colour Cosmetics Stability	Enhancing the stability of pigments in formulations.	Longer-lasting makeup products.	Colorants, pigments	Foundations, blushes
Anti-Inflammatory Formulations	Delivering soothing agents for sensitive skin.	Reduced redness and irritation.	Aloe vera, chamomile	Creams, gels
Regenerative Treatments	Facilitating the delivery of growth factors and peptides for skin regeneration.	Promotes skin repair and rejuvenation.	Growth factors, stem cells	Regenerative creams, serums
Personal Care Products	Application in deodorants and personal hygiene products for enhanced efficacy.	Improved odour control and freshness.	Antiperspirants, fragrances	Deodorants, body washes

Materials Research Foundations 173 (2025) 441-469 https://doi.org/10.21741/9781644903452-17

5. Advantages of Mesoporous Materials in Cosmetic Formulations

Mesoporous materials are beneficial to the use of cosmetics for several reasons that are related to product performance and consumer experience. Most importantly, they can protect numerous active components, which are very often vulnerable to the influence of processes like heat, light, or moisture [71]. The implementation of mesoporous features allows for such components to be minimised transforming degradation and creating a better product. What is more, mesoporous materials incorporated into a system could extend the durability of the test products so that they could be active for a long period. This stability is a plus for the producers since it minimizes waste and is a plus for the end users as they get products that are reliable for use over some time. Figure 3 and Table 3 show the advantages and disadvantages of Mesoporous Materials in Cosmetic Formulations.

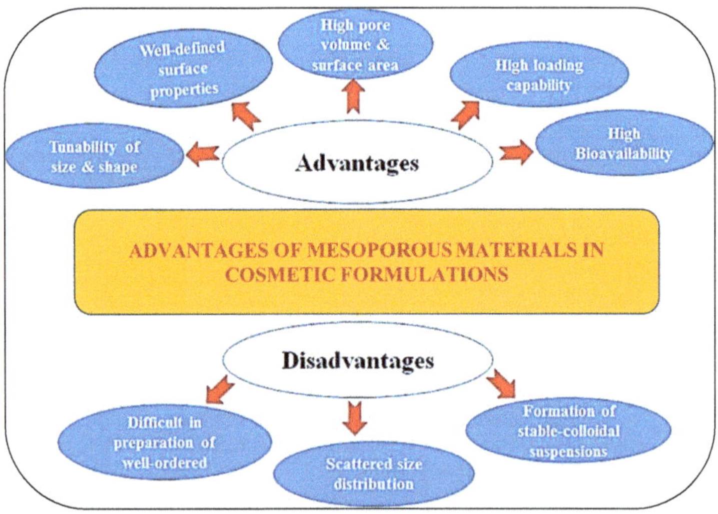

Figure 3. Advantages and Disadvantages of Mesoporous Materials in Cosmetic Formulations.

In addition to this, mesoporous materials also facilitate deeper penetration into the skin, increasing their efficacy by allowing active agents to better permeate the skin. This also improves the effectiveness of the wavelength active as it can penetrate thick layers of the skin where active ingredients can perform better. In the end, the adaptation of mesoporous materials could help minimize irritancy or sensitization of skin from most cosmetic formulations. The usage of mesoporous materials, when active ingredients are released from the formulation, it is controlled so that adverse effects will not occur. In conclusion, such materials in the enhancement of cosmetic products should be welcome as they improve product stability, efficacy and safety for users.

Materials Research Foundations 173 (2025) 441-469 https://doi.org/10.21741/9781644903452-17

Table 3. Advantages of Mesoporous Materials in Cosmetic Formulations.

Advantage	Description	Benefit to Formulation	Impact on Product Performance	Consumer Benefit	Example Ingredient/Application
Enhanced stability of active ingredients	Encapsulation of sensitive ingredients reduces degradation	Prolongs activity	Preserves efficacy over time	Longer-lasting products	Vitamin C
Prolonged shelf-life	Protects ingredients from environmental degradation	Increases shelf life	Delays oxidation and degradation	Reduced waste	Antioxidants
Improved skin penetration	Enhanced delivery of active agents into deeper skin layers	Improves absorption	Maximizes efficacy	Enhanced product effectiveness	Retinoids
Controlled release	Gradual and sustained release of ingredients	Maintains consistent effectiveness	Prolonged action over time	Longer-lasting effects	Moisturizers, UV filters
Reduction in skin irritation	Reduced exposure to high concentrations of active agents	Minimizes irritation	Improved tolerability for sensitive skin	Less risk of allergic reactions	Anti-aging agents like AHAs and BHAs
Targeted delivery	Directs active ingredients to specific skin areas	Increases precision	Optimizes the use of key ingredients	Focused treatment of problem areas	Whitening agents
Versatility in formulation	Compatibility with various cosmetic forms	Increases flexibility for product types	Enables use in creams, gels, lotions, etc.	Customizable product textures	Sunscreen, moisturizers
High surface area	Increased contact area for active agents	Enhances interaction with skin	Greater ingredient efficacy	Higher performance of actives	Antioxidants, enzymes
Enhanced moisture retention	Helps retain the skin's natural moisture	Prevents dehydration	Better skin hydration	Softer, more moisturized skin	Moisturizing creams
Stabilization of volatile ingredients	Encapsulation of volatile compounds	Prevents loss of activities	Maintains potency of formulations	Increased product performance	Essential oils

Materials Research Foundations 173 (2025) 441-469 https://doi.org/10.21741/9781644903452-17

	to prevent evaporation				
Customizable pore size	Ability to tailor pore size for specific ingredients	Optimizes ingredient loading	Enhances efficiency of encapsulation	Better controlled formulation properties	Vitamin complexes
Enhanced UV protection	Improves stability of UV filters	Increases UV protection efficiency	Prolonged protection under sun exposure	Safer, more effective sun care products	Sunscreens
Compatibility with natural ingredients	Supports natural and organic product development	Enhances clean label formulations	Boosts natural product stability	Consumer preference for natural options	Botanical extracts, essential oils
Improved texture and spreadability	Smooth, even application due to nanostructure properties	Enhances user experience	Easier, more pleasant product application	Improved satisfaction with product feel	Foundations, creams
Anti-ageing benefits	Slow release of anti-ageing actives for sustained effect	Prolongs the activity of key anti-aging agents	Reduces the appearance of fine lines	Smoother, younger-looking skin	Retinol, peptides
Enhanced antioxidant efficacy	Protects and releases antioxidants gradually	Prolonged antioxidant protection	Reduces oxidative stress on the skin	Better protection against environmental damage	Vitamin E, Coenzyme Q10
Boosted efficacy of natural actives	Helps enhance the bioavailability of plant-based activities	Increases the potency of natural ingredients	Maximizes benefits of natural formulations	More effective natural and organic products	Green tea, aloe vera
Improved colouration in decorative cosmetics	Ensures uniform distribution of pigments	Enhanced pigment stability	Vibrant and long-lasting colour	Brighter, more consistent makeup performance	Lipsticks, foundations
Anti-microbial properties	Some mesoporous materials possess intrinsic antimicrobial activity	Improves product preservation	Prevents microbial contamination	Safer, longer-lasting products	Skin cleansers, antimicrobial creams

Materials Research Foundations 173 (2025) 441-469
https://doi.org/10.21741/9781644903452-17

Temperature stability	Protects formulations from temperature fluctuations	Prevents melting or separation of ingredients	Maintains integrity across temperature ranges	Better product resilience	Creams and emulsions
Eco-friendly material options	Mesoporous silica and other materials can be sustainably sourced	Supports sustainable product development	Reduces environmental impact	Eco-conscious product choices	Organic and green beauty products
Minimal invasive delivery	Enables non-invasive application of active agents	Reduces the need for more aggressive treatments	Gentler skin treatments	Comfort and safety for users	Hydration masks, anti-ageing serums
Enhanced product transparency	Improves clarity in gel and serum formulations	Aesthetic appeal of clear formulations	Visually attractive cosmetic products	Popular for transparent gels and serums	Serums, facial gels
Cost-effectiveness	Allows for more efficient use of expensive activities	Reduces the amount of activity needed	More affordable high-performance products	Affordable skincare solutions	Anti-ageing creams
Customizable release profiles	The release can be tailored for immediate or delayed effects	Adaptable to various product needs	Enhanced performance customization	Versatile, user-tailored product experience	Time-release moisturizers, treatments

6. Safety and Biocompatibility of Mesoporous Materials

Considering their application in cosmetics, the safety and biocompatibility of mesoporous materials are among the primary concerns when they are incorporated into a product. The high surface area, adjustable pore sizes and chemical properties make such materials a lot of merits to the improvement of cosmetic products. Nevertheless, this has to be done through safe means which involve doing all the necessary toxicological studies, long-term studies of biocompatibility, and following the set laws and standards regarding the use of substances in cosmetics [72].

The first step in almost every evaluation of mesoporous materials for skin contact and use in formulations is the toxicological evaluation of these materials. This includes the toxicity of materials up to cellular and organism levels. Cytotoxicity tests are standard practice: these are performed to determine if there are any adverse effects of mesoporous materials on skin cells that may cause irritation and other inflammatory effects. Studies indicate that at the amounts used for example in formulations, cosmetic ingredients such as mesoporous silica and mesoporous titanium dioxide possess relatively high cytotoxicity. The low size of these particles is an advantage in the delivery of the active compounds but at the same time creates fears regarding their penetration across the skin and distribution in the body. Nevertheless, most studies indicate that most of the non-porous particles, especially those that are above 100nm are not absorbed into the skin thus the chances of causing systemic toxicity are very low [73].

Besides the evaluations oriented towards the short term, assessment of long-term effects and biocompatibility is also required to address concerns about the safety of using mesoporous materials for prolonged and repeated applications [74]. It also includes the evaluation of the long-term effect of these materials on surrounding biological tissues. For example, due to its potential for drug delivery and skin care products, mesoporous silica has been the subject of research, with the vast majority of literature noting very good biocompatibility, where the materials used do not cause side effects such as inflammation, and allergy or chronic toxicity. However, the biocompatibility of mesoporous materials may differ because of different methods of synthesis, technochemistry, and sizes of the particles. That is why there is a need for constant advancement of these materials and their applications to guarantee that they are still safe. There is also an attempt to develop studies concentrating on diable polymers with mesoporous structures, which are less likely to pose prolonged dangers since they can be degraded in the body or the environment [75].

The other important issue is allowing the mesoporous materials incorporated in cosmetics to comply with the regulations and standards of the respective authorities. As for cosmetic products, and especially for products intended for application on the skin, a high level of safety efficacy and quality is required in every region of the world. American Food and Drug Administration (FDA) and European Medicines Agency (EMA) – cover cosmetic ingredients including mesoporous substances and stipulate very high safety, efficacy and labelling standards. Such prohibitions normally state that extensive information on the chemical makeup, dimensions and surface characteristics of the material as well as evidence clearance of biotoxicity must be provided.

7. Challenges, Limitations, and Future Perspectives in Mesoporous Materials for Cosmetics

As much as mesoporous materials offer great prospects in cosmetics, there are still several obstacles and constraints that will require additional research and creativity. One of the major limitations is the issue of scalability. While cosmetics produced in mesoporous form are efficient and effective, the actual process involved in the making of controlled mesoporous materials including their micropore tweaks, is complex and costly, thus hampering production at commercial levels. Maintaining product efficacy and safety requires addressing finished product quality and uniformity across batches, but this may be difficult to achieve with modern nanomaterials.

Another limitation relates to biocompatibility and safety issues. Although mesoporous substances such as silica are generally safe in studies, it is still worrying to contemplate extended periods of exposure, especially for substances with well-defined nanoscale particles. A few of the mesoporous substances may even enter through the skin, or produce unexpected effects within the cells. Toxicological safety assessment along with long-term studies on lower age groups should be done before their application as users would be using such products more readily and shortly.

In addition, other challenges tend to come from the fact that some moves have to be made due to regulatory constraints. This is because cosmetic products that contain nanomaterials have to meet strict safety regulations that differ from one region to another. It ensures that the regulatory

Materials Research Foundations 173 (2025) 441-469
https://doi.org/10.21741/9781644903452-17

framework on the use of nanotechnology in cosmetics, which is dynamic, will lead to significant delays in the commercialization of new mesoporous-based products. Table 4 discusses challenges, Limitations, and Future Perspectives in Mesoporous Materials for Cosmetics.

Table 4. Challenges, Limitations, and Future Perspectives in Mesoporous Materials for Cosmetics.

Challenges	Limitations	Production Scalability	Regulatory Issues	Environmental Impact	Future Perspectives
Scaling up production for industrial use	High production costs	Difficulty in large-scale manufacturing	Stringent safety regulations for cosmetics	Potential environmental accumulation of materials	Innovations in mesoporous material design
Cost considerations	Limited commercial feasibility	Ensuring batch-to-batch uniformity	Evolving regulatory frameworks	Lack of biodegradable materials	Sustainable and green synthesis approaches
Difficulty in achieving consistent quality	High synthesis complexity	Tailoring pore size for diverse applications	Complex approval processes for nanomaterials	Concerns about nanoparticle persistence	Integration of nanotechnology in future cosmetics
Limited stability with some active ingredients	Compatibility with other cosmetic agents	Challenges in functionalizing material surfaces	Regional differences in safety standards	Environmental contamination due to nano-release	Potential for personalized cosmetic products
Unpredictable interaction with biological systems	Sensitivity to formulation conditions	Optimizing surface modifications	Lack of standardized safety assessments	Inadequate disposal or recycling methods	Targeted drug and active agent delivery systems
Concerns over toxicity of nanomaterials	Potential for skin penetration issues	Reducing agglomeration in formulations	Risk of delayed approvals	Impact on aquatic ecosystems	Enhanced skin delivery and penetration techniques
Limited long-term biocompatibility studies	Difficulty in determining long-term safety	High energy consumption in production	Need for clearer regulatory guidelines	Limited eco-friendly disposal options	Improved bioavailability and efficacy
High energy consumption in production	Limited availability of functionalized variants	Durability under diverse cosmetic conditions	Non-standardized testing protocols	Long-term degradation issues	Nanomaterial-based skin treatments
Difficulties in functionalizing surfaces	Low commercial availability	Achieving reproducibility in material properties	Variable safety testing between countries	Potential for harmful byproducts	Advancement in biodegradable mesoporous materials
Complex synthesis procedures	Limited compatibility with sensitive ingredients	Increased production costs due to complexity	Lengthy regulatory approval times	Long-term ecological impact of nanomaterials	Evolution of eco-friendly cosmetic solutions

Conclusion

The modern distinctive structural, biocompatible and multifunctional characteristics of mesoporous materials make them suitable for a great range of applications in cosmetics. The most remarkable advantages of these materials for cosmetic preparations are their high surface area, uniform pore diameter and controllable porosity. Such materials will primarily allow the encapsulation and sustained release of active substances such as vitamins, antioxidants, moisturizers, UV filters and the like, which increase the stability and effectiveness of these materials and ensure the shelf life of the aa products. Given sustained release, mesoporous materials promote deeper penetration into the skin and longer lasting benefits thus enhancing the performance and effectiveness of the product.

Reducing skin irritation and sensitization has become one of the most important pros of the use of mesoporous materials, which makes it an ideal candidate for sensitive skin care products. In addition, they enhance cosmetic formulations by offering protection to sensitive active ingredients from storage degradation as well as from the environment. In addition, the demonstrated application of these materials in colour cosmetics, anti-ageing therapies and sunscreens indicates the extensive field of usage for these materials. Notwithstanding these merits, several challenges and limitations persist that need to be dealt with. The cost of production as well as the inability to scale up the manufacturing process greatly impede the prospects of mass production in developing industries. Also, the risks of approval for the use of nanomaterials in cosmetic products have been equally responsible for the delays in commercializing these products.

From the societal perspective, there are also concerns about the use of nanomaterials during their life cycle, which means that biosynthetic or fuss-free green chemical approaches ought to be employed instead. Cosmetics application of mesoporous materials would appear to be attractive because of the recent advancement of material designing and functionalization as well as the green approaches to synthesis. This enables the evolution of next-women products that complement the specific needs of different skin types. This makes such cosmetic products become individualized. Due to the availability of research data on nanotechnology as well as biocompatibility, it is clear that the use of mesoporous materials in the cosmetics industry will be enhanced, particularly in skin delivery systems aiming at designing effective and long-lasting cosmetic products.

References

[1] A. Ribeiro, M. Estanqueiro, M. Oliveira, J.S. Lobo, Main benefits and applicability of plant extracts in skin care products, Cosmetics. 2 (2015) 48-65. https://doi.org/10.3390/cosmetics2020048

[2] M.A. Farage, K.W. Miller, P. Elsner, H.I. Maibach, Intrinsic and extrinsic factors in skin ageing: A review, Int. J. Cosmet. Sci. 30 (2008) 87-95. https://doi.org/10.1111/j.1468-2494.2007.00415.x

[3] J. Krutmann, A. Bouloc, G. Sore, B.A. Bernard, T. Passeron, The skin aging exposome, J. Dermatol. Sci. 85 (2017) 152-161. https://doi.org/10.1016/j.jdermsci.2016.09.015

[4] R.H. Dahal, D.S. Shim, J. Kim, Development of actinobacterial resources for functional cosmetics, J. Cosmet. Dermatol. 16 (2017) 243-252. https://doi.org/10.1111/jocd.12304

[5] A. Long, J. Crowther, K. Beach, J. Neil, Skin topometry changes demonstrate anti-aging efficacy of a topically applied cosmetic product via in vivo testing, J. Am. Acad. Dermatol. 56 (2007) 224. https://doi.org/10.1016/j.jaad.2006.10.184

[6] R. Ganceviciene, A.I. Liakou, A. Theodoridis, E. Makrantonaki, C.C. Zouboulis, Skin anti-aging strategies, Derm.-Endocrinol. 4 (2012) 308-319. https://doi.org/10.4161/derm.22804

[7] R.A. Baxter, Anti-aging properties of resveratrol: Review and report of a potent new antioxidant skin care formulation, J. Cosmet. Dermatol. 7 (2008) 2-7. https://doi.org/10.1111/j.1473-2165.2008.00354.x

[8] S. Mukherjee, A. Date, V. Patravale, H.C. Korting, A. Roeder, G. Weindl, Retinoids in the treatment of skin aging: An overview of clinical efficacy and safety, Clin. Interv. Aging. 1 (2006) 327-348. https://doi.org/10.2147/ciia.2006.1.4.327

[9] B. Fubini, A. Hubbard, Reactive oxygen species (ROS) and reactive nitrogen species (RNS) generation by silica in inflammation and fibrosis, Free Radic. Biol. Med. 34 (2003) 1507-1516. https://doi.org/10.1016/S0891-5849(03)00149-7

[10] M. Valko, D. Leibfritz, J. Moncol, M.T.D. Cronin, M. Mazur, J. Telser, Free radicals and antioxidants in normal physiological functions and human disease, Int. J. Biochem. Cell Biol. 39 (2007) 44-84. https://doi.org/10.1016/j.biocel.2006.07.001

[11] L.A.P. Huy, H. Hua, C.P. Huy, Free radicals, antioxidants in disease and health, Int. J. Biomed. Sci. 4 (2008) 89-96. https://doi.org/10.59566/IJBS.2008.4089

[12] B. Uttara, A.V. Singh, P. Zamboni, R. Mahajan, Oxidative stress and neurodegenerative diseases: A review of upstream and downstream antioxidant therapeutic options, Curr. Neuropharmacol. 7 (2009) 65-74. https://doi.org/10.2174/157015909787602823

[13] E.B. Kurutas, The importance of antioxidants which play the role in cellular response against oxidative/nitrosative stress: Current state, Nutr. J. 15 (2016) 71. https://doi.org/10.1186/s12937-016-0186-5

[14] O.A. Fawole, N.P. Makunga, U.L. Opara, Antibacterial, antioxidant and tyrosinase-inhibition activities of pomegranate fruit peel methanolic extract, BMC Complement. Altern. Med. 12 (2012) 1-11. https://doi.org/10.1186/1472-6882-12-200

[15] M. Kim, T. Shibata, S. Kwon, T.J. Park, H.Y. Kang, Ultraviolet-irradiated endothelial cells secrete stem cell factor and induce epidermal pigmentation, Sci. Rep. 8 (2018) 4235. https://doi.org/10.1038/s41598-018-22608-y

[16] K.D. Hsu, H.J. Chen, C.S. Wang, C.C. Lum, S.P. Wu, S.P. Lin, K.C. Cheng, Extract of ganoderma formosanum mycelium as a highly potent tyrosinase inhibitor, Sci. Rep. 6 (2016) 32854. https://doi.org/10.1038/srep32854

[17] S. Mondal, A. Thampi, M. Puranik, Kinetics of melanin polymerization during enzymatic and nonenzymatic oxidation, J. Phys. Chem. B. 122 (2018) 2047-2063. https://doi.org/10.1021/acs.jpcb.7b07941

[18] C. Corinaldesi, G. Barone, F. Marcellini, A. Dell'Anno, R. Danovaro, Marine microbial-derived molecules and their potential use in cosmeceutical and cosmetic products, Mar. Drugs. 15 (2017) 118. https://doi.org/10.3390/md15040118

[19] A. Herman, A.P. Herman, B.W. Domagalska, A. Młynarczyk, Essential oils and herbal extracts as antimicrobial agents in cosmetic emulsion, Indian J. Microbiol. 53 (2013) 232-237. https://doi.org/10.1007/s12088-012-0329-0

[20] S.F. Nabavi, A.D. Lorenzo, M. Izadi, E.S. Sánchez, M. Daglia, S.M. Nabavi, Antibacterial effects of cinnamon: From farm to food, cosmetic and pharmaceutical industries. Nutrients. 7 (2015) 7729-7748. https://doi.org/10.3390/nu7095359

[21] U. Klaschka, Natural personal care products-analysis of ingredient lists and legal situation, Environ. Sci. Eur. 28 (2016) 8. https://doi.org/10.1186/s12302-016-0076-7

[22] G.A. Rivera, M. Llompart, M. Lores, C.G. Jares, Preservatives in cosmetics: Regulatory aspects and analytical methods, Anal. Cosmet. Prod. (2018) 175-224. https://doi.org/10.1016/B978-0-444-63508-2.00009-6

[23] L.H. Lee, K.G. Chan, J. Stach, E.M.H. Wellington, B.H. Goh, Editorial: The search for biological active agent(s) from actinobacteria. Front. Microbiol. 9 (2018) 824. https://doi.org/10.3389/fmicb.2018.00824

[24] R. Xu, N. Shang, P. Li, In vitro and in vivo antioxidant activity of exopolysaccharide fractions from Bifidobacterium animalis RH, Anaerobe. 17 (2011) 226-231. https://doi.org/10.1016/j.anaerobe.2011.07.010

[25] J.A. Frank, C.I. Reich, S. Sharma, J.S. Weisbaum, B.A. Wilson, G.J. Olsen, Critical evaluation of two primers commonly used for amplification of bacterial 16S rRNA genes, Appl. Environ. Microbiol. 74 (2008) 2461-2470. https://doi.org/10.1128/AEM.02272-07

[26] S.H. Yoon, S.M. Ha, S. Kwon, J. Lim, Y. Kim, H. Seo, J. Chun, Introducing EzBioCloud: A taxonomically united database of 16S rRNA gene sequences and whole-genome assemblies, Int. J. Syst. Evol. Microbiol. 67 (2017) 1613-1617. https://doi.org/10.1099/ijsem.0.001755

[27] National Center for Biotechnology Information. https://www.ncbi.nlm.nih.gov/, 2024 (accessed 25 March 2024).

[28] S. Kumar, G. Stecher, K. Tamura, MEGA7: Molecular evolutionary genetics analysis version 7.0 for bigger datasets, Mol. Biol. Evol. 33 (2016) 1870-1874. https://doi.org/10.1093/molbev/msw054

[29] H.M. Shin, H.K. Byung, Y.C. Eun, S.H. Jung, S.K. Yeong, R.M. Kyung, Y. Kim, Suppressive effect of novel aromatic diamine compound on nuclear factor-κB-dependent expression of inducible nitric oxide synthase in macrophages, Eur. J. Pharmacol. 521 (2005) 1-8. https://doi.org/10.1016/j.ejphar.2005.07.013

[30] D. Zhao, Q. Huo, J. Feng, B.F. Chmelka, G.D. Stucky, Nonionic triblock and star diblock copolymer and oligomeric surfactant syntheses of highly ordered, hydrothermally stable, mesoporous silica structures, J. Am. Chem. Soc. 120 (1998) 6024-6036. https://doi.org/10.1021/ja974025i

[31] H.D. Brightbill, R.L. Modlin, Toll-like receptors: Molecular mechanisms of the mammalian immune response, Immunology. 101 (2000) 1-10. https://doi.org/10.1046/j.1365-2567.2000.00093.x

[32] Y. Emre, C. Hurtaud, T. Nübel, F. Criscuolo, D. Ricquier, A.M.C. Doulcier, Mitochondria contribute to LPS-induced MAPK activation via uncoupling protein UCP2 in macrophages, Biochem. J. 402 (2007) 271-278. https://doi.org/10.1042/BJ20061430

[33] W. Sosroseno, I. Barid, E. Herminajeng, H. Susilowati, Nitric oxide production by a murine macrophage cell line (RAW264.7) stimulated with lipopolysaccharide from Actinobacillus actinomycetemcomitans, Oral Microbiol. Immunol. 17 (2002) 72-78. https://doi.org/10.1046/j.0902-0055.2001.00091.x

[34] S. Ahmed, S.A. Sulaiman, A.A. Baig, M. Ibrahim, S. Liaqat, S. Fatima, S. Jabeen, N. Shamim, N.H. Othman, Honey as a potential natural antioxidant medicine: An insight into its molecular mechanisms of action, Oxid. Med. Cell. Longev. 2018 (2018) 8367846. https://doi.org/10.1155/2018/8367846

[35] S.J. Padayatty, A. Katz, Y. Wang, P. Eck, O. Kwon, J.H. Lee, S. Chen, C. Corpe, A. Dutta, S.K. Dutta, et al. Vitamin C as an antioxidant: Evaluation of its role in disease prevention, J. Am. Coll. Nutr. 22 (2003) 18-35. https://doi.org/10.1080/07315724.2003.10719272

[36] D.E. Pratt, Natural antioxidants from plant material, in: M.T. Huang, C.T. Ho, C.Y. Lee (Eds.), Phenolic Compounds in Food and Their Effects on Health II, ACS Symposium Series, American Chemical Society: Washington, DC, USA, 1992, pp. 54-71. https://doi.org/10.1021/bk-1992-0507.ch005

[37] D.M. Kasote, S.S. Katyare, M.V. Hegde, H. Bae, Significance of antioxidant potential of plants and its relevance to therapeutic applications, Int. J. Biol. Sci. 11 (2015) 982-991. https://doi.org/10.7150/ijbs.12096

[38] L. Wei, N. Hu, Y. Zhang, Synthesis of polymer-mesoporous silica nanocomposites, Materials. 3 (2010) 4066-4079. https://doi.org/10.3390/ma3074066

[39] S.P. Chaudhari, A. Gupte, Mesoporous silica as a carrier for amorphous solid dispersion, J. Pharm. Res. Int. 16 (2017) 1-19. https://doi.org/10.9734/BJPR/2017/33553

[40] F.L. Li, H.J. Zhang, Synthesis of hollow sphere and 1D structural materials by sol-gel process, Materials. 10 (2017) 995. https://doi.org/10.3390/ma10090995

[41] E. Blanco, L. Esquivias, R. Litrán, M. Piñero, M.R.D. Solar, N.D.L.R. Fox, Sonogels and derived materials, Appl. Organomet. Chem. 13 (1999) 399-418. https://doi.org/10.1002/(SICI)1099-0739(199905)13:5<399::AID-AOC825>3.0.CO;2-A

[42] S.P. Chaudhari, A. Gupte, Mesoporous silica as a carrier for amorphous solid dispersion, J. Pharm. Res. Int. 16 (2017) 1-19. https://doi.org/10.9734/BJPR/2017/33553

[43] A.M. Putz, W. Kunzhou, A. Len, J. Plocek, P. Bezdicka, G.P. Kopitsa, T.V. Khamova et al. Mesoporous silica obtained with methyltriethoxysilane as co-precursor in alkaline medium, Appl. Surf. Sci. 424 (2017) 275-281. https://doi.org/10.1016/j.apsusc.2017.04.121

[44] W.K. Tan, H. Muto, G. Kawamura, Z. Lockman, A. Matsuda, Nanomaterial fabrication through the modification of sol-gel derived coatings, Nanomaterials. 11 (2021) 181. https://doi.org/10.3390/nano11010181

[45] A. Len, G. Paladini, L. Románszki, A.M. Putz, L. Almásy, K. László, S. Bálint et al. Physicochemical characterization and drug release properties of methyl-substituted silica

xerogels made using sol-gel process, Int. J. Mol. Sci. 22 (2021) 9197. https://doi.org/10.3390/ijms22179197

[46] O.A. Kamanina, A.S. Evgeniya, P.V. Rybochkin, V.A. Arlyapov, A.N. Vereshchagin, V.P. Ananikov, Preparation of hybrid sol-gel materials based on living cells of microorganisms and their application in nanotechnology, Nanomaterials. 12 (2022) 1086. https://doi.org/10.3390/nano12071086

[47] S.P. Chaudhari, A. Gupte, Mesoporous silica as a carrier for amorphous solid dispersion, J. Pharm. Res. Int. 16 (2017) 1-19. https://doi.org/10.9734/BJPR/2017/33553

[48] A.M. Putz, K. Wang, A. Len, J. Plocek, P. Bezdicka, G.P. Kopitsa, T.V. Khamova et al. Mesoporous silica obtained with methyltriethoxysilane as co-precursor in alkaline medium, Appl. Surf. Sci. 424 (2017) 275-281. https://doi.org/10.1016/j.apsusc.2017.04.121

[49] R.E. Morsi, R.S. Mohamed, Nanostructured mesoporous silica: Influence of the preparation conditions on the physical-surface properties for efficient organic dye uptake, R. Soc. Open Sci. 5 (2018) 172021. https://doi.org/10.1098/rsos.172021

[50] S. Sadjadi, M.N. Jamal, A survey on application of mesoporous materials in chemistry, Curr. Chem. Lett. 8 (2019) 69-86. https://doi.org/10.5267/j.ccl.2019.2.001

[51] H.Q. Qutaish, Synthesis and Characterisation of Mesoporous Transition Metal Oxides Based on Soft-Templating Method, University of Wollongong, Australia, 2019.

[52] C. Chircov, A. Spoială, C. Păun, L. Crăciun, D. Ficai, A. Ficai, E. Andronescu, Ş.C. Turculeţ, Mesoporous silica platforms with potential applications in release and adsorption of active agents, Molecules. 25 (2020) 3814. https://doi.org/10.3390/molecules25173814

[53] N. Pal, J.H. Lee, E.B. Cho, Recent trends in morphology-controlled synthesis and application of mesoporous silica nanoparticles, Nanomaterials. 10 (2020) 2122. https://doi.org/10.3390/nano10112122

[54] R. Szymanska, P. Pospisil, J. Kruk, Plant-derived antioxidants in disease prevention, Oxid. Med. Cell. Longev. 2016 (2016) 1-2. https://doi.org/10.1155/2016/1920208

[55] M.P.N. Rao, M. Xiao, W.J. Li, Fungal and bacterial pigments: Secondary metabolites with wide applications, Front. Microbiol. 8 (2017) 1113. https://doi.org/10.3389/fmicb.2017.01113

[56] Y. Wang, Y. Wu, Y. Wang, H. Xu, X. Mei, D. Yu, Y. Wang, W. Li, Antioxidant properties of probiotic bacteria, Nutrients. 9 (2017) 521. https://doi.org/10.3390/nu9050521

[57] R.K. Ko, G.O. Kim, C.G. Hyun, D.S. Jung, N.H. Lee, Compounds with tyrosinase inhibition, elastase inhibition and DPPH radical scavenging activities from the branches of Distylium racemosum Sieb. et Zucc, Phytother. Res. 25 (2011) 1451-1456. https://doi.org/10.1002/ptr.3439

[58] P. Manivasagan, J. Venkatesan, K. Sivakumar, S.K. Kim, Pharmaceutically active secondary metabolites of marine actinobacteria, Microbiol. Res. 169 (2014) 262-278. https://doi.org/10.1016/j.micres.2013.07.014

[59] J. You, K.B. Roh, Z. Li, G. Liu, J. Tang, S. Shin, D. Park, E. Jung, J. You, K.B. Roh, et al. The antiaging properties of Andrographis paniculata by activation epidermal cell stemness, Molecules. 20 (2015) 17557-17569. https://doi.org/10.3390/molecules200917557

[60] N. Oliver, M. Sternlicht, K. Gerritsen, R. Goldschmeding, Could aging human skin use a connective tissue growth factor boost to increase collagen content?, J. Investig. Dermatol. 130 (2010) 338-341. https://doi.org/10.1038/jid.2009.331

[61] G.O. Gular, Studies on antioxidant properties of the different solvent extracts and fatty acid composition of Hyoscyamus reticulatus L, J. Food Biochem. 36 (2012) 532-538. https://doi.org/10.1111/j.1745-4514.2011.00564.x

[62] Z.A. Elagbar, R.R. Naik, A.K. Shakya, S.K. Bardaweel, Fatty acids analysis, antioxidant and biological activity of fixed oil of annona muricata L. seeds, J. Chem. 2016 (2016) 3-4. https://doi.org/10.1155/2016/6948098

[63] A. Aktumsek, G. Zengin, G.O. Guler, Y.S. Cakmak, A. Duran, Assessment of the antioxidant potential and fatty acid composition of four Centaurea L. taxa from Turkey, Food Chem. 141 (2013) 91-97. https://doi.org/10.1016/j.foodchem.2013.02.092

[64] T. Nakatsuji, M.C. Kao, J.Y. Fang, C.C. Zouboulis, L. Zhang, R.L. Gallo, C.M. Huang, Antimicrobial property of lauric acid against propionibacterium acnes: Its therapeutic potential for inflammatory acne vulgaris. J. Investig. Dermatol. 129 (2009) 2480-2488. https://doi.org/10.1038/jid.2009.93

[65] M. Otto, Staphylococcus colonization of the skin and antimicrobial peptides, Expert Rev. Dermatol. 5 (2010) 183-195. https://doi.org/10.1586/edm.10.6

[66] T. Yanagisawa, T. Shimizu, K. Kuroda, C. Kato, The preparation of alkyltriinethylaininonium-kaneinite complexes and their conversion to microporous materials, Bull. Chem. Soc. Jpn. 63 (1990) 988-992. https://doi.org/10.1246/bcsj.63.988

[67] J.S. Beck, J.C. Vartuli, W.J. Roth, M.E. Leonowicz, C.T. Kresge, K.D. Schmitt, C.T.W. Chu, D.H. Olson, E.W. Sheppard, S.B. McCullen, et al. A new family of mesoporous molecular sieves prepared with liquid crystal templates, J. Am. Chem. Soc. 114 (1992) 10834-10843. https://doi.org/10.1021/ja00053a020

[68] B.G. Trewyn, I.I. Slowing, S. Giri, H.T. Chen, V.S.Y. Lin, Synthesis and functionalization of a mesoporous silica nanoparticle based on the sol-gel process and applications in controlled release. Acc. Chem. Res. 40 (2007) 846-853. https://doi.org/10.1021/ar600032u

[69] D. Zhao, J. Feng, Q. Huo, N. Melosh, G.H. Fredrickson, B.F. Chmelka, G.D. Stucky, Triblock copolymer syntheses of mesoporous silica with periodic 50 to 300 angstrom pores, Science. 279 (1998) 548-552. https://doi.org/10.1126/science.279.5350.548

[70] L.S.D. Silva, A.L. Moretti, G.G. Cher, P.A. Arroyo, Influence of crystallization and ageing time on the reproducibility of mesoporous molecular sieve SBA-15, Rev. Matér. 24 (2019). https://doi.org/10.1590/s1517-707620190002.0695

[71] N. Knežević, J.O. Durand, Large pore mesoporous silica nanomaterials for application in delivery of biomolecules, Nanoscale. 7 (2015) 2199-2209. https://doi.org/10.1039/C4NR06114D

[72] M.L. Ojeda, J.M. Esparza, A. Campero, S. Cordero, I. Kornhauser, F. Rojas, On comparing BJH and NLDFT pore-size distributions determined from N2 sorption on SBA-15 substrata, Phy. Chem. Chem. Phy. 5 (2003)1859-1866. https://doi.org/10.1039/b300821e

[73] T. Arshad, H.M.S. Khan, N. Akhtar, H. Hanan, M.D. Hussain, M. Kazi, Structural elucidation and development of azelaic acid loaded mesoporous silica nanoparticles infused gel: Revolutionizing nanodrug delivery for cosmetics and pharmaceuticals, Heliyon. 10 (2024). https://doi.org/10.2139/ssrn.4721424

[74] S. Saijo, M. Shibata, C. Izawa, Enhancement of the adsorbed amount of ferulic acid on mesoporous silica for use as an ultraviolet radiation absorber in cosmetics, J. Jpn. Soc. Colour Mater. 96 (2023) 108-112. https://doi.org/10.4011/shikizai.96.108

[75] N. Chen, S. Yao, M. Li, Q. Wang, X. Sun, X. Feng, Y. Chen, Nonporous versus mesoporous bioinspired polydopamine nanoparticles for skin drug delivery, Biomacromolecules. 24 (2023) 1648-1661. https://doi.org/10.1021/acs.biomac.2c01431

Materials Research Foundations 173 (2025) 470-495 https://doi.org/10.21741/9781644903452-18

Chapter 18

Mesoporous Materials for Diagnostics Applications

Melkamu Biyana Regasa

Department of Chemistry, College of Natural and Computational Sciences, P.O.Box 395, Nekemte, Ethiopia

Abstract

Mesoporous materials emerged as highly functional material systems to solve the problems of medical issues in health diagnostics. This might be because of these materials' tunable structure, broad surface areas, advanced cargo capability, voluminous chemical availability, low cost, low toxicity, high performance for visualization and treatment, and outstanding possibility to incorporate the desirable functions for diagnosis. Recently, tremendous investigations were performed along the potential applications of mesoporous materials in diagnosis like therapeutics, theranostic, medical imaging, point-of-care tests and clinical laboratory testing. This chapter discusses the applications of mesoporous materials in the advancement of precise drug transportation, improving the accuracy of bioimaging techniques, theranostic, point-of-care test and clinical laboratory testing.

Keywords

Mesoporous Materials, Health Diagnosis, Theranostic, Bioimaging, Point-of-Care-Test, Therapeutics

Contents

Materials Research Foundations 173 (2025) 470-495
https://doi.org/10.21741/9781644903452-18

1. Introduction

It is common to come across advanced functional materials such as mesoporous materials (either natural or synthetic) in our daily life. Materials which vacant voids or pores in the size range of 2-50 nanometers (nm) diameter are termed as mesoporous materials based on the IUPAC [1]. These permeable materials contain holes to an extrinsic fluid in the form of open holes, closed holes, through holes, or blind holes with structure or shape such as rounded, inkbottle, cone-forms, or slit-forms holes [2,3]. Mesoporous materials are a type of substance defined by a network of holes or pores with less than or equal to 50 nm. Inorganic (oxides, such as SiO_2, Al_2O_3, TiO_2 and ZrO_2), organic materials, organic-inorganic composites and mesporous carbon materials which usually have cubic or hexagonal structures are some examples of mesoporous materials [4]. From the different kinds of mesoporous materials, mesoporous silica (MSN) and metal oxides are well-studied and commonly used as support or substrate [5]. These materials with various holes have attracted great interest owing to the special possessions and wider application areas they have demonstrated. This might be due to their unique features such as consistent and adjustable voids, flexible surface structure, broad surface area, ample pore size, low toxicity and possibility for surface engineering [6,7].

The utilization of tiny molecules to identification is common and its therapeutical operations are limited by the unwanted side effects and fast secretion from the body part. Mesoporous materials have proven to be effective transporters for diagnosis, anticancer, anti-inflammatory, neurological medications, medical imaging and drug carriers [6-9]. Functional mesoporous materials come forth as a speedy grow up practical application for addressing the barriers of methods based on small molecules and enhancing clinical outcomes. Therefore, this chapter brings an insight into the applications of mesoporous materials in health diagnosis.

2. Synthesis and characterization methods

The preparation of mesoporous materials can be achieved by using different synthetic routes such as soft template, hard template and sol-gel based methods [1]. There are various synthesis and characterization techniques as depicted in Figure 1.

2.1 Soft-templating synthesis

The method involves the application of wetting agents (e.g. surfactants and block copolymers) as templates and inorganic materials as starting material. The self-organization between organic and inorganic materials forms mesoporous materials while the templates are used for the structure and pore-volume during preparation [10]. First, micelles are formed by the templates followed by the addition of inorganic materials. Second, the network between the organic and inorganic is formed by hydrolyzing the precursor then condensation and polymerization occur to form the mesoporous

Materials Research Foundations 173 (2025) 470-495
https://doi.org/10.21741/9781644903452-18

matrix. Next, further networking and adjusting pore size and nucleation is done via hydrothermal treatment after reaction is completed. Finally, the template release is performed via solvent extraction or calcination to obtain mesoporous materials for the specific requirements of various applications [4,11].

2.2 Sol-Gel Processes

Porous materials with different pore sizes are formed by the precursor materials that hydrolyze and participate in condensation reactions. This method is the type of bottom-up approach in nanotechnology and is among the most established protocols to prepare mesoporous materials [12,13]. Mesoporous inorganic nanomaterials like glasses and ceramics in various forms such as nanogels, films, fibers, and powders are synthesized by the sol-gel method [14,15]. It offers some advantages including low reaction temperature and adjustable parameters during synthesis processes which make this method more interesting [6].

2.3 Hard-templating synthesis

The method also called nanocasting and is another simple method to synthesize mesoporous materials [16]. It involves replicating the structure of a sacrificial template material, such as colloidal particles or porous membranes, to create mesoporous materials [17]. The hard-templating method involves four important steps [18]: template preparation to obtain avoid materials with tunable structure, precursor introduction, polymerization and destructive distillation, and liberation of the guide (template) to obtain porous materials with hollow structures and rigid particles [19]. Few mesoporous materials with uniform arrangements and adaptable ratio of the various pores are possible to prepared using hydrothermal artificial methods apart from its low freedom over materials structures [20].

2.4 Other methods

There are also other methods such as in situ-templating, template free and reticular chemistry methods that are used to prepare mesoporous materials. The in situ-templating involves the phase separation of precursors' chemical species during thermal or chemical treatment. Another method is template free strategy. It is template free method, and the mesoporous materials nanoparticles formation is based on interparticle spaces between neighboring nanosize building blocks. However, in the reticular chemistry approaches, molecular building blocks are used to produce covalent–organic frameworks or metal–organic frameworks with crystalline pore walls. In addition to these, other methods such as electrochemical [21], hydrothermal [22], valorization [23], and calcination [24] are also used to synthesize mesoporous materials.

2.5 Characterization techniques

The synthesis efficiency, chemical composition, functional units, pore-structures of the mesoporous materials is confirmed through characterization using various techniques. These techniques include Brunauer-Emmett-Teller (BET), infrared spectroscopy (IR), X-ray diffraction (XRD), thermogravimetric analysis (TGA), scanning electron microscopy (SEM), nuclear magnetic resonance (NMR), transmission electron microscopy (TEM), N_2 adsorption, and [25,26].

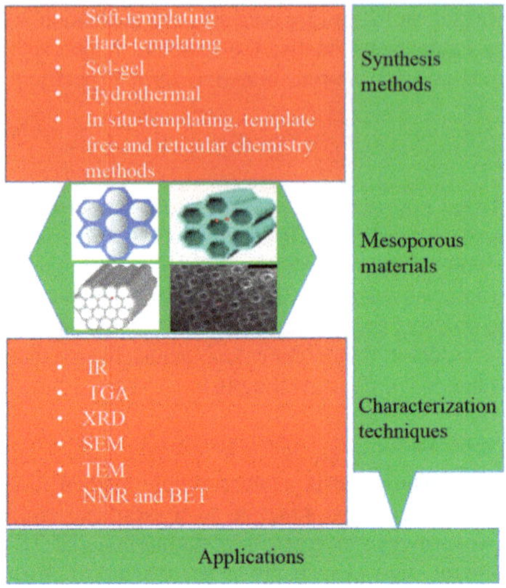

***Figure 1**. Mesoporous materials preparation and characterization methods*

XRD is used to identify the crystallographic symmetry, SEM is employed for morphological and particle size analysis, N_2 adsorption method is used to determine the holes volume and artifact area determination, NMR analysis offers the carbon and hydrogen content of the surfaces, BET gives surface area, IR gives functional units and TGA obtains thermal stability [27]. This implies that chemical and physical analysis provides an insight about the crystalline nature, morphology and structure of the mesoporous materials [28].

3. Types of mesoporous materials

Depending on their precursors, mesoporous materials can be grouped into carbon materials (obtained from biomass and petrochemicals), metal organic frameworks (MOFs), inorganic, mesoporous polymers and amorphous glasses which show highly small pore volume arrangement in the range of 0.5–2 nm [4]. The nature of the pores which can be either amorphous or crystalline framework that may be cylinderical or cage type geometry also determines the category of the mesoporous materials. Thus, most nanoporous materials might be mesoporous, macroporous, and microporous, depending on the magnitude of voids [29]. Generally, there are non-silica and silica-based materials with voids (Table 1). Also, ordered mesoporous materials such as silicas, organosilica compounds, sulfides, metal oxides, carbonitriles, nitrides, and nitrides, zeolites or pure metals are few examples available vastly [30,31].

Table 1*. Classes of mesoporous materials based on their compositions.*

Types	Composition	Examples	References
Inorganic	Metal, metalloid and nonmetal	Metal oxides (SiO_2, Al_2O_3, TiO_2, Fe_3O_4 and ZrO_2), metal Phosphonates etc	[16,32]
Organic	Carbon, hydrogen, oxygen etc	Carbon nanotube, grapheme, carbon spheres, fullerene, activated carbon etc	[29,33]
Organic-inorganic	Carbon, hydrogen, metal etc	Organosilica, benzene-Silica, Ti(IV) metal–organic framework, thiophene-silicas	[34-36]
Mesoporous Carbon	Carbon, hydrogen, oxygen etc	Carbon nanotubes, fullerene, graphene oxide etc	[29,37,38]

Mesoporous materials have wider application areas such as adsorption, drug delivery, hydrogen energy storage, separation technology, catalysis and sensors due to their regulated composition, larger surface areas and broad hole volumes [39,40]. Both organic and inorganic porous materials such as microporous (contain void radius less than 1 nm), mesoporous (have void radius from 1-25 nm), and macroporous (contain void radius greater than 25 nm) [41] exist as natural and synthetic materials. From the various types of porous materials, the inorganic based ones like silica nanoparticles are studied very well because of their high adaptability in the handling of construction and material possession [42]. Mesoporous silica nanoparticles (MSN) can have hexagonal, cubic, lamellar, cubic cages and cages like structures [42]. There are also magnetic mesoporous nanomaterials containing magnetic Fe_3O_4 nanoparticles [32] and carbon-based materials are common [29].

4. Properties of mesoporous materials

The structure and void size/geometry are the main factors affecting the features of materials with voids. The void geometry can be either three dimensional (3D) in the form of cage, or cylindrical in the form of two- or three-dimensional structures [38]. These structural dimensions are made up of the ellipsoidal or spherical cages that are interconnected into three dimensions to facilitate the mass transfer of active agents. On the other hand, mesoporous materials with cylindrical structures have homogeneous pore size and show potential applications in drug carriers, adsorption, and catalysis [14]. Mesoporous materials typically have large surface areas, ordered surfaces simple to adjust, flexible pore volume and void transmission, and low toxicity, materials stability, and modest structure [32]. Moreover, mesoporous materials have adaptable void size and surface chemistry. The void size and superficial chemistry of mesoporous materials can be tailored by adjusting synthesis parameters, allowing for meticulous body process over their characteristics and applications. Mesoporous materials often exhibit a well-defined and uniform pore structure, which is advantageous for applications requiring high selectivity and efficiency [6,30].

In conclusion, due to their unique features, mesoporous materials are applicable in health diagnostics [43]. Thus, the integration of mesopoorous materials in different diagnostics methods including therapeutics, medical imaging, point-of-care testing, **theranostics, and** clinical laboratory tests are described in the subsequent sections.

 Materials Research Forum LLC
https://doi.org/10.21741/9781644903452-18

5. Mesoporous materials in diagnostics

5.1 Therapeutics

The development of porous nanomaterials highly advanced the drug transportation in the body systems to advance the therapeutic technology and lessen the side effects [44]. The bioavailability and the characteristics of medicine are the essential causes that influence the treatment efficiency [1]. Direct patient management through efficient and improved drug delivery systems is possible by using nanocarriers for effective patient treatment (Figure 2). Mesoporous material application in therapeutics for better treatment of various diseases including cancer becomes important [45,46].

The drug is added in to the mesoporous materials and closed with capping agents. Upon the removal of the cape with the help of stimuli and the drug is released at specific site [47,48]. The surface chemical modification of mesoporous materials improves both the bioavailability of drugs and increases their water solubility [49,50]. The tumor tissues are known with vigorous reactions (metabolism and inflammation) and experience high temperature, thus temperature-responsive systems can be introduced. The temperature-responsive mesoporous materials such as glycerophosphate, paraffin wax, and poly (N-isopropyl acrylamide) (PNIPAAm) were synthesized. These materials have phase-transition temperatures greater than that of patient body temperature [51,52].

Similarly, pH-responsive systems were developed to treat cancer. For this purpose, mesoporous polymers such as aromatic amines, calcium carbonate, and polyacrylic acid (PAA) that are excitable to low pH through the use of pH sensitive linkers including imine, hydrazine, steric, and acetal linkers [53,54]. Organic mesoporous materials were prepared from polymer nanocomposite (polyacrylic acid-co-poly-itaconic acid) as a bilayer supported on MSN [55] and supramolecular polypseudorotaxane made of chitosan [56] as a pH-sensitive drug carrier. In another work, a pH-responsive hydrazone linkage was prepared for medicine transportation for the management of cancer using MSN functionalized by two ligands [57]. More interestingly, both temperature and pH sensitive medicine transportation system was developed by preparing silica from rice husk to treat breast cancer [58].

Enzyme-sensitive mesoporous materials-based drug delivery method is also common as therapeutic technology. In this work, magnetic mesoporous materials were used to deliver drug at specific sites using magnetic field [59-61].

Light is also another environmental stimuli used to trigger drug delivery. In this approach, photosensitizers are used in therapy based photothermal technology. The drug is released after thermal heating to damage cancer cells selectively. Materials like MSNs combined with gold nanorods (AuNRs) [62-64], graphene nanosheets [65], copper chalcogenides (Cu_9S_5) [66], single-walled carbon nanotubes (SWNTs) [67,66], and selenium-coated tellurium nanoheterojunctions [68] are used in chemotherapeutic processes.

Cancer is the deadliest disease originated by the abnormal and rampant cell growth in the body. To date, there are different treatment methods which are supposed to increase the efficacy and endurance rates of cancer long-suffering [69,70]. In line with this, many inorganic and organic based mesoporous materials have been formulated for drug transportation applications to treat cancer illness. These developments improved the drug cargo ratio, outflow and the uncontrolled

Materials Research Foundations 173 (2025) 470-495
https://doi.org/10.21741/9781644903452-18

liberation of the nanocarriers [71-73]. Many efforts have made to lessen the severity of anti-cancer drugs through the localization of the medicines into the malignant tumor cells location [74]. The presence of various functional units (e.g. silanol groups) highly facilitates the drug loading and release efficiency [75].

Mesoporous platinum [76], polydopamine modified ferrite nanoparticles and oxaliplatin [77], and gold nanocage [78] were also developed and implemented in cancer treatment.

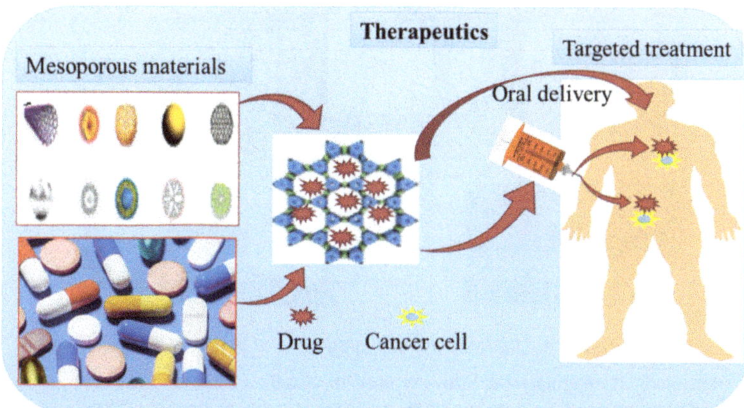

Figure 2. *Applications of mesoporous materials in targeted disease therapy.*

Hydrogel and nanogel based on pullulan and chitosan [72] and chitosan and dopamine modified with boronic acid [79] were made use of in doxorubicin drug transportation for malignant tumor management. The detection and treatment of tumours were performed using metallic nanoparticle [80], magnetic nanoparticles (Fe_3O_4@TiO_2–ZrO_2@mSiO_2) [81], Mn-based theranostic nanoplatforms [82], and manganese oxides [83].

Photodynamic and photothermal therapy based treatments were developed the the application of MSNs [84], 2 devinyl-2-(1-hexyloxyethyl)-pyropheophorbide (HPPH) [84] and carboxy aluminum phthalocyanine with small Pb nanosheets [85,84] to reduce cervical cancer cells viability by 65%. Moreover, various mesoporous materials such as hollow mesoporous carbon functionalized with ZnO quantum dots [86], polysorbate-coated MSN [87], oxidized mesoporous carbon nanospheres modified with peptide [88], MSN [89], and zwitterionic nanogels [90] were developed and applied for reference point drug transportation for different diseases mainly cancer.

5.2 Medical imaging

In medical sciences, bioimaging is a very important and compelling means that is widely used for the detection of diseases [91]. Different medical imaging methods, like magnetic resonance imaging (MRI), X-ray, computed tomography (CT), positron emission tomography (PET), and ultrasound supply elaborate anatomical reference and structural message about internal structures and organs (Figure 3).

Materials Research Foundations 173 (2025) 470-495
https://doi.org/10.21741/9781644903452-18

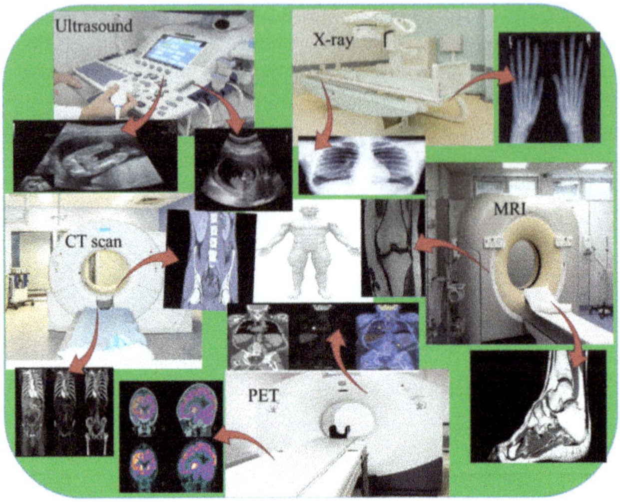

Figure 3. *Different medical imaging used in diagnosis.*

The different medical imaging modalities are used to visualize abnormalities, tumors, fractures, and other pathological conditions, aiding in diagnosis and treatment planning [92]. Presently, there is growing sake in the improvement of contrast agents for different imaging modalities [91].

Mesoporous materials are being used in medical imaging, CT scan, MRI, optical imaging, and ultrasound and PET for health diagnostics [50,93,94]. This implies that these functional materials play an important role in medical examination due to their unique characteristics such as enhanced surface area, tunable pore magnitude, and ample pore volume. These characteristics make them idealistic options for different imaging utilization (Table 2).

Porous metallic nanomaterials were applied as CT demarcation agents [97] and as X-ray radiation shielding including barium-doped PVC/Bi_2WO_6 composites [98], WO_3-nanoparticles [99], tungsten carbide, and polyester added to artificial marble materials and WO_3 Nanoparticles [100].

Table 2. *Some examples for applications of mesoporous materials in medical imaging*

Type	Diagnosis	Mesoporous materials	Reference
Ultrasound	Stomach, renal pelvis, body fluid vessels, mammary gland, excretory organ, muscular tissue, clappers and joints	Mesoporous silica nanoparticles-silver nanoparticles, carbon-doped TiO_2 nanoparticles, organosilica–Fe_3O_4 Hybrid	[95,96]
X-ray	Skeletal system problems, breast cancer and digestive issues (through barium swallows and enemas	Metallic nanoparticles, barium-doped PVC/Bi_2WO_6 composites, WO_3-nanoparticles, ungsten carbide, bismuth sulfide@mesoporous silica core-shell nanoparticles	[97,98-101]
MRI	Neural structure and skeletal structure abnormalities, breast, prostate, liver abnormalities, spinal injuries, blood flow	MSN, Fe_3O_4 nanoparticles, magnetic-MSN, hybrid protein-coated magnetic core-MSN, Gadolinium-Coated MSN	[96,102-105]
CT scan	Interior organs, bones, soft tissue and blood vessels within the body	Gadolinium and iron oxide nanoparticles, gadolinium-labeled dendrimer-entrapped gold nanoparticles	[106,104]
PET scan	Tumours, heart disease, brain disorders, assess the entire body or one specific area	Gadolinium-labeled dendrimer-entrapped gold nanoparticles, carbon nanotubes	[105,107]

Various mesoporous materials such as gadolinium-labeled dendrimer-entrapped gold nanoparticles [108], and carbon nanotubes [107], and gadolinium engineered, magnetic alloy nanoparticles [108] were successfully integrated into PET and CT imaging techniques. Furthermore, different types of mesoporous materials particularly MSN [102,109] and Fe_3O_4 nano-size particles are integrated into MRI to obtain highly resolved image [96].

In cancer related pathology, the engineering of composition, and properties of MSNs using gadolinium led to superior resolution [102,105]. In addition to gadolinium, manganese oxides [110,111], superparamagnetic iron oxide nanoparticles (SPION) [112], Dy DOTA [113] and binuclear ^{1}H and ^{19}F spectroscopy [114] can be used. Also, mesoporous silica nanoparticles-silver and carbon-doped TiO_2 nanoparticles showed good performance in detecting breast cancer cells [95] using ultrasound technique.

5.3 Point-of-care testing

Point-of-care testing (POCT) means examination carried out close to the patient or at the point of care, typically outside of traditional laboratory settings. It provides advantages such as small-

Materials Research Foundations 173 (2025) 470-495 https://doi.org/10.21741/9781644903452-18

volume samples, small reagent requirement, portability, highly sensitive, ease of use, low cost, and quick turnaround times over central laboratory testing [115,116]. POCT devices are portable, rapid, and easy to use, enabling real-time diagnosis and treatment monitoring [35]. Glucose meters, pregnancy tests, infectious disease rapid tests, and cardiac biomarker assays are few examples of POCT devices (Figure 4).

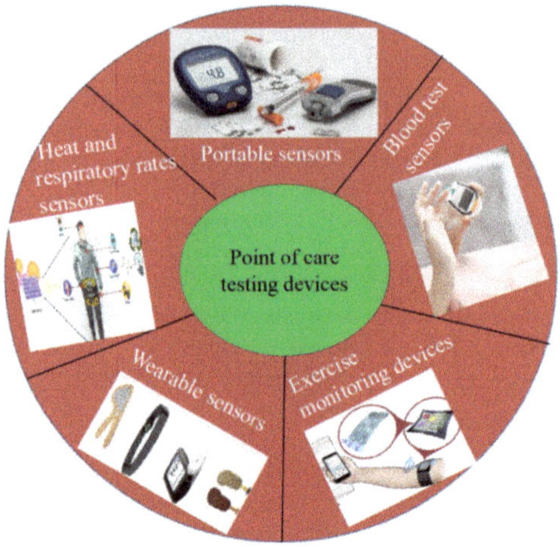

Figure 4. *Various sensors used in point-of-care testing devices*

POCT demands high performing techniques to obtain reliable and precise data. However, most of the time, it is difficult to find devices with such capabilities to address the prevailing demands [117]. Thus, mesoporous materials play a vital role in advancing POCT, offering several advantages that address the challenges of rapid and accurate diagnostics at the point of need. These advantages include enhanced analyte capture, reduced non-specific binding, and improved signal-to-noise ratios. By integrating mesoporous materials with microfluidic systems or paper-based devices, fast and delicate sensing of biomarkers or pathogens is performed with minimal sample volumes and instrumentation. The combination of enzyme and magnetic mesoporous materials avoided the interference from matrix and improved the detection of diseases [32].

Mesoporous materials integrated with natural recognition elements can serve as platforms for biosensors [118]. Biosensors based on mesoporous materials offer higher sensitivity, rapid outcome, and selectivity, making them appropriate for detecting various biological analytes, including proteins, nucleic acids, and pathogens [102]. For example, an electrochemiluminescence sensor combined with microfluidic chip was developed using Ru (bpy)$^{32+}$ loaded MSN for real-time H-FABP analyt measurement with detection limit of 0.72 ng/mL and linear ranges from 1 to100 ng/mL [119]. In another work, MSN was employed to develop a portable fluorescence sensor to detect reactive protein (CRP) in the linear ranges from 0.1-50 ng/mL and limit of detection of 0.05 ng/mL [120]. The biomarker luteinizing hormone (LH) level is directly related to the female ovulation cycle. A very sensitive microfluidic immunosensor was developed using

Materials Research Foundations 173 (2025) 470-495
https://doi.org/10.21741/9781644903452-18

fiber to LH to achieve detection limit and linear rage of 0.125 μIU/mL and 0.001–100 mIU/mL), respectively [16].

5.4 Theranostics

The word theranostics is the combination of two approaches namely diagnosis and therapy. In this approach, the paired radiopharmaceuticals are used to detect and target cancer cells. Most of the time, radiation emitting substances (radiotracers) are used to selectively detect and target [121]. The radiotracers are lined to other substances to image and perform therapeutics [122].

Simultaneous diagnose and treatment of patients by theranostic or theranostics is important in modern health care. So far small molecules were commonly used in diagnostic and therapeutic procedures. However, such technologies suffer from unwanted side effects and fast excretion. Recently, mesoporous materials or nanoparticles emerged as a innovative technology for simultaneous detection and treatment [8]. Furthermore, surface modification of mesoporous materials with targeting ligands can render them biocompatible, allowing for safe drug administration in biological systems. By incorporating diagnostic agents into mesoporous carriers, these systems enable simultaneous identification and treatment of malignant tumor [123,124]. Moreover, doxorubicin loaded MSN encapsulated iron oxide nanocomposites were synthesized for targeted delivery. These nanocomposites exhibited the capability for target-hunting drug delivery to the site due to its optical properties and thus considered as ideal for an anticancer drug and imaging investigation [125].

Nanotherapeutic applications of mesoporous materials for the treatment of cancer and much other pathology become common nowadays due to their interesting properties (Figure 5). In this method, mesoporous materials act as nanocarriers to pave the way for precision medicine [126]. This can be achieved through the preparation and application of mesoporous materials with required properties [69]. The application of DNA damage response inhibitors target improved the poor response to the radium-223 application in the administration of pathological process-resistant prostate cancer radiation therapy [127]. The anticancer agent doxorubicin loaded on to MSN-capped with a pH-sensitive gatekeeper aiming to minimize the side effects and enhancing its efficacy [128].

MSN composites show better performances and even greater versatility in theranostic applications [86,129]. Advancement of theranostic continued with the assistance of MSN, quantum dots, iron oxide nanoparticles, gold nanoparticles, and carbon nanotubes [130,93]. MSN-based materials (e.g. mesoporous bioactive glasses) modified with metal ions were used for therapeutic drugs [131]. Mesoporous materials mainly silica-based nanocarriers are successfully used in the delivery of proteins with therapeutic effect. As an example, anticancer proteins [132,133], anticancer peptides [134,135], immunostimulating peptides [136], hormones and growth factors [137] and antibacterial peptides [138] were effectively delivered to the site where treatment would be more effective. Moreover, MSN [102], magnetic MSN [103], modular hydrogel MSN [139], magnetic MSN shell nanocomposite coated with protein [140], and porous organosilica–Fe_3O_4 hybrid nanocapsules [96] were prepared and successfully applied for theranostic applications.

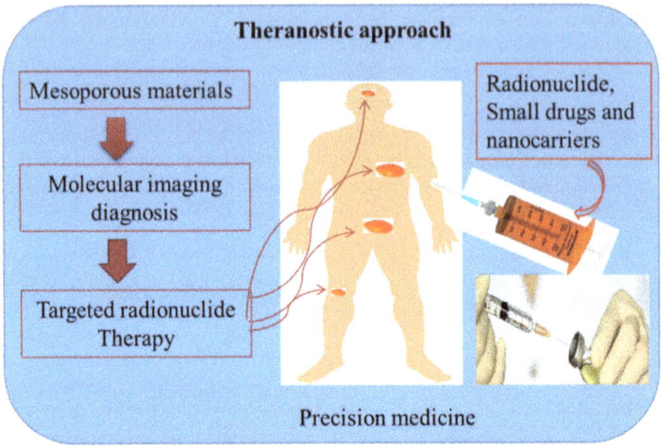

Figure 5*. Mesoporous materials-based theranostic approach*

5.5 Clinical Laboratory Tests

Clinical laboratory tests involve analyzing biological samples (e.g. blood, urine, tissue, etc.) to detect and quantify specific biomarkers, metabolites, or pathogens associated with diseases or conditions [141]. Common laboratory tests include blood chemistry panels, complete blood counts (CBC), enzyme-linked immunosorbent assays (ELISA), polymerase chain reaction (PCR), and genetic testing. Mesoporous materials are increasingly finding applications in clinical laboratory tests, offering unique advantages because of their tailored pore construction, high surface area, and tunable properties (Figure 6) [142].

Sample preparation and purification plays a key role in the clinical laboratory tests. With this regard, mesoporous materials can be functionalized to selectively capture and purify target analytes from complex biological samples such as blood, urine, or saliva. Their large surface area and void volume enable efficient adsorption and separation of biomolecules, facilitating sample preparation for downstream analysis [143]. New technologies based on mesoporous materials were introduced for sample preparation in clinical tests. For example, MSN was used sample fractionation before mass spectrometry detection of disease biomarker [144]. Hexagonal mesoporous silica [145], and aminopropyl hexagonal mesoporous silica [146], multi-functional polystyrene [129], and other mesoporous materials [147] were successfully applied in solid phase extraction for peptidomic profiling of clinical specimens.

Mesoporous materials are widely used in the sample isolation and enrichment [148]. MSN and others are widely used as sensing layers or substrate for immoblizing biorecognition elements (e.g. enzymes, antibodies or cells). In both cases the materials play role of facilitating electron transfer at the interface for optical or electrochemical sensors [149]. The detection of disease makes the treatment more effective. Bovine tuberculosis diseases are among the deadliest diseases that need early treatment. Hence, MSNs are employed in the modification of electrochemical impedance

Materials Research Foundations 173 (2025) 470-495 https://doi.org/10.21741/9781644903452-18

immunosensor for the sensing of bovine interferon-γ at picogram (pg) level (linear range of 0.001-10.0 ng/mL and detection of limit of 0.62 pg/mL) [150].

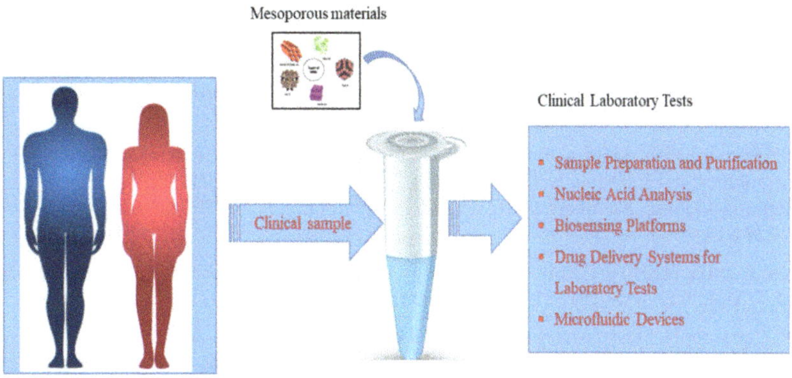

Figure 6. *Applications of mesoporous materials in clinical laboratory tests*

In addition, mesoporous silica nanoparticles were prepared and applied to remove immunoglobulin G (IgG) from serum for the in vitro IgE sensing for patients' hypersensitivity reaction to amoxicillin [151]. Fluorescence sensor based on cleavable silica nanocapsule (SNCs) nanocomposite was developed to detect Covid-19 nucleocapsid protein (N protein) biomarker. The optical sensors performed in linear ranges of 0.01 to 5 ng/mL and limit of detection of 2 pg/mL [152]. Interestingly, MSNs are compatible with different analytical techniques to develop sensors and other detectionmethods [37].

Conclusion and future perspectives

Mesoporous materials are materials with void spaces, unique properties and versatile functionalities, devising them important materials for a broad scope of uses. There are different kinds of mesoporous materials like inorganic, organic, and organic-inorganic materials. These materials emerged as versatile platforms for developing advanced diagnostic technologies with enhanced sensitivity, specificity, and functionality. These materials offer versatile platforms for medical imaging, providing opportunities for enhanced contrast, targeted delivery, bio-imaging, clinical laboratory tests, point-of-care tests, and theranostics, thereby advancing the field of diagnostic imaging and therapy.

Ongoing research focuses on developing novel synthesis methods, exploring new materials, and improving their characteristics for particular utilization. Furthermore, significant attention ought to be given to improving sensitivity, multiplexing capabilities, rapid assay development, stability, and biocompatibility, making them suitable platforms for advancing the field of diagnostic immunology and improving patient care.

References

[1] I.A. Spiridon, I.D. Căruntu, I. Spiridon, R. Brăescu, Insight into potential biomedical application of mesoporous materials, Pharmaceutics. 14 (2022) 2382. https://doi.org/10.3390/pharmaceutics14112382

[2] R. Diab, N. Canilho, I.A. Pavel, F.B. Haffner, M. Girardon, A. Pasc, Silica-based systems for oral delivery of drugs, macromolecules and cells, Adv. Colloid Interface Sci. 249 (2017) 346-362. https://doi.org/10.1016/j.cis.2017.04.005

[3] L. Giorno, E. Piacentini, F. Bazzarelli, Macroporous, Mesoporous, and microporous membranes. In ; E. Drioli, L. Giorno, (Eds.), Encyclopedia of Membranes, Springer: Berlin/Heidelberg, Germany, 2016, pp. 1-2. https://doi.org/10.1007/978-3-642-40872-4_2244-1

[4] J.A. Cecilia, R.M. Tost, M.R. Millan, Mesoporous materials: From synthesis to applications, Int. J. Mol. Sci. 20 (2019) 3213. https://doi.org/10.3390/ijms20133213

[5] N. Carlsson, H. Gustafsson, C. Thörn, L. Olsson, K. Holmberg, B. Åkerman, Enzymes immobilized in mesoporous silica: A physical-chemical perspective, Adv. Colloid Interface Sci. 205 (2014) 339-360. https://doi.org/10.1016/j.cis.2013.08.010

[6] D.B. Emrie, Sol-gel synthesis of nanostructured mesoporous silica powder and thin films, J. Nanomater. 2024 (2024) 16. https://doi.org/10.1155/2024/6109770

[7] M. Bayoumia, M.M. Kamil, F.A. Rahi, N.Z. Kbah, A.A. Ma'aen, Advanced mesoporous silica nanoparticles: Synthesis, characteristics, and applications, J. Med. Sci. 2 (2023) 43e48. https://doi.org/10.55810/2789-9128.1019

[8] M. Vallet-Regí, Our contributions to applications of mesoporous silica nanoparticles. Acta Biomater. 137 (2022) 44-52. https://doi.org/10.1016/j.actbio.2021.10.011

[9] J.M. Rosenholm, C. Sahlgren, M. Linden, Multifunctional mesoporous silica nanoparticles for combined therapeutic, diagnostic and targeted action in cancer treatment, Curr. Drug Targets. 12 (2011) 1166 - 1186. https://doi.org/10.2174/138945011795906624

[10] K.W. Kim, B. Park, J. Kim, C. Jo, J.K. Kim, Recent progress in block copolymer soft-template-assisted synthesis of versatile mesoporous materials for energy storage systems, J. Mater. Chem. A. 11 (2023) 7358-7386. https://doi.org/10.1039/D2TA09353G

[11] H.N. Lokupitiya, A. Jones, B. Reid, S. Guldin, M. Stefik, Ordered mesoporous to macroporous oxides with tunable isomorphic architectures: Solution criteria for persistent micelle templates, Chem. Mater. 28 (2016) 1653-1667. https://doi.org/10.1021/acs.chemmater.5b04407

[12] F. Trindade, M.J. Politi, 2-Sol-gel chemistry-deals with sol-gel processes, in: R. Bacani, M.J. Politi, F. Trindade, E.R. Triboni (Eds.), Nano Design for Smart Gels, Elsevier, 2019, pp. 15-34. https://doi.org/10.1016/B978-0-12-814825-9.00002-3

[13] N.I. Vazquez, Z. Gonzalez, B. Ferrari, Y. Castro, Synthesis of mesoporous silica nanoparticles by sol-gel as nanocontainer for future drug delivery applications, Bol. Soc. Esp. de Cerám. Vidr. 56 (2017) 139-145. https://doi.org/10.1016/j.bsecv.2017.03.002

[14] Y. Han, L. Zhang, W. Yang, Synthesis of mesoporous silica using the sol-gel approach: Adjusting architecture and composition for novel applications, Nanomaterials. 14 (2024) 903. https://doi.org/10.3390/nano14110903

[15] A.E. Danks, S.R. Hall, Z. Schnepp, The evolution of 'sol-gel' chemistry as a technique for materials synthesis, Mater. Horiz. 3 (2016) 91-112. https://doi.org/10.1039/C5MH00260E

[16] K. Xue, B. Lin, S. Huang, B. Cai, W. Lai, T. Zhan, Y. Liang, C. Zhang, A dry chemistry, ultrasensitive microfluidic fiber material-based immunosensor for electrochemiluminescence point-of-care testing of luteinizing hormone, Sens. Actuators B: Chem. 393 (2023) 134259. https://doi.org/10.1016/j.snb.2023.134259

[17] S.M. Savić, K. Vojisavljević, M. Počuča-Nešić, K. Živojević, M. Mladenović, N.Z. Knežević, Hard template synthesis of nanomaterials based on mesoporous silica, Metall. Mater. Eng. 24 (2018) 225-241. https://doi.org/10.30544/400

[18] W. Xin, Y.H. Song, Mesoporous carbons: Recent advances in synthesis and typical applications, RSC Adv. 5 (2015) 83239-83285. https://doi.org/10.1039/C5RA16864C

[19] M. Marcos-Hernández, D. Villagrán, 11-Mesoporous composite nanomaterials for dye removal and other applications, in: George Z. Kyzas, Athanasios C. Mitropoulos (Eds.), Micro and Nano Technologies, Composite Nanoadsorbents, Elsevier, 2019, pp 265-293. https://doi.org/10.1016/B978-0-12-814132-8.00012-5

[20] J. Zhang, B. Shen, J. Chen, L. Chen, Y. Meng, J. Feng, A dual-sensitive mesoporous silica nanoparticle based drug carrier for cancer synergetic therapy, Colloids Surf. B: Biointerfaces. 175 (2019) 65-72. https://doi.org/10.1016/j.colsurfb.2018.11.071

[21] S. Daneshjou, B. Dabirmanesh, F. Rahimi, S. Jabbari, K. Khajeh, Catalytic parameters and thermal stability of chondroitinase ABCI on red porous silicon nanoparticles, J. Biotechnol. 324 (2020) 83-90. https://doi.org/10.1016/j.jbiotec.2020.09.020

[22] S. Azizi, J. Soleymani, S. Shojaei, N. Shadjou, Synthesize of folic acid functionalized dendritic fibrous nanosilica and its application as an efficient nanocatalyst for access to direct amidation of carboxylic acids with amines, J. Nanostructures. 10 (2020) 671-681.

[23] E. Elimbinzi, S.S. Nyandoro, E.B. Mubofu, J.C. Manayil, A.F. Lee, K. Wilson, Valorization of rice husk silica waste: Organoamine functionalized castor oil templated mesoporous silicas for biofuels synthesis, Microporous Mesoporous Mater. 294 (2020) 109868. https://doi.org/10.1016/j.micromeso.2019.109868

[24] X.Q. Wu, Z.D. Shao, Q. Liu, Z. Xie, F. Zhao, Y.M. Zheng, Flexible and porous TiO2/SiO2/carbon composite electrospun nanofiber mat with enhanced interfacial charge separation for photocatalytic degradation of organic pollutants in water, J. Colloid Interface Sci. 553 (2019) 156-166. https://doi.org/10.1016/j.jcis.2019.06.019

[25] S. Kumar, M. Malik, R. Purohit, Synthesis methods of mesoporous silica materials, Mater. Today Proc. 4 (2017) 350-357. https://doi.org/10.1016/j.matpr.2017.01.032

[26] S.A. Jadhav, I. Miletto, V. Brunella, D. Scalarone, G. Berlier, Porous silica particles: Synthesis, physicochemical characterization and evaluation of suspension stability, Phys. Chem. Ind J. (2017) 1-11. https://doi.org/10.3144/expresspolymlett.2017.11

[27] W. Yang, S. Shirazian, R. Soltani, M.H. Zare, Bio-originated mesosilicate SBA-15: Synthesis, characterization, and application for heavy metal removal, npj Clean Water. 7 (2024) 49. https://doi.org/10.1038/s41545-024-00340-7

[28] I. Nowak, A. Feliczak-Guzik, Mesoporous materials: Materials, technological, and environmental applications, Int. J. Mol. Sci. 24 (2023) 9197. https://doi.org/10.3390/ijms24119197

[29] G. Sharma, Biogenic carbon nanostructured materials for detection of cancer and medical applications: A mini review, Hybrid Adv. 5 (2024) 100166. https://doi.org/10.1016/j.hybadv.2024.100166

[30] P.S. Shinde, P.S. Suryawanshi, K.K. Patil, V.M. Belekar, SA. Sankpal, S.D. Delekar, S.A. Jadhav, A brief overview of recent progress in porous silica as catalyst supports, J. Compos. Sci. 5 (2021) 75. https://doi.org/10.3390/jcs5030075

[31] A.H. Chowdhury, N. Salam, R. Debnath, S.M. Islam, T. Saha, Design and fabrication of porous nanostructures and their applications, in: Y.B. Pottathara, S. Thomas, N. Kalarikkal, Y. Grohens, V. Kokol (Eds.), Micro and Nano Technologies, Nanomaterials Synthesis, Elsevier, 2019, pp 265-294. https://doi.org/10.1016/B978-0-12-815751-0.00008-0

[32] J. Li, J. Huang, Y. Jiang, L. Wu, Y. Deng, Magnetic large-mesoporous nanoreactors enable enzymatically regulated background-free and persistently signalling diseases diagnosis. Adv. Funct. Mater. 33 (2023) 2212317. https://doi.org/10.1002/adfm.202212317

[33] Y. Kuwahara, D.Y. Kang, J.R. Copeland, N.A. Brunelli, S.A. Didas, P. Bollini, C. Sievers, T. Kamegawa, H. Yamashita, C.W. Jones, Dramatic enhancement of CO2 uptake by poly(ethyleneimine) using zirconosilicate supports. J. Am. Chem. Soc. 134 (2012) 10757-10760. https://doi.org/10.1021/ja303136e

[34] S. Liu, X. He, X. Hu, Y. Puab, X. Mao, Porous nanomaterials for biosensing and related biological application in in vitro/vivo usability, Mater. Adv. 5 (2024) 453. https://doi.org/10.1039/D3MA00498H

[35] F.R. Mansour, S.F. Hammad, I.A. Abdallah, A. Bedair, R.M. Abdelhameed, M. Locatelli, Applications of metal organic frameworks in point of care testing, Trends Anal. Chem. 172 (2024) 117596. https://doi.org/10.1016/j.trac.2024.117596

[36] E.B. Cho, J. Park, M. Jaroniec, Structural stability of Si-C bonds in periodic mesoporous thiophene-silicas prepared under acidic conditions, J. Phys. Chem. C. 117 (2013) 21441-21449. https://doi.org/10.1021/jp408385t

[37] Y.Q. Almajidi, R.H. Althomali, K. Gandla, H. Uinarni, N. Sharma, B.M. Hussien, M.S. Alhassan, Romero-Parra, R.M., Bisht, Y.S. Multifunctional immunosensors based on mesoporous silica nanomaterials as efficient sensing platforms in biomedical and food safety analysis: A review of current status and emerging applications, Microchem. J. 191 (2023) 108901. https://doi.org/10.1016/j.microc.2023.108901

[38] L. Xu, Z. Lin, Y. Chen, Z. Fan, X. Pei, S. Yang, X. Kou, Y. Wang, Z. Zou, D. Xi, P. Yin, G. Su, M. Zhou, J. Dai, L. Pan, Y. Zhao, Carbon-based cages with hollow confined structures for efficient microwave absorption: State of the art and prospects, Carbon. 201 (2023) 1090-1114. https://doi.org/10.1016/j.carbon.2022.10.015

[39] S. Jarmolińska, A. Feliczak-Guzik, I. Nowak, Synthesis, characterization and use of mesoporous silicas of the following types SBA-1, SBA-2, HMM-1 and HMM-2, Materials. 13 (2020) 1-33. https://doi.org/10.3390/ma13194385

[40] N. Pal, A. Bhaumik, Soft templating strategies for the synthesis of mesoporous materials: Inorganic, organic-inorganic hybrid and purely organic solids, Adv. Colloid Interface Sci. 189 (2013) 21-41. https://doi.org/10.1016/j.cis.2012.12.002

[41] M. Thommes, K. Kaneko, A.V. Neimark, J.P. Olivier, F.R. Reinoso, J. Rouquerol, K.S.W. Sing, Physisorption of gases, with special reference to the evaluation of surface area and pore size distribution (IUPAC Technical Report), Pure Appl. Chem. 87 (2015) 1051-1069. https://doi.org/10.1515/pac-2014-1117

[42] B. Xu, S. Li, R. Shi, H. Liu, Multifunctional mesoporous silica nanoparticles for biomedical applications, Signal Transduct. Target. Ther. 8 (2023) 435. https://doi.org/10.1038/s41392-023-01654-7

[43] T.T.H. Thi, V.D. Cao, T.N.Q. Nguyen, D.T. Hoang, V.C. Ngo, D.H. Nguyen, Functionalized mesoporous silica nanoparticles and biomedical applications, Mater. Sci. Eng. C. 99 (2019) 631-656. https://doi.org/10.1016/j.msec.2019.01.129

[44] M.K.M. Esfahani, S.E. Alavi, P.J. Cabot, N. Islam, E.L. Izake, Application of mesoporous silica nanoparticles in cancer therapy and delivery of repurposed anthelmintics for cancer therapy, Pharmaceutics. 14 (2022) 1579. https://doi.org/10.3390/pharmaceutics14081579

[45] M. Ghaferi, M.K.M. Esfahani, A. Raza, S. Al Harthi, Ebrahimi Shahmabadi H., Alavi S.E. Mesoporous silica nanoparticles: Synthesis methods and their therapeutic use-recent advances, J. Drug Target. 29(2021) 131-154. https://doi.org/10.1080/1061186X.2020.1812614

[46] Y. Gao, D. Gao, J. Shen, Q. Wang, A review of mesoporous silica nanoparticle delivery systems in chemo-based combination cancer therapies, Front. Chem. 8 (2020) 598722. https://doi.org/10.3389/fchem.2020.598722

[47] M. Moritz, M. Geszke-Moritz, Mesoporous materials as elements of modern drug delivery systems for anti-inflammatory agents: A review of recent achievements, Pharmaceutics. 14 (2022) 1542. https://doi.org/10.3390/pharmaceutics14081542

[48] C. Han, H. Huang, Y. Dong, X. Sui, B. Jian, W. Zhu, A comparative study of the use of mesoporous carbon and mesoporous silica as drug carriers for oral delivery of the water-insoluble drug carvedilol, Molecules. 24 (2019) 1770. https://doi.org/10.3390/molecules24091770

[49] S. Porrang, S. Davaran, N. Rahemi, S. Allahyari, E. Mostafavi, How advancing are mesoporous silica nanoparticles? A comprehensive review of the literature, Int. J. Nanomed. 17 (2022) 1803-1827. https://doi.org/10.2147/IJN.S353349

[50] S. Jafari, H. Derakhshankhah, L. Alaei, A. Fattahi, B.S. Varnamkhasti, A.A. Saboury, Mesoporous silica nanoparticles for therapeutic/diagnostic applications, Biomed Pharmacother. 109 (2019) 1100-1111. https://doi.org/10.1016/j.biopha.2018.10.167

[51] M. Ashrafizadeh, H. Saebfar, M.H. Gholami, K. Hushmandi, A. Zabolian, P. Bikarannejad, M. Hashemi, S. Daneshi, S. Mirzaei, E. Sharifi, A.P. Kumar, H. Khan, H.S Hossein, M.

Vosough, N. Rabiee, V.K. Thakur, P. Makvandi, Y.K. Mishra, F.R. Tay, Y. Wang, A. Zarrabi, G. Orive, E. Mostafavi. Doxorubicin-loaded graphene oxide nanocomposites in cancer medicine: Stimuli-responsive carriers, co-delivery and suppressing resistance, Expert Opin. Drug Deliv. 19 (2022) 355-382. https://doi.org/10.1080/17425247.2022.2041598

[52] L. Liu, H. Ma, J. Yu, Y. Fan, Fabrication of glycerophosphate-based nanochitin hydrogels for prolonged release under in vitro physiological conditions, Cellulose. 28 (2021) 4887-4897. https://doi.org/10.1007/s10570-021-03819-5

[53] B.K. Poudel, Z.C. Soe, H.B. Ruttala, B. Gupta, T. Ramasamy, T.K. Thapa, M. Gautam, W. Ou, H.T. Nguyen, J.H. Jeong, S.G. Jin, H.G. Choi, C.S. Yong, J.O. Kim, In situ fabrication of mesoporous silica-coated silver-gold hollow nanoshell for remotely controllable chemo-photothermal therapy via phase-change molecule as gatekeepers, Int. J. Pharm. 548 (2018) 92-103. https://doi.org/10.1016/j.ijpharm.2018.06.056

[54] M. Karimi, H. Mirshekari, M. Aliakbari, P. Sahandi-Zangabad, M.R. Hamblin, Smart mesoporous silica nanoparticles for controlled-release drug delivery, Nanotechnol Rev. 5 (2016) 195-207. https://doi.org/10.1515/ntrev-2015-0057

[55] Z. Zhang, W. He, J. Zheng, G. Wang, J. Ji, Rice husk ash-derived silica nanofluids: Synthesis and stability study, Nanoscale Res. Lett. 11 (2016) 502. https://doi.org/10.1186/s11671-016-1726-9

[56] A. Pourjavadi, Z.M. Tehrani, S. Jokar, Chitosan based supramolecular polypseudorotaxane as a pH-responsive polymer and their hybridization with mesoporous silica-coated magnetic graphene oxide for triggered anticancer drug delivery, Polymer. 76 (2015) 52-61. https://doi.org/10.1016/j.polymer.2015.08.050

[57] M. Mladenović, I. Morgan, N. Ilić, M. Saoud, MV. Pergal, G.N. Kaluđerović, N.Z. Knežević, pH-Responsive release of ruthenium metallotherapeutics from mesoporous silica-based nanocarriers, Pharmaceutics. 13 (2021) 460. https://doi.org/10.3390/pharmaceutics13040460

[58] L. Ma'mani, S. Nikzad, H. Kheiri-Manjili, S. Al-Musawi, M. Saeedi, S. Askarlou, A. Foroumadi, A. Shafiee, Curcumin-loaded guanidine functionalized PEGylated I3ad mesoporous silica nanoparticles KIT-6: practical strategy for the breast cancer therapy, Eur. J. Med. Chem. 83 (2014) 646-54. https://doi.org/10.1016/j.ejmech.2014.06.069

[59] K. Vaghasiya, E. Ray, A. Sharma, O.P. Katare, R.K. Verma, Matrix metalloproteinase-responsive mesoporous silica nanoparticles cloaked with cleavable protein for "self-actuating" on-demand controlled drug delivery for cancer therapy, ACS App. Bio. Mater. 3 (2020) 4987-4999. https://doi.org/10.1021/acsabm.0c00497

[60] Y. Wang, L. Wang, L. Guo, M. Yan, L. Feng, S. Dong, J. Hao, Photo-responsive magnetic mesoporous silica nanocomposites for magnetic targeted cancer therapy, New J. Chem. 43 (2019) 4908-4918. https://doi.org/10.1039/C8NJ06105J

[61] N. Asadi, N. Annabi, E. Mostafavi, M. Anzabi, R. Khalilov, S. Saghfi, M. Mehrizadeh, A. Akbarzadeh, Synthesis, characterization and in vitro evaluation of magnetic nanoparticles modified with PCL-PEG-PCL for controlled delivery of 5FU, Artif. Cells Nanomed. Biotechnol. 46 (2018) 938-945. https://doi.org/10.1080/21691401.2018.1439839

Materials Research Forum LLC
https://doi.org/10.21741/9781644903452-18

[62] J. Wu, D.H. Bremner, S. Niu, D. Li, R. Tang, L.M. Zhu, Multifunctional A7R peptide-modified hollow mesoporous silica@ Ag2S nanotheranostics for photoacoustic/near-infrared fluorescence imaging-guided tumor-targeted chemo-photothermal therapy, J. Biomed. Nanotechnol. 15 (2019) 1415-1431. https://doi.org/10.1166/jbn.2019.2729

[63] C. Ong, B.G. Cha, J. Kim, Mesoporous silica nanoparticles doped with gold nanoparticles for combined cancer immunotherapy and photothermal therapy, ACS App. Bio. Mater. 2 (2019) 3630-3638. https://doi.org/10.1021/acsabm.9b00483

[64] L. Zhang, Y. Li, C.Y. Jimmy, Chemical modification of inorganic nanostructures for targeted and controlled drug delivery in cancer treatment, J. Mater. Chem. B. 2 (2014) 452-470. https://doi.org/10.1039/C3TB21196G

[65] X. Liu, X. Wu, Y. Xing, Y. Zhang, X. Zhang, Q. Pu, M. Wu, J.X. Zhao, Reduced graphene oxide/mesoporous silica nanocarriers for pH-triggered drug release and photothermal therapy, ACS App. Bio. Mater. 3 (2020) 2577-2587. https://doi.org/10.1021/acsabm.9b01108

[66] G. Song, Q. Wang, Y. Wang, G. Lv, C. Li, R. Zou, Z. Chen, Z. Qin, K. Huo, R. Hu, J. Hu, A low-toxic multifunctional nanoplatform based on Cu9S5@ mSiO2 core-shell nanocomposites: Combining photothermal-and chemotherapies with infrared thermal imaging for cancer treatment, Adv. Funct. Mater. 23 (2013) 4281-4292. https://doi.org/10.1002/adfm.201203317

[67] J. Liu, C. Wang, X. Wang, L. Cheng, Y. Li, Z. Liu, Mesoporous silica coated single-walled carbon nanotubes as a multifunctional light-responsive platform for cancer combination therapy, Adv. Funct. Mater. 25 (2015) 384-392. https://doi.org/10.1002/adfm.201403079

[68] S. Chen, C. Xing, D. Huang, C. Zhou, B. Ding, Z. Gou, Z. Peng, D. Wang, X. Zhu, Y. Cao, Eradication of tumor growth by delivering novel photothermal selenium-coated tellurium nanoheterojunctions, Sci. Adv. 6 (2020) eaay6825. https://doi.org/10.1126/sciadv.aay6825

[69] S. Barui, V. Cauda, Multimodal decorations of mesoporous silica nanoparticles for improved cancer therapy, Pharmaceutics. 12 (2020) 527. https://doi.org/10.3390/pharmaceutics12060527

[70] Y. Kawabata, K. Wada, M. Nakatani, S. Yamada, S. Onoue, Formulation design for poorly water-soluble drugs based on biopharmaceutics classification system: Basic approaches and practical applications, Int. J. Pharm. 420 (2011) 1-10. https://doi.org/10.1016/j.ijpharm.2011.08.032

[71] V. Negri, J. Pacheco-Torres, D. Calle, P. López-Larrubia, Carbon nanotubes in biomedicine, Top. Curr. Chem. (Cham). 378 (2020) 15. https://doi.org/10.1007/s41061-019-0278-8

[72] Y. Wang, Y. Xie, K.V. Kilchrist, J. Li, C.L. Duvall, D. Oupický, Endosomolytic and tumor-penetrating mesoporous silica nanoparticles for sirna/mirna combination cancer therapy, ACS Appl. Mater. Interfaces. 12 (2020) 4308-4322. https://doi.org/10.1021/acsami.9b21214

[73] S. Pinel, N. Thomas, C. Boura, M. Barberi-Heyob, Approaches to physical stimulation of metallic nanoparticles for glioblastoma treatment, Adv. Drug. Deliv. Rev. 138 (2019) 344-357. https://doi.org/10.1016/j.addr.2018.10.013

[74] M. Vallet-Regí, M. Colilla, I. Izquierdo-Barba, M. Manzano, Mesoporous silica nanoparticles for drug delivery: Current insights, Molecules. 23 (2018) 47. https://doi.org/10.3390/molecules23010047

[75] N. Vadia, S. Rajput, Applications of mesoporous material for drug delivery, in: Importance & Applications of Nanotechnology, MedDocs Publishers, 2019.

[76] B. Fu, M. Dang, J. Tao, Y. Li, Y. Tang, Mesoporous platinum nanoparticle-based nanoplatforms for combined chemo photothermal breast cancer therapy, J. Colloid Interface Sci. 570 (2020) 197-204. https://doi.org/10.1016/j.jcis.2020.02.051

[77] N. Alamer, A. Meshkini, L. Khoshtabiat, A. Behnamsani, Synergizing effects of chemodynamic therapy and chemotherapy against breast cancer by oxaliplatin-loaded polydopamine/BSA@copper ferrite, J. Drug Deliv. Sci. Technol. 72 (2022) 103391. https://doi.org/10.1016/j.jddst.2022.103391

[78] H. Sun, J. Su, Q. Meng, Q. Yin, L. Chen, W. Gu, Z. Zhang, H. Yu, P. Zhang, S. Wang, Y. Li, Cancer cell membrane-coated gold nanocages with hyperthermia-triggered drug release and homotypic target inhibit growth and metastasis of breast cancer, Adv. Funct. Mater. 27 (2017) 1604300. https://doi.org/10.1002/adfm.201604300

[79] E.B. Kang, G.B. Lee, I. In, S.Y. Park, pH-Sensitive fluorescent hyaluronic acid nanogels for tumor-targeting and controlled delivery of doxorubicin and nitric oxide, Eur. Polym. J. 101 (2018) 96-104. https://doi.org/10.1016/j.eurpolymj.2018.02.016

[80] H.S. Tuli, R. Joshi, G. Kaur, V.K. Garg, K. Sak, M. Varol, J. Kaur, S.A. Alharbi, T.A. Alahmadi, D. Aggarwal, K. Dhama, V.S. Jaswal, S. Mittal, G. Sethi, Metal nanoparticles in cancer: From synthesis and metabolism to cellular interactions, J. Nanostructure Chem. 13 (2022) 321-348. https://doi.org/10.1007/s40097-022-00504-2

[81] W. Li, Z. Cao, R. Liu, L. Liu, H. Li, X. Li, Y. Chen, C. Lu, Y. Liu, AuNPs as an important inorganic nanoparticle applied in drug carrier systems, Artif. Cells Nanomed. Biotechnol. 47 (2019) 4222-4233. https://doi.org/10.1080/21691401.2019.1687501

[82] S. Liang, G. Liao, W. Zhu, L. Zhang, Manganese-based hollow nanoplatforms for MR imaging-guided cancer therapies, Biomater. Res. 26 (2022) 32. https://doi.org/10.1186/s40824-022-00275-5

[83] G. Yang, L. Xu, Y. Chao, J., Xu, X. Sun, Y. Wu, R. Peng, Z. Liu, Hollow MnO2 as a tumor-microenvironment-responsive biodegradable nano-platform for combination therapy favoring antitumor immune responses, Nat. Commun. 8 (2017) 902. https://doi.org/10.1038/s41467-017-01050-0

[84] P.L. Abbaraju, Mesoporous Silica Nanoparticles for Biomedical Application, Aust Institute for Bioengineering & Nanotechnology Institution, The University of Queensland, https://espace.library.uq.edu.au/ view/UQ:681687/2017 (accessed July 20, 2024).

[85] Z.X. Zhao, Y.Z. Huang, S.G. Shi, S.H. Tang, D.H. Li, X.L. Chen, Cancer therapy improvement with mesoporous silica nanoparticles combining photodynamic and photothermal therapy, Nanotechnology. 25 (2014) 285701. https://doi.org/10.1088/0957-4484/25/28/285701

[86] S. Feng, Y. Mao, X. Wang, M. Zhou, H. Lu, Q. Zhao, S. Wang, Triple stimuli-responsive ZnO quantum dots-conjugated hollow mesoporous carbon nanoplatform for NIR-induced dual model antitumor therapy, J. Colloid Interface Sci. 559 (2020) 51-64. https://doi.org/10.1016/j.jcis.2019.09.120

[87] B.S. Fateh, M. Hamidi, A. Maleki, Z. Karami, H. Mohamadpour, M.R.S. Zanjani, Polysorbate-coated mesoporous silica nanoparticles as an efficient carrier for improved rivastigmine brain delivery, Brain Res. 1781 (2022) 147786. https://doi.org/10.1016/j.brainres.2022.147786

[88] S. Wang, W. Wu, Y. Liu, C. Wang, Q. Xu, Q. Lv, R. Huang, X. Li, Targeted peptide-modified oxidized mesoporous carbon nanospheres for chemo-thermo combined therapy of ovarian cancer in vitro, Drug Deliv. 29 (2022) 1947-1952. https://doi.org/10.1080/10717544.2022.2089298

[89] D. Gkiliopoulos, I. Tsamesidis, A. Theocharidou, G.K. Pouroutzidou, E. Christodoulou, E. Stalika, K. Xanthopoulos, D. Bikiaris, K. Triantafyllidis, E. Kontonasaki, SBA-15 mesoporous silica as delivery vehicle for rhbmp-2 bone morphogenic protein for dental applications, Nanomaterials. 12 (2022) 822. https://doi.org/10.3390/nano12050822

[90] Z. Wang, X. Li, X Zhang, R. Sheng, Q. Lin, W. Song, L. Hao, Novel contact lenses embedded with drug-loaded zwitterionic nanogels for extended ophthalmic drug delivery, Nanomaterials. 11 (2021) 2328. https://doi.org/10.3390/nano11092328

[91] A. Khorasani, D. Shahbazi-Gahrouei, A. Safari, Recent metal nanotheranostics for cancer diagnosis and therapy: A review, Diagnostics. 13 (2023) 833. https://doi.org/10.3390/diagnostics13050833

[92] P. Sowmiya, T. Stalin Dhas, D. Inbakandan, N. Anandakumar, S. Nalini, K.S.U. Suganya, R.R. Remya, V. Karthick, C.M. Vineeth Kumar, Optically active organic and inorganic nanomaterials for biological imaging applications: A review, Micron. 172 (2023) 103486. https://doi.org/10.1016/j.micron.2023.103486

[93] E. Aznar, M. Oroval, L. Pascual, J.R. Murguia, R. Martinez-Manez, F. Sancenon, Gated materials for on-command release of guest molecules, Chem. Rev. 116 (2016) 561-718. https://doi.org/10.1021/acs.chemrev.5b00456

[94] W.X. Mai, H. Meng, Mesoporous Silica Nanoparticles: A multifunctional nano therapeutic system, Integr. Biol. 5 (2013) 19-28. https://doi.org/10.1039/c2ib20137b

[95] K. Ponsanti, B. Tangnorawich, N. Ngernyuang, C. Pechyen, Synthesis of mesoporous silica nanoparticles (MSNs)/silver nanoparticles (AgNPs): Promising hybrid materials for detection of breast cancer cells, J. Mater. Sci. Mater. Electron. 33 (2022) 7515-7527. https://doi.org/10.1007/s10854-022-07891-w

[96] M. Ma, F. Yan, M. Yao, Z. Wei, D. Zhou, H. Yao, H. Zheng, H. Chen, J. Shi, Template-free synthesis of hollow/porous organosilica-fe3o4 hybrid nanocapsules toward magnetic resonance imaging-guided high-intensity focused ultrasound therapy, ACS Appl. Mater. Interfaces. 8 (2016) 29986-29996. https://doi.org/10.1021/acsami.6b10370

Materials Research Forum LLC
https://doi.org/10.21741/9781644903452-18

[97] N. Aslan, B. Ceylan, M.M. Koç, F. Findik, Metallic nanoparticles as X-ray computed tomography (CT) contrast agents: A review, J. Mol. Struct. 1219 (2020) 128599. https://doi.org/10.1016/j.molstruc.2020.128599

[98] L. Gholamzadeh, H. Sharghi, M.K. Aminian, Synthesis of barium-doped PVC/Bi2WO6 composites for X-ray radiation shielding, Nucl. Eng. Technol. 54 (2022) 318-325. https://doi.org/10.1016/j.net.2021.07.045

[99] H. Al-Ghamdi, H.M. Hemily, I.H. Saleh, Z.F. Ghataas, A.A. Abdel-Halim, M.I. Sayyed, S. Yasmin, A.H. Almuqrin, M. Elsafi, Impact of WO3-nanoparticles on silicone rubber for radiation protection efficiency, Materials. 15 (2022) 5706. https://doi.org/10.3390/ma15165706

[100] M.N. Azman, N.J. Abualroos, K.A. Yaacob, R. Zainon, Feasibility of nanomaterial tungsten carbide as lead-free nanomaterialbased radiation shielding, Radiat. Phys. Chem. 202 (2022) 110492. https://doi.org/10.1016/j.radphyschem.2022.110492

[101] L. Li, Y. Lu, C. Jiang, Y. Zhu, X. Yang, X. Hu, Z. Lin, Y. Zhang, M. Peng, H. Xia, c. Mao, Actively targeted deep tissue imaging and photothermal-chemo therapy of breast cancer by antibody-functionalized drug-loaded X-ray-responsive bismuth sulfide@mesoporous silica core-shell nanoparticles, Adv. Funct. Mater. 28 (2018) 1704623. https://doi.org/10.1002/adfm.201704623

[102] X. Wang, M. Zheng, F. Raza, Y. Liu, Y. Wei, M. Qiu, J. Su, Recent developments in mesoporous silica nanoparticles for tumor theranostic applications, Curr. Pharm. Des. 28 (2022) 151-164. https://doi.org/10.2174/1381612827666211111152839

[103] M.S. Laranjeira, T.P. Ribeiro, A.I. Magalhães, P.C. Silva, J.A.M. Santos F.J. Monteiro, Magnetic mesoporous silica nanoparticles as a theranostic approach for breast cancer: Loading and release of the poorly soluble drug exemestane, Int. J. Pharm. 619 (2022) 121711. https://doi.org/10.1016/j.ijpharm.2022.121711

[104] Z. Li, J. Guo, M. Zhang, G. Li, L. Hao, Gadolinium-coated mesoporous silica nanoparticle for magnetic resonance imaging, Front. Chem. 10 (2022) 837032. https://doi.org/10.3389/fchem.2022.837032

[105] C. Li, L. Zhao, L. Jia, Z. Ouyang, Y. Gao, R. Guo, S. Song, X. Shi, X. Cao, 68Ga-labeled dendrimer-entrapped gold nanoparticles for PET/CT dual-modality imaging and immunotherapy of tumors, J. Mater. Chem. B. 10 (2022) 3648-3656. https://doi.org/10.1039/D2TB00378C

[106] Z.J. Wang, Q. Li, L.L. Tan, C.G. Liu, L. Shang, Metal-organic frameworks mediated assembly of gold nanoclusters for sensing applications, J. Anal. Test. 6 (2022) 163-177. https://doi.org/10.1007/s41664-022-00224-0

[107] B.T. Cisneros, J.J. Law, M.L. Matson, A. Azhdarinia, E.M. Sevick-Muraca, L.J. Wilson, Stable confinement of positron emission tomography and magnetic resonance agents within carbon nanotubes for bimodal imaging, Nanomedicine. 9 (2014) 2499-2509. https://doi.org/10.2217/nnm.14.26

[108] S.W. Chou, T.H. Chen, Y.C. Fa, Y.Y. Yang, C.Y. Lin, K.L. Kuo, S.T. Wu, E.C. Lin, J.A. Annie Ho, J.K. Hsiao, gadoliniumengineered magnetic alloy nanoparticles for magnetic

resonance T1/T2 dual-modal and computed tomography imaging, Chem. Mater. 34 (2022) 10050-10058. https://doi.org/10.1021/acs.chemmater.2c02537

[109] R. Huang, Y.W. Shen, Y.Y. Guan, Y.X. Jiang, Y. Wu, K. Rahman, L.J. Zhang, H.J. Liu, X. Luan, Mesoporous silica nanoparticles: Facile surface functionalization and versatile biomedical applications in oncology, Acta Biomater. 116 (2020) 1-15. https://doi.org/10.1016/j.actbio.2020.09.009

[110] X. Cai, Q. Zhu, Y. Zeng, Q. Zeng, X. Chen, Y. Zhan, Manganese oxide nanoparticles as MRI contrast agents in tumor multimodal imaging and therapy, Int. J. Nanomed. 14 (2019) 8321-8344. https://doi.org/10.2147/IJN.S218085

[111] K. Deka, K.A. Guleria, D. Kumar, J. Biswas, S. Lodha, S.D. Kaushik, S. Dasgupta, P. Deb, Exclusive T2 MRI contrast enhancement by mesoporous carbon framework encapsulated manganese oxide nanoparticles, Curr. Appl. Phys. 20 (2020) 89-95. https://doi.org/10.1016/j.cap.2019.10.010

[112] W. Xu, S. Zhang, Q. Zhou, W. Chen, VHPKQHR peptide modified magnetic mesoporous nanoparticles for MRI detection of atherosclerosis lesions, Artif. Cells Nanomed. Biotechnol. 47 (2019) 2440-2448. https://doi.org/10.1080/21691401.2019.1626411

[113] X.Y. Zheng, J. Pellico, A.A. Khrapitchev, N.R. Sibson, J.J. Davis, Dy-DOTA integrated mesoporous silica nanoparticles as promising ultrahigh field magnetic resonance imaging contrast agents, Nanoscale. 10 (2018) 21041-21045. https://doi.org/10.1039/C8NR07198E

[114] D. Zhuang, H. Zhang, G. Hu, B. Guo, Recent development of contrast agents for magnetic resonance and multimodal imaging of glioblastoma, J. Nanobiotechnol. 20 (2022) 284. https://doi.org/10.1186/s12951-022-01479-6

[115] Y. Hou, C.C. Lv, Y.L. Guo, X.H. Ma, W. Liu, Y. Jin, B.X. Li, M. Yang, S.Y. Yao, Recent advances and applications in paper-based devices for point-of-care testing, J. Anal. Test. 6 (2022) 247-273. https://doi.org/10.1007/s41664-021-00204-w

[116] H. Wang, Y. Wang, W. Wang, Y. Zhang, Q. Yuan, J. Tan, Materials-functionalized point-of-care testing devices for pathogen detection, Acc. Mater. Res. 4 (2023) 1083-1094. https://doi.org/10.1021/accountsmr.3c00151

[117] Z. Zhang, P. Ma, R. Ahmed, J. Wang, D. Akin, F. Soto, B.F. Liu, P. Li, U. Demirci, Advanced point-of care testing technologies for human acute respiratory virus detection, Adv. Mater. 34 (2022) 1-39. https://doi.org/10.1002/adma.202103646

[118] M.B. Behyar, A. Nilghaz, M. Hasanzadeh, N. Shadjou, Recent progresses and challenges on mesoporous silica nanoparticles for DNA-based biosensors and diagnostics, TrAC Trends Anal. Chem. 178 (2024) 117846. https://doi.org/10.1016/j.trac.2024.117846

[119] L. Zhu, W. Fu, B. Zhu, Q. Feng, X. Ying, S. Li, J. Chen, X. Xie, C. Pan, J. Liu, C. Chen, X. Chen, D. Zhu, An integrated microfluidic electrochemiluminescence device for point-of-care testing of acute myocardial infarction, Talanta. 262 (2023) 124626. https://doi.org/10.1016/j.talanta.2023.124626

[120] J. Guo, S. Chen, S. Tian, K. Liu, X. Ma, J. Guo, A sensitive and quantitative prognosis of C-reactive protein at picogram level using mesoporous silica encapsulated core-shell up-

Materials Research Forum LLC
https://doi.org/10.21741/9781644903452-18

conversion nanoparticle based lateral flow strip assay, Talanta. 230 (2021) 122335. https://doi.org/10.1016/j.talanta.2021.122335

[121] B.J. Burkett, D.J. Bartlett, P.W. McGarrah, A.R. Lewis, D.R. Johnson, K. Berberoğlu, M.K. Pandey, A.T. Packard, T.R. Halfdanarson, C.B. Hruska, G.B. Johnson, A.T. Kendi, A review of theranostics: Perspectives on emerging approaches and clinical advancements, Radiol Imaging Cancer. 5 (2023) e220157. https://doi.org/10.1148/rycan.220157

[122] K. Bannik, B. Madas, M. Jarzombek, A. Sutter, G. Siemeister, D. Mumberg, S. Zitzmann-Kolbe, Radiobiological effects of the alpha emitter Ra-223 on tumor cells, Sci. Rep. 9 (2019) 18489. https://doi.org/10.1038/s41598-019-54884-7

[123] A. Rahikkala, S.A.P. Pereira, P. Figueiredo, M.L.C. Passos, A.R.T.S. Araújo, M.L.M.F.S. Saraiva, H.A. Santos, Mesoporous silica nanoparticles for targeted and stimuli-responsive delivery of chemotherapeutics: A review, Adv. Biosyst. 2 (2018) 1800020. https://doi.org/10.1002/adbi.201800020

[124] B. Darvishi, L. Farahmand, K. Majidzadeh-A, Stimuli-responsive mesoporous silica nps as non-viral dual sirna/chemotherapy carriers for triple negative breast cancer, Mol. Ther. Nucleic Acids. 7 (2017) 164-180. https://doi.org/10.1016/j.omtn.2017.03.007

[125] K. Hemant, P. Balaram, K. Jitender, K. Pramod, In vitro and bioimaging studies of mesoporous silica nanocomposites encapsulated iron-oxide and loaded doxorubicin drug (DOX/IO@Silica) as magnetically guided drug delivery system, Curr. Pharm. Biotechnol. 24 (2023) 1297-1306. https://doi.org/10.2174/1389201023666220428084920

[126] R.R. Castillo, D. Lozano, M. Vallet-Regí, Mesoporous silica nanoparticles as carriers for therapeutic biomolecules, Pharmaceutics. 12 (2020) 432. https://doi.org/10.3390/pharmaceutics12050432

[127] V.L. Dunne, T.C. Wright, F.D.C.G. Liberal, J.M. O'Sullivan, K.M. Prise, Synergistic activity of DNA damage response inhibitors in combination with radium-223 in prostate cancer, Cancers. 16 (2024) 1510. https://doi.org/10.3390/cancers16081510

[128] S. Sargazi, U. Laraib, M. Barani, A. Rahdar, I. Fatima, M. Bilal, S. Pandey, R.K. Sharma, G.Z. Kyzas, Recent trends in mesoporous silica nanoparticles of rode-like morphology for cancer theranostics: A review, J. Mol. Struct. 1261 (2022) 132922. https://doi.org/10.1016/j.molstruc.2022.132922

[129] J. Chen, W. Han, J. Chen, W. Zong, W. Wang, Y. Wang, G. Cheng, C. Li, L. Ou, Y. Yu, High performance of a unique mesoporous polystyrene-based adsorbent for blood purification, Regen. Biomater. 4 (2017) 31-37. https://doi.org/10.1093/rb/rbw038

[130] Y. Pan, P. Xue, S. Liu, L. Zhang, Q. Guan, J. Zhu, X. Tian, Metal-based hybrid nanoparticles as radiosensitizers in cancer therapy, Colloid Interf. Sci. Commun. 23 (2018) 45-51. https://doi.org/10.1016/j.colcom.2018.01.004

[131] H. Zhu, K. Zheng, R. Aldo Boccaccini. Multi-functional silica-based mesoporous materials for simultaneous delivery of biologically active ions and therapeutic biomolecules, Acta Biomater. 129 (2021) 1-17. https://doi.org/10.1016/j.actbio.2021.05.007

[132] X. Liu, F. Wu, Y. Ji, L. Yin, Recent advances in anti-cancer protein/peptide delivery, Bioconjug. Chem. 30 (2019) 305-324. https://doi.org/10.1021/acs.bioconjchem.8b00750

[133] T. Karpinski, A. Adamczak, Anticancer activity of bacterial proteins and peptides, Pharmaceutics. 10 (2018) 54. https://doi.org/10.3390/pharmaceutics10020054

[134] J. Xie, W. Xu, Y. Wu, B Niu, X. Zhang, Macroporous organosilicon nanocomposites co-deliver Bcl2-converting peptide and chemotherapeutic agent for synergistic treatment against multidrug resistant cancer, Cancer Lett. 469 (2020) 340-354. https://doi.org/10.1016/j.canlet.2019.10.018

[135] Y. Wu, P. Ge, W. Xu, M. Li, Q. Kang, X. Zhang, J. Xie, Cancer-targeted and intracellular delivery of Bcl-2-converting peptide with functional macroporous silica nanoparticles for biosafe treatment, Mater. Sci. Eng. C. 108 (2020) 110386. https://doi.org/10.1016/j.msec.2019.110386

[136] E. Tenland, A. Pochert, N. Krishnan, K.U. Rao, S. Kalsum, K. Braun, I. Glegola-Madejska, M. Lerm, B.D. Robertson, M. Lindén, G. Godaly, Effective delivery of the anti-mycobacterial peptide NZX in mesoporous silica nanoparticles, PLOS ONE. 14 (2019) e0212858. https://doi.org/10.1371/journal.pone.0212858

[137] A. Zakeri Siavashani, M. Haghbin Nazarpak, F. Fayyazbakhsh, T. Toliyat, S.J.P. McInnes, M. Solati-Hashjin, Effect of amino-functionalization on insulin delivery and cell viability for two types of silica mesoporous structures, J. Mater. Sci. 51 (2016) 10897-10909. https://doi.org/10.1007/s10853-016-0301-1

[138] Q. Yu, T. Deng, F.C. Lin, B. Zhang, J.I. Zink, Supramolecular assemblies of heterogeneous mesoporous silica nanoparticles to co-deliver antimicrobial peptides and antibiotics for synergistic eradication of pathogenic biofilms, ACS Nano. 14 (2020) 5926-5937. https://doi.org/10.1021/acsnano.0c01336

[139] M. Gerstenberg, C.M. Stürzel, T. Weil, F. Kirchhoff, M. Lindén, Modular hydrogel−mesoporous silica nanoparticle constructs for therapy and diagnostics, Adv. NanoBiomed Res. 2 (2022), 2100125. https://doi.org/10.1002/anbr.202100125

[140] M. Ménard, F. Meyer, C. Affolter-Zbaraszczuk, M. Rabineau, A. Adam, P.D. Ramirez, S. Bégin-Colin, D. Mertz, Design of hybrid protein-coated magnetic core-mesoporous silica shell nanocomposites for MRI and drug release assessed in a 3D tumor cell model, Nanotechnology. 30 (2019) 174001. https://doi.org/10.1088/1361-6528/aafe1c

[141] P. Mario. Why C-reactive protein is one of the most requested tests in clinical laboratories?, Clin. Chem. Lab. Med. 61 (2023) 1540-1545. https://doi.org/10.1515/cclm-2023-0086

[142] H.M. Alshekhabobakr, S.O AlSaqatri, N.M. Rizk, Laboratory test utilization practices in hamad medical corporation; role of laboratory supervisors and clinicians in improper test utilization; A descriptive pilot study, J. Multidiscip. Healthc. 15 (2022) 413-4 https://doi.org/10.2147/JMDH.S320545

[143] L. Zhao, H. Qin, R. Wu, H. Zou, Recent advances of mesoporous materials in sample preparation, J. Chromatogr. A. 1228 (2012) 193-204. https://doi.org/10.1016/j.chroma.2011.09.051

[144] R. Savino, R. Terracciano, Mesopore-assisted profiling strategies in clinical proteomics for drug/target discovery, Drug Discov. Today. 17 (2012) 143-152. https://doi.org/10.1016/j.drudis.2011.10.002

[145] R. Terracciano, M. Preianò, G. Maggisano, C. Pelaia, R. Savino, Hexagonal mesoporous silica as a rapid, efficient and versatile tool for MALDI-TOF MS sample preparation in clinical peptidomics analysis: A pilot study, Molecules. 24 (2019) 2311. https://doi.org/10.3390/molecules24122311

[146] F. Farjadian, S. Azadi, S. Mohammadi-Samani, H. Ashrafi, A.A. Azadi, A novel approach to the application of hexagonal mesoporous silica in solid-phase extraction of drugs, Heliyon. 4 (2018) e00930. https://doi.org/10.1016/j.heliyon.2018.e00930

[147] K. Liang, H. Wu, T.Y. Hu, Y. Li, Mesoporous silica chip: Enabled peptide profiling as an effective platform for controlling bio-sample quality and optimizing handling procedure, Clin. Proteom. 13 (2016) 34. https://doi.org/10.1186/s12014-016-9134-9

[148] J. Yao, N. Sun, C. Deng, Recent advances in mesoporous materials for sample preparation in proteomics research, TrAC Trends Anal. Chem. 99 (2018) 88-100. https://doi.org/10.1016/j.trac.2017.11.016

[149] M. Mladenović, S. Jarić, M. Mundžić, A. Pavlović, I. Bobrinetskiy, N.Z. Knežević, Biosensors for cancer biomarkers based on mesoporous silica nanoparticles, Biosensors. 14 (2024) 326. https://doi.org/10.3390/bios14070326

[150] Y. Xia, H. Chen, R. Liu, F. Shi, C. Ren, J. Li, J. Zhao, X. Chen, Z. Yang, Mesoporous SiO2 sphere-based electrochemical impedance immunosensor for ultrasensitive detection of bovine interferon-γ, J. Anal. Test. 7 (2023) 295-303. https://doi.org/10.1007/s41664-023-00255-1

[151] L.P. Juan, M. Cristina, A.M. Pérez-Moreno, R. Jurado-Escobar, G. Bogas, D. T.D. Fernández, M.I. Montañez, C. Mayorga, M.J. Torres, Influence of pore size in protein G'-Grafted mesoporous silica nanoparticles as a serum pretreatment system for in vitro allergy diagnosis, Adv. Healthc. Mater. 12 (2023) 2203321. https://doi.org/10.1002/adhm.202203321

[152] S. Hamd-Ghadareh, A. Salimi, A. Vaziry, Ultrasensitive ratiometric fluorescence bioassay for accurate detection of covid-19-specific nucleocapsid protein in clinical serum samples using modified cleavable mesoporous sio2 satellite-enriched carbon dots, ACS Biomater. Sci. Eng. 9 (2023) 5279-5292. https://doi.org/10.1021/acsbiomaterials.3c00391

Keyword Index

About the Editor

Dr. Inamuddin is working as an Assistant Professor at the Department of Applied Chemistry, Aligarh Muslim University, Aligarh, India. He obtained a Master of Science degree in Organic Chemistry from Chaudhary Charan Singh (CCS) University, Meerut, India, in 2002. He received his Master of Philosophy and Doctor of Philosophy degrees in Applied Chemistry from Aligarh Muslim University (AMU), India, in 2004 and 2007, respectively. He has extensive research experience in multidisciplinary fields of Analytical Chemistry, Materials Chemistry, and Electrochemistry and, more specifically, Renewable Energy and Environment. He has worked on different research projects as a project fellow and senior research fellow funded by the University Grants Commission (UGC), Government of India, and the Council of Scientific and Industrial Research (CSIR), Government of India. He has received the Fast Track Young Scientist Award from the Department of Science and Technology, India, to work in the area of bending actuators and artificial muscles. He has also received the Sir Syed Young Researcher of the Year Award 2020 from Aligarh Muslim University. He has completed four major research projects sanctioned by the University Grant Commission, Department of Science and Technology, Council of Scientific and Industrial Research, and Council of Science and Technology, India. He has published 216 research articles in international journals of repute and nineteen book chapters in knowledge-based book editions published by renowned international publishers. He has published 200 edited books with Springer (U.K.), Elsevier, Nova Science Publishers, Inc. (U.S.A.), CRC Press Taylor & Francis Asia Pacific, Trans Tech Publications Ltd. (Switzerland), IntechOpen Limited (U.K.), Wiley-Scrivener, (U.S.A.) and Materials Research Forum LLC (U.S.A). He is a member of various journals' editorial boards. He has served as Associate Editor for journals (Environmental Chemistry Letter, Applied Water Science and Euro-Mediterranean Journal for Environmental Integration, Springer-Nature), Frontiers Section Editor (Current Analytical Chemistry, Bentham Science Publishers), Editorial Board Member (Scientific Reports-Nature)and Review Editor (Frontiers in Chemistry, Frontiers, U.K.) He has also guest-edited various special thematic issues for the journals of Elsevier, Bentham Science Publishers, and John Wiley & Sons, Inc. He has attended as well as chaired sessions at various international and national conferences. He has worked as a Postdoctoral Fellow, leading a research team at the Creative Research Initiative Center for Bio-Artificial Muscle, Hanyang University, South Korea, in the field of renewable energy, especially biofuel cells. He has also worked as a Postdoctoral Fellow at the Center of Research Excellence in Renewable Energy, King Fahd University of Petroleum and Minerals, Saudi Arabia, in the field of polymer electrolyte membrane fuel cells and computational fluid dynamics of polymer electrolyte membrane fuel cells. He is a life member of the Journal of the Indian Chemical Society. His research interest includes ion exchange materials, a sensor for heavy metal ions, biofuel cells, supercapacitors and bending actuators.

Dr. Arwa Alrooqi is an Assistant Professor in the Department of Chemistry, Faculty of Science, at Al Baha University, Saudi Arabia. She has published several research articles in internationally recognized journals. Dr. Alrooqi is actively engaged in multidisciplinary research, focusing on the synthesis and engineering of organic semiconductors and nanomaterials, their characterization, and their application in smart windows.

Dr. Amal Alshehri is an Assistant Professor of Physical Chemistry in the Department of Chemistry, Faculty of Science, at Al Baha University, Saudi Arabia. She earned her Ph.D. with a focus on nanomaterials, catalysis, and adsorption. Her research expertise encompasses green synthesis approaches for nanomaterials, water treatment, heterogeneous catalysis, photocatalysis, and the selective oxidation of cyclohexane.Dr. Alshehri has published several research articles in internationally recognized journals, contributing to fundamental advancements in the fields of nanomaterials and catalysis.

Dr. Mohammad A. Jafar Mazumder has been serving as a Professor of Chemistry at King Fahd University of Petroleum & Minerals (KFUPM), Saudi Arabia. He has extensive experience in designing, synthesizing, and characterizing various organic compounds, ionic and thermo-responsive polymers for corrosion, water treatment, and biomedical applications. Dr. Jafar Mazumder obtained his B.Sc (Hons.), M.Sc (Chemistry) from Aligarh Muslim University, India, MS (Chemistry) from KFUPM, Saudi Arabia, and Ph.D. in Chemistry (2009) from McMaster University, Canada.

In more than 20 years of academic research, Dr. Jafar Mazumder has had the opportunity to work with several international collaborative research groups and has exposed himself to a broad range of research areas. Dr. Jafar Mazumder secured 8 US patents, published more than 85 articles in peer-reviewed journals, 37 conference proceedings, 9 book chapters, and co-edited 4 books with Springers and Trans Tech publications. He is awarded as a Fellow of the Royal Society of Chemistry and Chartered Chemist, Association of Chemical Profession of Ontario, Canada. Besides, Dr. Jafar Mazumder is a member of the American Chemical Society (ACS), Canadian Society for Chemistry (CSC), Canadian Biomaterial Society (CBS), and a life member of the Bangladesh Chemical Society (BCS). In his academic career, he was awarded numerous national and international scholarships and awards including the prestigious Indian Council for Cultural Relations (ICCR) Scholarship from Govt. of India for undergraduate studies in India, Aligarh Muslim University undergraduate & graduate Gold medal, and certificate of excellence from the Ministry of Human Resource Development, Govt. of India, and MITACS postdoctoral fellowship (Canada) for pursuing postdoctoral research in Chemical and Biomedical Engineering.

Currently, Dr. Jafar Mazumder is actively involved in several ongoing university (KFUPM), government (KACST, NSTIP), and client (Saudi Aramco) funded projects in the capacity of principal and co-investigators. His current research interest includes the design, synthesis, and characterization of various modified monomers and polymers for potential use in the inhibition of mild steel corrosion in oil and gas industries and the preparation of multilayered polyelectrolyte coated membranes for the removal of heavy metals and organic contaminants from aqueous water samples. The long-term scientific goal of Dr. Jafar Mazumder is not merely to make science fun and entertaining for people. It is to engage them with a multidisciplinary scientific mission at a deeper level to create a space through which they can interact with scientific ideas, develop connections between science, engineering, and biology, and thoughts of their own to contribute to society. He feels this goal and engaging personality make him a pleasant person to work with and help inspire his co-workers in any professional setting.

www.ingramcontent.com/pod-product-compliance
Ingram Content Group UK Ltd.
Pitfield, Milton Keynes, MK11 3LW, UK
UKHW020659041026
15590UKWH00001B/367